Ethics and the Military Profession

The Moral Foundations of Leadership
NROTC Third Edition

Edited by:

George R. Lucas, Jr.
Class of 1984 Distinguished Chair in Ethics
Stockdale Center for Ethical Leadership
U.S. Naval Academy & Professor of Ethics and Public Policy
Naval Postgraduate School

W. Richard Rubel
Captain, U.S. Navy (Retired)
Distinguished Military Professor of Ethics
U.S. Naval Academy

Learning Solutions

New York   Boston   San Francisco
London   Toronto   Sydney   Tokyo   Singapore   Madrid
Mexico City   Munich   Paris   Cape Town   Hong Kong   Montreal

Cover photographs courtesy of the U.S. Navy

Copyright © 2011, 2010, 2008, 2007, 2005, 2004 by George Lucas and Rick Rubel
Copyright © 2002, 2001, 2000, 1999 by Pearson Custom Publishing
Copyright © 1998 by Simon & Schuster Custom Publishing
All rights reserved.

This copyright covers material written expressly for this volume by the editor/s as well as the compilation itself. It does not cover the individual selections herein that first appeared elsewhere. Permission to reprint these has been obtained by Pearson Learning Solutions for this edition only. Further reproduction by any means, electronic or mechanical, including photocopying and recording, or by any information storage or retrieval system, must be arranged with the individual copyright holders noted.

All trademarks, service marks, registered trademarks, and registered service marks are the property of their respective owners and are used herein for identification purposes only.

Pearson Learning Solutions, 501 Boylston Street, Suite 900, Boston, MA 02116
A Pearson Education Company
www.pearsoned.com

Printed in the United States of America

1 2 3 4 5 6 7 8 9 10 V312 15 14 13 12 11 10

000200010270649965

RG/LP

PEARSON

ISBN 10: 0-558-91640-6
ISBN 13: 978-0-558-91640-4

DEDICATION

For
Dr. Paul E. Roush, Ph.D.
Colonel, U.S. Marine Corps (Retired)
USNA Class of 1957

Copyright Acknowledgments

Grateful acknowledgment is made to the following sources for permission to reprint material copyrighted or controlled by them:

"The Ring of Gyges," by Plato, reprinted from *The Republic,* Book 2, Crofts Classics Series, translated and edited by Raymond Larson (1979), reprinted by permission of Harlan Davidson, Inc.

"Why Ethics is So Hard," by Thomas B. Grassey (1998).

"Relativism and Objectivism: Are There Universal Values?" by Burton F. Porter, reprinted from *Philosophy Through Fiction and Film* (2003).

"A Higher Moral Standard for the Military," by J. Carl Ficarrotta, reprinted by permission of the author.

"The Moral Foundations of Military Service," by Martin L. Cook (spring 2000), reprinted by permission of the author.

"Constitutional Ethics," by Paul E. Roush, reprinted by permission of the author.

"The American Professional Military Ethic," by Anthony E. Hartle, reprinted from *Moral Issues in Military Decision Making*, second edition (2004), by permission of University of Kansas Press.

"Letter from Birmingham City Jail," by Martin Luther King, Jr., the Estate of Martin Luther King, Jr., c/o Writer's House. Copyright © 1963 by Martin Luther King, Jr., copyright © renewed 1991 by Coretta Scott King.

Excerpt from "Religion and Morality: Exploring the Connections," by Martin L. Cook, reprinted by permission of the author.

"Practical Reasoning and Moral Casuistry," by Albert R. Jonsen, reprinted from *The Blackwell Companion to Religious Ethics*, edited by William Schweiker (2005), John Wiley & Sons, Inc.

"Utilitarianism," by Louis P. Pojman, reprinted from *Ethics: Discovering Right and Wrong* (2002), by permission of Gertrude Pojman.

"The Ones Who Walk Away from Omelas," by Ursula K. Le Guin, reprinted from *Omelas* (1973, 2001), by permission of the author and the author's agents, the Virginia Kidd Agency, Inc. Copyright © 1973, 2001 by Ursula K. LeGuin. First appeared in *New Directions 3*. From the author's collection *The Wind's Twelve Quarters*.

Excerpts from "Ordinary Rational Knowledge of Morality to Philosophical," by Immanuel Kant, reprinted from *Groundwork of the Metaphysic of Morals*, translated by H.J. Paton (1948), condensed and edited by Lawrence Lengbeyer, Thomson Publishing Group.

"A Simplified Account of Kant's Ethics," by Onora O'Neill, reprinted from *Matters of Life and Death*, edited by Tom Regan (1986), McGraw-Hill Companies.

"On A Supposed Right to Lie because of Philanthropic Concerns," by Immanuel Kant (1799). First appeared in *Berlinische Blaetter* (September 1799), published by Biester.

"The Moral Virtues," by Aristotle, reprinted from *The Ethics of Aristotle*, translated by J.A.K. Thomson (1976), by permission of Penguin Books, Ltd.

"Habit and Virtue," by Aristotle, reprinted from *Nicomachean Ethics*, translated by Terence Irwin (1985), by permission of Hackett Publishing Company, Inc.

"Courage," an excerpt from *The Nicomachean Ethics*, by Aristotle, reprinted from *Introduction to Aristotle*, edited by Richard McKeon (1974), by permission of Michael McKeon.

"Friendship," an excerpt from *The Nicomachean Ethics,* by Aristotle, reprinted from *Nicomachean Ethics*, translated by J.E.C. Welldon (1897).

"Love: War's Ally and Foe," by J. Glenn Gray, reprinted from *The Warriors: Reflections on Men in Battle* (1959), by permission of Charlotte Gray Martin and Lisa Gray Fisher. Copyright © 1987 by Ursula A. Gray.

Excerpts from *Summa Theologica, Ethics and Natural Law*, by St. Thomas Aquinas, translated by Fathers of the English Dominican Province (1911), published by Burns, Oates & Washbourne Ltd.

Excerpt from "The Ethics of Natural Law," by C.E. Harris, reprinted from *Applying Moral Theories*, Third Edition (1997), by permission of Wadsworth, a part of Cengage Learning, Inc.

"Law and Order in International Society" and "Wars of Anticipation," by Michael Walzer, reprinted from *Just and Unjust Wars: A Moral Argument with Historical Illustrations* (2006), by permission of Basic Books, a member of the Perseus Books Group.

"From jus ad bellum to jus ad pacem: Re-thinking Just War Criteria for Irregular Warfare," by George R. Lucas, Jr., reprinted from *Ethics and Foreign Intervention*, edited by Dean K. Chatterjee and Donald E. Scheid (2003), by permission of Cambridge University Press.

"The Concept and Practice of Jihad in Islam," by Michael G. Knapp, reprinted from *Parameters: Journal of the U.S. Army War College* (2002).

"Jus in bello—Just Conduct in War," by Brian Orend, reprinted from *The Morality of War* (2006), Broadview Press.

"Guerrilla War," by Michael Walzer, reprinted from *Just and Unjust Wars: A Moral Argument with Historical Illustrations* (2006), by permission of Basic Books, a member of the Perseus Books Group.

"Just War Criteria and the New Face of War: Human Shields, Manufactured Martyrs, and Little Boys with Stones," by Michael Skerker, reprinted from the *Journal of Military Ethics*, March 2004, by permission of the Taylor & Francis Group.

"Perspectives on Intervention: The Case of Somalia—A Response to Professor Lucas," by Anthony C. Zinni, reprinted from *Perspectives on Humanitarian Military Intervention* (2001), by permission of Berkeley Public Policy Press, Institute of Governmental Studies at the University of California, Berkeley.

"Why Warriors Need a Code," by Shannon E. French, reprinted from *The Code of the Warriors: Exploring Warrior Values, Past and Present* (2003), by permission of Rowman & Littlefield Publishers, Inc.

Excerpts from *On Liberty*, by John Stuart Mill (1859).

Excerpt from *War and Self Defense*, by David Rodin, edited by Lawrence Lengbeyer (2003), reprinted by permission of Oxford University Press.

Excerpts from *A Theory of Justice*, by John Rawls (1971), Belknap Press of Harvard University Press.

"Distributive Justice," by Julian Lamont and Christi Favor, reprinted from the *Stanford Encyclopedia of Philosophy*, edited by Edward N. Zalta (fall 2008). Originally published by Stanford University Press.

"Crime and Punishment," by R.A. Duff, reprinted from *Routledge Encyclopedia of Philosophy*, vol. 2 (1998), by permission of Taylor & Francis Books UK.

Excerpt from "Billy Budd," by Herman Melville, reprinted from *Billy Budd: Sailor*, edited by Harrison Hayford and Merton Sealts (1972).

Introduction to "Billy Budd," by Christina and Fred Sommers, reprinted from *Vice and Virtue in Everyday Life*, Fourth Edition (1997), by permission of the authors.

"Is the 'Whole Truth' Attainable?" by Sissela Bok, reprinted from *Lying: Moral Choice in Public and Private Life* (1978), Pantheon Books, a division of Random House, Inc.

Excerpts from *Courage Under Fire: Testing Epictetus's Doctrines in a Laboratory of Human Behavior*, by James B. Stockdale (1993), reprinted by permission of Hoover Institution Press. Copyright © 1993 by the Board of Trustees of the Leland Stanford Junior University.

Excerpts from *The Enchiridion* by Epictetus.

"Roman Stoicism," by Shannon E. French, excerpted and reprinted from *The Code of the Warrior: Exploring Warrior Values, Past and Present* (2003), by permission of the author.

"A Vietnam Experience, Duty," by James B. Stockdale, adapted from his Address to the Class of 1983, United States Military Academy.

"Admiral James B. Stockdale's Leadership Model," by Paul E. Roush, reprinted by permission of the author.

DEPARTMENT OF THE NAVY
NAVAL SERVICE TRAINING COMMAND
2601A PAUL JONES STREET
GREAT LAKES, ILLINOIS 60088-2845

1533
Ser N00/1949
October 4, 2010

Congratulations on beginning your capstone Naval Science course! As you approach the end of your time as college students, you must increasingly look ahead to your future as military officers. The lessons you learn in this course will forge a more solid foundation for the significant tasks you will face as a military leader.

As Naval officers, your actions and decisions will be under close scrutiny by your subordinates, fellow officers, and the public you serve. This course will help prepare you to make the right decisions for the right reasons. By examining the actions of others, discussing the ethical decision process, and the potential consequences of your decisions, you will build a valuable framework for "what is right" when faced with ethical dilemmas.

The Naval Services' core values of honor, courage and commitment must play a daily role in your ethical decisions. You will find that it routinely takes courage and commitment to stand up and make the hard ethical decision rather than looking the other way or making the popular decision. Perhaps Oliver Wendell Holmes said it best "Not always right in all men's eyes but faithful to the light within." I ask you to apply yourself fully in this curriculum to set the light within burning brightly enough to give you the courage to lead your Sailors and Marines with the very best ethical leadership possible. Our profession demands no less!

Sincerely,

David. F. Steindl
Rear Admiral, U.S. Navy
Commander Naval Service
Training Command

CONTENTS

Introduction to the Third Edition: "The Age of Hybrid War" xiii
   George R. Lucas, Jr.

I. Why Study Ethics? 1

On the Eve of Battle 9
   George R. Lucas, Jr.

The Ring of Gyges 13
   Plato (c. 360 B.C.E.)

Why Ethics is so Hard 15
   Capt Thomas B. Grassey, U.S. Naval Reserve (Retired)

II. The Moral Framework of Military Service 21

A. The Relativity of Moral Beliefs and Situations 25

Relativism and Objectivism: Are There Universal Values? 29
   Burton F. Porter

A Higher Moral Standard for the Military 33
   Maj. J. Carl Ficcarotta, U.S. Air Force (Retired)

B. The U.S. Constitution and the Moral Foundations of Military Service: Conflicts of Principles and Loyalties 45

The Moral Foundations of Military Service 47
   Martin L. Cook

Constitutional Ethics 57
   Col. Paul E. Roush, U.S. Marine Corps (Retired)

The American Professional Military Ethic 63
   Anthony E. Hartle, Colonel, U.S. Army (Retired)

Letter from Birmingham City Jail 71
   The Rev. Dr. Martin Luther King, Jr.

REFERENCE: The Constitution of the United States 81

C. Religion and Military Ethics 97

Religion and Morality: Exploring the Connections 99
Martin L. Cook

Practical Reasoning and Moral Casuistry 105
Albert R. Jonsen

Abraham's Obedience Tested 111
Genesis 22:1-12

III. Traditions of Moral Reasoning in Western Culture 113

A. Utilitarianism and the Greatest Good 119

Utilitarianism 123
John Stuart Mill

Utilitarianism 133
Louis P. Pojman

The Ones who Walk Away from Omelas 139
Ursula K. Le Guin

B. Kantian Ethics and the Basis of Duty 143

from Groundwork of the Metaphysic of Morals 149
Immanuel Kant

A Simplified Account of Kant's Ethics 161
Onora O'Neill

On a Supposed Right to Lie because of Philanthropic Concerns 167
Immanuel Kant

C. Aristotle and the Ethics of Virtue and Character 169

The Moral Virtues 173
Aristotle

Habit and Virtue 179
Aristotle

Courage 183
Aristotle

Friendship 187
Aristotle

The Warriors: Reflections on Men in Battle 191
J. Glenn Gray

D. The Tradition of Natural Law 195

from Summa Theologica, Ethics and Natural Law 199
St. Thomas Aquinas

from The Ethics of Natural Law 203
C. E. Harris

Natural Law and the Principle of Double Effect: Six Hypothetical Cases..... 211
George R. Lucas, Jr.

IV. The Moral Role of the Military Professional in International Relations 219

A. The Justification for Going to War (*Jus ad Bellum*) 225

Is it Always Sinful to Wage War? 231
St. Thomas Aquinas

Law and Order in International Society 233
Michael Walzer

Wars of Anticipation 241
Michael Walzer

From jus ad bellum to *jus ad pacem*: Re-thinking Just War Criteria for Irregular War 249
George R. Lucas, Jr.

The Concept and Practice of Jihad in Islam 261
Michael G. Knapp

B. The Moral Code of the Warrior (*Jus in Bello*) 271

Jus in bello—Just Conduct in War 273
Brian Orend

Guerrilla War 287
Michael Walzer

Just War Criteria and the New Face of War: Human Shields, Manufactured Martyrs, and Little Boys with Stones 299
Michael Skerker

Ethical Issues in the Use of Military Force in Irregular Warfare 309
George R. Lucas, Jr.

Perspectives on Irregular Warfare: The Case of Somalia 315
General Anthony C. Zinni, U.S. Marine Corps (Retired)

Why Warriors Need a Code 323
Shannon E. French

V. Upholding Truth, Enforcing Justice, Defending Liberty and Rights ... 335

A. Liberty and Rights ... 339

On Liberty ... 343
John Stuart Mill

War and Self-Defense ... 359
David Rodin

B. Justice ... 367

Excerpts from A Theory of Justice ... 373
John Rawls

Distributive Justice ... 383
Julian Lamont and Christi Favor

Crime and Punishment ... 389
R. A. Duff

from Billy Budd ... 393
Herman Melville

C. Upholding the Truth ... 399

Lying: Moral Choice in Public & Private Life ... 401
Sissela Bok

VI. Moral Leaders and Moral Warriors: Vice Admiral James B. Stockdale and Stoic Philosophy ... 413

Courage Under Fire ... 417
Vice Admiral James B. Stockdale, U.S. Naval Service (Retired)

The Enchiridion ... 427
Epictetus

Roman Stoicism ... 439
Shannon E. French

A Vietnam Experience, Duty ... 451
Vice Admiral James B. Stockdale

Vice Admiral James B. Stockdale's Leadership Model ... 457
Col. Paul E. Roush, U.S. Marine Corps (Retired)

VII. Epilogue: The Hiding Places of Memory ... 461

The Hiding Places of Memory ... 463
George R. Lucas Jr.

Introduction to the Third Edition
"The Age of Hybrid War"
George R. Lucas, Jr.

"Hybrid war" is a relatively new designation from the U.S. Department of Defense, representing the latest attempt to classify the ever-changing nature of armed conflict in the 21st century. "Hybrid war," or "irregular war" (IW), or "unconventional" war envisions a mix of elements of conventional armed conflict between states with elements of insurgency, civil unrest, guerrilla warfare, terrorism, counterinsurgency, genocide and humanitarian intervention, peace-keeping and stability operations.[1] This is a tall order for any nation's military. But like it or not, this is the face of conflict in our lifetime, and it is essential that members of our armed forces be fully prepared to cope with it.

By whatever label we chose to name it, the new era of "irregular" or "unconventional" warfare is often (and I think, mistakenly) characterized primarily by the substantial changes and challenges in military tactics and military technology it portends—the so-called "revolution in military affairs" (RMA). On the one hand, asymmetric warfare, including a variety of types of terrorist attacks involving rudimentary technology, has largely replaced the massive confrontations between opposing uniformed combat troops employing the high-tech and heavy-platform weaponry that previously characterized conventional war. On the other hand, new kinds of "emerging technologies," including military robotics and unmanned weapons platforms, non-lethal weapons, and cyber-attacks, ever-increasingly constitute the tactical countermeasures adopted by established nations in response.

The era of irregular or unconventional war, however, also requires another revolution—a cultural sea-change that has been much harder to identify, and much slower in coming. This radical cultural shift entails re-conceptualizing the nature and purpose of war-fighting, and of the war-fighter. This relatively neglected dimension of RMA requires new thinking about the role of military force in international relations, as well as about how nations raise, equip, train, and, most especially, about the ends for which these nations ultimately deploy, their military forces.

The new era of IW requires that military personnel develop a radically-altered vision of their own roles as warriors and peace-keepers, as well as appropriate recognition of the new demands IW

---

[1] In the Joint Operating Concepts of the U.S. Department of Defense, for example, an earlier designation of "Military Operations Other than War" (MOOTW) has since been replaced by a somewhat more refined, tripartite distinction between "irregular" warfare (IW), major combat operations (MCO), and SSTRO: "stabilization, security, transition and reconstruction operations" (such as humanitarian intervention and disaster relief). IW itself is defined as "A violent struggle among state and non-state actors for legitimacy and influence over the relevant populations. IW favors indirect and asymmetric approaches, though it may employ the full range of military and other capabilities, in order to erode an adversary's power, influence, and will." As I mention above, the most recent "term of art" to describe the complex interplay between conventional and unconventional operations is "hybrid" war. These distinctions are somewhat artificial and for the most part, poorly drawn and even more poorly described. The message, however, is clear: the U.S. and other allied and coalition governments will face criminal and terrorist attacks of unconventional sorts, employing a variety of tactics and weapons, aimed largely at civilian populations rather than enemy governments or their military forces, and aimed at creating insecurity among the populations that are destabilizing to their respective governments. The joint doctrine is aimed at defining and formulating strategic responses to a variety of such conflicts. http://www.dtic.mil/futurejointwarfare/concepts/iw_joc1_0.pdf

imposes upon the requisite expertise, knowledge, and limits of acceptable conduct for members of the profession of arms. Finally, the new era of IW requires that those institutions and personnel who educate and train future warriors be much more effective in helping them understand, and come to terms with, their new identity, the new roles they will be expected to play in the international arena, and the new canons of professional conduct appropriate to those roles. This book is intended to address those challenges.

I.

On Friday, September 4, 2009, two U.S. F-15E fighters dropped bombs on two fuel trucks near Kunduz, Afghanistan that were believed to have been hijacked by Taliban insurgent forces. The strike was ordered from a NATO command center by a German military officer, Col. Georg Klein. Despite his belief that the individuals surrounding the trucks were the armed, Taliban hijackers themselves, subsequent investigations clearly showed that a large number of local civilians, including children, were killed in the attack, more than 100 victims in all.[2] The Taliban hijackers, recognizing they were "sitting ducks," had abandoned the fuel trucks to the inhabitants of a nearby village, who were eagerly trying to scavenge the precious fuel.

The event prompted a controversy in Germany, already nervous about its first postwar deployment of military troops to foreign soil. Here is the sequence of events, excerpted from a report later that month in Germany's leading international news magazine, *Der Spiegel*:

> The crews manning the F-15 fighters first asked German Colonel Klein and his forward air controller in Kunduz whether they should first fly their jets at low altitude over the tankers. Such a "show of force" would have given the Taliban fighters and civilians the opportunity to flee. Klein apparently turned down the request, thereby "omitting" one of the escalation levels which, according to NATO procedures, need to precede an air strike.
>
> The U.S. pilots then asked whether the situation posed an "imminent threat." Klein, through his forward air controller, responded with a terse "confirmed." Klein's forward air controller or air commander, was a master sergeant, code-named "Red Baron." The American pilots also asked Klein's air commander twice whether German forces had had "troops in contact" with the enemy. The response, once again, was: "Confirmed." In truth, however, it appears that German forces from the Kunduz base had not been deployed to carry out reconnaissance of the situation in the riverbed where the tanker trucks were. The fact that the tankers had been stuck in the riverbed for hours meant they probably posed no acute threat to the base.
>
> In the absence of enemy contact or an acute threat, Klein lacked the authority to order the air strike by himself. If a commander's own forces are not under acute threat, he is required to consult with ISAF headquarters in Kabul before ordering an air strike. And if there is a risk of civilian casualties, then an air strike can only be authorized by NATO's Joint Force Command in the Dutch town of Brunssum. The charges are serious. They suggest that Klein and the "Red Baron" may have not told the pilot the truth and that the air strike was ordered on the basis of false information.[3]

My point in raising this case is not to add further to the criticism aimed at the time, and since, at the conduct of Germany's, or of any single, NATO member-nation's forces. Rather, it is to link this instance with others like it in a disturbing pattern. The problem seems to be not simply that there are allegations of inappropriate use of force by NATO troops in Afghanistan. The deeper problem seems to be that such instances point to the lack of a common and widely shared interpretation, for example,

---

[2] *Washington Post*, Saturday 5 September 2009: http://www.washingtonpost.com/wp-dyn/content/article/2009/09/04/AR2009090400543.html?sid=ST2009090400002.

[3] Der Spiegel, International Edition, 24 September 2009: http://www.spiegel.de/international/world/0,1518,k-6948,00.html.

of what concepts like "proportionality," "military necessity," "discrimination" (or what international treaties sometimes term "distinction") and non-combatant immunity actually mean among the different national forces that make up the NATO coalition in this conflict. Those disagreements lead to inconsistent applications of force, to avoidable instances unwanted civilian casualties, which, in turn, jeopardize the NATO coalition forces' goals in their conduct of their counterinsurgency campaign against the Taliban in Afghanistan.

The concepts cited above are essential in understanding American and allied policy in the midst of their present "irregular" or "hybrid" wars in Afghanistan and Iraq. Angry German citizens, for example, had labored under the false impression that their own troops participating in the Afghan conflict were merely engaged in "peace-keeping" activities, and were not empowered or authorized to use deadly force. Their misunderstanding of their own public policy failed to recognize, as a frustrated former Dean of the U.S. Army War College remarked, "that, in Afghanistan, there is, at present, no peace to keep!"[4]

These concepts are essential to understanding the rules of engagement and the status of forces agreements under which these troops currently operate in these countries—and indeed, under which militaries are likely to operate in any conflict for the foreseeable future. These concepts are essential components of international humanitarian law, and the "Law of Armed Conflict" (LOAC), to whose historic treaties and conventions the U.S. and its allies are, for the most part, signatories. But most essentially, the concepts cited above are *moral* concepts, derived from centuries of moral discussion about the justifiable entry into, and conduct of, armed conflict, and about the *ethical ideals of the profession of arms* itself. To say that these essential concepts are "not well understood," let alone that they might not be "taken seriously" by our nation's representatives and our military allies in these conflicts is tantamount to saying that we fail to understand why we are there, what we are doing, or what it will take to prevail in these tragic conflicts abroad.

The new era of "irregular warfare," in particular, highlights the urgent need for military personnel to understand and abide by these essential pillars and core values of the military profession itself.[5] Our reliance on coalition operations, as well as joint-forces maneuvers among and between the members of different branches of military services, highlights the need for a widely-shared understanding of the common goals we seek, and the common constraints under which we are entitled to pursue them. Many have referred to this problem as one of "ethical inter-operability," the need for the same kind of consensus and understanding regarding key moral and professional knowledge that we routinely strive for in joint technical and tactical operations in other respects.[6]

II.

It might be thought, however, that this entire matter is one of policy and international law, rather than ethics. Treaties and conventions set forth the basic principles of *jus in bello*, the proper conduct of war, or LOAC, setting forth proper regard for noncombatant immunity, discrimination, proportional use of force balancing military necessity against other values and concerns, such as the welfare of the civilian population, proper treatment of detainees, and the like. "Status of Forces" agreements, or "Rules of Engagement" (ROE), moreover, are the work of military lawyers, translating those broad legal principles into specific, and sometimes labyrinthian guidance concerning constraints placed upon military personnel conducting combat operations in the field.

---

[4] *Viva voce* remark of Col. Jeff McCausland, Ph.D., U.S. Army (retired), senior fellow at the Carnegie Council for Ethics in International Affairs, and also at the Stockdale Center for Ethical Leadership, at a seminar on Afghanistan at the Council's headquarters in New York City (October 22, 2009).

[5] See G. R. Lucas, Jr., " 'This is Not Your Father's War': Confronting the Moral Challenges of 'Unconventional' War," NATO 60th Anniversary Banquet Address. *Journal of National Security Law and Policy*, 3, no. 2 (2009), 331–342.

[6] See G.R. Lucas, Jr., "Pirates and PMCs: Internationalism and Ethical Interoperability," *International Journal of Applied Philosophy*. 23 no. 1 (Spring 2009), 87–94 [American Philosophical Association Symposium on President-elect Obama and U.S. Foreign Policy, 29 December 2008].

Frankly, if I were to point diagnostically to the grave failure of military ethics, it would be here, in allowing this perverse understanding of LOAC and *jus in bello* to prevail. This is an entirely external, and minimalist notion of the proper conduct of war, focused entirely upon what is permitted or prohibited, rather than upon what is sought or intended. Moreover, it reduces required moral reasoning entirely to the level of procedural questions about an often bewildering array of rules and regulations. All too often this results in the brigade or battalion commander asking the JAG (military lawyer) for guidance about ethics. The Commanding Officer's question, "Can we do this?" becomes a question about what the law permits, *when it is the CO himself who should be offering guidance about professional ethics*: "*Should* we do this?"

Our present attitude about ethics and leadership, in sum, is the moral equivalent of the military obsession with tactics (means), when what is wanted and needed is the moral equivalent of strategic thinking (ends). *Ethics, finally, is nothing more or less than strategic thinking about ends*, about goals and purposes, and about the relation of military means toward those moral and political ends. Military leaders, as professionals, require both a vision of the goal or purpose of military operations themselves, and a vision of what their own military personnel individually and collectively represent or stand for in the attainment of that goal. It is that strategic, ethical vision that this book endeavors to inform and inspire in our nation's future military leaders.

For my part, when midshipmen inquire somewhat skeptically about what, in this book, is termed *jus in bello*—the long historical discussion about "just conduct of war" and about that often confusing array of laws, treaties, ROEs, and the like—I reply that the entire subject can be reduced to a discussion between four members of a special forces team that took place on a mountainside in southern Afghanistan in late June of 2005.[7] It was, by the report of the lone witness to the conversation, not a relaxed discussion of differing views of procedure or principle over a cup of herbal tea. It was, rather, an intense argument, reflecting sharply contrasting differences of opinion, and wildly divergent, largely untrained intuitions about the prudent, the right, and the Good—raw and untrained moral intuitions that each member of the team brought to bear at that fateful moment of decision.

SEAL Team 10's heated discussion that summer morning was, finally, an argument over a *moral dilemma* that they encountered unexpectedly that day. They disagreed over what they might be permitted to do, even required to do, to ensure the success of their mission. They disagreed about what they decided to do (and just as importantly, what they decided *not* to do), and about what that agonizing decision meant to them, and ultimately what that decision (and countless dilemmas and decisions like it) mean to us all. Importantly, their reported conversation that morning had nothing whatever to do with alcohol, drunkenness, sexual harassment, or the other sorts of "character concerns" that routinely pass for "ethical issues" in the military.

The argument among the four members of SEAL Team 10 was over what to do with two adult goat herders and a fourteen year-old boy who, with their herd of 100 goats, had accidentally stumbled upon the Special Forces team, who had been inserted that morning into the rugged and isolated terrain in search of a notorious Taliban guerilla leader. To be sure, the discussion focused upon what was permitted, and what might be punishable under the law. But it also apparently focused upon something more: about who these "innocent civilians" actually were, whether they had any independent moral standing that deserved consideration, even at considerable risk to the armed combatants who detained them.

Theirs was a discussion, that is to say, about *professional military ethics*, much more than a debate over arcane terms of international law. And though they initially disagreed violently over the proper concerns to be considered and conclusions to be reached, they finally came to an agreement that, whatever the law of nations might conceivably permit or condone in this instance, they simply did not have the right to enforce a death sentence upon local inhabitants, who were otherwise going about their

---

[7] I first discussed this case and its instructive importance in "Inconvenient Truths: Moral Challenges to Combat Leadership in the new Millennium," 20th Annual Joseph Reich, Sr. Memorial Lecture, U.S. Air Force Academy (November 7, 2007): http://www.usna.edu/Ethics/Publications/Inconvenient%20Truths%20for%20USAF%20Academy.pdf. A more detailed version of the case by Capt. Rick Rubel, "Life or Death: SEAL Team 10 in Afghanistan" can be found in the accompanying case study book.

routine business in their own land, and who had unfortunately, inadvertently, stumbled upon the SEAL team's position. That misfortune, SEAL Team 10 concluded, awkwardly, and at great peril to themselves, did not warrant a death sentence.

And so they let the goatherds go, and bore the risks of their mission themselves, rather than transferring that risk to these locals. That decision, by all accounts, cost three of the four their lives. And so the lone survivor,[8] and many others who have studied and reviewed this situation, believe in hindsight that the SEAL team made a mistake, and that they should have executed these locals summarily on the spot, rather than subjecting themselves to additional (and what turned out to be very real) risk of harm and death.

I believe that conclusion to be precisely the wrong one to draw. And to their credit, SEAL Team 10 did not draw it. They paid a terrible price for their principled decision, and it is not surprising that one of those who helped reached that conclusion now questions his own judgment out of guilt over the loss of his comrades. But he was not wrong then, and we would be wrong now, to attribute their loss to a mistake in moral judgment. Instead, we should respectfully honored these fallen for having the courage to reflect upon their overall mission, and the purpose for their presence.

Their CO, Lt Michael "Murph" Murphy, apparently reminded them of this, quite appropriately. If they killed the goatherds, they could be found out, he reportedly remarked, and charged with war crimes. But even more importantly, their behavior in offloading their risky mission onto the shoulders of largely innocent bystanders would invariably compromise their mission, and cast their own roles as military professionals into disgrace and disrepute. In order merely to increase their own chances of survival, they would, by such logic, be compelled to betray the larger purpose that had brought them to this land. They were finally, if grudgingly, unwilling to purchase their lives at the price of their professional honor and integrity.

That, I think, was precisely the right conclusion to reach in such circumstances. To think otherwise is not merely to be confused about the professional military ethic, but is to fundamentally betray it. But that is clearly one person's opinion about an admittedly complicated and troubling case. The point of the present exercise is not to accept that opinion, or any opinion uncritically, but instead to develop the resources to form moral judgments for oneself, especially when confronted with dire and demanding circumstances.

There is an understandable tendency to want to avoid taking on that burden. Thus, it is often asserted by military personnel in the U.S. that "we are warriors, whose job it is to kill people and break things when ordered to do so, until someone else orders us to stop." That is what a respected colleague rightly describes, at the very end of a book on this subject, as ethics education designed to fit troops to fight "old fashioned" wars.[9] In the present instance, such a characterization of the 21st-century warrior is not quite correct, and not nearly up to the challenges military personnel now routinely face in "irregular" or "hybrid" wars. In the new era of "irregular war," the troubling truth about military personnel, instead, is that "you are warriors whose job it is to intercede between prospective victims, and others who are killing <u>them</u> and breaking <u>their</u> things, and order those <u>others</u> to stop!"

III.

This volume on military ethics and moral philosophy, now in its third edition, explores nothing more or less than the validity of this conclusion about what those Navy SEALS encountered, and what they did that day. It provides exhaustive resources to explore the background necessary to understand their actions, and evaluate their conclusions for oneself. This volume of philosophical readings is meant to accompany a companion volume, *Case Studies in Ethics and the Military Profession*, also edited by Captain William "Rick" Rubel, USN (retired) and myself. These two volumes, encompassing military theory

---

[8] See Petty Officer Marcus Lutrell, *Lone Survivor: The Eyewitness Account of Operation Redwing and the Lost Heroes of SEAL Team 10* (New York: Little, Brown & Co., 2007). The references I cite in my Reiff Lecture offer additional, conflicting journalistic accounts from interviews closer to the occurrence itself.

[9] Don Carrick, one of the editors, in his own concluding essay for *Ethics Education in the Military* (London: Ashgate Press, 2008).

and practice, comprise the course of study in ethics required of all midshipmen at the United States Naval Academy during their second year of study. There is also a companion "instructor's guide," the current edition of which is available in electronic format, prepared by Captain Rubel.[10]

The course is grounded upon genuine moral dilemmas encountered in the profession of arms, like those described above. This book, and the course it supports, are meant to address the military's "professional ethic" by aiding students and officer candidates to explore the justifiable resort to the use of military force to resolve international conflicts, provide security, and enforce the law. Most importantly, this volume helps readers to examine the restraint and good moral judgment required of those empowered with the use of deadly force for these justifiable public purposes. This text provides a wide range of philosophical and professional readings and resources for reflecting on an array of moral dilemmas that arise in the course of carrying out one's military duty, focused on a number of key themes and issues in professional military ethics. The companion book of case studies provides ample, updated circumstances for young officers and officer candidates to ponder as they prepare to pursue their careers as members of America's unrivaled military forces. For, along with unrivaled power comes sobering responsibility, and the demand to discharge one's chosen role of public service and public defense with honor, courage, and integrity. This book is intended to help its readers fulfill that challenging responsibility.

The term "professional" is rightly thought to convey a sense of public recognition, admiration, and trust. For any individual to be so designated and recognized is a mark of honor and prestige in society. Thus it is a cause for pride and celebration when a physician completes a lengthy internship and residency following graduation from medical school, or when a young engineer or lawyer successfully completes licensure or bar exams and assumes their place within his or her chosen profession. This is certainly true, perhaps even more so, in the case of the military officer, swearing an oath and accepting his or her commission in the nation's armed forces.

This pride of place and public honor is accorded, however, precisely because most people expect, and have a right to expect, a combination on the part of professionals of the highest technical and intellectual competence with the moral elements of public service, integrity, and self-sacrifice demanded of the members and practitioners of that profession. Thus, society generally acknowledges the profound moral values and obligations attendant upon joining a profession—be it medicine, engineering, law, journalism, the clergy, or the military—upon whose service its own welfare ultimately depends, and consequently, members of society pay homage to (and again, have a right to expect) high standards of personal character and conduct from each of that profession's members as "professionals."

In this context, military officers in service to their country, engaged in upholding and defending the universal moral values of individual liberty, justice, the rule of law, and respect for human rights, are rightly understood as "professionals," providing each is willing to understand, reflect upon, and accept the obligations of public service attendant on his or her professional service. These volumes of theory and practice are intended to meet this goal by acquainting both senior military officers, and midshipmen and cadets in training, with the moral foundations of military service, with the moral principles each will be expected to uphold as a representative of their organization or service, and also with the range of moral dilemmas or conflicts each is likely to encounter in the course of his or her career.

Finally, it is well to observe that professional education, in any setting, involves a unique wedding of theory and practice, of intellectual reflection and extensive, engaged experience. Everyone knows

---

[10]The first edition of this guide was prepared by Prof. Frank Chessa of the University of Nevada-Reno (at the time a graduate student in ethics at Georgetown University) and myself in 1997, using resources developed by our military faculty engaged in teaching the first pilot versions of the course. It was considerably improved in a second edition by Professor Susan Dwyer, a member of our faculty who is now professor of philosophy and director of the graduate program in ethics at the University of Maryland Baltimore County. Subsequent editions of the guide were prepared annually by myself, Captain Corky Vasquez, USN, and Professor Shannon E. French, who considerably expanded the use of multimedia video and film segments, and prepared teaching guides and discussion questions for these. Captain Rubel subsequently revised, improved, fully integrated and migrated the entire guide to our current USNA web site on "Blackboard," to support distance education training and to provide resources for military officers teaching a course in ethics for the first time. The guide in electronic form may be obtained free of charge by contacting him at rubel@usna.edu.

that it is not possible to study medicine or engineering, for example, exclusively from a textbook, or to comprehend the values and obligations of those professions merely on the basis of intellectual reflection alone. The same is certainly true, perhaps even more so, for military service. All three, as with other professions, are activities that engage the whole person, and commit the individual professional to a course of practice, dedication, and habituation in the means and methods of the profession whose mastery may encompass a lifetime.

Mere involvement in unreflective activity associated with the profession of arms, however, is not itself sufficient to produce a fully competent and trustworthy military officer and leader. In the past, the military services have been especially prone to believe such a fallacy, and to denigrate knowledge in favor of performance. Rather than celebrating with amusement the lackluster academic performance of many who rise to positions of authority in the military, perhaps we would do well to heed their own sober and rueful reflections on how much better they might have performed, and how much more they might have accomplished, had they attended equally both to theory *and* to practice, both to academic study and to practical achievement, when they had the opportunity. Moral ideals have great power, but only when they are recognized and valued within organizations, and put into practice by men and women of courage and dedication who collectively constitute those organizations. This volume, encompassing representative readings on traditions of moral reasoning, together with its companion volume of case studies grounded in military practice, aim to address this important need in professional military education.

Acknowledgements

As mentioned, this is the third edition of what has become the single most widely used textbook on military ethics in the world, now in use at both the U.S. Air Force and Naval Academies, and at more than 100 ROTC programs throughout the U.S. and abroad. It has been a privilege over nearly two decades to participate in the development of the core ethics course at the Naval Academy, and in the process, to help bring the various editions of this accompanying text to fruition.

Early on in this collective effort, a generous gift from Mr. William K. Brehm and Dr. Ernst Volgenau of the USNA Class of 1955 enabled the Academy to establish a Visiting Distinguished Chair in Ethics during the first six years of our program, from 1997 through 2003. This gift enabled us to bring a nationally-distinguished ethics scholar to the Academy each year for a residency of two years as we sought to organize our course, recruit permanent faculty, and build our program. The visiting chair was held, in succession, by Professor Nancy Sherman (Georgetown University), Douglas MacLean (University of North Carolina), and finally Roger Wertheimer (University of California-Irvine). For each, especially prior to the events of "9/11" and America's subsequent wars in Afghanistan and Iraq, the direct encounter with military personnel, and with the largely unaddressed problems of professional ethics in a military context, was a new and novel experience, largely without precedent in their own prior teaching, research, and writing. It has been heartening to see the extent to which our visiting distinguished faculty have since recognized the importance of this subject, and made substantial contributions to our understanding of it based upon their experiences in residence here.

Subsequently, the Class of 1984 resurrected the Distinguished Chair in Ethics, lodging it in the Stockdale Center for Ethical Leadership, where it has been my privilege to succeed these eminent colleagues as the first permanent appointee to the Chair. I wish to take this opportunity to thank both the Class of 1984, and the Director of the Stockdale Center, Col. Art Athens (USMCR, retired) for their support of my work in this field.

The present organization of our course, and contents of these two volumes, are now almost entirely due, however, to the subsequent ongoing work of our permanent Ethics faculty who succeeded me at the Naval Academy: Professors Shannon E. French (now Inamori Professor of Ethics and Director of the Inamori Center for Ethics at Case-Western Reserve University), Lawrence Lengbeyer (who prepared several of the edited classics for inclusion in the present edition), Deane-Peter Baker, Michael Skerker, Patricia Cook (now teaching ethics at the Naval Postgraduate School), David Garren, and Christopher Eberle. Their sustained work behind the scenes over several years in organizing the course, conducting educational seminars for our volunteer officer-instructors, and their consistent excellence in the classroom and as colleagues are what made our course, and these two textbooks, possible. Each has contributed both selected readings and, in some cases, original materials for our use, reflecting their wonderfully diverse areas of scholarly expertise.

This volume is enriched through the inclusion of several significant essays on essential topics written by my long-time colleague and friend, Dr. Martin L. Cook, who currently holds the Vice Admiral James B. Stockdale Chair in Ethics at the School for Operational and Strategic Leadership in the Naval War College (Newport, R.I.). Previously, Cook was Elihu Root Professor of Ethics at the Army War College (Carlisle, PA), and associate department chair in philosophy at the U.S. Air Force Academy. He is universally regarded as one of the nation's premier authorities in military ethics, and, in addition, is the only subject matter expert in that field to have taught in all of the military service cultures, a fact that lends additional authenticity to his insights.

Finally, the present volume remains dedicated to the one individual who, more than any other, is responsible both for the form and the very existence of the course from which it grew: Dr. Paul E. Roush. Readers will note a number of his essays included in this volume, and several cases also authored by him in the companion volume of case studies. There he is identified as Colonel Paul E. Roush, United States Marine Corps and U.S. Naval Academy Class of 1957 (see http://www.mcu.usmc.mil/mcu/Bios/Roush.htm). As a midshipman, he led his school's football team to victory over nationally-ranked Notre Dame University. As a member of both the 2nd and 3rd Marine Divisions, he served with bravery and distinction, and was a combat-decorated veteran of the War in Vietnam. Later, as Assistant Naval Attaché to the United States Embassy in Moscow, he personally provided the protection of his Marine guard to religious dissidents seeking political asylum and freedom in the former Soviet Union, an incident that sparked international attention and concern.

Upon retirement, and with advanced degrees from American University and the Johns Hopkins School for Advanced International Studies, Dr. Roush then served for many years as professor of leadership studies in what is now known as the Department of Leadership, Ethics and Law at the U.S. Naval Academy. He was an unswerving champion of equal opportunity for men and women in military service, and a principal figure in the initial development of the Naval Academy's core ethics course, for which he served as the first course coordinator. He championed, with equal courage and steadfastness, the notion that such a course must always represent both a marriage of theory and practice, and a collaboration in both design and teaching between civilian academic scholars and subject-matter experts, and senior military officers with line and command experience. His vision continues in force to this day. He remains—as did his own hero and mentor, Vice Admiral James B. Stockdale—an exemplar of the resolve, steadfastness, moral courage, compassion, and devotion to moral duty that this volume and the course it supports aim to instill in our nation's future leaders. *Semper Fi!*

Part I
Why Study Ethics?

Ethics and the Military in America

The purpose of any introductory course in ethics is to develop each individual's intellectual capacities for making well-reasoned ethical decisions, and for explaining or justifying those decisions to others.

Within military service, this particular sort of competence is especially important. In the United States, for example, the several military services and the nation they serve are said to be founded alike on a common set of basic moral principles and values. These principles and values are set forth in key documents pertaining to America's Founding, such as the Declaration of Independence and the Constitution. The broad principles of liberty, rights, equality under the rule of law, separation of church and state, and the separation of political powers (including civilian control of the military) are taken to provide a framework that guides, and to an important degree constrains, the individual military officer's personal moral and political deliberations and actions.

If this assumption is correct, then it is incumbent upon each member of the U.S. military to have a reasonable degree of familiarity with these documents, with the traditions of moral and political reflection they sustain, and also with the expectations or constraints they lay upon his or her personal life and professional service. Thus, a survey of the main features of military law, and a course in government and the Constitution are frequently among the requirements of any junior officer's preparatory education.

Two centuries of Supreme Court rulings, however, testify to the difficulties of achieving consensus on the proper interpretation and application of those moral and political principles, even as they testify to the lively and ongoing public interest and concern in their meaning and proper application in support of democracy and the rule of law. It is not easy, that is to say, simply to derive one's personal responsibilities and a consistent code of conduct straightforwardly from the provisions of such documents.

Part of the practical difficulty is that the Constitution, the Declaration of Independence, and other documents outlining the foundations of our political order for the most part offer very general, abstract principles about our fundamental political relations as citizens. They do not specify the particularities of the endless variety of complex moral issues that military officers, not to mention individual citizens generally, are likely to confront during the course of their lives and careers. Moreover, the attempt by Congress to clarify and codify some of the more important of these specific responsibilities for men and women in military service—the Uniform Code of Military Justice—is at once a lengthy and complex document in its own right, and likewise doesn't pretend to cover every conceivable circumstance.

Hence, a great deal more hard work, individual study, and reflection are required beyond merely a cursory examination of our nation's founding documents, or even of the specific provisions of military law, in order to prepare each member of the armed services for his or her public responsibilities. To that end—in addition to the study of law and Constitutional government—each of the undergraduate service academies, most of the university-based programs of the Reserve Officer Training Corps (ROTC), as well as Officer Candidate School programs, and even the nation's senior military war colleges, also require the study of ethics and moral philosophy in some form. This text, and its companion volume of case studies in military ethics, are intended as resources in support of the teaching of ethics in these various settings.

The Frustrations of Ethics

That said, these texts and cases cannot, and do not, provide a straightforward recipe for resolving all moral dilemmas. Neither the editors, nor the many contributors (past and present) assume that there is some single, clear formula or algorithm, acceptable to all reasonable persons of good will, that unfailingly identifies the right answer for every difficult moral choice. This is often a hard limitation to accept.

It seems to support a common complaint about the study of ethics: namely, that *"in ethics, there are never any answers, so why must we study it?"*

It would be vastly more convenient if centuries of moral reflection and debate could be "boiled down" to a few essential principles—such as "be a person of integrity," or "do the right thing" or "respect truth and loyalty," or "never leave a comrade behind"—of the sort that might fit conveniently on a plastic wallet card, or serve as a catchy billboard slogan to entice new recruits. Leaders in the military, business, and political sectors often long for such a clear-cut system that they might follow themselves, and pass along to their colleagues, co-workers, or subordinates.

Yet every attempt on the part of organizations or "professional ethicists" to achieve such an economy of expression seems to be doomed to failure. The "crash course" or the "one-day" or "weekend seminar" in essentials of professional ethics ends up invariably providing little more than a comic and superficial caricature of moral reasoning. Likewise, the understandable desire to condense all of moral reasoning into a few catchy "bullets" on a multimedia computer slide invariably ends up making important moral principles (such as the several worthy moral ideals listed above), or the traditional values of venerable organizations (such as the U.S. Naval Service's emphasis on "honor, courage, and commitment"), seem somehow trite or trivial, instead of profound. In the end, the "shortcut approach" to ethics just doesn't seem able to capture the moral complexities of life in the organization, or to do justice to the history and rich moral heritage of the organization itself.

Even if there *were* a single, unified moral "theory," discerning right from wrong conduct in each and every case, or even if we could develop a single, all purpose Ethical Algorithm or decision-guiding formula or checklist to guide our practical behavior, its proper interpretation and application in particular cases would still be ripe for endless debate. Whether our fundamental moral framework is traced to the U.S. Constitution, or to passages from Holy Scripture, or alternatively to a set of professional core values, it would nevertheless remain the case that honest and sincere persons, while swearing to abide by these codes, would still ferociously oppose one another's understanding and application of these general principles to particular cases. Hence, no simple, abbreviated handbook of "Ethical Rules" can be a substitute for the skills of judicious deliberation. And these may only be honed through hard work, reading, study, discussion, reflection, and ultimately through practice and application. These two volumes of moral theory and practice aim to provide the resources for such reflection.

What Might We Gain from the Study of Ethics?

Many students, especially those required to take an "ethics course" as part of their professional preparation in medicine, engineering, or business, come to their first class honestly wondering what any such ethics course could teach them that they need to know and don't already know. It is not uncommon for them to feel certain that their moral beliefs are just fine the way they are—perhaps because their faith in God's word as a source of moral guidance is firm, or perhaps, quite differently, because they think everyone's beliefs are their own private affair, and consequently no one has the right to tell anyone else what to believe, let alone how to behave.

In fact, the situation is even more paradoxical than it first appears. Serious ethical reflection cannot begin, and cannot be taught, unless the inquirers and students, as well as the teachers, already have a basic understanding of the essential ethical concepts and can make moral discriminations and judgments and reason about them. They must already know enough for them to be held morally responsible for their decisions and actions. But if that's already the case, what is to be learned?

The answer is: *quite a lot*, far more than could be covered in one semester.

To begin with: it is hardly surprising that many cadets or midshipmen share moral attitudes in common with their age group, or the population at large, that are nonetheless rejected by the several branches of the United States Armed Forces. Right off the bat, this generates a serious moral conflict that must be resolved. For example: many midshipmen or cadets arrive at their respective service academies unprepared for the insistent demand among military organizations for rigorous truthfulness, even when it means exposing the misdeeds of one's friends. Placing truth over loyalty to friends may seem to young people both wrong in principle, and difficult if not impossible to live by in fact. Yet this is the expectation placed upon them from the outset as part of their orientation and education to military life. Likewise, prevailing attitudes toward sexual mores and "fraternization," or "sharing" work on examinations, or many other serious matters, may run counter to the expectations placed upon a uniformed officer in service to his or her country, even if such practices otherwise appear to be widely tolerated, even when condemned in the larger society.

The situation is likely to be quite different, however, for senior officers teaching ethics for the first time in a service academy, or studying the subject for the first time in a war college. After twenty or more years of professional service such individuals will be fully imbued with the need for truthfulness, integrity, and circumspect professional behavior, and may reject the widespread moral relativism so characteristic of our larger culture. Still, they may find themselves equally at a loss to engage rationally and analytically in a topic or subject like ethics that they firmly believe to be a matter of intuition and innate character (and so not a subject that can be taught or learned).

Finally, and most seriously: both groups, younger professionals-in-training and older, more experienced professionals, may share the view of many people who think that, in war, whatever helps win is permissible, as though when waging war we inhabit an amoral world where only victory matters. A hardly more advanced idea is that, if one's cause is just, then whatever helps win is justified. *Our Armed Forces as organizations, however, do not tolerate that attitude.* By treaty, and by the venerable tradition of the warrior's honor alike, our military is bound to respect substantial ethical constraints on its conduct in war, constraints that may make victory more difficult, dangerous and costly. Our conception of the proper conduct of war has grown more restrictive over time as our sense of justice, decency and humanity has matured. Officers and leaders, at any age, and in all leadership situations, must understand what those restrictions are, and come to see them as worth upholding. They must come to understand the moral constraints on the use of force as reflecting universal moral values to which they themselves would otherwise voluntarily subscribe, rather than as arbitrary and irrational constraints, unreasonably imposed from sources outside the profession.

Such constraints on the exercise of force and on conduct during combat will appear far less arbitrary and inconsistent if all of us come to understand more adequately the moral rationale for those restrictions. Otherwise, the temptations to transgress them can be as powerful as the passions of war. There would be no need for the legal restrictions if there weren't such powerful motives to transgress them. Those temptations can prove irresistible when the restrictions are learned by rote and regarded as unreasonable impediments to effective military action. The study of ethics and moral philosophy, especially in the context of wartime cases, helps dispel this misunderstanding and reinforce our nation's commitment to use force sparingly, and only insofar as is necessary, for the purpose of defending liberty, protecting human rights, and enforcing the rule of law among nations and peoples.

The Role of Philosophy in Morality

An appreciation of the ethics of the military service is only one part of what needs to be imparted. Perhaps more important (and certainly more difficult) is the development of general ethical reasoning and decision-making skills. As Socrates discovered over two millennia ago, it is only after we try to reflect on our moral beliefs in a rigorous way that we begin to realize how deficient our understanding of ethics really is. Few people are naturally motivated to engage in intellectually serious reflection on their moral beliefs, which entails both the ability to articulate their beliefs and the reasons for holding them clearly and coherently, and also the ability to respond intelligently to sustained critical analysis. Yet such reflection is vitally important, and we need always to be sharpening our ability to undertake it. Despite our certainties, that is to say, we just might, on occasion, be mistaken! It is important to examine our most strongly held convictions from time to time, if for no other reason than to ensure we don't become the next public example of misplaced confidence or arrogance.

In other respects, the historical figure of Socrates provides a superb illustration of the central purposes of this text and of the course of study it advocates and supports. In his *Apology* to the Athenian assembly, Socrates explicitly cited his own military service as the source of his understanding of what it is to remain steadfast, to continue to do one's duty in the face of adversity and never be willing (as he put it) to abandon his assigned post. He then explained how, during his later life, he came to understand his "assigned duty" from the gods to serve Athens as a teacher, and to goad his fellow citizens toward greater self-examination and toward the improvement of their moral lives, even if they did not always wish to undertake that improvement. This divinely-assigned "military duty" he likewise would not abandon, even under the threat of death.

Centuries later, the great Roman emperor, warrior, and Stoic philosopher, Marcus Aurelius, would quote the inspiring words from Socrates's *Apology* in order to inspire his own legions for battle:

> "... wherever a man has placed himself thinking it is the best place for him, or has been placed by a commander, there in my opinion he ought to stay and to abide the hazard, taking nothing into the reckoning, either death or anything else, before the baseness of deserting his post." (Plato, *Apology* 28d)

In contrast to teachers in our present era, Socrates was not a purveyor of elaborate, abstract "theories," for which he exhibited no patience whatsoever. Even less was he a vain or pretentious advocate of philosophy as the special domain of "experts." Quite the contrary, his method of inquiry—of teaching and learning—was grounded firmly in actual practice, and he carried it out daily with, among, and (he argued) *for the sake of* ordinary men and women engaged in the pursuit of their common life.

Thus, he cross-examined the devoutly religious about the nature of piety, and questioned the leading military figures of Athens about their understanding of courage. Their piety, authority, or military experience alone were never, upon cross-examination, found to be sufficient in and of themselves. All of these individuals, like all of us, were reasonably content with what they were doing, and reasonably "self-certain" that they knew right from wrong. And, like them, it is only upon forcible reflection—even while engaged in doing what is right, or when acting courageously or piously—that we discover ruefully that we have no very clear idea what we are about, or why what we are doing is really important. As a result, in a crisis or under challenging circumstances, we have no resources upon which to fall back for guidance. This, finally, is what a good course in ethics, rooted in case-studies, will help us develop.

The Readings

"On the Eve of Battle" is a short story that incorporates the actual, recorded text of a brief, inspirational speech given by a British commanding officer to his troops in Kuwait just prior to the onset of the March, 2003 war in Iraq. Citizens from many nations have argued very publicly and vociferously about the morality *of* that war at the time, and since. But what are we to make of Lieutenant Colonel Collins' remarks about what we might term "ethics *in* warfare?" What are the moral values and organizational traditions he invokes, and why do warriors (not to mention the rest of us) find such language so inspiring? Does this speech in fact embody noble and honorable moral truths, which inspire directly as a result of their universal recognition and claim on us? Or, as the young Marine officer listening to the speech wonders, can such lofty ideals sometimes disguise other, less worthy political motives?

The next reading, from Plato's *Republic*, directly addresses the second of these two options, one that we might call moral *realism*. Realism in ethics, as in political science, is the concept that actions are always self-interested and amoral, and that morality is merely a fancy way to dress up this depressing fact. Thus, Socrates' friend, Glaucon, offers the view that "justice" in social institutions—or what we might call "righteousness" or integrity in the case of individuals—is really an elaborate and misleading masquerade for actions that are purely self-interested. Morality is thus only an outward convention, an appearance disguising mere self-interest, and not a valid expression of an individual's true character. Since, on Glaucon's hypothesis, our true nature is always to act in our self-interest, if we found ourselves freed from the constraints that public accountability and fear of reprisal place upon our actions, we would discover ourselves capable of doing almost anything, no matter how "immoral" such behavior might then appear.

This is a hard view to countenance: what we think of as "morality" or good individual character may really be only the result of enlightened self-interest, a kind of amoral *realpolitik*. So, is Glaucon correct? And are the noble ideals that Lieutenant Colonel Collins invokes merely, in the end, a form of empty rhetoric designed to deceive or to manipulate others into fighting bravely?

In particular, do the stories and illustrations that Glaucon offers as evidence for his position really prove his point? This is the hard part about ethics: evaluating evidence, and assessing the quality and validity of arguments offered in support or criticism of a particular position. This is distinguished from a debate competition in important respects, in that debates are often won by the individual who speaks the loudest, the fastest, or assembles the most apparently relevant evidence in rapid-fire succession.

Philosophical argument should not be decided by such tactics, but instead, by the relevance and quality of evidence produced, and by the logical rigor or strength of the arguments put forth.

Glaucon's position—that morality is nothing more than a lofty appearance disguising self-interested behavior—is a very old position, even though it is often offered up by contemporary cynics as the most enlightened and modern of views. To that end, we might pay close attention to Glaucon's third example, in which he offers a choice between two kinds of lives: one truly righteous, but unrecognized; the other morally corrupt, but otherwise quite prosperous and successful. Which life would we choose, if those were the only choices?

Finally, Captain Thomas B. Grassey's "Why Ethics Is So Hard," provides a military officer's rationale for the study of ethics. Military officers need not be philosophers, he argues, but they are often required, as leaders and decision-makers, to consider and react to a range of practical dilemmas with a strong moral component. (What Grassey has in mind, in fact, are the kinds of actual scenarios captured in our accompanying volume of case studies in military ethics.) Moral decisions are, he laments, invariably made about complex issues in the context of shifting historical circumstances, and in the absence of sufficient time to reflect, or of complete information to consider. That makes moral deliberation difficult, and clear answers hard to come by.

Ethics is also "hard," more ominously, because morality is not only about individuals and their character, but about the values and organizational coherence of the institutions within which individuals operate. The style of organizational leadership, the reward system, and the traditions of the institution may erode, rather than nurture and reward, the ability and confidence of even most morally upright of individuals. Each of us must prepare *in advance*, as best we can, to confront our inevitable dilemmas in less-than-perfect organizational settings by equipping ourselves with as much knowledge, insight, and moral courage as we can develop.

On the Eve of Battle
*George R. Lucas, Jr.**

In Memoriam: Kylan Jones-Huffman, Lieutenant USNR
USNA Class of '94
KIA 21 August 2003 outside Al Hillah, Iraq, with the 1st Marine Expeditionary Force.

On Tuesday, March 18, 2003, Captain John Erskine, U.S. Marine Corps, First Marine Expeditionary Force, watched the sun set over the desert along the northwestern border of Kuwait. Sunset was the one time this bleak and forbidding landscape seemed to come alive with a peculiar kind of fragile beauty, as the bleached-white and barren terrain suddenly seemed to glow like a living fire in the rays of the setting sun.

Not many miles from where he stood, the legendary patriarch, Abraham, was said to have lived many thousands of years ago. According to stories sacred to Jews, Christians, and Muslims alike, Abraham had been commanded by God to set out from the city of Ur on a journey across this very desert in search of a new land in which to found a nation. In a few more hours, Erskine fully expected that he and his comrades would likewise be commanded to set out on the modern highway, stretching out before him like a thin ribbon in the sand into Iraq, heading toward Basra, turning west toward Nasiriyah, past the birthplace of Abraham, and on toward Baghdad, on a mission of a very different sort.

Erskine and his company had been bivouacked in this location for several weeks, together with British Army troops from the Royal Irish Brigade, preparing for a possible military strike into Iraq. The endless waiting, and training for an uncertain mission, had begun to wear on their nerves. They were eager for something to happen, one way or another. Only a few hours before, he and members of his company had listened to the long-anticipated speech by the U. S. President to the nation, issuing a final ultimatum for Iraq's leader, President Saddam Hussein, to abandon his office and go into exile, or face a punitive military invasion by U.S.-led coalition forces. His Commander-in-Chief had explained the pending military mission as a war of law enforcement, authorized in accordance with terms set forth in a number of resolutions passed by the United Nations—resolutions with which Iraq had refused to comply. Erskine and his comrades, however, knew the matter at hand was far more complex than this, and he could not put the nagging uncertainties out of his mind. In the fading light, he tried to put his thoughts in order.

Captain Erskine harbored no illusions about Iraq's leader. He had been a young teenager when, over a decade earlier, Saddam and his then-massive army had moved into Kuwait to capture its rich oil fields, prompting an international outcry and, ultimately, a swift, massive and destructive war that drove the Iraqi army out of Kuwait. He knew of the man's brutality, of the many questions concerning

---

*"Captain John Erskine" is a composite figure based upon embedded news reports from the battlefront, intended to capture their collective experience while respecting the privacy of those actually depicted in news accounts and interviews at the time. The speech of Lieutenant Colonel Tim Collins to the Royal Irish brigade on the eve of the Iraq war was reported in the *London Times* by Ben MacIntyre on March 22, 2003.

the legitimacy of his rule, and of his suspected links to terrorism (including bounty payments to the families of Palestinian suicide bombers in Israel). Erskine had little doubt about the authenticity of reports claiming that Saddam's military possessed terrifying biological and chemical (and possibly also nuclear) weapons of mass destruction in clear violation of U. N. sanctions, despite the failures of U.N.-appointed weapons inspectors to find any concrete evidence of them during the past few weeks. It gave him genuine satisfaction to contemplate the prospect of putting this dangerous and ruthless tyrant out of commission, once and for all.

But the world was filled with bad men and even worse governments. Rebel militaries in countries like Liberia, Sierra Leone, Congo, and Algeria had committed unspeakable atrocities against unarmed and helpless civilian non-combatants in those countries in an effort to terrorize these people into accepting their authority. Shouldn't we, thought Erskine, or *someone*, do something about that as well? An even more ruthless tyrant in North Korea, the dictator Kim Jong Il, pursued reckless policies of militarization and mismanagement that drove millions of his own people to the brink of starvation, while making no secret of the fact that he actually possessed weapons of mass destruction, including nuclear weapons, and was fully prepared to use them against South Korea, Japan—even boasting that he could fire missiles with nuclear warheads at the western coast of the United States. Erskine's own grandfather was a combat-decorated Army veteran in the Korean War in the 1950s. The blood of American servicemen had been spilled in an effort to bring peace and protect freedom in that country already. If we were now concerned about the threat of Saddam, shouldn't we be even more concerned about Kim Jong Il?

Erskine shook his head to clear it. While he couldn't help having his own opinions about these questions, he acknowledged that there are times when such matters were simply not his or any other junior officer's to decide. As a member of the military profession, he told himself, he had taken an oath of loyalty to support and defend the Constitution. The Constitution, at least as he understood it, placed responsibility for using military force on the shoulders of democratically-elected civilian leaders in a complicated decision process. While those processes might not be always infallible or beyond reproach, they certainly contained no provisions for "second-guessing" by a junior officer. As a U.S. Marine, moreover, he was fond of telling his friends and family that he was proud to go wherever he was ordered to go, and to serve the causes his nation deemed significant. "Me and my guys," he would proudly assert, "are the 'pointy end of the spear.' It's our job to go where the elected leadership tell us and 'kill people and break things' until we are told to stop!" This, it now seemed to him, was one of those times.

Besides, Erskine remembered, he had been at home, on leave with his family for his daughter's third birthday, on the morning of September 11, 2001, scarcely eighteen months ago. He had just happened to have the morning news on the television around 8:45 when the first reports of the dramatic terrorist attacks on the Pentagon and the World Trade Center began to be broadcast. He, like millions of other Americans, had watched in horror and stunned disbelief as, in short succession, both of the immense, burning towers crumbled to the ground, killing thousands. He had been disappointed in the ensuring months that he was not personally deployed to Afghanistan to help search out and capture or destroy the perpetrators of that atrocity. "If our Commander-in-Chief, Secretaries of Defense and State," he thought, "and senior members of the intelligence community all believe that Saddam may have had something to do with this, and may now be manufacturing and selling weapons of mass destruction to these criminals, while starving and brutalizing his own people in defiance of the international community, that's good enough for me!"

His thoughts were interrupted by the stirring of personnel in the British encampment nearby. He walked over for a closer look, nodding to the sentry on duty, whom he had come to know quite well. In the days and weeks preceding, he had often had time to trade jokes and barbs with "the Brits" in the neighboring camp, and was quietly grateful for their steadfast support of his country's efforts to combat terrorism. He recognized their C.O., a tough, brazen, cigar-chomping Lieutenant Colonel named Tim Collins, gathering everyone within earshot to listen to his briefing. Erskine caught him in mid-sentence:

". . . the enemy should be in no doubt that we are his Nemesis, and that we are bringing about his rightful destruction. There are many regional commanders who have stains on their souls and they are stoking the fires of Hell for Saddam. As they die they will know their deeds have brought them to this place. Show them no pity. But those who do not wish to go on that journey, we will not send."

"We go to liberate, not to conquer. We will not fly our flags in their country. We are entering Iraq to free a people, and the only flag that will be flown in that ancient land is their own. Don't treat them as refugees, for they are in their own country. I have known men who have taken life needlessly in other conflicts. They live with the mark of Cain upon them. If someone surrenders to you, then remember they have that right in international law, and ensure that one day they go home to their family. The ones who wish to fight, well, we aim to please. If there are casualties of war, then remember, when they woke up and got dressed in the morning they did not plan to die this day. Allow them dignity in death. Bury them properly, and mark their graves."

"You will be shunned unless your conduct is of the highest, for your deeds will follow you down through history. Iraq is steeped in history. It is the site of the Garden of Eden, of the Great Flood, and the birth of Abraham. Tread lightly there. You will have to go a long way to find a more decent, generous, and upright people than the Iraqis. You will be embarrassed by their hospitality, even though they have nothing."

"There may be people among us who will not see the end of this campaign. We will put them in their sleeping bags and send them back. There will be no time for sorrow. Let's leave Iraq a better place for us having been there. Our business now is north."

Erskine walked back to his own encampment afterwards, inspired, and yet also sombered, by the British soldier's speech. "This is not my father's, or my grandfather's, war," he thought, shaking his head. He had never heard a senior officer speak in quite this way: tough, proud, confident, but also full of high moral ideals and compassion. Instinctively he grasped the "lucky marble" that he always carried in his belt pouch. His daughter, Amanda, had given it to him. It was clear, with a deep blue center. She had called it "the big blue marble" after a kid's show she loved to watch. The "big blue marble" in that show was the earth, with all its countries, oceans, and peoples. "My daddy's somewhere out on that marble, taking care of us" she would tell her friends. To Erskine, however, the beautiful blue color matched perfectly the color of Amanda's eyes.

Suddenly, as happens in the desert, night was fast upon him. One moment the trailing edge of red sun had just dipped below the horizon, and, in almost an instant, it was dark. Erskine gazed at the startlingly-bright stars just beginning to appear overhead. In the clear, dry air of this ancient place, their light shown with a special brightness. Erskine wished his thoughts were as clear as the bright light of those stars above. By their light, according to another famous legend, a trio of astronomers in ancient Babylon had navigated their way toward a tiny village further west, known as Bet Lehem, in search of a savior. Tomorrow, Erskine thought, he would make his way, this time navigating by global positioning satellites, toward Baghdad, the legendary capital of that ancient empire from which those three "wise men" had first set out. His mission was salvation of a different sort: to put an end to the rule of a cruel tyrant, and to set free his captive population. Erskine sighed. He hoped they would appreciate the gesture, and that not too many of them, or of his own troops, would have to die in the process.

On the morning of March 19, 2003, British troops and the U.S. Marines of the First Marine Expeditionary Force encountered fierce resistance on the outskirts of Basra. An exploding mine overturned an armored troop carrier in Erskine's convoy, injuring some of his men. While providing protection and rendering aid to Marines caught in the withering cross-fire, Captain John Erskine was struck multiple times by enemy fire and killed, among the first casualties of the second Gulf war. Over seven years later, coalition forces still remained on deployment in postwar Iraq.

The Ring of Gyges
Plato (c. 360 B.C.E.)
Excerpt from *The Republic,* Book 2
Plato

Glaucon: "I'll begin with my first subject then: the nature and origin of justice.

"People say that injustice is by nature good to inflict but evil to suffer. Men taste both of its sides and learn that the evil of suffering it exceeds the good of inflicting it. Those unable to flee the one and take the other therefore decide it pays to make a pact neither to commit nor to suffer injustice. It was here that men began to make laws and covenants, and to call whatever the laws decreed 'legal' and 'just.' This, they say, is both the origin and the essence of justice, a thing midway between the best condition—committing injustice without being punished—and the worst—suffering injustice without getting revenge. Justice is therefore a compromise; it isn't cherished as a good, but honored out of inability to do wrong. A real 'man,' capable of injustice, would never make a pact with anyone. He'd be insane if he did. That, Socrates, is the popular view of the nature of justice and of the conditions under which it develops.[1]

"That men practice justice unwillingly out of inability to do wrong may be seen by considering a hypothetical situation: Give two men, one just and the other unjust, the opportunity to do anything they want and then observe where their desires take them. We'll catch the just man ending up in the same place as the unjust. This is because human nature always wants more, and pursues that as a natural good. But nature has been diverted by convention and forced to honor equality.

"Our two men would have the opportunity I mentioned if we gave them the power once given to the ancestor of Gyges, the famous king of Lydia. They say he was once a shepherd serving the man who was then king. One day a great earthquake opened a chasm in the place where he was pasturing sheep. Astonished, he climbed down into the fissure and saw, among other fabulous things, a hollow bronze horse with windows in it. He peeped inside and saw the body of a man, seemingly larger than life and wearing only a golden ring. He took the ring and left. Later he wore it to the monthly meeting of shepherds, at which they made their reports to the king. As he was sitting there with the others he happened to turn the setting of the ring toward him. Suddenly he became invisible, and the others began to speak of him as though he were gone. Amazed, he turned the setting away from him again and reappeared. After further experiments had convinced him that the ring indeed had the power to make him visible and invisible at will, he contrived to become a messenger to the king. He seduced the queen, with her help murdered the king, and usurped the throne of Lydia.

"Now if there were two such rings and we gave one to our just man and one to the unjust, no one, they say, would have the iron will to restrain himself when he could with impunity take what he liked from the market, slip into houses and sleep with anyone he wanted or kill whomever he wished, free

---
[1] The first presentation in literature of the "social contract" theory of justice.

people from jail, and like a god among mortals, do whatever he pleased. In that situation there'd be no difference between our two men; both would act alike. So a person may use this as evidence that no one is willingly just, since whenever a man thinks he can get away with injustice he does it. Justice is practiced only under compulsion, as someone else's good—not our own. Everyone really believes that injustice pays better than justice; rightly, according to this argument. Because if a man had the power to do wrong and yet refused to touch other people's property, discerning men would consider him a contemptible dolt, though they'd openly praise him and deceive one another for fear of being harmed.

"We'll be able to make a proper judgment on these two ways of life only if we contrast the perfectly just man with the perfectly unjust. Make each a perfect specimen of his kind. The unjust man should act like a skilled craftsman. An accomplished doctor or navigator distinguishes between what's possible and impossible in his field and attempts the one but ignores the other. And if he botches something he knows how to straighten it out. So with our accomplished criminal: he should discriminate nicely and get away with his crimes. If caught he's a bungler. And injustice's highest perfection is to *seem* just without *being* so. Deprive our perfectly unjust man of nothing, therefore, but lend him utter perfection: Let him contrive the greatest reputation for justice while committing the most heinous crimes, and if he should bungle something, grant him the ability to straighten it out; give him the persuasion to sway juries when informers denounce him, and the wealth, influence, courage, and vigor to use force when force must be used.

"Alongside this paragon of perfidy let us place our perfectly just man—a man noble and simple, desiring, in Aeschylus's words, 'not to *seem,* but to *be* good.'[2] All right, let's take away seeming. A reputation for justice will bring him honor and rewards and make it uncertain whether he practices justice for its own sake or for the rewards. So strip him of everything but justice and make him the exact opposite of our other specimen. Let him win the worst reputation for injustice while leading the justest of lives, to test him and see if his justice holds up against ill repute and its evil effects. And let him persevere until death, seeming unjust while being just, so we may examine the extremes of justice and injustice and decide which makes the happier life."

"Uncanny, Glaucon!" I cried. "You've scoured these two types for our judgment as though they were bronzes."

"I do my best, Socrates. Now with these two men before us it shouldn't be hard to tell what sort of life awaits each. And if it sounds uncouth, Socrates, don't blame me, but the champions of injustice. They say our just man will be whipped, racked, chained, and after having his eyes burnt out and suffering every torment, be run up on a stake and impaled, and so learn that one ought 'not to *be,* but to *seem* just.' Aeschylus's words rightly apply to the unjust man who, since he doesn't live by opinion but pursues what clings to the truth, desires 'not to *seem,* but to *be un*just:'

> out of the deep furrow of his mind
> reaping a crop of faultless designs.[3]

Because he *seems* just, they say he'll rule in his city, marry into the family of his choice, form partnerships with anyone he wants, match his daughters with whomever he wishes, and from his lack of aversion to committing injustice reap advantage and gain. He'll get the better of his enemies by defeating them in lawsuits, public and private, and so grow rich and able to help his friends and harm his enemies. With ostentatious sacrifices, dedications, and gifts, he'll serve the gods and the men of his choosing far better than the just man, and one can reasonably expect he'll be more loved by the gods. Thus, Socrates, both gods and men, they say, see to it that the unjust man enjoys a better life than the just."

---

[2] *Seven against Thebes* 592.

[3] Aeschylus, *Seven against Thebes* 593–94.

Why Ethics Is So Hard
Capt Thomas B. Grassey, USN*
James B. Stockdale Professor of Ethics,
Naval War College

Naval officers are practical, concerned with what should be done. Ethics is therefore interesting, for it is preeminently practical. The purpose of ethical consideration is to decide what ought to be done now, in this situation, all things considered.

Ethics has an ancient, even pre-historic, connection with the bearing of arms. Those who wielded the weapons of death were regarded as having sacred powers, a responsibility that placed them with those who healed the sick and interceded with the gods. Priests, doctors, and warriors were three easily recognized classes in many societies.

In our own day, General Sir John Winthrop Hackett famously observed, "The military life, whether for sailor, soldier, or airman, is a good life. The human qualities it demands include fortitude, integrity, self-restraint, loyalty to other persons, and the surrender of the advantage of the individual to the common good.... What the bad man cannot be is a good sailor, or soldier, or airman."[1] This often has been recognized by service leaders, past and present. Honor, courage, and commitment are at the core of our profession.

But ethics rests on theories about the individual, society, the universe, knowledge, and life. Because such matters are beyond the normal ken of naval officers and most other professionals, we should not be surprised when their complexity manifests itself in confusing implications about our duties. Indeed, perhaps the most important point to acknowledge is that *ethics is hard, not easy*. Why is this so?

The Perspective of the Individual

Happily, in most situations we easily distinguish right from wrong. In fact, we usually make decisions without any awareness of their "ethical" aspects; it simply never occurs to us that we could steal a shipmate's wallet or lie to our commanding officer. With properly formed characters, the reinforcement of good habits enhances the efficiency of our thinking, and we pursue goals with no conscious reference to standards of right and wrong. This is the most common reason why we imagine ethics to be obvious, straightforward, and easy.

Everyone, however, regularly confronts an inescapable fact of moral complexity: life does not always offer a choice between good and evil. Sometimes we realize that our options are compounds with elements we normally would not accept in our actions. We grow aware of each alternative's

---

*I wish to thank all the officers who participated since 1994 in the Naval War College elective "Ethics and the Military"; they significantly contributed to and clarified the ideas offered here.

[1] Sir John Winthrop Hackett, "The Military in the Service of the State," in Malham W. Wakin (ed.), *War, Morality, and the Military Profession* (2d ed.) (Boulder: Westview, 1986), 118–119.

negative features, and we hesitate. Thus, a midshipman who is offered a preview of a final examination may weigh the goods and evils of cheating against those of not cheating.

A second factor is that morality changes, so one must stay informed about ethical matters. This need not signify acceptance of moral relativism, for our knowledge of right and wrong develops through cultural growth just as our knowledge of nature advances through science. Aristotle, St. Paul, and Thomas Jefferson thought that nothing in "the heavens" could crash to Earth, and that slavery is morally acceptable; we know that both views are wrong. If even the greatest thinkers of any age can be in error about such important facts, we must be humble about the possibility—the inevitability—of errors in what we believe about science and ethics. But this is intellectual honesty, not relativism. Thus, just as many naval officers were wrong about the practice of flogging that Congress forbade in 1850, and others were wrong about racial segregation in the armed forces that President Harry Truman prohibited with Executive Order 9981 in 1948, and while not every change must be for the better, we are experiencing changes in how society views alcohol, adultery, women's role in the military, and homosexuality. Numerous leaders failed to recognize before 1991 that America would no longer tolerate the antics at Tailhook conventions; by the time they did, it was too late.

Two related factors also make ethics hard: the pressure of time, and the limits of knowledge. Sometimes deadlines are recognized long before the moment of decision, while other times our decision must be made almost immediately upon becoming aware of a problem. In either case, we do not have enough time to avoid or eliminate the moral problem before we must decide what to do about it—whether to "gun deck" the training records, whether to stay with the aircraft on fire until it no longer endangers those on the ground.

Because ethics is based on knowledge, what we should do depends on what we know. Usually we feel we know enough to decide correctly. Frequently, however, our uncertainty is significant: we would decide one way if our beliefs were confirmed, another if they were refuted. The greatest difficulties arise when we seek to understand human motivations, including our own. As we contemplate non-judicial punishment for a petty officer who damaged government property, it matters decisively whether the cause was carelessness or clumsiness. If the expense was major and our superior is irate, we might also question whether our judgment is prejudiced toward punishment in order to be seen "holding someone accountable." How does one decide correctly in situations where critical factors are not known, and may never be ascertained?

The final aspect of an individual's prominence in ethics is the influence of emotions. As Ovid reflected: "I see the better way, and approve it; I follow the worse."[2] Our intellect and will are not always in harmony, even on mundane matters. Yet this considerably complicates ethics, for we must recognize in others what we know about ourselves: occasionally we do what we know to be "the worse," and we truly do not know why.

Although the power of emotions—anger, greed, lust, fear, vanity, despair, love, joy, hope, courage, and the like—is commonly acknowledged to affect a person's self-control, it may also distort thinking. How regularly do we see an individual with lengthy service experience and a record of outstanding performance fall victim to incredibly poor judgment, traceable to one or another passion? When a mid-grade pilot performs an unsafe takeoff in front of his family, a colonel submits an inflated travel expense claim, or a flag officer maintains an adulterous relationship with an enlisted subordinate, it is hardly the result of a sudden impulse overwhelming clear thinking. Rather, the officer's thought processes are so distorted that what others see as obvious truths are jumbled if not hidden from the perpetrator. When Admiral Jeremy M. Boorda wrote in a suicide note that "what I am about to do is not too smart," but then went ahead anyhow, he demonstrated that the thinking of our most able and senior leaders is subject to emotions and passions which complicate assessments about ethics. Was what he did morally wrong? Yes. Was he therefore immoral?

Perspectives on Organizations

Too many civilian officials and officers are under the impression that professional ethics is exclusively about individual choices in particular cases; they mistakenly assume that improvement in military ethics is tantamount to development of personal virtue. This is deleterious for two reasons: what it ignores, and what it implies.

---

[2]Ovid, *Metamorphoses*, vii.20.

To focus one's attention on the individual and on specific cases is to neglect the more consequential and difficult topics of institutional policies and practices, formal and informal codes of conduct, and people's organizational roles. It is, in short, to miss the essence of the naval services. When we put on our uniform, issue orders, and engage the enemy, we do so not as private persons but as members of an organized armed force, agents of our nation. All of the rules and procedures, laws and customs, duties and authorities that constitute the Navy and Marine Corps exist indifferently to the particular persons who are in the service at any given moment. What really matters is the organization.

If Herman Wouk was so much as obliquely correct in saying "The Navy is a machine invented by geniuses, to be run by idiots," the critical questions concern the machine.[3] Policies and practices shape most of the moral environment of our professional lives; codes of behavior provide much of the remainder; and individual decisions—while highly visible—define what is left. For example, when "body count" became the measure of effectiveness in Vietnam, after-action reports exaggerated the number of enemy dead, and (more tragically) any body counted. If commanding officers' evaluations are closely correlated with their units' readiness figures, that practice puts pressure throughout the service for false reports. When mixed-gender crews are assigned to ships and units that deploy, many people experience greater moral challenges than they had faced before. When a service, in little over a decade, shifts from a largely unmarried force to one in which family members outnumber service personnel, its policies on child care, housing, health services, schooling, time away from home, and geographic bachelor arrangements acquire greater moral significance. Poor organizational policies, such as those that hurt retention, will be felt by individuals in ways they identify as personal moral quandaries: how, for instance, to meet increased recruiting quotas without compromising standards.

Formal and informal codes have nearly as much influence in shaping the ethics of our lives, yet it is easy to overlook the innumerable features of service culture which must be learned. Leading senior non-commissioned officers, taking care of the troops before yourself, going through shellback initiation, stashing booze in your stateroom on "the boat," and believing that "it's only wrong if you get caught" are customs we may have encountered in our careers. But a service education addresses more than individual behavior: one learns in the Pentagon how the service plays the budget game; in joint assignments how large commands "act purple"; in dealings with Congress how political goals are attained; in interactions with contractors how capitalism works; and in operating with allied forces how national objectives are pursued. No one can rise in rank without mastering service culture, which has enormous ethical import.

The error of equating professional ethics with personal virtue is dangerous also for what it implies: that one is moral if one's personal behavior is pure, and immoral if it is not. Senior officers are thought to be lying, yielding to "political correctness," or putting their own advancement ahead of the needs of the service when they testify in their official capacities or make comments to the press about the progress of this program, the success of that project, the adequacy of the service budget, or the wisdom of an administration policy. Service members and others—failing to appreciate the complexity of organizational practices yet being bombarded with exhortations about the importance of ethics and core values (understood as personal virtues such as never lying)—condemn those leaders as hypocrites who do not stick up for their service and subordinates.

Clearly we are not suggesting that practices, codes, and customs are immune from moral appraisal—quite the contrary. Yet unless we understand the context of an officer's actions, we are apt to invoke individual standards in judging the performance of an organizational function. Just as we do not "tell the truth" when we open negotiations for the purchase of a used automobile (a routine practice with understood rules), a service leader might not "tell the truth" in Congressional or media interviews (also routine practices with understood rules). To assess the morality of such behavior, we must judge the practice itself as well as the specific performance.

Because we are members of many organizations, each with its policies and practices, customs and codes of conduct, we all live multiple roles; we are spouses, parents, friends, service members, church congregants, alumni, and volunteers. The fact that competing roles create moral dilemmas is familiar. At any given moment, we could be tugged from many directions, and deciding what is the right thing to do requires understanding much about the various organizations, assessing the practices or codes that are contending for our attention, and evaluating our own preferences. Such deliberations are sometimes far from easy.

---

[3]Herman Wouk, *The Caine Mutiny* (Garden City, N.Y.: Doubleday, 1951).

Some Thoughts on Theory

Let us look briefly at the foundation on which ethical judgments stand, because a basic understanding of the theoretic support for our particular judgments and decisions will help us to see why they may be perplexing.

We should note right away that ethics can be intended, at one extreme, to specify bare minimums of performance. At the opposite extreme, ethics can be a call to strive for the highest ideals. Roughly put, the difference is this: compliance with Defense Department ethics regulations may keep you out of jail, but it will not suffice to get you into heaven. Yet some discussions about service ethics appear oblivious to the difference in purpose that might be desired, an error in theory that complicates our deliberations. For instance, do the general precepts of honor and telling the truth deprive one of the Constitutional right not to incriminate oneself?

The proliferation of legal requirements, prohibitions, and procedures in modern life, including our professional activities (note the growing role of the judge advocate in operational matters), could lead us to imagine that what is legal is also moral. It is symptomatic that the Defense Department's "ethics regulations" are written by lawyers, and a JAG officer invariably is designated the "command ethics counsellor." However, the lawyer's business is with what is lawful, not with what is moral.

Second, and of immense importance in thinking about how to make ethical decisions, we should be alert to the fact that all of us employ at least three radically different approaches in our deliberations. The first invokes absolute rules, such as "Enemy prisoners shall not be tortured." These can be extremely complicated and precise, as our legal system illustrates. Another focuses on the consequences of actions, and directs that one should do whatever brings about the best overall result. A third asks about intentions and character; what is this person's purpose?

Disturbingly, although all of these approaches seem essential for sound ethical thought, they sometimes appear to be plainly inconsistent. For example, given a looming deadline of a terrorist threat to explode a nuclear weapon in an American city, should we torture a captured terrorist to extract vital information we are certain he knows? Our intentions would be noble, and we could save millions of lives, but only by breaking the rule not to torture prisoners.

Of small consolation, yet nevertheless important to note, is that disagreements reign and perplexities abound even among scholars and moral theorists. Can modern war be just? Are economic embargoes ethical? Is nuclear deterrence morally acceptable? Should women be assigned to combat? Is assassination of an enemy head of state permissible? Arguments on such questions are currently filling scholarly journals, with little consensus on what is correct.

Nor are such disputes confined to applications. Experts are divided about many of the most fundamental conceptual matters, such as how to answer the question "Why should I be moral?" Therefore, we in the military can accept with some comfort the idea that our struggles with professional ethics are reasonable, even warranted. However, the fact that the theories upon which our decisions ultimately stand are complex, conflicting, sometimes contradictory, and often inconclusive should warn us against the illusion that our profession's ethics can be uncomplicated and straightforward.

A final thought on theory. This is what Robert Pirsig calls "the high country of the mind." "Few people travel here," he warns, because "one has to become adjusted to the thinner air of uncertainty, and to the enormous magnitude of questions asked, and to the answers proposed to these questions. The sweep goes on and on and on so obviously much further than the mind can grasp one hesitates to go near for fear of getting lost in them and never finding one's way out."[4] Most of us choose not to wrestle too often with questions such as: What is the meaning of life? Is there a God? What determines whether an act is moral? How can I know what is true? What is worth dying for? Rather, we prefer to live with answers we accepted years ago. Yet if we are serious about professional ethics, we must acknowledge that all of our present convictions and beliefs depend, in the end, on our answers to such questions.

---

[4] Robert M. Pirsig, *Zen and the Art of Motorcycle Maintenance* (New York: William Morrow, 1974), 127.

The Bottom Line

Naval officers need not be philosophers, so the study of theoretical matters may be left to scholars. But we are members of a profession which has its own distinguished ethical code, and in the performance of our duties we are bound to uphold that code. We are obligated, therefore, to reflect upon its character and practical directives.

The ethical code of the American military profession is specific, complex, and binding on all who wear the uniform. Senior leaders of the Navy and Marine Corps are especially obliged to recognize that it entails much more than vague pieties about "being a good person, " "doing the right thing, " and "practicing our core values." They must exemplify that code in their own performance of duties, articulate it coherently, and ensure that all under their command are educated in its precepts. Equally important, they must not demean their subordinates by trivializing the difficulty of meeting its high standards.

The bottom line of our profession's ethics is that we may have to lay down our lives in the service of our nation. No discussion of military ethics should lose sight of that fact. It is a hard truth, one not to be obscured by banal assurances that those who maintain their service's ethical code will become more popular, get promoted early, receive the best assignments, and end up being much happier and more successful people. None of those is necessarily true (nor automatically false). What is true is that our profession's ethics compels every service member's attention because its implementation is his or her responsibility, and nothing about it is easy.

An ethical decision is one that determines what should be done now, in this situation, all things considered. It is comprehensive—"*all* things considered"—and thus, at that moment of action, the ultimate authority on what to do. Because it requires us to consider everything that is relevant, ethical action particularly calls upon judgment, sensitivity, prudence, imagination, creativity, foresight, broad-mindedness, and wisdom. None of these capacities can usefully be taught, although through experience several of them can be learned.

What may be worse is that for none of those skills is our own self-evaluation trustworthy. We seldom can recognize when our judgment is misleading us, when we are lacking sensitivity or imagination, when we are not looking far enough ahead, or when we are being downright foolish. Since our internal sense is unreliable, we each have a professional responsibility to get external assistance to improve our moral deliberations. In general, good reading, reflection and discussion, and the study of persons who model right behavior can enhance our capacity to make moral decisions. Specifically, professional mentors and trusted friends are invaluable, particularly during the decision process when they can be perceptive and frank with us. Each of us can provide that help to others, just as we need it ourselves.

Intrinsic to being an officer is adherence to the military ethic. The nature of that commitment, the extent of its requirement, and the ways by which one fulfills it must be taught by all to all, and enforced by all on all. It is not easy.

Part II
The Moral Framework of Military Service

The ethical principles and ideals that properly govern military life derive, for the most part, from the principles and ideals that properly govern human life in general. The function, situation and the particular activities characteristic of military life, however, are quite unique. It is therefore crucial that military officers have a clear, articulate understanding of the moral implications of the position they assume.

First and foremost, that position establishes military officers as the managers of the terrifying and unimaginably destructive activity of warfare. Unlike the duties of other professionals—doctors, engineers, lawyers, teachers, or the clergy, for example—the distinctive function of military leaders *inherently* involves them in the contemplation and commission of morally problematic acts: in the extreme case, the military is responsible for deploying lethal force in a manner that results in killing and injuring people, destroying their property, and disrupting their lives, in order to achieve some presumably desirable political end. That is why the activity is morally problematic on its face.

That moral problem does not disappear, but it is considerably altered, if we can somehow show that the "desired political ends" of such destruction can also be plausibly linked to the defense of the state and its citizens from genuine threat or harm, or to participation in the protection of the lives, liberties, and rights of vulnerable peoples elsewhere. In that case, the use of force would not be merely "desirable" from some self-interested political perspective. It would also be, at least in principle, *morally justifiable*.

It is very important to recognize that political leaders almost never resort to simple statements of self-interest in explaining their recourse to war. Instead, they invariably find it necessary to *appeal (or to attempt to appeal) to moral principles in order to justify* their actions. This tactic demonstrates the importance of moral principles in the lives of ordinary citizens, and it therefore follows that our general responsibility, whether as citizens or as military officers, is to understand the nature of such appeals, and to be able to determine whether or not they are even plausible, let alone valid. None of us (as the philosopher Kant will demonstrate) is put on this earth merely to serve as an unwitting means to the immoral schemes of others!

Warfare is the distinctive, and the preeminent purpose of the military, but surprisingly, the conduct of war is hardly the military's sole (or, in normal circumstances, even its chief) activity. The several military services in this nation, for example, exist and serve essential functions during times of peace as well as in war, including helping to maintain peace, deterring aggression and lawlessness in the international arena, and providing humanitarian assistance, all while maintaining their preparedness for war.

These activities may not be so inherently problematic as the conduct of war itself, but all manner of ethical issues may arise in carrying out such activities. In these cases, the military services function as large, complex social organizations, invariably encountering many intricate moral dilemmas that are also found to arise in the operations of a business organization, government agency, or any other diverse collection of human beings engaged in some common enterprise. So military organizations, like organizations in general, face generic moral dilemmas arising from their particular manner of treating the individual members of the organization, especially in the opportunities provided in the organizational hierarchy for entering, training, advancing, and obtaining rewards and recognition on the basis of merit. Members of the military profession will be obliged, like all of us, to cope with issues of race, gender, and personal sexual orientation, to interact with and represent (and sometimes advocate) their activities truthfully and faithfully to their immediate clients and to the wider public, distribute their resources, services, and benefits fairly and equitably within and beyond the organization, and so forth. Each of these activities can, on occasion, be fraught with moral dilemmas.

Even though the ethical issues characterizing military service and military organizations may resemble those that civilian organizations confront—racial prejudice, sexual harassment, equality of opportunity and fairness in the distribution of rank, salary, and work assignments, for example—these moral dilemmas may still come to take on a uniquely different cast when encountered inside the context of a military organization. On occasion it may at least seem that military personnel are held to a different, or even a "higher" moral standard than civilians in wrestling with racial or sexual conflicts, for example, or in coping with dishonest or corrupt behavior on the part of individual members of a specific branch of military service. This may strike some members of the profession as an odd state of affairs: unrealistic, unreasonable, and perhaps even unfair. Why should military officers who fail in the discharge of their moral obligations be scrutinized more closely, or condemned more harshly, than anyone else?

This is a difficult question, all the more difficult unless we obtain a better understanding of what the moral foundations of military service are. That is the aim of this unit. Surely the distinctive function of the military may necessitate its operating, even in peacetime, on somewhat different principles than civilian organizations. In particular, while other professions operate on fairly generic and universal moral principles, *the nation's armed forces are uniquely dedicated to serving the nation and protecting its way of life. No other organization or profession is charged with that particular responsibility.* In that sense, the moral foundations of military service in the United States and in other democratic nations are grounded in the ethical heritage of the nation it serves. In this unit, accordingly, we focus on some of the ways in which the distinctive nature and purpose of the U.S. military serves as the basis for the principles and ideals that properly guide its officers.

A. The Relativity of Moral Beliefs and Situations

The oath sworn by military officers at the time of their commissioning commits them to moral principles and values that may explicitly dictate the kinds of actions they are either obligated or permitted to undertake in a great many circumstances. In a great many other situations, however, the generalities of the oath of office can, at best, merely rule out some of the available options as unacceptable, without providing further specific guidance about what may or must be done. An officer is a leader, a decision-maker, and as such, is regularly responsible for making ethical decisions that are not straightforwardly dictated by the stated or generally accepted rules governing his or her office.

For example: is a senior officer who happens to be personally and conscientiously opposed to abortion permitted to exercise his or her authority of command to prevent or discourage a subordinate from obtaining this procedure during a personal crisis? To what degree, in another instance, ought a senior officer allow, let alone order, that those serving under his or her command to conform to unfamiliar moral customs and modes of behavior widely accepted in the foreign ports-of-call they may visit? What should be permitted, or commanded, when such customs conflict sharply with our own moral beliefs and principles? Would it be acceptable to expect that men and women would abide by stricter and more conservative dress codes than are customary in the U.S., or maintain a strict segregation of personnel based upon gender or race that would be unacceptable (not to mention, illegal) in our own country? Even more seriously, would an officer be permitted to participate, or to turn a blind eye to the participation by those under his command, in sexual practices involving minors, for instance, that would be judged both illegal and immoral in our nation, simply because such behavior is tolerated in the foreign locale?

Faced with the bewildering variety of moral customs and practices encountered throughout the world, it is tempting to seek refuge in the thought that there really is no morally right or wrong decision. *Moral Relativism* is the idea that the truth of a moral judgment or principle is dependent upon and relative to its acceptance by some person or group of persons. Thus, confronted with a conflict of personal beliefs (as in the abortion case), the officer ought simply to follow his own beliefs. Or, perhaps, quite differently, he or she should adhere to and advocate his or her own personal beliefs, while learning simply to *tolerate* the differing beliefs of others. *But which is it*, and if we are to cultivate tolerance, how widely should our moral latitude extend? Would it cover toleration of the sexual enslavement of minors, for example, or the violent and brutal suppression of political opponents? Should we stand silently by and tolerate (let alone advise or participate in) the torture of such opponents, simply because it is the practice of the host country?

In any of these instances, *according to the concept of moral relativism*, no one is in a privileged position of saying what is, in fact, right or wrong—we are only at best able to state with some certainty what we ourselves *believe* to be right or wrong. In the case of a conflict between accepted moral values (for example, about the rights of women, or of minors, or of ethnic minorities, or the treatment of political opponents) in our society versus another, relativism seems to say that no criticism of one society by another would be objectively valid. Relativists might further recommend, in keeping with the value of tolerance, that we adopt the moral values and practices of the society in which we find ourselves, as captured in the familiar and well-worn proverb: *"When in Rome, do as the Romans do!"*

Thus, it would seem that our commanding officer visiting a foreign port-of-call would be acting correctly in ordering those under his or her command, or at least in permitting them, to follow the rules, laws, customs, and moral practices of the society they are visiting. Yet this officer would still be confronted with the question, how far should such toleration extend? Doesn't our toleration, or participation, in the practices of other cultures depend upon what those practices are? To the conventional advice, "when in Rome, do as the Romans do," perhaps we should reply, *"Well, that depends upon what the Romans are doing!"*

Thus, once again: ought the members of the crew or command to engage in practices that our society would find morally reprehensible (and very likely also illegal) just because these practices are tolerated or approved of in the society in which they are resident? Ought members of the crew or company be willing to subject themselves to constraints on their behavior that are not merely strange or inconvenient, but humiliating and personally degrading (not to mention illegal in our own society) just to accommodate hospitably to the conditions of the culture they are visiting? Should a commanding officer reign in the behavior of subordinates under his or her command when these individuals seem to be following the time-honored naval custom that "what goes on during deployment stays on deployment?" The experience of senior line officers is chock full of such encounters, and such dilemmas, often graphic and deeply troubling, about which there seems to be no clear guidance or resolution. Indeed, moral relativism as a theory seems to predict that no such guidance would be forthcoming, *because there is no perspective from which such conflicts could be decisively resolved.*

Certainly, everyone's beliefs are influenced by their culture, their parents, peers, clergy, the press, and myriad other sources of influence. So people from the same culture tend to have similar beliefs, and people from different cultures tend to have different beliefs. This much the Greek historian, Herodotus, observed many centuries ago.[1]

But all of these observations are true with respect to *all* our beliefs, scientific and moral alike. And as with our other beliefs, as we develop into adults we acquire the capacity to evaluate our beliefs, to reason about them, and refine them by that reasoning. Because of that capacity we become responsible for our beliefs, and for our decisions and actions based on those beliefs. Thus, we cannot simply take refuge in cheating on exams or tax returns, or lying to cover up for the misdeeds of our friends, merely because such behavior seems widespread and tolerated. Neither ought we merely to add to the sum total of human misery in the world by, for example, engaging in sexual acts with minors, or with young adults sold, enslaved, or indentured as prostitutes in foreign ports of call, simply because we can get away with it. We are required to face the additional question of whether such behaviors, whether they are customary or tolerated, are also *morally justified*. Contrary to the claims of relativists, this seems an independent question—one which, when mirrored in eyes of the weary and broken victims of such oppression, is less easily dismissed.

The Readings

Burton Porter gives an account of *moral or ethical relativism*. Porter contrasts this view with what he calls "objectivism," the belief that there are, in the end, at least some universally recognized and acknowledged moral principles. He cites some examples of these, and faults advocates of relativism for having been so fascinated by cultural *differences* as to entirely overlook a number of *broad and widespread moral similarities* or areas of common consensus.[1] He finds fault with the reasoning from such facts to the (unwarranted) conclusion that such facts decisively demonstrate that no universal moral values are possible. This kind of familiar argument is nonetheless mistaken. He concludes by considering some of the paradoxical positions that might arise, were we to try to adopt and live consistently by such an account of morality.

---

[1] Indeed, this understanding is captured in the words we use to label the subject we are currently studying. Readers may be familiar with a widespread tendency to jump back and forth between the use of the terms "ethics" and "morality" for this subject. That is because the Latin word *more* is a translation of the Greek term *ethos*, that Aristotle, Herodotus and others used to describe the habits, customs, and social practices of a given society or culture. Our modern usage differs slightly, by convention only: philosophers prefer to use the word "ethics" to characterize *the study of* "morality," including the study of a wide variety of universal moral theories and principles (we will consider several of these subsequently in this book) as well as of the particular settled customs and practices of specific societies.

While it may on occasion be difficult to discern a single right, or "objective" moral perspective, it does not follow that "anything goes" and all moral views are equally valid. Rather, there may be a very few distinct, but compatible, approaches to moral reasoning, each of which is valid to a degree, while other views (such as selfishness or cruelty) are definitely found to be deficient. This view is described as *moral pluralism*, according to which we might expect to discover two or three, but not an infinite variety, of legitimate ways to think about moral problems or discern right from wrong conduct. (We consider the most important of these distinct, but compatible approaches to moral reasoning in Unit III of this text.)

This chapter concludes with the troubling question about relativism for military personnel raised at the outset: Are there *two standards* of conduct in our society, one for military personnel, and another (less strict) moral standard for everyone else in our own society? In this particular instance, the question is whether moral standards may differ, not because the agents' beliefs differ, but because the agents' objective situations differ. Here the issue is whether the commitment to serve as a military officer obligates military personnel not merely to perform the duties officially assigned to them, but also *to live a morally exemplary life*. Many military officers uphold the basic importance of holding military personnel to an exemplary standard, in part as an example to society generally, and in part as an essential element in the cultivation of military leadership. Ficarrotta, by contrast, explores the possible meanings of such a double standard, and finds all plausible accounts of what this "higher standard" might consist of to be largely unjustifiable.

Relativism and Objectivism: Are There Universal Values?
Burton Porter [*]

One primary question in ethics is whether our values only reflect our culture or whether they have an Objective basis. According to the *relativist* view, when we make a value judgment we are only stating the attitudes and prejudices of our society. To the objectivist, we may be identifying something genuinely right or wrong, good or bad. For example, when we blame someone for lying, the relativist sees this as an expression of our culture's disapproval; our particular society favors honesty and condemns dishonesty. The objectivist would argue that we are stating an objective truth: that lying, by its very nature, is wrong; society depends upon mutual trust.

Some relativists point out that our judgments are also a matter of personal taste as well as a reflection of societal norms. What is right to one person may be wrong to another; values are an individual matter. John may believe that sex is only permissible within marriage whereas Bill may think premarital sex is perfectly acceptable. Just as we have different tastes in food we have different sets of values, and no one can be judged wrong either for liking broccoli or for being sexually promiscuous. As the Romans said, "*de gustibus non est disputandum*"; about taste there can be no dispute.

All standards are relative to a particular person or society, and they have no general validity outside that context. Therefore, to deliberate about which values are really worthwhile is a pointless exercise. Every thing is relative to the individual, the culture, the time, and the place.

Relativists often argue for their theory by citing the variety of value systems across the world, each different and each supported by people who believe themselves right. In some cultures, a man gains esteem by having several wives while in other cultures polygamy is considered immoral; in some societies, drugs are taken for pleasure and insight, while in others using drugs is a crime; in some places, a woman must cover her body completely in a burka with netting over her eyes, while in others, women can wear attractive clothes and bikinis on the beach. In the past, many cultures practiced slavery whereas today it is considered immoral, and war was thought heroic, but now we regard it as tragic. Old age used to be respected and even venerated, but currently youth is celebrated as the ideal, just as piety and wisdom were once highly valued, but now we prize wealth and status. Because of the multiplicity of cultural perspectives the relativist concludes that morality is a matter of history and geography. At the extreme, the relativist theory maintains that values are a matter of opinion. Whatever a person thinks is right becomes right because the person thinks so. We tell one another, "It's all a matter of opinion, it's how you feel," or "What gives you the right to judge?" or "Who's to say?" The implication is that whatever people believe to be true is true for them, and no outside judgment can prove them wrong. As Shakespeare declares in *Hamlet*, "There is nothing either good or bad, but thinking makes it so."

The impulses behind relativism seem to be admirable. First, there is the desire for *tolerance* and open-mindedness toward other people's ideas-including those that are different from our own. In a

---

[*] From Burton F. Porter, *Philosophy Through Fiction and Film* (NY: Prentice-Hall, 2003).

democracy, everyone has a right to his or her opinion as well as a right to be heard, and we should not presume that our ideas are the only correct ones. Such an attitude smacks of arrogance and righteousness. Furthermore, we should be wary of people who are sure they are right because such certainty can result in inquisitions, witch burnings, crusades, pogroms, purges, ethnic cleansing, and so forth.

A second source of relativism lies in our wish to maximize our *freedom*. If there are correct moral principles then we are compelled to acknowledge them, whether we like it or not, whereas if right and wrong depend upon how we feel then we have a great deal of personal control. We are then free to choose our values, and all ethical decisions become a function of our preferences. We do not choose that which is valuable but whatever we choose becomes valuable because we've chosen it.

A third motive for accepting relativism is our *uncertainty* in today's world as to which values are worth accepting and defending. History has proven us wrong too often with regard to political doctrines, social theories, or religious beliefs, so we have lost confidence in the truth of our ideas. Furthermore, our awareness of diverse values in a multicultural world casts doubt on any one theory of what is right or good.

Our uncertainty is increased by the scientific approach to knowledge, which has nearly eclipsed every other way of knowing. As science operates, only empirical statements are capable of being verified, which implies that all value judgments are a matter of opinion. Added to this are specific scientific findings such as Einstein's theory of relativity, which takes space and time as relative phenomena. Although relativity theory only applies to physics, people have taken it as evidence for the relativity of ethical values as well.

It appears, then, that every value judgment we make should be tentative and qualified. The sense of being sure about what is right has now been lost, and we are acutely conscious that every moral statement we make is potentially false. Furthermore we want to give ourselves maximum freedom in deciding how to live, and, in addition, to be tolerant toward other people's choices. Therefore it seems correct to say that everything is relative.

Persuasive as the case for relativism might be, many philosophers accept the opposite position of *objectivism*. According to the objectivist theory, we can identify particular acts as right, others as wrong, and certain purposes in living as more desirable than others. Societies do not create values but can reach an understanding about them, and individuals do not invent values but discover them. When we make a moral judgment we are not revealing something about ourselves or our culture's attitudes but expressing insight into the moral nature of an act.

For example, the judgment that stealing is wrong tells us something about the inherent nature of stealing. To take someone else's property, especially something they have worked hard to acquire, is to cause them injury. Therefore it is wrong, not just for us but for anyone. Whether we are in Africa, Europe, or America, in the fourth, the sixteenth, or the twenty-first century, we should not take what does not belong to us.

The objectivist, therefore, believes that we as human beings should follow certain standards of behavior because they are right in themselves. We should not be self-righteous, of course, and assume we know what those standards are, but we can have confidence that such standards exist. That gives direction to our search, and through rational discussion we can hope to get closer to the truth of things.

The objectivist also rejects many of the specific arguments used by the relativist. For instance, the objectivist points out that although values differ between cultures, that does not imply that all values are relative. The differences can be attributed to one society being more backward or enlightened than another, seeing values more dimly or clearly. To take an analogy from science, the fact that people thought the earth was flat at one time, and round at another, does not mean that each idea is right. Rather, people came to understand that although they believed the earth was flat, it really is round. In the same way, people have come to realize that women should be treated with absolute equality, that enslaving people is wrong in any society, and that we should respect minority rights. Prejudice and discrimination is not wrong in some cultures and right in others; it is wrong whenever and wherever it occurs.

In addition, the objectivist argues that the diversity of values between cultures may be more apparent than real. A wide area of moral agreement exists between cultures across the earth. For example, one society may condone a husband killing his wife's lover, another may condemn it, and this seems like a major difference. However, both societies will probably have laws prohibiting murder and a strong belief in the protection of human life. They will differ only in their definition of murder, that is, taking life unjustifiably. In the same way, one society may be hostile to strangers, another warm and welcoming, but both may believe in the value of hospitality. The difference is that one applies the rules of hospitality only to families within the society, the other extends them to outsiders as well.

Objectivists will sometimes cite the golden rule as a major example of a cross-cultural value. In Christianity we read, "Whatsoever ye would that men should do unto you, do ye even so to them"; in Judaism, "What is hateful to yourself, don't do to your fellow man"; in Buddhism, "Hurt not others with that which pains oneself"; and in Hinduism, "Do naught to others which if done to thee would cause pain." Confucianism tells us," What you don't want done to yourself, don't do to others"; Zoroastrianism says, "Do not do unto others all that which is not well for oneself"; Sikhism declares, "Treat others as thou wouldst be treated thyself"; and even Plato advises himself, "May I do unto others as I would that they should do unto me." Perhaps all societies do have a substratum of shared values.

The objectivist also points out a variety of logical criticisms showing that relativism is self-contradictory. For one thing, the relativist claims that the statement, "Everything is relative" is really true, but if everything is relative then nothing is really true, including that statement. It may be true relative to one's culture or according to one's tastes, but to say it is objectively true contradicts the theory itself.

Plato identifies another type of contradiction in a dialogue called the *Theatetus*. Here Socrates says to Protagoras, "and the best of the joke is that he acknowledges the truth of their opinions who believe his own opinions to be false for he admits that the opinions of all men are true." In other words, if everyone is right, then the person who thinks you are wrong must be right.

Finally, there is the self-contradiction with regard to tolerance (which was mentioned as one of the supports for relativism). Relativists claim that their position has the virtue of fostering tolerance because no value is considered really worthwhile. However, by extolling tolerance, relativists are assuming it possesses value. The relativists thereby give the game away, for tolerance at least is considered objectively valuable.

A Higher Moral Standard for the Military
Maj. J. Carl Ficarrotta, USAF (Retired)

It is commonly believed that, in some sense, military professionals are bound by a higher moral standard. This belief is especially prevalent inside the military. Even though there are occasional (perhaps inevitable) moral failures, there are nevertheless numerous internally promulgated codes and public espousals that enunciate such a belief.[1] Many commanders exhort their troops to moral goodness and chastise them when they fall short.[2] Military education frequently includes courses on the demands of professional ethics.[3] Indeed, from the top down, part of the background noise of professional military life are these higher expectations, and a belief that somehow, this line of work is one with a special moral status, special moral problems, and special moral demands.

In this article, I want critically to address what, at least generally, this higher moral standard might amount to. I want briefly to offer a more concrete interpretation of what we might mean by a higher standard. I'll then explore what reasons there might be for believing military professionals are bound by one. While my posture is a skeptical one, I still think there *are* arguments that make a partial and important case for some unique and especially strict military obligations. But I do not think we will be able fully to justify a more robust (and I think more commonly endorsed) conception of higher demands on military behavior and character.

What Might We Mean By a Higher Moral Standard?

There are at least two ways we might elucidate the idea of a higher moral standard for the military. First, we could mean there are *unique* moral obligations for military professionals that most other people simply do not have. For example, we might think military professionals (but not people in general) are morally obligated to follow the orders of their superiors or be courageous in the face of physical

---

[1] As for codes that require higher standards of ethical conduct, the most obvious example is the Uniform Code of Military Justice, especially Article 133. Charles R. Myers, in reviewing the history of such codes, has concluded that military professionals are held, under the law at least, "to unusually high ethical standards" (*USAFA Journal of Legal Studies*, Volume 5, 1994/95, p. 15). As for public espousals, take as a recent example the claim by former Navy Secretary James H. Webb, Jr. that "there is no substitute for an insistence on ethics, loyalty, accountability and moral courage" ("Webb Raps 'Self-Serving' Navy Leadership" in *Navy Times*, 6 May 96). When failures occur, corrections often go beyond mere chastisement. The media regularly reports on military professionals that have been relieved and punished under the UCMJ for moral failures including lying, disobedience, sexual harassment, adultery, and theft.

[2] The former Air Force Chief of Staff, General Ronald R. Fogleman, has insisted that "our standards must be higher than those that prevail in society at large" and that members of the military must "always exhibit the utmost in principled behavior, off-duty as well as on" (from a letter released to the public early in 1996 though the Air Force News Service, Washington DC).

[3] Examples include, but may not be limited to, the curriculums of all the service academies, various ROTC programs, officer training schools, and resident professional military education for NCOs, senior NCOs, and company grade, field grade, and senior grade officers.

danger. Call this the "uniqueness" interpretation. Second, we could mean military people have good reasons for being bound *more strictly* to the moral standards that apply to everyone. Here, we would ask military professionals more insistently to be moral, and would find them more blameworthy should they fail. Along these lines, we might say honesty is something we want from everyone, but that it is especially important for military people to be honest. Call this the "strictness" interpretation.

Having offered these two meanings for consideration, a few preliminary remarks are in order to head off possible confusion. First, these two meanings are not mutually exclusive (so we might mean some combination of both); nor am I claiming them to be exhaustive of the possibilities (so there might be other meanings of the phrase I'm not addressing). Second, I think these two meanings might apply just as well either to what counts as moral *behavior* or what counts as a morally good *character*. Obviously, character and behavior are tightly interrelated, but moral theorists sometimes disagree about the place and role each of these properly occupies in the structure of our moral thinking. It is a disagreement I think we can fruitfully bracket for my purposes. Higher standards, if we find them, might bind in terms of either behavior or character or both. And as it turns out, the arguments for higher standards I'll be examining move freely (without suffering) between these two objects of moral evaluation. Third, keep in mind that I'll be addressing *moral* standards for the military professional as opposed to standards of some other type (e.g., legal standards, standards of etiquette, standards of prudence, etc.).[4] Last, nothing in what follows suggests that military professionals are not bound *at least* by the same moral standards which bind us all, or that we should relax the enforcement of morality, either in or out of the military. The only thing at issue is whether military professionals are bound by higher standards.

There are several lines of argument that might lend some support to claims that military professionals are bound by one or both of these understandings (uniqueness and strictness) of a higher moral standard: arguments that start with unique military situations, arguments that pay attention to the military function as such, and arguments that concentrate on the role of the military and its relationship to the larger society. These lines of argument, while distinct, share a good deal in common and overlap somewhat in both their approaches and their conclusions.

Unique Situations, Contexts and Problems

This much seems to me uncontroversial. The military profession, and the conducting of military operations, puts people in unique situations and contexts that pose unique and particularly pressing moral problems. Anyone taking the moral point of view will immediately notice them. To varying degrees, this is true of many—maybe even most—lines of work. Doctors, lawyers, clergy, businesswomen, whatever, find themselves faced with unique situations and contexts that create moral problems which simply wouldn't come up very often in other endeavors.

Keeping this in mind gives us one possible way to make sense of how and why the military professional is bound by a higher moral standard. We could examine all the special situations, contexts and problems we encounter in the military, and try to puzzle out the right way, morally speaking, to think about them. For instance, in a military operation, we no doubt judge it a moral obligation to do whatever we can to avoid hurting innocents. Or we might judge that because military officers have extraordinary authority over their subordinates, they ought to take extraordinary care in looking out for their subordinates' welfare when issuing orders. This way of thinking lends some support to the "uniqueness" interpretation, and could lead us to suppose that the higher moral standard is merely an enumeration of the unique moral demands placed on military professionals because of the unique situations, contexts and moral problems they face in their work.

Importantly, on this view the unique moral demands would bind *anyone* who happened to be similarly situated. Of course, military professionals are far more likely than other people actually to find

---

[4]We might wonder at this point what makes a standard a *moral* one. This is a large question that would go beyond the scope and focus of this paper, and I hope I can put that issue aside for another day. I think we can in this study rely on a pre-theoretical, common sense notion of what counts as a moral standard, even if that amounts to nothing more than an implicit list of standards we are disposed to characterize as moral. The focus here, as we shall see, is on whether we can find special reasons for binding the military professional to such standards we already recognize as moral ones.

themselves in these contexts. But on this account, the reason the military professional is morally required to do this or that is not primarily because of who or what he *is*. Rather, it is primarily *the situations or the contexts* in which the military professional finds himself that generate the moral requirements. This or that would be required of any person in the same situation. Likewise, we could make similar arguments for unique and hence higher moral standards in almost any endeavor. A doctor, for example, might be bound by a higher standard of helping the sick. Of course it is plausible that anyone who happens to be able to help a sick person has some (perhaps) limited moral obligation to do so, but the doctor is uniquely situated in that she is most often in a position to help. She is, in this sense we're considering, bound by a higher standard.

So this is one way we could understand and justify a higher moral standard for the military professional. The approach will generate a long list of (general and specific) morally appropriate responses to situations military professionals are likely to face. To follow orders of appropriately appointed superiors, not to kill or injure non-combatants, to attend conscientiously to one's military duties and the like, would all be part of what binds the military professional more or less uniquely and would hence collectively constitute the higher moral standard.

At least as far as it goes, what this approach establishes must be right. There *are* unique situations, contexts and problems, and these do generate unique moral demands. Still, this is rather a thin construal of a higher moral standard for military professionals, and is as notable for it what it *doesn't* establish as for what it does. To begin, one might be inclined to think that invoking a higher standard for someone means (in some sense anyway) that a person is bound to do *more* than any similarly situated person would be bound. This thin approach—as I have developed it so far—does not establish such a requirement; and this may point up an inadequacy for this way of understanding a higher standard, depending on how important we think the requirement is. But more importantly, a higher moral standard so thinly construed says nothing directly about what the military professional may or may not do *outside of the military context*. If a military professional fails to pay his taxes, cheats on his wife, lies to his friend, whatever, I may be as disappointed in him as anyone else (for he was bound by the same moral standards that apply to us all). But I may not be *especially* disappointed in light of the higher standard (so construed and justified), because this standard was generated from and applies only to situations and contexts that are unique to the military.

Maybe using this general unique situations approach, we could also say some something about a higher standard understood in terms of the "strictness" interpretation. We've seen that, at the very least, a military professional is obligated by the *same* moral standards as everyone else. Morality, in general, always makes its special and insistent claims on each of us, simply in virtue of the fact that we are human beings. But given the morally tough situations that come up in the military line of work, maybe military professionals ought to attend *more* carefully to these common moral standards, and indeed not succumb to the temptation to comport their behavior and character in accordance with *lower* standards. Anscombe was exactly right to warn us about the dangers of commonplace "pride, malice and cruelty" and to point out how quickly warfare can become injustice, and how easily the military life can become a bad life.[5]

So when we consider the moral dangers and temptations of military service, and survey the extraordinarily bad things that can happen when the military professional is not strict and courageous in upholding moral standards, we may rightly worry. Combat no doubt can be corrosive to the moral personality. If we are concerned to minimize the immorality that can be, and too often is, found in war, we will see good reasons to be on guard. The military professional, then, ought to be especially strict and morally steadfast, and not yield to the extraordinary stresses that might easily lead him to violating moral principles that bind us all. Hence we might have a rough argument for binding the military professional in accordance with the "strictness" interpretation of a higher standard.

This way of thinking about things seems to avoid the first difficulty we noticed with the thinner approach (which concentrated only on the "uniqueness" interpretation). That is, it seems to leave room for us to demand *more* of the military professional than we would of someone else similarly situated.

---

[5] Elizabeth Anscombe, "War and Murder" in Wakin, *War Morality, and the Military Profession*, 2d Ed, ed. Malham M. Wakin (Boulder: Westview Press, 1986), 286.

Specifically, since military professionals know well the moral danger they might face, we might think they are bound—on this view—to be stronger, more disciplined, and have more moral courage in facing the temptations to do wrong in wartime. However, none of this addresses the second worry we noticed. Because these demands of strictness come from the special moral dangers present in military situations and contexts, we still are not in a position to say anything directly about the military professional's conduct and character *outside of the military context*.

In spite of these worries, I think by starting with the unique situations, contexts and problems faced by the military professional, we get a nice start in sharpening our understanding and justification of being bound by a higher moral standard, for both the "uniqueness" and the "strictness" interpretations. Still, I'm sure this way of understanding a higher moral standard for the military fails to capture all, or even most, of what many people are thinking when they invoke such a standard. If we hope to establish more, we must turn to some other ways of approaching the question, ways that might establish a thicker, more demanding version of the standard.

The Functional Line

Hackett has claimed that a bad person "cannot be . . . a good soldier, or sailor, or airman."[6] Wakin and others seem to agree with this claim.[7] These thinkers base their conclusion on an argument I'll call the functional line. They acknowledge the "unique" moral situations and demands placed on the military professional, which we just explored. But they furthermore think that there are certain rather *general* demands on the character and behavior of military professionals, mostly of the strictness variety, that flow directly from the military function itself.[8] For example, military units cannot function well, especially in combat environments, if the members of the unit are not scrupulously honest with each other. Also, military folk simply will not be able to do their jobs if they are not, to a certain degree, selfless. Otherwise, they wouldn't be willing to tolerate even the ordinary hardships of military life, much less be willing to risk their lives. Similar arguments can be made for the virtues of courage, obedience, loyalty, and conscientiousness. Hence if one thinks (for whatever reason) that it is important to have a military that functions as well as it can, one also is committed for these same reasons to thinking military professionals are more strictly bound to exhibiting these functionally necessary virtues and behaviors.

Notice that the functional line might be applied in some measure to any enterprise, especially cooperative ones. To the degree that any undertaking is important, then we at once have special reasons for more strictly binding those engaged in the enterprise to general moral standards that are necessary for its success. And cooperative enterprises typically depend very heavily on observing a number of moral standards. For instance, commerce would likely fail if the honesty of the participants dropped below a certain level. Hence insofar as, and to the degree that, commerce is important, we have reasons to be strict about honesty in a commercial setting. Identical arguments can be run for a large number of other enterprises (for example, fire fighting or police work). In each of these cases, we could argue for varying degrees of higher moral standards appropriate to participants in the enterprises.

But the application of the functional argument to the military is particularly apt, and establishes particularly strict and broad versions of a higher moral standard, for several reasons. First, few undertakings require the level and intensity of cooperation that is demanded by the military function. So moral standards, the observance of which are needed for cooperation, become particularly important for the military professional. Second, there are other demands of the military function that, while not

---

[6]Sir John Winthrop Hackett, "The Military in the Service of the State," in *War, Morality, and the Military Profession*, 2d Ed, ed. Malham M. Wakin (Boulder: Westview Press, 1986), 119.

[7]Malham M. Wakin, "The Ethics of Leadership: I," and "The Ethics of Leadership: II" in *War, Morality, and the Military Profession*, 191, 208, *passim*.

[8]I understand "the military function" for these purposes very narrowly, to mean something like fighting battles and wars. We might also assume a more normatively loaded characterization of the function, so as to include things like "defending the innocent" or "fighting for the right." This would probably make it easier to justify "strictness" in enforcing many moral standards, but I would rather assume less rather than more and see where it takes us.

directly or primarily concerned with cooperation per se, are also facilitated by clearly moral standards. The needs for bravery, selflessness, and conscientiousness come to mind as examples. These functional requirements need not be related directly to cooperation (though they might be), yet they also generate special reasons for being strict with what amount to moral standards. So the military function seems to make broader moral demands than many other undertakings, in that the military function makes a greater number of these strict demands on behavior and character. Third, failure in the military context likely will issue in tremendously bad consequences, whether considered morally or otherwise. When the military person violates *functionally* grounded moral rules, there is potential for disaster we just do not see in many lines of work.

If all this is right, then we have found some good reasons to think that military professionals have not only some obligations not normally encountered by others (as we saw in the unique situations approach), but that there are special reasons to be *strict* in enforcing many general obligations that apply to us all. I think the main idea here is right. But I also think we should be careful not to conclude too much from the functional line. All this argument leads us to are higher standards *in the military context*. Military people must be scrupulously honest with each other when there is some military issue at hand. They must be selfless when it comes to the demands of military work. They must be courageous when there is some military task to be performed.

What the functional line does *not* establish is that the military professional has special reasons to be "good" through and through. The argument gives a soldier who would never even think about lying in his unit no *special* reason not to lie to his spouse or cheat on his income tax. The military function will be no worse off if a sailor always put the needs of the service above her own, but still gives nothing to charity. As long as a pilot is courageous in combat or in dealing with his fellow professionals, he might just as well be a coward with a burglar or his father or his wife. We might well be (and should be) disappointed about these non-military moral failures, but the functional line doesn't give us *special* reasons to be strict outside the military context.

Now one might be inclined to think that what I'm imagining is not possible. Either people are honest or they're not, selfless or not, brave or not. This kind of functionalist would think virtues or character traits are not something we can easily exercise in one context and then fail to exercise in another. Hence, if that is true, then for functionalist reasons, the military professional ought to be held to higher standards of honesty, selflessness, or courage in every context, through and through.[9] Otherwise, failures will invariably bleed through into military life. So when a military professional, say, cheats on his taxes, or lies to a salesman, I would still have a special, functionally grounded reason for being particularly disappointed.

I do not think this works. Clearly, perfectly ordinary human beings are capable of forming extremely complicated, situation sensitive dispositions. Do not almost all of us easily internalize habits of etiquette that alternately allow and prohibit us to do all sorts of things depending on the context?[10] Likewise, given a normal human psychology, I see no reason to think we cannot form complex, situation-sensitive moral dispositions. Indeed, it would be very surprising if there were *not* moral dispositions sensitive to contexts and that take account of what might be at stake. I take it as obvious that there is some sense in which there can be honor among thieves. And the ugly truth is, history is full of examples of effective military professionals (who must have had the requisite functionally grounded moral qualities) who were—all things considered—very bad people indeed.[11]

---

[9] We could take this even farther, and believe that a person is either *good* or they are not. On this more radical view, reminiscent of a Platonic-style unity of the virtues, any moral failing whatever is reason to suspect other moral failings are forthcoming. We would be committed to thinking, for example, that a person who lies on their income taxes could not be relied on to be brave in battle.

[10] Do I really need to cite an example? Belching and passing wind, among other things, fit the bill here.

[11] Nazi Germany provides many particularly well-known examples of these sorts of context-sensitive dispositions. Stories abound of concentration camp guards and doctors who, while brutal with prisoners, were tender with their families and otherwise more than decent human beings. And the attitude conveyed in this famous passage from a speech by Himmler is instructive: ". . . we must be honest, decent, loyal, and comradely to members of our own blood and to nobody else. . . . Whether 10, 000 Russian females fall down from exhaustion while digging an anti-tank ditch interests me only in so far as the anti-tank ditch for Germany is finished" in the Office of United States Chief of Counsel for Prosecution of Axis Criminality, *Nazi Conspiracy and Aggression,* Volume 4, (Washington, DC: US Government Printing Office, 1946), 559.

Of course, whether or not (or to what extent) we have the moral-psychological capabilities I am postulating is an empirical matter. One functionalist thinker (Wakin) has conceded to me in conversation that we might well produce some examples of people who are complex in the way I am suggesting. Still, he maintains that people of globally good character are far more *likely* to possess the functionally grounded military virtues. It is this likelihood he thinks, even though we occasionally see those of 'split personality,' that justifies our functionally grounded desire that military professionals be good through and through.

There is more we might say in defense of the 'good through and through' functional view. Perhaps those willing to indulge in moral shortcomings in non-military contexts are revealing something important about their global motivational structures and value priorities. Someone willing to lie, cheat, steal, etc., must, on this view, have a certain sort of disregard for other human beings. They must be willing to rank, in a morally inappropriate way, their own needs and wants over the needs and wants of others. This disregard will not suddenly disappear in military contexts. Hence, goes the argument, those not globally good cannot be counted on to be good in the military context. So if I have functional reasons for a higher standard in the military context, I also have reasons to be strict in the other contexts: failure in the other contexts reveals that a person is missing something crucial for making the higher standard work where it is needed.

This argument may not be substantially different from Wakin's 'more likely to be good' claim. Indeed, I think it has more or less the same force, and the same limitations. This argument still relies on there being some connection (whether it be invariant or merely a reliable correlation) that reaches across contexts. It assumes that what one values, how one thinks, what one does, and what habits one forms will remain constant and not change with context. This is once again an empirical claim, but one that I do not find compelling on its face.

I am reluctant to do any more armchair psychology than I have already done, and point to this question of context sensitivity as an important area for further research by those competent to carry it out. But to the extent that people can and typically do form complex, context-sensitive moral dispositions and motivational structures, I think the functional line is weakened in its attempt to make more general demands on the military professional's behavior and character.

We might try another twist on this functional approach. If the military professional has the *appearance* of being moral through and through, the "more moral" image might contribute to military effectiveness in some way. Appearing more moral might make the military professional more effective in getting money and other kinds of support with those in the public who are morally minded. Indeed, garden variety moral *failures* by military professionals might erode public trust of the military, which could in turn impact money and support. Perhaps this twist will give us the reasons we need for extra strictness outside the immediate military context.

I do not think this twist gets us very far. First, I am sure we were not exploring reasons military members have for merely *looking* good, but instead were trying to establish that they have special reasons for *being* moral. To make out the stronger conclusion using this argument we would have to add a further premise. Specifically, we would have to say one cannot *appear* good without actually *being* good. And I take that to be false. We could weaken the added premise (and also to a degree the strength of the conclusion), and claim that the best, easiest, or most reliable way of appearing good is to be good. But I take even the weaker premise to be dubious. Even though appearing good without being good requires a special skill in the settings we are considering, it is not a rare or difficult skill to acquire.

This line of thinking also rests on a controversial empirical claim as to the relationship between one's willingness to support military professionals and one's beliefs about their more general moral behavior and character. Here once again, the strength of the argument may turn in large part on the results of empirical studies of this relationship. Will beliefs about "extra" moral uprightness really result in more support? Will beliefs about specific moral failings actually lead to a more general lack of trust? I think it is prudent, without the results of careful studies, to withhold judgment on these questions. And even if we assume these relationships do obtain, we would still need to finish the argument by showing that the *degree* of trust, support or confidence lost would be sufficient to impact the military function. That, finally, is where all functionalist arguments must bottom out.

The demands of leadership in the military might provide some other functional justifications for a higher standard outside the military context. First, morally upright troops might be more inclined to follow military leaders they believe are exceptionally moral. Of course, we still need to establish the empirical connections between perceptions of moral uprightness and the effectiveness of the leader. Do beliefs about a leader's moral conduct *outside* the professional context have effects on the ability to lead? Besides, if we base the strict adherence to moral obligations solely on what it takes to get troops to follow leaders, it might turn out that the military professional has just as compelling reasons on occasion to be especially *bad*. A less than morally upright soldier might identify *more* readily with a leader that shares his vices, and be *less* inclined to follow a leader that he views as "moralistic" or a "goody-two-shoes."

We should not, however, overlook the powerful effect military leadership can have on those led. Misconduct by a superior might create a climate within a military unit that would make similar misconduct in the ranks more likely. All military leaders, especially commanders, have an ability to 'set the tone' and their own conduct has a strong motivational and educative effect on those led. So assume there are standards that bind everyone. If failures by a military leader can lead to failures by others, then I have good reasons to be especially strict with leaders. I think this argument works, and establishes a 'strictness-style' higher standard for leaders, both in and out of the military context. We should keep in mind, though, that the argument would apply to anyone in a position to function as a role model for others. Also keep in mind it only works to the degree that a military professional is functioning as a leader. Those in the highest positions of leadership, to the extent that they are functioning as role models, ought to feel the weight of this argument most. To the extent that a military professional is not occupying a position of leadership, this argument does much less to establish a higher standard.

Overall the functional line does give us some special reasons to be morally strict with the military professional, but I think the strongest demands are made only in the military context. The argument does not get us a knight in shining armor. Indeed, the higher moral standards for the military professional established by the functional line (bravery in combat, honesty in the military context, selflessness to one's country, etc.) are ones that even a Nazi could and would endorse.[12]

Demands of the Role

This next argument I will explore is a lot like the functional line. Call it the "role-based" argument for a higher moral standard. On this view it is not just that the military function—narrowly defined as fighting—demands more of the military professional. There is also a more or less well-defined *role* one occupies in the military structure and in society at large when one is a military professional. Perhaps the military role carries with it unique and stricter moral demands that include, but go beyond, what can be generated by functionalism alone.

Take as an illustration of this role-based idea the moral standards concerning the behavior of police officers. A police officer is morally bound to do something about a crime in progress, while ordinary citizens are not always expected to step in. The unique obligation flows immediately from the *role* the police officer is filling. A parent is bound to care for his or her children in ways others are not morally required to do. The obligations are attached to the roles. So if one assumes a role in society (rather than pretending to assume it) this frequently carries with it some very definite moral baggage. As long as you are not a charlatan or a con man, you take on either unique obligations or stricter obligations (or both) because you *agree* to them by assuming the role.[13] Indeed, these various expectations and understandings concerning one's behavior and character as an occupant of a role are part of what it is for something to *be* a role.

Now I do not want to claim that the only role-based moral obligations are those driven by the brute expectations of society. There may be other sorts of reasons for higher standards attaching to roles that

---

[12]Again, we can draw on Himmler as a source, for he extols the "virtues of the SS Man: Loyalty . . . Obedience . . . Bravery . . . [and] Truthfulness" in *Nazi Conspiracy and Aggression*, 556–7.

[13]To be sure, there are both explicit and implicit elements to such agreements. Actually assuming the roles is most often (but not always) done explicitly. What the role requires is often part of an implicit, but no less clear, understanding.

I have not thought of or explored. Nor do I want to claim that society's brute expectations are always ones we should meet—they might, after all, be unreasonable expectations. My thought here is merely that if a society has certain expectations, and I voluntarily assume this publicly understood role, then I have at least *prima facie*, honesty-based reasons for meeting those expectations.

Now consider the military professional. If one voluntarily assumes that role[14] then there are certain standards of behavior and character to which one at once agrees. What are they, exactly? As an easy starting point, certainly an obligation to attend honestly and conscientiously to every day military duties comes with the package. A similar demand is made by almost every role. If called, doing one's best in combat seems uncontroversially to be an obligation attached to the role. We should also assume that the explicit oaths that demand obedience to superiors and loyalty to the constitution are part of the public understanding of the military professional's role-based obligations. There may be more. But when someone assumes the military role, unless he is a fraud, he at once assumes some moral obligations which are attached to the role (whatever those obligations turn out to be).

Of course, one might ask why the military professional should not be a fraud. Fair enough, and we might be able to conjure some special reasons military people have not to be frauds in regard to their role. But that is bigger game than I am stalking here. I am happy at this point to explore what kind of complex, role-based moral obligations we can deduce from a more simple moral obligation like not being a fraud. It is a *higher* standard that we are trying to establish.

So, if we assume the role as it is generally understood in the society, and couple it with a prohibition of fraud, I think we can establish the uncontroversial moral standards I have already mentioned. But have we found an argument for higher standards, particularly higher standards that go beyond the demands of functionalism? The obligations I have listed (attending to duty, fighting when called, obeying superiors, and being loyal) are not anywhere near exhaustive of the moral possibilities, do not ask more in degree or in kind than functionalism, and as stated, may be obligations we all have. How much more does the military role require? For this line of argument to do its work, something about the military role and our publicly shared understanding of it would need to point toward unique and stricter demands that go beyond functionalism, toward the strict obligation to be good through and through.

I am skeptical. Indeed, there are at least two worries about taking this any further. First, we can wonder straight away if role-based expectations for the military professional in our (or any) culture actually *do* go beyond the uncontroversial demands I have already listed. If they do not, then we would have no basis inside this role-based strategy to invoke any higher standards. It would be as if we told a doctor that she should not cheat on her spouse *because* she was a professional. I do not think this makes sense. Granted, a doctor has some special reasons not to lie to her patients about their medical conditions *precisely because* she is filling the role of a doctor, since this role carries with it a special component of trust in the doctor-patient relationship. But if it is wrong for her to cheat on her spouse, it is because infidelity would be wrong for anyone, not because she is a doctor. If there is no special expectation attached to the role, then there is no justifiable criticism based on such an expectation.

So does the military profession have special moral expectations attached to it as a role in society, expectations that should lead the military professional to be morally "better" than the functionalist would require, maybe even good through and through, in virtue of her role? It is not easy for me to answer this question with certainty, but I would judge the answer is no. When a military person neglects his children, writes a bad check, cheats on his taxes, whatever, I object morally and legally. But I think the grounds of these objections are standards I would apply to anyone, and there is not a sense that the military person has let me down specifically in regard to his role. Once again, though, I would expect a more definitive answer to come from actual empirical investigation (perhaps a sociological study of exactly what constitutes our publicly shared understanding of these roles).

---

[14]The voluntariness might not be essential, but the role-based case for unique or stricter moral obligation seems stronger to me when someone voluntarily undertakes the role. If this does not work to bind the military professional to a higher moral standard, the strategy would be hopeless for obligating a draftee.

I suppose I might be judging wrongly about the content of these role-based expectations. But that leads me to my second worry about taking this strategy any further. If the culture actually does expect the military professional to be more morally upright than others in many or most ways, and believes this is inherent in the role, *should* this be part of their expectations? Sure, reasonable expectations coupled with assuming the role generates obligations. But *is* it reasonable to expect as much from the military professional as we are supposing here? I think that if some people believe the role demands more than functionalism, their conception of the role is an unjustifiable one. I do not know of a better way to justify the reasonableness of our role-based expectations than grounding the expectations in the functions themselves. And indeed, the shapes these functions take are not arbitrary. The traditional professions, for example, are tethered to several important and perennial human needs (for health, justice, and defense) and the professional roles are conditioned by the function of best providing for those needs. Now we have already seen that the military function, even broadly understood, only makes certain limited demands in the moral sphere. So, I think, a functionally ungrounded claim that military professionals are bound by a higher moral standard is unreasonable, and should carry no weight as part of the foundation for this role-based strategy. Thus it makes the question of *actual* public expectations for the role an interesting empirical question but, without some supporting rationale, not of too much use for our purposes.

But why assume we need an argument in support of the expectations? Assume (controversially) that the public simply expects military professionals to meet a higher moral standard, and this has nothing to do with their thinking about the function or their legitimate understanding of the role. They are paying military salaries, so if this is what they want, however overly demanding, and for whatever reason, this is how the military should be. Given the brute expectation, the professional would be cheating the taxpayer if he took the job pretending to be especially morally upright, but not really taking the higher moral standard aspect of it seriously.

There is a great deal wrong with being willing to abandon the requirement for *reasonable* expectations. I'll mention, but not explore, how arbitrary and unfair it would be to take this view. How could we consistently hold one group on the public payroll (the military) accountable for higher, non-functionally grounded, moral standards, but not all the others on the public payroll (various civil servants and politicians at almost every level)? And I will set aside wondering once again if there *is* such an expectation in the public at large. It is still by my lights doubtful that there is a brute demand in our culture for the military to be *more* moral than the rest of us in non-functionally grounded contexts.

Maybe worst of all for this idea is that, given we stipulated these were not reasonable, functionally grounded expectations, we leave ourselves open to a couple of disquieting possibilities. First of all, here we say the *only* reason military professionals have for being bound by some higher moral standard is that this is what the public expects in the job, and hence they agreed to these things when they took the job. What then would keep these higher standards from disappearing in the future? If we uncritically base the obligation on brute public sentiment, history teaches us that this sentiment can change, and not always for the better. Second, I fear this kind of thinking might lead too many military professionals to think (however wrongly) that brute public sentiment is the sole (or at least "trumping") source of *all* their moral obligations. Then, in another place or time, we might hear the specious moral argument that the public *wants* Jews killed, and they are paying military salaries, so the military professional is obligated to do it.[15] No, if there is a higher moral standard based on something beyond function or a functionally shaped role, we had better have a good reason for thinking so. And "just because the public says" is not, by itself, good enough.

Group Image

While I would not rest my case on the public's brute expectations for the military, the public image of the military is not morally irrelevant. To see this, consider a commander required to discipline some of her troops for writing bad checks to merchants off base. In addition to the appropriate punishments, the commander could correctly admonish such offenders for the bad effects their actions had on the

---

[15] I would not bring up this incredible possibility if similar bad arguments were not so frequently made.

image of the military with local merchants. Because the military constitutes a readily identifiable group, many kinds of misconduct by the few can lead to bad consequences for the many. Some people might form, however hastily, general opinions about how they should view all military professionals.

So the fact that it is easy to identify someone as a member of the military, together with the tendency of some people to form generalizations based on thin evidence, gives us yet another special reason for the military member to be moral.[16] One person's misconduct or lack of character hurts her fellow professionals. Maybe here we have a reason for a higher moral standard outside the military context.

There seems to be something to this. But as with our other arguments, I think we should be sensitive to its limitations. First, (as we saw in our earlier twist on the functional line) it insists directly only on good image and not on genuine goodness. To establish genuine moral strictness we would need to believe genuine moral goodness was causally necessary (weaker version: causally closely related) to good image. Insofar as this is true, the argument provides reasons for a higher standard; to the degree that it is false, the more we would be inclined to say that the argument establishes only a case for good appearance and that the real crime is getting caught. These issues parallel the ones we examined in the twist on the functional line.

Also, even if the argument works, it only establishes higher moral standards the breach of which hurts other service members because of the resulting bad image. So to even get started, we would need to have enough people being caught to *create* the bad image. If only one soldier in 10 years, for example, commits murder, and this has no appreciable effect on the image of soldiers in general, then this image argument does not give me a special reason to be disappointed in the soldier. Besides, even if such an image does take hold, a bad image does not always hurt the members of the affected group. Segments of the population might disapprove of what they perceive of as misconduct in a group without this doing serious, positive harm to the members of the group. Pretend there is a widespread (even if mistaken) belief that military professionals are heavy drinkers, have foul mouths, are sexually promiscuous, and do not have proper regard for their health. Are we sure this will cause harm to the group as a whole, harm sufficient to provide a special reason for all members of the military to refrain from these behaviors? If the harm does not result, then the argument fails. Moreover, the types of misconduct that would elicit this kind of societal response (a generalization that results in harm to the group) could and probably would vary from place to place and time to time. Hence what the higher standard requires would vary as well. So all in all, while group image considerations give us some reasoning in support of a higher standard, it is not what I would call a firm foundation—what it requires would be tentative and variable.

Last, it is also interesting to notice that the argument would apply to any readily identifiable group. If this line of thinking is correct, then doctors, lawyers, racial groups, women, men, members of any group really, also have special reasons under certain circumstances not to misbehave publicly. After all, the image problem can affect any of these groups as well. In fact, all of us belong to one or more of these readily identifiable groups. This being the case, I am not sure how we can make sense of the resulting standard being a higher one, particularly if we thought that meant it was a standard that bound just, or especially, the military professional.

Conclusion

I do not think there is any simple and single answer to the question of what should count as a higher moral standard, and whether we have good reasons for thinking military professionals are bound by one. A number of different arguments point to a loose collection of unique military obligations and some special reasons for being strict with the same obligations that bind us all. But these higher standards mostly are restricted to the military context. Even if we stretch what counts as relevant to military

---

[16]As was the case with fraud in the role-based argument, we are here depending on deriving a complex set of obligations from a simple and presumably uncontroversial one (not hurting one's fellows). It is once again not a knock down argument, but I think it at least counts as a *reason*.

duties and responsibilities to the broadest extent plausible, the higher standards we can truly justify are not as extensive as may be commonly thought. We should conclude that military professionals are bound by *many* unique and/or especially strict moral standards, but these extra demands do not encompass all of morality. We ask an awful lot of military professionals, particularly in the moral sphere. But outside of functionally driven contexts, I claim we have little or no basis for asking them more insistently than others to be moral, or blaming them to a greater extent than we blame others for the same offenses. I do not think we can justifiably ask them to be saints.

B. The U.S. Constitution and the Moral Foundations of Military Service: Conflicts of Principles and Loyalties

Martin L. Cook's article, "The Moral Foundations of Military Service" describes a pivotal tension within the ethics of the military. By history and ancient tradition, on the one hand, the warrior is committed to a professional code of conduct that emphasizes universal moral values, including recognizing the immunity of non-combatants, treating enemy prisoners of war with restraint, using deadly force only for morally justifiable ends, and only to the degree necessary to attain those ends.

Sadly, he notes, decades of destructive wars once were fought in early modern Europe allegedly in behalf of other "universal moral ideals" (the religious wars of the Reformation and Counter-Reformation). This terrible experience led exhausted and disillusioned modern Europeans (and by extension, Americans) to the view that—universal ideals notwithstanding—nations must be perceived as sovereign entities with rights of territorial integrity, permitting the persons who live within their boundaries to pursue their common lives and interests as they see fit without outside interference.

The resulting "Peace of Westphalia" (1648) constitutes a watershed in military ethics, Cook claims, for, ever since, the warrior has come to be seen merely as the servant of his or her individual nation-state, willing to kill and willing to die in its defense, or in the pursuit of its political interests. But this is not a very inspiring moral notion, especially when compared with the military as exemplifying universal moral ideals, as above. How can the mere survival or the pursuit of the self-interests of states constitute a morally legitimate foundation for military service?

The morality of being a servant of the state would, at the very least, depend heavily upon the moral stance and moral conduct of the state one served, leaving the universal moral values that once constituted the common "code of the warrior" in limbo. We will return to this peculiar dilemma of contemporary warfare in Part IV, when discussing the justification for the use of military force in international relations.

Col. Paul E. Roush's "Constitutional Ethics" offers a moral framework that organizes an array of potentially conflicting loyalties within the world of the military officer serving *this* country, the United States of America.

Specifically, in the U.S., the military officer's oath and commission commit each officer to pledge loyalty, not to a specific leader, political party, or even a way of life, but instead, and for the first time in history, to a Founding Document. Within that document, the *Constitution of the United States*, broad principles are set forth that define and sustain the framework for a common, democratic rule of law among peoples who otherwise represent a bewildering and sometimes conflicting array of languages, ethnicities, religious beliefs, and political commitments. By pledging to "support and defend the Constitution" without "mental reservation," each officer becomes honor-bound to the universal moral principles embodied in that founding document, and to the governmental structure and chain of command legitimated by it, within which, by implication, all those different peoples, together with their languages, cultures, religious beliefs and political allegiances, *are to be protected with even-handed impartiality*.

In order for this system to work, however, military personnel must sustain the trust of the general public by showing themselves willing to prioritize their own loyalties and moral commitments. A warrior's true loyalties are properly structured by the function of the military organization he or she serves.

The officer's oath is an acceptance of a commitment to sacrifice: sacrifice, if necessary, of self for the sake comrades-in-arms, of comrades in turn for the sake of the unit (the "ship"), and sacrifice, in turn, of the unit for the sake of the service and its larger mission. That mission, however defined, must itself always represent (and must never conflict with) the support and defense of the Constitution and of the concept of democratic government, equality, and the rule of law it embodies. This is the military officer's "Constitutional paradigm," the foundation of public service to the State.

As U.S. Army Col. Anthony E. Hartle observes, simply understanding this complex hierarchy of loyalties, however, does not guarantee that conflicts of loyalties will never occur. There may be times when the officer is tempted to put his own, or her own friends' or family's interests above the demands of the service, or loyalty to the mission or even the Constitution itself. When the conflicts are relatively trivial (such as a desire not to be bound by lawful orders, or restrictive codes of professional conduct), the options are rather straightforward: obey, or else resign from the service. In particular, Roush argues, it is never an option to ignore the "mental reservation" clause and choose to disobey normal and reasonable orders while remaining in the military service. Even the most elementary forms of such self-serving disobedience represent, in principal, a profound betrayal of trust, the living of a lie bordering on treason, and may ultimately prove of great harm to the officer's individual character and professional caliber.

In Martin Luther King's renowned "Letter from a Birmingham Jail," by contrast, we read about profound disagreements between the structure, laws, and order imposed by the State (even in a democracy), and the "higher" demands of moral conscience and what we will later study as "natural law." Even the most benevolent and well-intentioned democratic government can be guilty of moral error. King, however, sets forth a very stringent burden of proof on the resistor or protestor against the rule of law: disobedience must be a last resort, when all attempts to remedy perceived injustice have failed. Disobedience must be undertaken publicly, faithfully, with an acceptance of the lawful consequences of such disobedience (trial, imprisonment, or other punishment) for the sake of calling attention of the majority to the presumed error of its ways.

For the military officer tempted to ignore or disobey an order, or living in profound dissatisfaction with an organizational policy imposed by civilian authorities, there is an important lesson to be learned in this example. The officer must not be merely inconvenienced or annoyed by the order or policy. For those relatively trivial kinds of conflicts of loyalties, the obligation is clear: obey the order, conform to the policy, "get used to it, get over it," or resign from the service. In contrast, for an officer genuinely concerned on the basis of personal conscience (or perhaps religious belief) about the possible errors of his or her chain of command, or of the morality or even legality of orders given or policies imposed, Roush derives from King's essay a demanding and exacting procedure, the "Hierarchy of Loyalties" and the "Four Prerequisites," that must strictly be followed as a guide to individual conduct in cases of conflict.

The Moral Foundations of Military Service
Martin L. Cook

The topic of morality and military service is a very broad one. For purposes of clarity, let's first group the issues into two areas: ethics *of* military service and ethics *in* military service.

By "ethics *of* military service," I mean the moral basis of the military profession itself. The introductory portion of this essay will address the moral meaning of soldiering and what it means to choose to devote one's life to national defense. Then we will turn to the ethic internal to the military profession—"ethics *in* military service."

Much has been written in recent years about the moral culture of the American military. Some see the unique (and some would say superior) moral demands of military service as a resource for national moral renewal—a kind of "light to the nation" in a time when the nation is sorely in need of confidence in the integrity of its leaders. Others view with alarm the supposed "culture gap" they see emerging between the American military and the civilian society it serves. The question of ethics in the military profession concerns the unique moral demands of military service. Apart from an inherent cultural conservatism of the military, what is it about military service that generates special moral obligations?

Ethics of Military Service

We begin, then, with the first topic, the ethics *of* military service. What is the moral basis of military service? Why is it morally legitimate to willingly assume the obligation to fight and die for one's country?[1]

Morally serious and thoughtful military officers feel a deep tension in the moral basis of their profession. On the one hand, there are few places in our society where the concepts of duty and service above self have such currency. High and noble ideals have a place in the military that they have in few other areas of modern American life. For many years now, polling data have shown that Americans respect and trust their military more than any other group in our society. Military service embodies some of the deepest values of human life and our society, and it produces character that inspires admiration and respect. Even a pacifist like the great American philosopher William James was deeply impressed by the "military virtues." He wrote a famous essay in which he speculated on what could be the "moral equivalent" of war and military service, and what else could produce such absolutely necessary virtues in our society.[2]

But while producing excellence of character and virtue, the military exists to serve the will of the political leadership of a particular state. The military will, at times, be employed for less-than-grand purposes in the service of that state. If ethics at its highest is about universal human values, such as the equal moral and spiritual value of every human being, how can that be manifested by serving to advance the interests of the very partial human community of a single nation?

---

[1] I am grateful to the US Air Force Academy, Department of Philosophy and Fine Art, which commissioned me to deliver their Seventh Annual Joseph A. Reich, Sr., Distinguished Lecture on War, Morality, and the Military Profession in 1994. Some of the argument which follows was first developed in that lecture.

[2] William James, "The Moral Equivalent of War," in *War and Morality,* ed. Richard Wasserstrom (Belmont, Calif.: Wadsworth, 1970), p. 7.

Clearly there is a tension between these highest and *universal* ethical ideals and the reality that the military serves particular states and their political leaders. If we believe Clausewitz's judgment that war is a continuation of politics by other means, the real purpose of military leadership is simply to serve the national interest.

Viewed in this perspective, all the rhetoric about the high moral purposes of military service constitutes a verbal smoke screen behind which lurks an unpleasant truth: It is *functional* to persuade individuals to think about military service in such moral terms, but such talk only makes it psychologically easier to evade the true reality that military people and organizations exist solely to serve the tribal interests of the state. And since states are engaged in a constant struggle to advance their interests and to diminish those of other states, there is little here to be seen as truly morally grand.

Of course we would probably disguise this reality by invocation of ideas of the "self-defense" of the state. But such talk is vague. We know the core meaning of "self-defense": self-defense is when we fend off someone who is attacking us personally or, extended to the state, when we resist a border incursion or protect the lives of fellow citizens in peril. In that narrow and relatively precise sense, all but absolute pacifists grant there is a right to self-defense.

But it requires considerable conceptual sleight-of-hand to extend the concept of self-defense to foreign interventions—whether humanitarian or imperial—and to balance-of-power wars. Only rarely do militaries (especially the US military) fight in wars that genuinely defend national political sovereignty and territorial integrity. Typically our wars serve something considerably broader and vaguer than strict self-defense would imply, something expressed in terms of national interests or important national values.

So we are now prepared to focus the fundamental question: What is the moral basis of states themselves that justifies our fighting to advance their interests? Certainly, one might argue, it is only human individuals who make moral claims on us, and the use of force and violence might be justified only in the defense of such individuals. So why should anyone be willing to kill and die for the state, an entity which is, after all, a relatively artificial construct, an abstraction?

What States Are and Why We Value Them

Our question was posed most sharply by St. Augustine 1600 years ago. He wrote at one of those few real crossroads of history, literally watching as the Roman Empire was collapsing around him, sensing that a new age of darkness was descending on the Western world. In his great work *The City of God*, Augustine reflected on the ruins of the Roman Empire. Romans of the old school had a ready explanation for the collapse of their civilization: the fall of Rome was the fault of the Christians.

For centuries Rome was secure in its political and military strength because it worshipped the civic gods of Rome. In return, those gods protected the Empire and sustained its armies. Indeed, for the Romans, much of religion had a primarily practical and civic function. So from the beginning Christians' appeal to universal and transcendent values that embraced all of humanity seemed politically dangerous and profoundly "un-Roman." From this pagan Roman perspective, a century of Christian rule had undermined those civic virtues and hence weakened the Roman character and will to fight.

It is Augustine's task, as he sees it, to refute the pagan charge. His point of departure is to question Roman assumptions about the glorious character of the state itself. He recounts the legends of the founding of Rome. Were not the legendary founders of the state, Romulus and Remus, suckled by wolves? Was the state not founded on murder and treachery? Was Imperial glory not built on a foundation of conquest and suppression of other peoples and nations?

In one of the most famous passages of Augustine's work, he offered his own view of the glory that was Rome:

> Remove justice, and what are kingdoms but gangs of criminals on a large scale? What are criminal gangs but petty kingdoms? A gang is a group of men under the command of a leader, bound by a compact of association, in which plunder is divided according to an agreed convention. If this villainy wins so many recruits from the ranks of the demoralized that it acquires territory, establishes a base, captures cities, and subdues peoples, it then openly arrogates to itself the title of kingdom, which is conferred on it in the eyes of the world, not by the renouncing of aggression but by the attainment of impunity. For it was a witty and truthful rejoinder which was given by a captured pirate to Alexander the

Great. The king asked the fellow, "What is your idea, in infesting the sea?" And the pirate answered.... "The same as yours, in infesting the earth! But because I do it with a tiny craft, I'm called a pirate: because you have a mighty navy, you're called an emperor."[3]

In Augustine's view there simply is no moral difference between states and bands of pirates. There is only the difference of scale, which can make the state seem grand while the robber band is simply evil. Both depend for their success on a kind of internal harmony and organization (what we might call "military virtues"), and both measure success by their ability to take and destroy the lives and property of others.

Because of this view of the state, Augustine counsels Christians to look to their true home not in the City of Man or the Earthly City, as he calls it, but in the City of God, a "city" of universal and transcendent value. Only in such a city can human beings find spiritual and moral rest; as he wrote in another work, "God, you have made us for yourself, and our hearts are restless until they find their rest in Thee."[4] Only there do universal human values find a true and lasting home.

But for the time being, Augustine said, we must live in both cities. In this life, and in this history, we must struggle amid the shades of gray of the state, of warfare, and of injustice, doing what we can to make things better and more peaceful than they would otherwise be, but not hoping for or expecting purity. We are to live, then, "between the times"—aware of the City of God, but not trying "before its time" to live as if we were its citizens exclusively.

It may be necessary to go to war in service of the *relatively good* state, which is all that stands between us and complete political and moral chaos. Augustine quite literally saw that chaos on the horizon as Rome fell and barbarian armies advanced on his own city in Africa. But such wars are to be entered "mournfully." They are justified because others have broken the peace, and the soldier who fights to restore the peace that had existed is the true peacemaker. But Augustine's mournful soldier is free of false hopes and unrealistic ideals. His wars will not create a City of God amid the shadows of the City of Man. At most they will maintain a kind of order and a kind of peace, resting on force and the threat of force.

This Augustinian line of thinking laid the foundation for the classic Christian and, later, the secular international-legal justification for participation in warfare: the Just (or Justified) War Theory. This theory worked out a place for moral conduct of soldiers intermediate between the pacifism of the early church and the amoralism of the nihilist's denial that moral categories apply to war at all.

Of course this tradition undergoes enormous elaboration and qualification as it wends its way through Western intellectual history, and through the changing political contexts in which it is worked out. For most of that next thousand years after Augustine, the Western world was a relative backwater compared to the stronger and more sophisticated civilizations of the East, first of the Eastern Roman Empire, and then of the new Islamic civilization centered in Baghdad.

But within the West, the ideal was a model of a unified Christian civilization, centered in Rome and under the authority of the bishop of Rome, the Pope. On this understanding, wars were justified as responses to disruptions of the order of that civilization.

So it is important in our thinking about fighting *in defense of states* to remind ourselves that the state as we know it is a fairly modern invention. For much of history, and for many cultures, *the state as we think about it* does not exist at all.

In the West it was not until the period after the Reformation that the nation-state, with its claims to sovereignty and territorial integrity, became the dominant institution. Prior to that, European nations and political leaders were subordinate in principle, and often in fact, to the ideal of a universal Christendom.

Similarly the Islamic world affirms in principle the unity of all Muslim peoples, and the ideal of gathering all these peoples in a single political order with a single political head. This Muslim civilization as a unified entity is set in contrast to the *dar al harb*, the world of conflict that lies outside the order of Islamic civilization.

In many parts of the world we discover daily that the boundaries of states on the map correspond poorly to people's senses of identity and belonging. Whether we're talking about failed states, or "tribes with flags" that are the reality in much of the world, the nation-state, founded on the civic equality of all citizens, is a recent and regionally specific development.

---

[3]Augustine, *The City of God*, Book IV, sect. 4.

[4]Augustine, *Confessions*, Book 1, 1(1).

The growth of the concept of states with rights to territorial integrity and political sovereignty, and the evolution of a world system taking that form of organization as fundamental, can be seen as an attempt to give a moral shape and definition to the realities of post-Reformation Europe. It became obvious as Europe exhausted itself in the religious wars following the Reformation that the last illusions of a unified Christian Empire were no longer thinkable. In place of the earlier ideal, laws and customs of international relations evolved to deal with those new realities, and particularly to put an end to perennial war over religious differences. At the end of the Thirty Years War, accepting the futility of restoring political and religious unity by force, the European states crafted the Peace of Westphalia in 1648.

In this new Westphalian international system, religion would no longer be a factor in determining alliances or granting or withholding citizenship. Nor would it be a cause of war. What resulted was the system of Westphalia in which Europe was organized into nation-states of differing religious systems. Here the rules of the game were that the internal matters of states were their own business.

From this specific set of historical circumstances we get the modern international system in which political sovereignty and territorial integrity of states are the highest values. The whole body of international law is founded on this idea of the sovereign state as an entity closely analogous to a free individual, able to do as he or she sees fit in matters that affect only individual welfare. Each free individual is at liberty to pursue the life and beliefs that seem to him or her most likely to lead to happiness, free from the interference of others.

The whole body of Christian and Medieval thought about just war is transposed in this new environment into a secular version of the theory. Here *jus ad bellum*, the reasons for going to war, are increasingly defined in terms of the defense of the twin principles of the new international system: territorial integrity and political sovereignty of states.

For the purpose of developing the moral foundations of military service, it is important to note that the military, too, comes to have a rather different conceptual framework in this model of the international system than it had in Medieval Europe or in the Islamic world.

Naturally, one would not defend the Crusades or the military aspects of the Islamic concept of *jihad* (religious struggle) to expand the realm of Islam. But we should note that those ideas place the activities of the warrior in a supposedly universal moral and religious frame. The soldier fights in the name of values believed to be universal and transcendent, not merely for a particular sovereign state.

In the context of the nation-state, by contrast, the role of the military is set in a much smaller, and probably more realistic, context. It is not fighting for God or universal humanity. Instead, it is defending a particular political and social order in the face of threats to it by other militaries in the service of other states.

It is an axiom of this new model of international order that all states *have equal moral claims* to territorial integrity and political sovereignty. Each state has the right to be free of aggression by others and to use its military in defense of those rights.

Occasionally the rhetoric is more grandiose, perhaps especially in American political discourse (the "war to end all wars," defending democracy or civilization, or defeating communism). But the "official rules" of the international system were built on the idea of the fundamental equal sovereignty of all states.

There is an implication in this for the morality of the military: In the Westphalian international system, military officers are moral equals as well, regardless of the state they serve. This is the classical modern European understanding of the moral foundation of officership—that all military officers are morally equal members of the profession of arms.

On this model, the moral demands on the military profession are great, but they are also delimited. The officer is obliged to serve the state with integrity and to conduct military operations in a professional manner, disciplining subordinates and ensuring that they conduct themselves within the bounds of the laws and customs of war. But it is *not* the moral responsibility of the officer to assess the moral worth of the state itself. Neither is the officer generally obliged to determine the justice of the war the state orders unless there is a compelling reason to do so.

But one may still ask our fundamental question: What is the moral basis on which officers of these particular states can justify killing and dying for their interests? Such states fall far short of being bearers of universal moral, religious, or political truth. Yet as ethical agents, soldiers too understand and appeal to universal human values.

Killing for One's Country

Fighting in defense of the interests of particular states requires a strong case be made for it. Let's look only at American history. Suppose someone said that the United States is built on the morally questionable foundation of conquest and territorial expansion. Further, American expansion and development rest on the destruction of an indigenous civilization by disease and war and by dishonorable and dishonest dealings with both Native American peoples and with Mexico. Suppose that person went on to say that racism is alive and well in American life, and that the relations between the sexes are far from fair and equal.

To that litany of charges, what can any well-informed and educated American say but, "True"? In a famous essay, Admiral James Stockdale, who had been the senior American Prisoner of War (POW) in Vietnam, discussed his observations on education and the POW experience.[5] He noted that a POW with little historical education or an uncritical patriotism in which America could do no wrong was at great risk. For that POW, the reality of a less-than-perfect American society could be exploited by an enemy. It could be used to shake his confidence and loyalty by making him see the nation's faults and flaws, perhaps for the first time.

Stockdale's point is a profound one: If one is to serve the state as a thinking military officer, one must serve the state *as it is*, not the fantasy state of America's highest ideals and ambitions. In this regard, Augustine's somber estimate of the state—of any state—is far closer to reality than the "alabaster cities" whose gleam is "undimmed by human tears" in our best national song.

Of course that litany of injustice, conflict, and conquest, with only regional variations, would be the story behind every other state in the world too, except in the worse cases of states whose borders are even more unnatural, imposed as they were by departing colonial powers. So we can now pose the question again: What about such states morally warrants a profession dedicated to serving their interests and killing and dying on their behalf?

Princeton philosopher Michael Walzer, in his fine book *Just and Unjust Wars*, attempts to work out why national loyalties should matter to people. True, existing states and their boundaries result from utterly irrational patterns of arbitrary map-making and histories of conquest. Still, in reasonably good states, the nation with its twin rights of territorial integrity and political sovereignty creates a "space" (both literal and metaphorical) where a group of people can attempt to work out a "common life." He explains this concept of common life as follows:

> Over a long period of time, shared experiences and cooperative activity of many different kinds shape a common life. The protection [of the state] extends *not only to the lives and liberties of individuals but also to their shared life and liberty,* the independent community they have made, for which individuals are sometimes sacrificed.[6]

On this model, one serves the state in order to protect that common life the officer shares with fellow citizens. One recognizes the complexity and often the moral ambiguity of the processes that give rise to that common life. But one also recognizes that the persistence and flourishing of that common life is a condition for human welfare and goods less tangible than life and property—the goods of shared memory, common symbols, shared suffering, defeat and victory, and history and culture built up over time together.

Walzer's idea of a common life provides a language to try to articulate why, in reasonably good states, it matters to be an American or a Japanese or an Egyptian, over and above the good of individual survival. It helps articulate what a well-grounded and honest patriotic feeling is about.

The foundation of this idea of a common life is Westphalian. It is applicable to every society possessed of sufficient historical continuity through time. If the moral basis of states is that they create and maintain the space within which a common life can flourish, it is obvious that states succeed in doing this to widely varying degrees. Walzer continues his argument:

---

[5]"The Moral World of Epictetus: Reflections on Survival and Leadership," in *War, Morality and the Military Profession*, ed. Malham M. Wakin (2d ed.; Boulder, Colo.: Westview Press, 1986), p. 20–21.

[6]Michael Walzer, *Just and Unjust Wars: A Moral Argument with Historical Illustrations* (2d ed.; Basic Books, 1977), p. 54.

> The moral standing of any particular state depends upon the reality of the common life it protects and the extent to which the sacrifices required by that protection are willingly accepted and thought worthwhile. If no common life exists, or if the state doesn't defend the common life that does exist, its own defense may have no moral justification.[7]

If we follow this line of argument, the defense of states rests on a universal foundation. The argument gives states the benefit of the doubt and assumes that they really do sustain the common life and welfare of their citizens. And I believe this is the right account. But note that it opens the door to judging and perhaps even intervening into the affairs of sovereign states if and when their defense of those values is fundamentally flawed. And it also challenges the moral equality of soldiers, linking the moral basis of their service to the moral character of the political order they serve.

Symbolically the change was marked by General Eisenhower's conduct at the end of World War II. When German General von Arnim was captured, he requested a meeting with General Eisenhower, a request completely reasonable if one assumes the moral equality of members of the profession of arms. Eisenhower refused, however, saying,

> The tradition that all professional soldiers are comrades in arms has . . . persisted to this day. For me, World War II was far too personal a thing to entertain such feelings. Daily as it progressed there grew within me the conviction that, as never before . . . the forces that stood for human good and men's rights were . . . confronted by a completely evil conspiracy with which no compromise could be tolerated.[8]

General Eisenhower's attitude marks a change from the idea of morally equal military professionals. It reflects an idea of military service set once again in the framework of universal moral questions about the nature of the states officers serve. It reflects a profound change in our thinking about states and war. It suggests the Westphalian pillars of state sovereignty and territorial integrity may not be as sacrosanct in the future as they have been. It suggests new thinking about warfare in the post-Cold War world.

Take the Gulf War as an example. No other war since World War II has so clearly matched the Westphalian paradigm: a sovereign state, internationally recognized, has its territorial integrity and political sovereignty directly and unambiguously attacked. That state requests help from the international community to restore it among the nations of the world. States respond; aggression is rolled back; Kuwaiti sovereignty is restored. This is the classic Westphalian story with a happy ending.

But, say the critics, the moral basis of the Gulf War is tainted. Despite all the rhetoric of international law and multinational coalitions, so the argument goes, really the war was about oil and economics. The implication seems to be that because there were important international economic interests involved in that war, the presence of such interests makes the motives impure. "No blood for oil!" went the chant. But would "blood for Kuwaiti sovereignty"—in the absence of oil—rally the enthusiasm of the critics?

For the sake of contrast, let us briefly examine the Kosovo intervention. Here, in extreme contrast to the Gulf War, it is very hard to make a case for a crucial US interest in Kosovo. Yet here the criticism is the opposite to that of the Gulf War: "No blood for the rights of foreigners when there is no national interest involved!" would probably be the cry.

These examples point to the horns of the Westphalian dilemma in its post–Cold War form: Is military power to be used in the service of national interests, wherever they are? If so, then claims to higher moral justifications are unnecessary and misguided. In other words, Clausewitz's word is the final word, and war really is just politics by other means.

Alternatively, is military power, freed from the fairly artificial and historically abnormal framework of the bipolar superpower world, now at last at liberty to serve the universal moral ends of promoting democracy, supporting human rights, and removing oppressors to the cheers of the

---

[7]Ibid.
[8]Cited in Walzer, p. 37.

oppressed? If so, how and why should national political leaders be willing to spend the blood and treasure of their individual nations in the service of the lives and rights of foreign nationals?

It seems we are now deeply ambivalent about these alternatives. Much of our national confusion about the role of US foreign policy generally, and of the purposes of military power specifically, results from this conflicting pair of models for thinking about the proper uses of military power.

In a profound speech to a Joint Session of the US Congress, South African President Nelson Mandela said the following:

> In an age such as this, when the fissures of the great oceans shall, in the face of human genius, be reduced to the narrowness of a forest path, much revision will have to be done of ideas that have seemed as stable as the rocks, including such concepts as sovereignty and the national interest. . . .
>
> If what we say is true, that manifestly, the world is one stage and the actions of all its inhabitants part of the same drama, does it not then follow that each of us . . . should begin to define the national interest to include the genuine happiness of others, however distant in time and space their domicile might be?[9]

Mandela's vision of the new world order has much to commend it. To a great degree it seems to reflect accurately the global convergence we daily witness around us. It reflects, too, the growing sense that the existing structures of international relations are increasingly inadequate to the tasks now facing them in the post–Cold War world. But it leaves much unresolved at the practical level. Are the armed forces and leaders of individual nation-states prepared to enlist in the service of such a vision? Can we ask the soldiers of our state, or of any state, to fight and die in the name of such a global vision of our interests?

For the moment, we're left with two answers to the question: Why serve the state? We have a Westphalian answer in defense of the common life of our nation. And we have a universalizing answer in terms of transcendent moral and political values. I would simply note that they are quite different. They correspond to differing understandings of the proper uses of military power. Different political leaders appeal to one or the other (or try, somewhat disingenuously, to combine them).

Both kinds of reasons will be adduced to justify a wide range of interventions and uses of force in the foreseeable future. For those readers who are future strategic leaders, the clearer you can become in your own thinking about this question, the better you will be able to advise political leaders on wise and morally persuasive courses of action.

Ethics in the Military

We turn now, and more briefly, to the question of ethics *in* the military. The profession of soldiering puts unique moral demands on military personnel. No other group in society is given as much latitude to define its own standards of conduct and talks so frequently and openly about the core values that define it. And in American history, no institution has been viewed as a bigger threat to American political values than the standing Army.

Recall President Eisenhower's farewell address. His strongest words of warning concerned the growing power of the military-industrial complex and the threat it posed to American political life.

At the time of the ratification of the Constitution, recall the great anxiety that the stronger federal government the Constitution proposed would allow a standing federal army. Alexander Hamilton, arguing for the Constitution, called a large standing army one of the "engines of despotism" (Federalist Paper No. 8), and argued strongly that only absolute civilian control by Congress over military matters would protect American liberty.

In 1943, General George Marshall gave one of the finest statements of the importance of the highest ethical standards in military service and of the special character of the American military in particular. It occurred in a conversation with Major General John Hilldring:

---

[9] US Congress, *Congressional Record*, 103d Cong., 2d sess., 6 October 1994, 140, Part 144, H11008.

In the spring of 1943, George Marshall called John Hilldring to his office. It was the middle of World War II. The Germans were still deep in Russia; the British and Americans had cleared North Africa, but had not yet invaded Sicily or Italy. The Japanese controlled most of the Western Pacific. Marshall was Chief of Staff of the US Army. Hilldring, a two-star general, had just been given the job of organizing military governments for countries to be liberated or conquered. Years afterward Hilldring reported what Marshall said to him:

"I'm turning over to you a sacred trust and I want you to bear that in mind every day and every hour you preside over this military government and civil affairs venture.... [We] have a great asset and that is that our people, our countrymen, do not distrust us and do not fear us. Our countrymen, our fellow citizens, are not afraid of us. They don't harbor any ideas that we intend to alter the government of the country or the nature of this government in any way. This is a sacred trust that I turn over to you today.... I don't want you to do anything, and I don't want to permit the enormous corps of military governors that you are in the process of training and that you are going to dispatch all over the world, to damage this high regard in which the professional soldiers in the Army are held by our people, and it could happen, it could happen Hilldring, if you don't understand what you are about."[10]

Opinion polls have shown for many years that the US military is the most highly respected professional group in the United States. That respect ebbs and flows a bit, of course, when scandals and conspicuous moral failures such as Aberdeen or Tailhook tarnish the military's image.

And there are big challenges ahead in terms of the growing gap between civilian and military cultures as fewer and fewer individuals in the general population and in the civilian branches of government have direct experience and understanding of the military. There is the risk that the military may come to view itself as too set apart from, and morally superior to, the civilian culture it is charged to protect.

That high regard in which American citizens hold soldiers and the military profession is, in Marshall's words, a truly sacred trust. It has real practical benefits, of course, in terms of arguing the military's case for resources and in making sure professional military advice is heeded in national security matters. But it is far more than that.

Only when the military articulates and lives up to it highest values can it retain the nobility of the profession of arms. Only when it retains a proper sense of its role in American democratic life does it retain the trust and respect Marshall spoke of. Only a military that daily lives out its values and feels its connection to the citizens is a military that engenders the respect and loyalty of the nation and keeps it from being feared.

Such respect and trust are the real foundations of morale, retention, and voluntary service. Young people may be drawn to military service by promises of education and training, by opportunities for adventure and travel. But they stay because in military service they come to see a kind of ideal human community, grounded in service to others and mutual respect.

Journalist Thomas Ricks has written eloquently of the power of military training on 18-year-old recruits. Ricks chronicles how those young people, drawn from some of the most unpromising neighborhoods and backgrounds in our society, are inspired and transformed by military training. For the first time in their lives, they experience and embrace a higher moral vision of life.[11]

For them as individuals, military service is a route out of what well might have been very unfulfilling lives. For our society, such young people form a nucleus of moral strength for the nation.

Only if the promise of moral community holds true for them as they experience the day-to-day routines of military service will such young people keep the vision and stay in military service. On issues of racial integration, of equal opportunity, of respectful treatment of every service member, the US military has been a guide to the nation at its best moments. Excellence in human relations and moral integrity can and does inspire the idealism of the young people who choose to be a part of this service.

---

[10]Richard Neustadt and Ernest May, *Thinking in Time* (New York: Free Press, 1986), p. 247.

[11]Thomas E. Ricks, "The Widening Gap between the Military and Society," *The Atlantic Monthly*, July 1997, pp. 66–78.

This alone points to the deep connection between ethics in the military and the more practical issues of recruitment and retention.

At a still larger level, ethics in the military is the absolute requirement if we are to retain the precious resource of the long American tradition of civil-military relations. There are few nations in the world with the same historical tradition and conviction that our citizens do not fear their military forces, and that they respect and admire them. The military's treatment of its own soldiers must reflect the highest degree of respect and fairness if that trust and respect is to be maintained.

Military operations must also be conducted in accordance with the values of the American people and international law if that trust is to be sustained. Some of us often complain about the CNN effect, but it is not all bad. Knowing that our military's conduct will be scrutinized closely by our citizens and by people all over the world insures that just war restraints are respected. It also forces military leaders to educate the American public to the limits of the possible in military operations.

As alluded to above, the connection between the American military and the general public is undergoing a particular strain unprecedented since the end of the Vietnam War. As the military becomes smaller, fewer and fewer civilians and civilian leaders have direct experience of military service. Consequently, there is the danger of a growing gap in values and mutual understanding. The costs to the military, the culture, and the nation will be great and long-lasting if some measure of mutual trust and respect is lost as a result.

In sum, the highest standards of ethical climate and conduct are essential to maintaining a healthy military service and to attracting and retaining the best and most talented of each new generation of Americans. The cultural gap needs explicit attention if it is not to undermine the trust and respect of the nation for its military. Only if the connection and trust between the populace and the military are maintained can military service remain the honorable and respected profession of arms that causes good people to enter service and to advance to senior levels of leadership. Issues centered on human relations, respect for others, command integrity, and the conduct of combat operations all raise ethical challenges that decisively affect that trust. Such issues will continue to pose significant challenges for current and future military leaders.

Constitutional Ethics
Col. Paul E. Roush, USMC (Retired)

Why do we need constraints? Shouldn't we, as the experts on the military, be entrusted to make our own decisions, without oversight by civilian government, concerning how to best defend the nation? Patrick Henry in a speech in 1788 answered the question about the need for constraints as follows, "And, Sir, would not all the world, from the Eastern to the Western hemisphere, blame our distracted folly in resting our rights upon the contingency of our rulers being good or bad. Show me that age and country where the rights and liberties of the people were placed on the sole chance of our rulers being good men, without a consequent loss of liberty." Faith in well-intentioned military commanders is not enough to ensure that farmers in Iowa and steel workers in Pennsylvania are participants and not just spectators in determining how their military is to be operated and employed. During the formative years of our constitution and government, there was a great fear of standing armies and the military in general. Placing so much power in the hands of a small minority was seen as inviting tyranny. There was much debate over who should control the military. The end result was that control was split between the different branches of government. A number of the Constitution's references to the military are in the form of constraints. Those constraints to the exercise of military force are found in four sources: the President, the Congress, the courts, and what the Constitution calls the Supreme Law of the Land.

Presidential Constraints

The President is declared in the Constitution to be the Commander-in-Chief of the armed forces. That means that no military commander has the right to employ his or her forces in ways that contravene the expressed desires of the President. The outstanding example of tension in this area involved President Truman firing General MacArthur during the Korean conflict when General MacArthur adamantly persisted on publicly arguing for offensive actions which the President had forbidden.

Congressional Constraints

The second constraint to unhindered activity on the part of the military is the Congress. The Constitution grants to the Congress a number of important powers. One is the power of the purse. The Congress represents the people from whom taxes are derived. In that capacity the elected representatives approve the purposes for which funds are to be made available and the amount to be allocated. The Constitution also grants to the Congress the power to make regulations governing the armed forces. This is a very broad-ranging power, indeed—one which touches the life of each sailor or marine every day. For example, the Congress legislated the Uniform Code of Military Justice (UCMJ), the legal system under which the military operates. The Congress also creates administrative regulations such as pay, retirement, and health care issues. Congress also enacts legislation which is not administrative in nature but has a direct impact on war-fighting. The Boland Amendment, which forbade providing funds for the Contras, the insurgents who were attempting to overthrow the Sandinista regime in

Nicaragua in the mid-to-late 1980s, is an example of such congressional legislation. When Lieutenant Colonel Oliver North apparently failed to comply with Boland, the course was set for confrontation between the legislative and executive branches of government. The War Powers Act, which places time limits on the use of force without congressional approval is another example of regulation which affects the freedom of the military to resort to force. The code of conduct was disseminated by the Congress to regulate the behavior of American prisoners of war. Every person in the military is under obligation to comply with every piece of legislation issued by the Congress to regulate the armed forces. There are no exceptions. Compliance means obedience to the spirit as well as the letter of the law at issue.

Judicial Constraints

The courts are less likely to intervene in the affairs of the military than is the case when they deal with civilian institutions. They have traditionally given the military very wide latitude. There are, nonetheless, many cases brought before the courts which could directly or indirectly affect some aspect of the way the military does business. For example, about twenty-five years ago, a judge in Washington, D.C. ruled that the Marine Corps could not legally prohibit dependents of Marines stationed in Okinawa, Japan from traveling to and living in Okinawa (at their own expense) during the time that the service member was stationed there. From then until the present that ruling has been the prevailing reality. It may well be that some persons in the military may take exception to various rulings issued from time to time by the courts. We are free, of course, to be happy or unhappy about such matters. What we are not free to do is act in opposition to such rulings, unless we are willing to risk very serious personal consequences.

The Supreme Law of the Land

The supreme law of the land is defined in the Constitution as comprising three parts. The first is all the provisions of the Constitution itself. Thus, to oppose either the right of the Congress to legislate the military, a clear provision of the Constitution, or to ignore or oppose the content of such congressional legislation, would be to oppose the supreme law of the land. The second part of the supreme law of the land is conventional international law. The meaning here is that the provisions of any formal agreements or treaties between the United States and some other country, when ratified, are binding upon the military. One example would be the Geneva Conventions of 1949, which have to do with treatment of prisoners and protection of non-combatants. Another would be the United Nations Charter. The third part of the supreme law of the land is customary international law. This pertains to situations that fall short of treaty status but where there is, nevertheless, consensus about the issue at hand and the President has formally elevated the issue to the level of customary international law by issuing a presidential proclamation or executive order to that effect. An example here would be the law of the sea, the provisions of which, with minor exception, are honored by the United States. Those provisions are spelled out by the President and are as binding upon the military as they would be if the Law of the Sea were a treaty to which we were signatory.

The Constitutional Paradigm

The question at issue is how to apply the above knowledge about the Constitution in a practical way. Clearly there is a need for a reference point which has utility in guiding the individual who wants to behave ethically. Without a reference point the individual in the military will operate in a frequently unethical world, a world in which societal mores may be as likely to obscure rather than clarify that which is right, good, and proper.

Four principles guide the practical application of the Constitution as an ethical reference point. THE PRINCIPLES MUST BE APPLIED SEQUENTIALLY.

The First Principle. There is a priority of loyalties for individuals in the Armed Forces that their oath obligates them to uphold. The hierarchy is as follows:

{ Constitution
Mission
Service
Ship or Command
Shipmate
Self

For a given individual, other loyalties may intervene, for example, those to family, to Supreme Being, etc. As long as the intervening loyalties do not conflict with the loyalties of the profession, the oath is still being complied with. The obligation to follow the hierarchy is the same for all persons who take the oath regardless of their personal beliefs or desires. Like Kant's moral obligations, the requirement to follow the Constitution is external to any individual.

The Second Principle. One must first resolve conflicting loyalties when they exist—then act. Reversing the order will not do. The military person is obligated to follow the loyalty hierarchy, but there will be times when there are conflicts. The conflict may be between following a Constitutional loyalty and a personal loyalty, for example, a legal order to kill might conflict with a belief that killing is immoral. The conflict might also be about wanting to place a lower loyalty above a higher loyalty, for example, falsifying records to make your department look good at an inspection. When it is apparent that a conflict exists, it must be resolved. Resolving implies taking some action. It does *not* mean merely coming to terms with the conflict in your own mind, *i.e.*, an "as long as I can look myself in the mirror" attitude. The Constitutional Paradigm is based on values external to your own and therefore must be resolved externally. In many cases, resolving the conflict will entail expressing the concerns to the chain of command or perhaps to several chains of command, depending upon the sources of the competing loyalty claimants. Thus, in the case of Iran-Contra, resolution of the conflict before *trading arms for hostages* would have required extensive exchange of views between the executive branch's National Security Council and members of Congress charged with applicable oversight functions. Eventually resolution might have required action by the courts or may have been achieved as a result of even more specific and more restrictive legislation by the Congress. The important point is that resolution would have clarified the priority of loyalties. If the ensuing priority of loyalties presented an untenable situation for one's conscience, then principles three and four are still available. The fundamental problem was the failure to resolve in advance of action. The decision to act first may have stemmed from the knowledge that attempts at resolution would have precluded taking the actions contemplated by Iran-Contra. That of course, is a by-product of living in our Constitutional system.

The Third Principle. The individual who cannot or will not follow principles one and two above should give serious consideration to finding a line of work in which those conflicts no longer apply, *i.e.*, resigning from the service. When officers accept a commission in the Armed Forces they also take an oath to the Constitution. If the oath is not being followed then the commission is, in a sense, being held under false pretenses. The very words of the commission obligate an officer to obey all the lawful orders of those appointed over him or her. Resigning, however, may not be as easy as taking off a uniform and walking out the door. Refusal to abide by the constitutional constraints may yield strict penalties, including being charged with the commission of offenses under the provisions of the UCMJ. For example, an officer in a combat situation who receives a legal order that he chooses not to obey may want to resign from the service after the heat of battle is past. That will not, however, absolve him of the consequences of his earlier actions in disobeying the legal order. In the words of the UCMJ, one who disobeys any order does so at his own peril.

The Fourth Principle. In very rare circumstances one could conceivably encounter a situation in which an obligation (an order, a law, etc.), while legal, is so offensive that one chooses to withhold the loyalty required by the first principle. If all efforts at resolving the loyalty aspects of the issue—

applying principle two—also fail, principle three would suggest resignation as the remaining option. It is, however, possible that the offended person believes that failure to confront the situation would be the greater evil. Under such circumstances one might decide to disobey a legal order or a law in order to attain a higher good; for example, saving lives that might otherwise be foolishly squandered. It is possible that a military doctor, if required by some future Congress to perform abortions, might choose not to obey in order to create a test case in the courts, even though such refusal might result in court martial with attendant severe punishment. Principle four is certainly not an option that should be exercised lightly. What follows is a series of prerequisites that should be applied *sequentially* before it is even contemplated.

Prerequisites for Principle Four

Prerequisite One. A number of prerequisites must be in place to apply principle four. The first is that the issue must be in response to a *fundamental violation of justice*. Please note above that in each cited example under principle four the circumstances were *non-trivial*. At issue were (1) foolish squandering in combat of the lives of the young people entrusted to the military leader, (2) continuation of patently unjust laws which fundamentally diminished and demeaned the dignity of millions of American citizens, and (3) deeply held views about the sanctity of life. By contrast, deliberately violating laws which govern the lives of midshipmen (amount of liberty granted, wearing of civilian clothes, attendance at military obligations, observance of reveille, etc.) would be limited to the first three principles because the circumstances are, by comparison, generally of insufficient gravity to merit application of principle four.

Prerequisite Two. Where possible, an attempt should be made to have the law or order changed by legal means before choosing the path of disobedience.

Prerequisite Three. All the cited examples of disobedience were or would be done *in public, with full awareness that the persons in authority knew or would know in advance that the order or law was going to be violated by a specific individual at a specified time and place*. In each case there was or would be full willingness to advise the authorities in advance that the law was going to be violated. Where time and circumstances permit, this action in the public arena is the third prerequisite. In a worst case scenario such as a combat situation in which such choices had to be made quickly it would be incumbent, in relying on principle four, to satisfy the requirement for public action by notifying one's senior of the nature of the action as soon as possible. Attempting to act secretly or in a manner to avoid being found out obviates the claim on principle four.

Prerequisite Four. Finally, note that the person choosing to violate the law or order, having first ensured that the authorities would know of the violation *was willing to accept the full legal consequences of the act*. This willingness to accept the presumably inevitable consequences of the act done in public is the fourth prerequisite for the appeal to principle four.

In an analogous situation at the Naval Academy, the midshipman who believed that the rules governing consumption of alcohol were grievously unjust and non-trivial would attempt to use his chain of command to have the rule changed. Assuming failure in that endeavor, he might advise his company officer that he was going to consume alcohol in his room at a specific time in order to test whether or not the rule could prevail. He would in the interim (as his case was being adjudicated) be willing to accept the full consequences of the act, including potential expulsion from the Naval Academy. Having made his point in public (*i.e.*, with foreknowledge of the authorities) and with full acceptance of the consequences of the act for the purpose of overcoming what he perceived as an egregious violation of justice posed by the alcohol rules, he could lay claim to principle four. If the outcome failed to substantiate his perception of grievous injustice then the law would prevail and the midshipman would bear the consequences. The application of the principle is radically different from the highly unprincipled approach of "you rate what you skate."

While the approach embodied in the Constitutional Paradigm is not an all-inclusive ethical norm, its consistent application would have resolved in a positive way many of the difficult situations which eventually had very negative consequences for individuals, for the military, and for the nation. The following examples address each of the constraints imposed on the military by the Constitution and show how the four principles above were applied by the individuals involved.

The Principles and Presidential Constraints

An example involving the constraints imposed by the President in the role of Commander-in-Chief is the case referred to above between President Truman and General MacArthur. The issue was the unexpected (by MacArthur) entry of the Chinese into the conflict after MacArthur had exceeded his authority in driving United Nations forces toward the Yalu River, which marked the boundary between Korea and China. Subsequently, MacArthur pushed hard and publicly for military victory in North Korea, apparently discounting the possibility of Soviet entry into the conflict. MacArthur had ignored the first principle by refusing to yield to the orders of the President. He violated the second principle by failing to resolve the conflict between his own assumption that war's only aim was victory and the orders of his President that a goal short of victory was in the interest of the nation and the world. He did not resign (violation of principle #3) and he attempted to evade accepting the consequences of his actions in subsequent speeches to Congress and to other groups (violation of principle #4).

The Principles and Congressional Constraints

An example of congressional legislation that should be familiar to all of you is Public Law 94-106, the legislation which required the assignment of women at the service academies. The oath all midshipmen take requires their full and unequivocal commitment to carrying out congressional mandates. The oath allows for *zero* mental reservation or purpose of evasion. Not only is there a requirement for compliance with the letter of the law; there is the requirement to work to make compliance with the spirit of the law the institutional norm. Reality has been otherwise. A certain proportion of male midshipmen have intentionally withheld from women the full inclusion which is their legal right and reasonable expectation. The consequences of that process are numerous. For the male perpetrators and those who tolerate the behavior it has meant legitimizing defiance of authority, diminishing one's commitment to ethical behavior, acquiring the habit of cover-up, and publicly degrading the naval profession. For the women who are denied full inclusion results have included loss of esteem and diminished effectiveness that accompanies invidious rejection (psychological exclusion on the basis of gender without regard to competency). For the Naval Academy there were diminishing of cohesion and teamwork and nurturing of attitudes which are precursors for Tailhook-like events. Gender bias is clearly a serious matter; for example, Tailhook has probably cost the Navy tens of millions of dollars, entailed enormous loss of trust among the American public and the Congress, and has caused some in Congress to evaluate the Navy in ways that cast doubt as to whether or not it can be trusted to develop new strategies for the post–Cold–War era. It would not be far amiss to say that those who diminish the readiness of the Navy by deliberately withholding full inclusion of women are guilty of a little bit of treason. Certainly they are routine participants in a clandestine form of mutiny against the commitment which their oaths to the constitution entailed.

The Principles and Treaties and Customary International Law

One of the more blatant examples of violations of treaties and customary international law involves the behavior of army Lieutenant William Calley and his platoon in the village of My Lai in South Vietnam. In that village Calley and his people assassinated helpless, unarmed old men, women, children and babies. The victims were shot down without justification in an hours-long orgy of senseless slaughter. Any orders issued by Calley to behave in that manner were illegal. All of the troops involved had an obligation to disobey the orders. The Geneva Convention of 1949 specifically prohibits acts by combatants which have as their intent bringing physical harm to non-combatants. The United States is signatory to the Geneva Convention. By virtue of ratification by the United States Senate, the Geneva Convention is a treaty. The Constitution grants to all treaties the status of supreme law of the land. The military person's oath to support and defend the Constitution obligates him or her to comply with the supreme law of the land. *The inexorable linkage from oath to all subsequent behavior is a fundamental fact of military life.* It cannot be otherwise if the nation is to retain the essential character of the relationship between the people (through their elected representatives) and the military. If the linkage goes, so does the trust between the citizen and the soldier. A military unconstrained by the provisions of the Constitution is a threat to the civil society.

It is sometimes asserted that military persons fight and die not for some abstract concept such as the Constitution, but for the close friend who is fighting alongside. That may be true, but it is also true that when the shooting had stopped at My Lai, the abstract concept had been translated into scores of actually-murdered (*i.e.,* not hypothetical or abstract) babies, children, mothers, wives, grandparents. It is also true that the actions at My Lai tore asunder whatever remained of the trust between citizen and soldier; hastened the withdrawal of support for the American effort in Vietnam and, in the final analysis, guaranteed that thousands of American lives had been squandered in a cause which became inextricably linked in the media and in the perception of the public with My Lai-type events. None of those consequences are abstract. We must learn sooner or later to do system thinking—to understand that act A produces consequence B, even though it was not apparent that such would be the case. That is precisely why the Tailhook incident was such a tragedy. A large number of persons in the military still do not understand that the consequences include a fundamental alteration in the relations between the Navy and the public and between the Navy and the Congress. In the American system the price of maintaining the trust relationship between citizen and soldier is scrupulous compliance with the letter and the spirit of the supreme law of the land.

The Principles and the Use of Force

One final example may be instructive in coming to an understanding of the extent to which the Constitution has a practical outworking in dealing with nearly intractable situations. In the August 1989 issue of the *Marine Corps Gazette* General Ray Davis, former Assistant Commandant of the Marine Corps, Division Commanding General in Vietnam, and winner of the Congressional Medal of Honor in the Korean conflict, wrote with great skill in an article entitled *Politics and War: Twelve Fatal Decisions that Rendered Defeat in Vietnam.* He provided a superb analysis of what had gone wrong. Unfortunately he advocated a course of action in the conclusion which cannot be justified in the light of the constraints levied on the military person by the Constitution. The following is quoted from the article.

> I join with those who proclaim "No more Vietnams," but for different reasons than most. I am not ready to surrender our role as protectors of freedom. Our friends must know that we do support their quest for freedom. "No more Vietnams" means to me that when we do launch military forces in that noble cause of freedom we must do so with an absolute desire to win. To commit our military forces and then withhold support is to betray those men who so bravely serve our country. When we go to war, we must go to win—that or stay at home.

Most persons in the military could probably agree with those sentiments. We tend to see the choices as he does—go to win or stay home. The painful experiences in Korea, and especially in Vietnam are burned into the very soul of those who took part in those conflicts. But wars are fought for political ends. The President, as in Korea, or the Congress, as in Vietnam, may decide on objectives short of military victory. At that point the ethical military officer will either submit to civilian authority or will resign. The alternative is to further diminish the trust relationship between citizen and soldier. The price is too high to pay.

Summary

The essential points in this paper include the following. First, the scope of the Constitution is far-reaching. A high proportion of the difficult ethical decisions which the military officer may be called upon to make are, in fact, amenable to resolution if one is willing to use the Constitution as an ethical reference point. Second, the constraints placed on the military are very straightforward. They include the power of the President as Commander-in-Chief, the Congress with its power to regulate, the courts as they rule on military-related issues, and the Supreme Law of the Land—the provisions of the Constitution, conventional international law, and customary international law. Third, there is a constitution-based ethical paradigm which can guide the military person through a decision process calculated to produce that which is right, good, and proper. The process includes commitment to the Constitution as one's highest loyalty, resolution of loyalty conflicts in advance of acting, resignation if the conflict cannot be resolved, and a willingness to accept the consequences of public disobedience if the first three principles of the paradigm cannot be squared with one's conscience.

The American Professional Military Ethic[1]
Anthony E. Hartle

The Sources

Not all young officers give serious thought to the professional ethics taught by the military education systems. Most simply accept the "rules" as established fact and either attempt to abide by them or choose to violate them for reasons of their own. After a certain period of time, those who make the military a career begin to identify themselves in terms of their role, which often makes objective analysis of "the rules" even more difficult and less frequent. That, at least, was true in my case. The only issue was that of ensuring that one knew and understood just what the rules were. The education at the Military Academy had placed great emphasis on the rules.

> Two years of combat service, however, generated numerous morally ambiguous situations, some of them deeply troubling, and the answers provided by the code as I understood it were sometimes incompatible with my intuitions of conscience. At some point in my experience, justification of the code became necessary. The nature of the conflict brought many soldiers to that point, I believe, and reactions to the experience varied widely. Disillusionment, cynicism, and resentment were not unusual ways of responding to the agonizing conflicts.

Men and women in uniform sometimes fail to recognize that being a member of a profession imposes moral obligations. We will explore those obligations in detail in this chapter, but we should remember that the profession of arms makes a vital contribution to society and to civilization. The philosopher Thomas Hobbes claimed that people need society to escape from the state of nature, where life is "solitary, poore, nasty, brutish, and short." In an imperfect world, armed forces are necessary to preserve society and the state. For while the state brings order to existence, and in Hobbes's view makes moral existence possible, international society remains a dangerous arena. Despite the role of the military in deterring war and defending with force when necessary, in the words of a thoughtful officer, "if the [armed forces] merely replicate the state of nature, as in societies run by warlords, [they cease] to be useful to society." The armed forces must reflect the demands of morality if they are to be consistently useful to society—indeed, if they are not to be a danger to it. Military officers need to understand that their commitment to professional activity brings a commitment to moral constraints. When we face enemies who accept no such constraints, that covenant can be difficult to maintain.

The Second Gulf War was fortunately brief. Had it continued for some time, fighting against an enemy that ignored the laws of war would have placed great pressure on the conduct of Coalition forces. In some instances, moral constraints would have been difficult to maintain.

The military services today seek to deepen understanding of professional commitment by explaining what it means to be an officer. The U.S. Army discusses the practice of being a commissioned leader in terms of four interrelated identities: servant to the nation, member of a profession, warrior, and leader of character. The oath of office that officers take upon commissioning sets the parameters for

---

[1] Anthony E. Hartle. "The American Professional Military Ethic." Chapter 4 in *Moral Issues in Military Decision Making*, 2nd ed., revised. (Lawrence, KS: University Press of Kansas, 2004), pp. 42-75, 244-247. Edited by Rick Rubel & Lawrence Lengbeyer.

the role of servant to the nation. The commission itself provides the foundation for service as a military professional, though the implications of being a member of a profession are complex. The qualities necessary to lead men and women in carrying out missions that impose both great responsibility and great danger shape what it means to be a leader of character. The skills and attributes necessary to be successful in combat establish what it means to be a warrior. These identities clarify the application of the professional code that guides and inspires the conduct of men and women in uniform.

My own experience suggests that a necessary step in preparing career professionals for the kinds of problems that soldiers face in combat is to make clear the code of the profession and its typical applications. The American professional military ethic has not been formally and systematically codified. The formal aspects of the code are found primarily in the oath of enlistment and the oath of commissioning (the wording of the commission actually awarded to officers), and the codified laws of war, though a variety of official publications contribute to the accepted guidelines for conduct. Informal elements of the ethic are taught, through professional socialization. For commissioned and noncommissioned officers, that process takes place most obviously in the structured programs of the military's professional development system. However, the day-to-day activities in military units and the examples set by superiors provide the most telling influences. The role of military leaders at all levels is a critical element in the overall process of professional socialization. For example, the official policy of the military services concerning sexual harassment may be quite clear and specific, but if unit commanders indicate that sexual harassment is an unimportant issue or that they will condone harassment, their perspective will dominate in the understanding of the informal code governing their unit's conduct.

The Oath of Office

Broad parameters are established in the oath of office itself:

> I do solemnly swear (or affirm) that I will support and defend the Constitution of the United States against all enemies, foreign and domestic; that I will bear true faith and allegiance to the same, and that I take this obligation fully, without any mental reservation or purpose of evasion; and that I will well and faithfully discharge the duties of the office on which I am about to enter. So help me God.

The oath requires two commitments. First, the Constitution is the object of allegiance, to be upheld whether its authority is challenged from within or from without. Secondly, the officer is committed to perform the duties required in the professional role. Though the commitment appears straightforward, many officers give insufficient consideration to the implications of the political and moral principles manifested in the Constitution. Because the nature of the duties required in the role is open-ended, the professional officer's explicit commitment to support and defend the Constitution becomes particularly important. Commitment to the Constitution entails a commitment to the values and principles represented by the Constitution. Accordingly, the substance of those principles and values is critical to an adequate understanding of the American professional military ethic.

Meaning of the Commitment to the Constitution

While it may be generally accepted that characteristic values endure for any particular society, it is just as clear that values do change, even if exceedingly slowly. If we consider the Constitution as the manifestation of the fundamental values of American society, we might be tempted to conclude that the values and principles that it represents change with regularity. Only through adaptation to new and unforeseeable circumstances could the Constitution continue as a viable blueprint for government.

Despite such changing views, certain fundamental principles and values are as clear today as they were when the Founding Fathers wrote the Constitution. "Questions of constitutional law involve matters of public policy which should not be decided merely because of the original meanings in the Constitution. They must be read as revelations of the general principles that are expansive and comprehensive in character. Those principles and purposes are what was intended to endure." The professional officer is particularly concerned with these principles. When officers commit themselves to the support and defense of the Constitution and acknowledge its fundamental authority as the basis of government, these basic principles and values constitute the object of their pledge. As one commentator points out, "The Constitution was not fixed for all time in 1789," but rather "is a set of fundamental ideas by which orderly change can take place in a stable society." The point is that whereas the circumstances in which constitutional law is applied may change, the guiding principles do not.

Rights and liberties are at the center of any analysis of the Constitution. The various political principles that structure American government (such as the allocation of powers between state and federal government, the separation of powers at the federal level, the checks and balances system, and representative legislation) all ultimately concern creating a system that protects the rights and liberties of individuals. Discussion of rights can become confusing and ambiguous because rights can take various forms, so we should note that the rights under scrutiny here are both legal and normative. We are all quite familiar with legal rights or positive rights, those for which there is both social recognition and legal protection. They are empirical and contingent. Normative rights are those for which there is a moral justification. Constitutional rights are those that the Framers of the Constitution and subsequent generations have held to be not only normative and legal but also universal and indefeasible. The Constitution has consistently been described as a document created to protect fundamental rights: "Logically, the document is less a willful assertion of power than an act of sovereign self-restraint in behalf of a hierarchy of values that would find us willing to adjust our notions of economic well-being and national security as needed to honor constitutional rights."

The concept of constitutionalism itself certainly belongs among the basic, unchanging principles represented by the central political document of the United States. The familiar preamble to the Constitution states that

> WE THE PEOPLE OF THE UNITED STATES, in order to form a more perfect union, establish Justice, insure domestic tranquility, provide for the common defense, and secure the blessings of liberty to ourselves and our posterity, do ordain and establish the Constitution of the United States of America.

The preamble makes it clear that the purpose of the Constitution is to secure liberty, justice, and the general welfare. The principle of constitutionalism holds that a written, comprehensible constitution that limits the power of government and of individuals and agencies in government is necessary for the maintenance of a civilized society in which citizens can enjoy liberty, justice, and equality. The body of the Constitution both establishes and limits the power of the specified branches of a republican form of government, and in doing so, reveals a profound wariness of the power of government. Because agencies of the government have command over extensive resources, the potential for the abuse of power is always present, as we know all too well from repeated incidents during the last three decades of the twentieth century. The "checks and balances" of the Constitution are an overt attempt to preclude or minimize such abuses. The principle of constitutionalism maintains that, in the interests of liberty, justice, and equality, the constitution must be the final authority in the affairs of the state, embodying the fundamental values of the society in which it functions and reflecting the ultimate source of authority in republican government—the people. As established in the U.S. Constitution, the people—and only the people—have the indefeasible right to change their government. An understanding of the principle of constitutionalism helps clarify the critical importance of an officer's oath to support and defend the values and principles represented in our founding document.

The principle of individual rights is another of the fundamental values manifested in that document. It concerns those rights of man that are not to be denied by the government itself or by the desires of the majority. The principle is reflected most obviously in the amendments to the Constitution and in the function of the Supreme Court, which is why nominations to the Court are so closely examined. The Bill of Rights, the first ten amendments to the Constitution, embodies the doctrine of natural rights, which was the "hard core of Revolutionary political theory." The Constitution forbids the majority or even the entire House and Senate to pass laws that impair the fundamental rights of individuals. The strength of this prohibition is clear in the First Amendment: "Congress shall make no law respecting an establishment of religion, or prohibiting the free exercise thereof; or abridging the freedom of speech, or of the press; or of the right of the people peaceably to assemble and to petition the Government for a redress of grievances." An examination of the wording of the first ten amendments supports the well-documented claim that the Bill of Rights was originally "intended to render certain rights immune from abridgement by legislative majorities."

In addition to specifying the rights of individuals to certain basic freedoms, the Constitution created the institution of the Supreme Court. It is not implausible to claim that the powers granted the Supreme Court are primarily for the purpose of protecting individual rights, even, to a certain extent, against the will of the majority. Through this institution, the mechanism of government and law has a "limited mandate to correct mistakes made by State and natural majorities." In so doing, the Court serves to protect the rights of individuals. The institution of the Supreme Court, of course, will eventually affect the will of the majority, but it ensures that action will be taken only after deliberate, reflective, considered study. Both the amendments to the Constitution and the institution of the Supreme Court embody the principle of individual rights within the framework of the Constitution.

During the 1990s, human rights emerged as a prominent factor in international affairs. United Nations interventions in Bosnia, Kosovo, East Timor, and other areas of conflict were in response to massive human rights violations. Recognition of human rights in the United States and around the world makes it necessary to understand how they fit into our discussion of the professional military ethic. Whereas the preceding discussion makes reference to natural rights and the equality of all persons before the law, our Constitution governs but one nation—the United States. It applies to the regulation and protection of American society and the conduct of American citizens. Increasingly, dialogue about rights occurs in the context of universal human rights.

Discussions abound in which such rights are assumed to be self-evident, beginning with the seminal human rights document, the 1948 Universal Declaration of Human Rights produced by the United Nations. In its most general form, the doctrine of human rights holds that there are certain individual rights that are obtained quite independently of any particular states, societies, or periods.

In any discussion of rights, specific elements become important: the holder of the right, the object of the right (that to which one has a claim), and the duty-bearers of the right (those against whom one makes the claim). Human rights are distinctive with regard to two of these three elements. Every human being qualifies as a right holder, and the objects of human rights include all other human beings and all human agencies. The identity and obligations of the duty-bearers vary depending on circumstances and relationships, but claims to at least noninterference apply to all.

Another core element of the concept of human rights is that human beings, the right holders, are all equally entitled to such rights because of their status as autonomous beings who are responsible for what they do. They have both equal moral worth and equal moral responsibility. Responsibility is important because what people merit in terms of treatment is a function of how they act, how they manage moral responsibility. When individuals—or states—act in ways that fail to respect the rights of others, they fail in their moral responsibility and they are accountable. Thus the murderer who has wrongfully taken the life of another person has failed in moral responsibility and is appropriately punished by society.

If we examine the nature of state sovereignty on the basis of the domestic analogy, famously applied by Michael Walzer, we can say that states have rights as well, by virtue of being states. As Walzer put it, "As with individuals, so with sovereign states: there are things that we cannot do to them, even for their own ostensible good."

The Peace of Westphalia, a treaty signed by the major European powers in 1648, inaugurated the modem state system in the Western world and heralded the principle of sovereignty that dominated world affairs until the end of the Cold War. The rights of states under this system "are summed up in law books as territorial integrity and political sovereignty." Walzer finds the foundation—and justification—for those rights in the rights of citizens: "The duties and rights of states are nothing more than the duties and rights of the men who compose them." But we can make an even stronger claim today: the legitimacy of the state, and thus its authority to exercise rights in the international community, is a function of its credibility in protecting the rights of its citizens and the corresponding support of the citizens for the state and its political structure. The fundamental moral and political rights of the citizens of a state are, in this view, everywhere the same. They are the human rights possessed by all human beings.

Whether we focus on natural rights, civil rights, or human rights, an independent judiciary and reliable law enforcement provide the means to secure individual rights and social order. If people are to live together in some regulated fashion, government must have the authority to restrict liberty, to regulate property, and to protect the pursuit of happiness—though such interference is permissible only under and within the rule of law. We find the principle of the rule of law clearly reflected in the Constitution and in its applications; one need look no further than the Due Process Clause of the Fifth and Fourteenth Amendments for evidence. Additional consideration reveals the interesting fact that while the President as Commander in Chief controls overwhelming military power, and while Congress controls the power of the purse, the power of the Supreme Court consists entirely in the principle of the rule of law established by the provisions of the Constitution. Thus, we can reasonably claim that individual rights secured by law constitute the central value reflected in the Constitution.

The moral rights with which the Constitution is concerned "constitute moral reasons for action of a special weight and urgency." They are "moral claims to kinds of individual needs and concerns which must be satisfied prior to other kinds of moral claims and which justify the use of force, other things being equal, in support of the moral urgency these claims involve." Through the Constitution, these moral rights become legal rights as well. In that form, they are protected from unjust actions by the majority. Accordingly, "[m]ajority rule is not the basic moral principle of the Constitutional order. The basic moral principle is the principle of greatest equal liberty. Majority rule is justified only to the extent that it is compatible with this deeper moral principle, which constitutes a standard of criticism for majority rule."

This conclusion, supported firmly by the content of the amendments to the Constitution, establishes the relationship between the concept of fundamental freedoms and moral rights. The latter take precedence within the constraint of greatest equal liberty for all. The principle of greatest equal liberty guides us in moving from the abstract principle of both the rule of law and of individual rights to actual practice.

Although interpretations and applications of constitutional law are perpetually in flux, firm ground exists for maintaining that the broad principles of constitutionalism, representative democracy, individual rights, the rule of law, and greatest equal liberty are fixtures in our national understanding of the Constitution. Also evident in our national history and implicit in the provisions of the Constitution, which authorizes the raising of armed forces, is the firm belief that these ennobling values are worth fighting for and that the use of force in their defense is fully justified. That is the soldier's purpose. When military members pledge their commitment to the support and defense of the Constitution, they commit themselves, by logical extension, to the principles and values that form the basis of its provisions.

The Commission

The commission provided to military officers begins with the following: "Know ye that reposing special trust and confidence in the patriotism, valor, fidelity, and abilities of [named officer] . . . ." The commission from the commander in chief continues, stating that "this officer is to observe and follow such orders and directions from time to time, as may be given by me, or by the future President of the United States of America." The fundamental law of the United States is the Constitution, so that the commission confirms the supremacy of the Constitution in the commitment of military officers. Were the President or any other superior to issue an unlawful order, military officers would be obligated by their role requirements to disobey it. That obligation finds its moral basis in the commissioning oath.

During President Nixon's last, bitter days in office, some people were concerned that the military might support an attempt by the President to remain in power despite the demands of Congress and the courts. Anyone who understood commitment of the officer corps to the Constitution, however, would not have taken the possibility seriously. An officer's loyalty is to the principles and values manifested in the Constitution, not to the person of the commander in chief.

The oath and the commission provide the foundation for the traditional idealistic code of the United States armed forces, the code I have been calling the professional military ethic. The Armed Forces Code of Conduct, promulgated after the Korean Conflict, also provides guidance for members of the military, but it has limited application because it concerns actions appropriate for those taken prisoner.

Duty—Honor—Country

The discussion that follows focuses on the "traditional idealistic code" against which institutional actions are measured. We will examine the standards of behavior that military professionals feel they ought to meet, not conduct that fails to meet such standards or institutional pressures of a particular period that contribute to the failure to meet them.

Duty. Duty incorporates the concepts of obedience and self-discipline. Whereas self-discipline would apply to most professions, it is of fundamental significance to the military professional, for the demands of duty can be particularly heavy. They may require the sacrifice of one's own life and those of others—an aspect of daily existence in a combat environment. The professional commitment is one of "ultimate liability." The requirement for both physical courage and the courage to make difficult decisions is implicit. In light of such demands, members of a military organization recognize that obedience is essential for effective functioning. As one long-time student of military sociology put it, "Integrity and instant obedience are the *sine qua non* of the military institution."

In the American tradition, the oath of commissioning indicates that an officer's duty is to the state and, in particular, to the Constitution. Duty assumes the subordination of personal desires to the requirement generated by the oath, that of defending the Constitution and through it the state. "As a member of a service, an individual accepts a series of narrowly defined duties to superiors and subordinates consistent with his responsibilities to uphold that oath."

Honor. For American military officers, honor connotes integrity, not military glory or prestige.

> Its underlying values are truth-telling, honesty, and integrity. Implicit in "honor" is a sense of trust within the officer corps. Subordinates must be able to trust their leaders implicitly. The trust must be mutual if the unity and cohesion which are so crucial to combat effectiveness are to be developed. Requirements of combat demand high standards of honor, integrity, loyalty, and justice. The same applies to the military institution as a whole in carrying out the heavy responsibilities entrusted to it by the host society.

Personal honor remains nonnegotiable. The following analysis states the position clearly:

> At the daily working level, an atmosphere of trust and confidence is essential for military organization to operate effectively. . . . Mutual confidence and esteem are essential to a unit's esprit de corps. . . high standards of personal integrity must be nurtured so that mutual confidence can survive long periods of stress.

> When orders imply substantial sacrifice and risk on the part of subordinates, they must have no lingering doubts of the commander's true motivations. To execute the orders effectively, they must accept his personal integrity without question.

> Of course, the most obvious and perhaps the unique requirement for high standards of honor in the military profession has to do with the necessity of accurate reporting in combat. The danger of unnecessary loss of life in such situations is too obvious to warrant elaboration. . . . The advice of the military professional to military or civilian superiors must accurately reflect current situations; otherwise, the consequences can be severe.

Country. The country is the objective to which the performance of duty and the maintenance of honor are devoted. The third element of the motto "Duty— Honor—Country" re-emphasizes the concept that no particular government administration or individual commands the allegiance of the military. The country itself (the state) is the beneficiary of the services of the armed forces. Further, members of the profession subordinate personal welfare to the welfare of the nation. This principle follows from the fundamental purpose of the armed forces. Since successful accomplishment of assigned missions often means that lives must be expended, the exigencies of the profession obviously demand that the means employed in combat, including the expenditure of human lives, are precisely that— means. Because of the criticality of the function of the organization, its welfare must have priority over the welfare of individual members. Here again, however, the American professional military ethic is modified from the straightforward requirements of military activity, in which mission goes before all (which is to say that the security of the state justifies all). If functional requirements alone provided the basis for moral decisions, the accomplishment of assigned missions would have priority over all other considerations. That, however, is not the case in the American military, given the laws of war and the values of American society.

The commissioning oath makes clear that the values of American society as exemplified in the Constitution give substance to the American professional military ethic. While human rights are not referred to as such in the document, it establishes the political system in which individual freedoms, representative democracy, and equality are to be optimized. The amendments to the Constitution reflect many of what we think of as human rights, and the specifics of the Bill of Rights can be justified in terms of the fundamental human rights highlighted in the Declaration of Independence. The commitment to the country is constrained by conceptions of morality and "guided by an overwhelming commitment to constitutional process."

Traditional Values

Within the context established by "Duty—Honor—Country," four more specific principles have developed that are also fundamental to the American professional military ethic. The principle of *professional competence* influences all professional activity in the American military. "The nature of the military profession, and the responsibilities of the profession to the society it serves, are such as to elevate professional competence to the level of an ethical imperative." To be capable of conscientiously striving to perform in accordance with the precepts of "Duty—Honor—Country," an officer must be competent to perform assigned tasks. The more capable officers are, the more successful they will be in living up to the other principles in the professional military ethic. Given the weight of responsibility shouldered by the military, no degree of competence short of the maximum possible can be declared acceptable in terms of the professional ideals.

Another fundamental principle is *civilian control of the military* by the elected representatives of the people. From this principle derives another that is far from unique to the American military but nonetheless basic to it. "The professional soldier is 'above politics' in domestic affairs." If the integrity of the military is to be beyond reproach, *professional officers must not directly involve themselves in domestic politics*. On the surface such a position appears somewhat naive. The military establishment is the largest institutional complex in the government of the United States, and the extension of the "military-industrial complex" in economic terms is most difficult to assess just because it includes so much. Functional influence, however, is one matter; overt participation by an individual is another. The code forbids the latter. While the distinction is one of degree in practice, the services maintain the distinction with surprising effectiveness. Certainly, the military as an institution traditionally observes political neutrality.

A fourth principle that is basic to the American professional military ethic concerns the *importance of the welfare of the individual soldier*, which goes beyond utilitarian concerns. Such a principle might be expected in a society that traditionally has placed great emphasis on individualism. This principle is firmly embedded in the ethic as well as explicitly stated in the law. Article 5947, Title 10, U.S. Code, states that "commanding officers and others in authority shall take all necessary and proper action to promote and safeguard the morale, physical well-being, and general welfare of the officers and enlisted men under their command and charge." Two aspects of combat in the Vietnam era also reflect this concern for the individual. It was not unusual in the jungles of the Cambodian border area or in the mountains of the central highlands to find American soldiers in the field eating ice cream flown in by helicopters. While such actions might reveal logistical mismanagement, they also reveal the extent to which the system attempted to "take care of" soldiers. The second aspect was tactical. The massive use of firepower reflected the concern (at times perhaps undue) with friendly casualties. Arguments continue concerning the degree to which manifestations of the concern about the welfare of uniformed members hindered the pursuit of military objectives, but the basis for that concern is firmly embedded in the American professional military ethic. In the southwest Pacific in 1944 during World War II, Japanese resistance on the coral island of Biak was intense, with the Japanese soldiers well protected in deep caves. One such emplacement was an area about 500 by 800 yards referred to as the Ibdi Pocket, defended by about 1,000 Japanese. Instead of ordering casualty-heavy attacks, the US commanders continuously shelled and bombed the Pocket from 1 June until 21 July. Although casualties were still significant, they were much lower than they would have been had the commanders not expended huge amounts of ordnance in reducing the enemy capabilities. In Vietnam, the attitude was epitomized by the dictum, "Spend bullets, not bodies." That policy was a function of public and Congressional relations at some level, of course, but "politics" was not the primary motive. Soldiers are members of American society who have value in their own right as persons. Lives of men and women in uniform are not to be risked without compelling cause.

The phrase "Duty—Honor—Country" thus represents the content of the American professional military ethic. Each term has particular connotations in the American context. The broad principles of civilian control of the military and that of political neutrality overarch these connotations, and the obligation to promote the physical and psychological welfare of the individual military member to the maximum extent possible within the context of mission accomplishment permeates all aspects of the motto. Lastly, in each of these principles, the concept of professional competence is required and assumed.

Letter from Birmingham City Jail
Dr. Martin Luther King, Jr. *

My dear Fellow Clergymen,

While confined here in the Birmingham city jail, I came across your recent statement calling our present activities "unwise and untimely." Seldom, if ever, do I pause to answer criticism of my work and ideas. If I sought to answer all of the criticisms that cross my desk, my secretaries would be engaged in little else in the course of the day, and I would have no time for constructive work. But since I feel that you are men of genuine good will and your criticisms are sincerely set forth, I would like to answer your statement in what I hope will be patient and reasonable terms.

I think I should give the reason for my being in Birmingham, since you have been influenced by the argument of "outsiders coming in." I have the honor of serving as president of the Southern Christian Leadership Conference, an organization operating in every southern state, with headquarters in Atlanta, Georgia. We have some eighty-five affiliate organizations all across the South—one being the Alabama Christian Movement for Human Rights. Whenever necessary and possible we share staff, educational and financial resources with our affiliates. Several months ago our local affiliate here in Birmingham invited us to be on call to engage in a nonviolent direct-action program if such were deemed necessary. We readily consented and when the hour came we lived up to our promises. So I am here, along with several members of my staff, because we were invited here. I am here because I have basic organizational ties here.

Beyond this, I am in Birmingham because injustice is here. Just as the eighth century prophets left their little villages and carried their "thus saith the Lord" far beyond the boundaries of their hometowns; and just as the Apostle Paul left his little village of Tarsus and carried the gospel of Jesus Christ to practically every hamlet and city of the Graeco-Roman world, I too am compelled to carry the gospel of freedom beyond my particular hometown. Like Paul, I must constantly respond to the Macedonian call for aid.

Moreover, I am cognizant of the interrelatedness of all communities and states. I cannot sit idly by in Atlanta and not be concerned about what happens in Birmingham. Injustice anywhere is a threat to justice everywhere. We are caught in an inescapable network of mutuality, tied in a single garment of destiny. Whatever affects one directly affects all indirectly. Never again can we afford to live with the narrow, provincial "outside agitator" idea. Anyone who lives in the United States can never be considered an outsider anywhere in this country.

You deplore the demonstrations that are presently taking place in Birmingham. But I am sorry that your statement did not express a similar concern for the conditions that brought the demonstrations into being. I am sure that each of you would want to go beyond the superficial social analyst who looks merely at effects, and does not grapple with underlying causes. I would not hesitate to say that it is unfortunate that so-called demonstrations are taking place in Birmingham at this time, but I would

---

*Reprinted by arrangement with the Heirs to the Estate of Martin Luther King, Jr., c/o Writer's House as agent for the proprietor.

say in more emphatic terms that it is even more unfortunate that the white power structure of this city left the Negro community with no other alternative.

In any nonviolent campaign there are four basic steps: (1) collection of the facts to determine whether injustices are alive, (2) negotiation, (3) self-purification, and (4) direct action. We have gone through all of these steps in Birmingham. There can be no gainsaying of the fact that racial injustice engulfs this community.

Birmingham is probably the most thoroughly segregated city in the United States. Its ugly record of police brutality is known in every section of this country. Its unjust treatment of Negroes in the courts is a notorious reality. There have been more unsolved bombings of Negro homes and churches in Birmingham than any city in this nation. These are the hard, brutal and unbelievable facts. On the basis of these conditions Negro leaders sought to negotiate with the city fathers. But the political leaders consistently refused to engage in good faith negotiation.

Then came the opportunity last September to talk with some of the leaders of the economic community. In these negotiating sessions certain promises were made by the merchants—such as the promise to remove the humiliating racial signs from the stores. On the basis of these promises Rev. Shuttlesworth and the leaders of the Alabama Christian Movement for Human Rights agreed to call a moratorium on any type of demonstrations. As the weeks and months unfolded we realized that we were the victims of a broken promise. The signs remained. Like so many experiences of the past we were confronted with blasted hopes, and the dark shadow of a deep disappointment settled upon us. So we had no alternative except that of preparing for direct action, whereby we would present our very bodies as a means of laying our case before the conscience of the local and national community. We were not unmindful of the difficulties involved. So we decided to go through a process of self-purification. We started having workshops on nonviolence and repeatedly asked ourselves the questions, "Are you able to accept blows without retaliating?" "Are you able to endure the ordeals of jail?" We decided to set our direct-action program around the Easter season, realizing that with the exception of Christmas, this was the largest shopping period of the year. Knowing that a strong economic withdrawal program would be the by-product of direct action, we felt that this was the best time to bring pressure on the merchants for the needed changes. Then it occurred to us that the March election was ahead so we speedily decided to postpone action until after election day. When we discovered Mr. Connor was in the run-off, we decided again to postpone action so that the demonstrations could not be used to cloud the issues. At this time we agreed to begin our nonviolent witness the day after the run-off.

This reveals that we did not move irresponsibly into direct action. We too wanted to see Mr. Connor defeated; so we went through postponement after postponement to aid in this community need. After this we felt that direct action could be delayed no longer.

You may well ask, "Why direct action? Why sit-ins, marches, etc.? Isn't negotiation a better path?" You are exactly right in your call for negotiation. Indeed, this is the purpose of direct action. Nonviolent direct action seeks to create such a crisis and establish such creative tension that a community that has constantly refused to negotiate is forced to confront the issue. It seeks so to dramatize the issue that it can no longer be ignored. I just referred to the creation of tension as a part of the work of the nonviolent resister. This may sound rather shocking. But I must confess that I am not afraid of the word tension. I have earnestly worked and preached against violent tension, but there is a type of constructive nonviolent tension that is necessary for growth. Just as Socrates felt that it was necessary to create a tension in the mind so that individuals could rise from the bondage of myths and half-truths to the unfettered realm of creative analysis and objective appraisal, we must see the need of having nonviolent gadflies to create the kind of tension in society that will help men to rise from the dark depths of prejudice and racism to the majestic heights of understanding and brotherhood. So the purpose of the direct action is to create a situation so crisis-packed that it will inevitably open the door to negotiation. We, therefore, concur with you in your call for negotiation. Too long has our beloved Southland been bogged down in the tragic attempt to live in monologue rather than dialogue.

One of the basic points in your statement is that our acts are untimely. Some have asked, "Why didn't you give the new administration time to act?" The only answer that I can give to this inquiry is that the new administration must be prodded about as much as the outgoing one before it acts. We will be sadly mistaken if we feel that the election of Mr. Boutwell will bring the millennium to

Birmingham. While Mr. Boutwell is much more articulate and gentle than Mr. Connor, they are both segregationists, dedicated to the task of maintaining the status quo. The hope I see in Mr. Boutwell is that he will be reasonable enough to see the futility of massive resistance to desegregation. But he will not see this without pressure from the devotees of civil rights. My friends, I must say to you that we have not made a single gain in civil rights without determined legal and nonviolent pressure. History is the long and tragic story of the fact that privileged groups seldom give up their privileges voluntarily. Individuals may see the moral light and voluntarily give up their unjust posture; but as Reinhold Niebuhr has reminded us, groups are more immoral than individuals.

We know through painful experience that freedom is never voluntarily given by the oppressor; it must be demanded by the oppressed. Frankly, I have never yet engaged in a direct action movement that was "well- timed," according to the timetable of those who have not suffered unduly from the disease of segregation. For years now I have heard the words "Wait! " It rings in the ear of every Negro with a piercing familiarity. This "Wait" has almost always meant "Never. " It has been a tranquilizing thalidomide, relieving the emotional stress for a moment, only to give birth to an ill-formed infant of frustration. We must come to see with the distinguished jurist of yesterday that "justice too long delayed is justice denied." We have waited for more than 340 years for our constitutional and God-given rights. The nations of Asia and Africa are moving with jetlike speed toward the goal of political independence, and we still creep at horse and buggy pace toward the gaining of a cup of coffee at a lunch counter. I guess it is easy for those who have never felt the stinging darts of segregation to say "Wait." But when you have seen vicious mobs lynch your mothers and fathers at will and drown your sisters and brothers at whim; when you have seen hate-filled policemen curse, kick, brutalize and even kill your black brothers and sisters with impunity; when you see the vast majority of your twenty million Negro brothers smothering in an airtight cage of poverty in the midst of an affluent society; when you suddenly find your tongue twisted and your speech stammering as you seek to explain to your six-year-old daughter why she can't go to the public amusement park that has just been advertised on television, and see tears welling up in her little eyes when she is told that Funtown is closed to colored children, and see the depressing clouds of inferiority begin to form in her little mental sky, and see her begin to distort her little personality by unconsciously developing a bitterness toward white people; when you have to concoct an answer for a five- year-old son asking in agonizing pathos: "Daddy, why do white people treat colored people so mean?"; when you take a cross-country drive and find it necessary to sleep night after night in the uncomfortable corners of your automobile because no motel will accept you; when you are humiliated day in and day out by nagging signs reading "white" and "colored"; when your first name becomes "nigger" and your middle name becomes "boy" (however old you are) and your last name becomes "John," and when your wife and mother are never given the respected tide "Mrs."; when you are harried by day and haunted by night by the fact that you are a Negro, living constantly at tiptoe stance never quite knowing what to expect next, and plagued with inner fears and outer resentments; when you are forever fighting a degenerating sense of "nobodiness"; then you will understand why we find it difficult to wait. There comes a time when the cup of endurance runs over, and men are no longer willing to be plunged into an abyss of injustice where they experience the blackness of corroding despair. I hope, sirs, you can understand our legitimate and unavoidable impatience.

You express a great deal of anxiety over our willingness to break laws. This is certainly a legitimate concern. Since we so diligently urge people to obey the Supreme Court's decision of 1954 outlawing segregation in the public schools, it is rather strange and paradoxical to find us consciously breaking laws. One may well ask, "How can you advocate breaking some laws and obeying others?" The answer is found in the fact that there are two types of laws: there are just and there are unjust laws. I would agree with Saint Augustine that "An unjust law is no law at all."

Now what is the difference between the two? How does one determine when a law is just or unjust?

A just law is a man-made code that squares with the moral law or the law of God. An unjust law is a code that is out of harmony with the moral law. To put it in the terms of Saint Thomas Aquinas, an unjust law is a human law that is not rooted in eternal and natural law. Any law that uplifts human personality is just. Any law that degrades human personality is unjust. All segregation statutes are unjust because segregation distorts the soul and damages the personality. It gives the segregator a false sense of superiority and the segregated a false sense of inferiority. To use the words of Martin Buber, the great Jewish philosopher, segregation substitutes an "I-it" relationship for the "I-thou" relationship, and ends up relegating persons to the status of things. So segregation is not only politically, economically and sociologically unsound, but it is morally wrong and sinful. Paul Tillich has said that sin is separation. Isn't segregation an existential expression of man's tragic separation, an expression of his awful estrangement, his terrible sinfulness? So I can urge men to disobey segregation ordinances because they are morally wrong.

Let us turn to a more concrete example of just and unjust laws. An unjust law is a code that a majority inflicts on a minority that is not binding on itself. This is difference made legal. On the other hand a just law is a code that a majority compels a minority to follow that it is willing to follow itself. This is sameness made legal. Let me give another explanation. An unjust law is a code inflicted upon a minority which that minority had no part in enacting or creating because they did not have the unhampered right to vote. Who can say that the legislature of Alabama which set up the segregation laws was democratically elected? Throughout the state of Alabama all types of conniving methods are used to prevent Negroes from becoming registered voters and there are some counties without a single Negro registered to vote despite the fact that the Negro constitutes a majority of the population. Can any law set up in such a state be considered democratically structured?

These are just a few examples of unjust and just laws. There are some instances when a law is just on its face and unjust in its application. For instance, I was arrested Friday on a charge of parading without a permit. Now there is nothing wrong with an ordinance which requires a permit for a parade, but when the ordinance is used to preserve segregation and to deny citizens the First Amendment privilege of peaceful assembly and peaceful protest, then it becomes unjust.

I hope you can see the distinction I am trying to point out. In no sense do I advocate evading or defying the law as the rabid segregationist would do. This would lead to anarchy. One who breaks an unjust law must do it *openly, lovingly* (not hatefully as the white mothers did in New Orleans when they were seen on television screaming, "nigger, nigger, nigger"), and with a willingness to accept the penalty. I submit that an individual who breaks a law that conscience tells him is unjust, and willingly accepts the penalty by staying in jail to arouse the conscience of the community over its injustice, is in reality expressing the very highest respect for law.

Of course, there is nothing new about this kind of civil disobedience. It was seen sublimely in the refusal of Shadrach, Meshach and Abednego to obey the laws of Nebuchadnezzar because a higher moral law was involved. It was practiced superbly by the early Christians who were willing to face hungry lions and the excruciating pain of chopping blocks, before submitting to certain unjust laws of the Roman Empire. To a degree academic freedom is a reality today because Socrates practiced civil disobedience. We can never forget that everything Hitler did in Germany was "legal" and everything the Hungarian freedom fighters did in Hungary was "illegal." It was "illegal" to aid and comfort a Jew in Hitler's Germany. But I am sure that if I had lived in Germany during that time I would have aided and comforted my Jewish brothers even though it was illegal. If I lived in a Communist country today where certain principles dear to the Christian faith are suppressed, I believe I would openly advocate disobeying these anti-religious laws. I must make two honest confessions to you, my Christian and Jewish brothers. First, I must confess that over the last few years I have been gravely disappointed with the white moderate. I have almost reached the regrettable conclusion that the Negro's great stumbling block in the stride toward freedom is not the White Citizen's Councilor or the Ku Klux Manner, but the white moderate who is more devoted to "order" than to justice; who prefers a negative peace which is the absence of tension to a positive peace which is the presence of justice; who constantly says, "I agree with you in the goal you seek, but I can't agree with your methods of direct action"; who paternalistically feels that he can set the timetable for another man's freedom; who lives by the myth of time and who constantly advised the Negro to wait until a "more convenient season." Shallow understanding from people of good will is more frustrating than absolute misunderstanding from people of ill will. Lukewarm acceptance is much more bewildering than outright rejection.

I had hoped that the white moderate would understand that law and order exist for the purpose of establishing justice, and that when they fail to do this they become dangerously structured dams that block the flow of social progress. I had hoped that the white moderate would understand that the present tension of the South is merely a necessary phase of the transition from an obnoxious negative peace, where the Negro passively accepted his unjust plight, to a substance-filled positive peace, where all men will respect the dignity and worth of human personality. Actually, we who engage in nonviolent direct action are not the creators of tension. We merely bring to the surface the hidden tension that is already alive. We bring it out in the open where it can be seen and dealt with. Like a boil that can never be cured as long as it is covered up but must be opened with all its pus-flowing ugliness to the natural medicines of air and light, injustice must likewise be exposed, with all of the tension its exposing creates, to the light of human conscience and the air of national opinion before it can be cured.

In your statement you asserted that our actions, even though peaceful, must be condemned because they precipitate violence. But can this assertion be logically made? Isn't this like condemning the robbed man because his possession of money precipitated the evil act of robbery? Isn't this like condemning Socrates because his unswerving commitment to truth and his philosophical delvings precipitated the misguided popular mind to make him drink the hemlock? Isn't this like condemning Jesus because His unique God- consciousness and never-ceasing devotion to his will precipitated the evil act of crucifixion? We must come to see, as federal courts have consistently affirmed, that it is immoral to urge an individual to withdraw his efforts to gain his basic constitutional rights because the quest precipitates violence. Society must protect the robbed and punish the robber.

I had also hoped that the white moderate would reject the myth of time. I received a letter this morning from a white brother in Texas which said: "All Christians know that the colored people will receive equal rights eventually, but it is possible that you are in too great of a religious hurry. It has taken Christianity almost two thousand years to accomplish what it has. The teachings of Christ take time to come to earth." All that is said here grows out of a tragic misconception of time. It is the strangely irrational notion that there is something in the very flow of time that will inevitably cure all ills. Actually time is neutral. It can be used either destructively or constructively. I am coming to feel that the people of ill will have used time much more effectively than the people of good will. We will have to repent in this generation not merely for the vitriolic words and actions of the bad people, but for the appalling silence of the good people. We must come to see that human progress never rolls in on wheels of inevitability. It comes through the tireless efforts and persistent work of men willing to be co-workers with God, and without this hard work, time itself becomes an ally of the forces of social stagnation. We must use time creatively, and forever realize that the time is always ripe to do right. Now is the time to make real the promise of democracy, and transform our pending national elegy into a creative psalm of brotherhood. Now is the time to lift our national policy from the quicksand of racial injustice to the solid rock of human dignity.

You spoke of our activity in Birmingham as extreme. At first I was rather disappointed that fellow clergymen would see my nonviolent efforts as those of the extremist. I started thinking about the fact that I stand in the middle of two opposing forces in the Negro community. One is a force of complacency made up of Negroes who, as a result of long years of oppression, have been so completely drained of self-respect and a sense of "somebodiness" that they have adjusted to segregation, and, of a few Negroes in the middle class who, because of a degree of academic and economic security, and because at points they profit by segregation, have unconsciously become insensitive to the problems of the masses. The other force is one of bitterness and hatred, and comes perilously close to advocating violence. It is expressed in the various black nationalist groups that are springing up over the nation, the largest and best known being Elijah Muhammad's Muslim movement. This movement is nourished by the contemporary frustration over the continued existence of racial discrimination. It is made up of people who have lost faith in America, who have absolutely repudiated Christianity, and who have concluded that the white man is an incurable "devil." I have tried to stand between these two forces, saying that we need not follow the "do-nothingism" of the complacent or the hatred and despair of the black nationalist. There is the more excellent way of love and nonviolent protest. I'm grateful to God that, through the Negro church, the dimension of nonviolence entered our struggle. If this philosophy had not emerged, I am convinced that by now many streets of the South would be flowing with floods of blood. And I am further convinced that if our white brothers dismiss us as "rabble-rousers" and "outside agitators" those of us who are working through the channels of nonviolent direct action and refuse to support our nonviolent efforts, millions of Negroes, out of frustration and despair, will seek solace and security in black nationalist ideologies, a development that will lead inevitably to a frightening racial nightmare.

Oppressed people cannot remain oppressed forever. The urge for freedom will eventually come. This is what happened to the American Negro. Something within has reminded him of his birthright of freedom; something without has reminded him that he can gain it. Consciously and unconsciously, he has been swept in what the Germans call the Zeitgeist, and with his black brothers of Africa, and his brown and yellow brothers of Asia, South America and the Caribbean, he is moving with a sense of cosmic urgency toward the promised land of racial justice. Recognizing this vital urge that has engulfed the Negro community, one should readily understand public demonstrations. The Negro has many pent-up resentments and latent frustrations. He has to get them out. So let him march sometime; let him have his prayer pilgrimages to the city hall; understand why he must have sit-ins and freedom rides. If his repressed emotions do not come out in these nonviolent ways, they will come out in ominous expressions of violence. This is not a threat; it is a fact of history. So I have not said to my people "get rid of your discontent." But I have tried to say that this normal and healthy discontent can be channelized through the creative outlet of nonviolent direct action. Now this approach is being dismissed as extremist. I must admit that I was initially disappointed in being so categorized.

But as I continued to think about the matter I gradually gained a bit of satisfaction from being considered an extremist. Was not Jesus an extremist in love—"Love your enemies, bless them that curse you, pray for them that despitefully use you." Was not Amos an extremist for justice—"Let justice roll down like waters and righteousness like a mighty stream. " Was not Paul an extremist for the gospel of Jesus Christ—"I bear in my body the marks of the Lord Jesus." Was not Martin Luther an extremist—"Here I stand; I can do none other so help me God." Was not John Bunyan an extremist—"I will stay in jail to the end of my days before I make a butchery of my conscience." Was not Abraham Lincoln an extremist- "This nation cannot survive half slave and half free." Was not Thomas Jefferson an extremist—"We hold these truths to be self-evident, that all men are created equal." So the question is not whether we will be extremist but what kind of extremist will we be. Will we be extremists for hate or will we be extremists for love? Will we be extremists for the preservation of injustice—or will we be extremists for the cause of justice? In that dramatic scene on Calvary's hill, three men were crucified. We must not forget that all three were crucified for the same crime—the crime of extremism. Two were extremists for immorality, and thusly fell below their environment. The other, Jesus Christ, was an extremist for love, truth and goodness, and thereby rose above his environment. So, after all, maybe the South, the nation and the world are in dire need of creative extremists.

I had hoped that the white moderate would see this. Maybe I was too optimistic. Maybe I expected too much. I guess I should have realized that few members of a race that has oppressed another race can understand or appreciate the deep groans and passionate yearnings of those that have been oppressed and still fewer have the vision to see that injustice must be rooted out by strong, persistent and determined action. I am thankful, however, that some of our white brothers have grasped the meaning of this social revolution and committed themselves to it. They are still all too small in quantity, but they are big in quality. Some like Ralph McGill, Lillian Smith, Harry Golden and James Dabbs have written about our struggle in eloquent, prophetic and understanding terms. Others have marched with us down nameless streets of the South. They have languished in filthy roach-infested jails, suffering the abuse and brutality of angry policemen who see them as "dirty nigger-lovers. " They, unlike so many of their moderate brothers and sisters, have recognized the urgency of the moment and sensed the need for powerful "action" antidotes to combat the disease of segregation.

Let me rush on to mention my other disappointment. I have been so greatly disappointed with the white church and its leadership. Of course, there are some notable exceptions. I am not unmindful of the fact that each of you has taken some significant stands on this issue. I commend you, Rev. Stallings, for your Christian stance on this past Sunday, in welcoming Negroes to your worship service on a non-segregated basis. I commend the Catholic leaders of this state for integrating Springhill College several years ago.

But despite these notable exceptions I must honestly reiterate that I have been disappointed with the church. I do not say that as one of the negative critics who can always find something wrong with the church. I say it as a minister of the gospel, who loves the church; who was nurtured in its bosom; who has been sustained by its spiritual blessings and who will remain true to it as long as the cord of life shall lengthen.

I had the strange feeling when I was suddenly catapulted into the leadership of the bus protest in Montgomery several years ago that we would have the support of the white church. I felt that the white ministers, priests and rabbis of the South would be some of our strongest allies. Instead, some have been outright opponents, refusing to understand the freedom movement and misrepresenting its leaders; all too many others have been more cautious than courageous and have remained silent behind the anesthetizing security of the stained-glass windows.

In spite of my shattered dreams of the past, I came to Birmingham with the hope that the white religious leadership of this community would see the justice of our cause, and with deep moral concern, serve as the channel through which our just grievances would get to the power structure. I had hoped that each of you would understand. But again I have been disappointed. I have heard numerous religious leaders of the South call upon their worshippers to comply with a desegregation decision because it is the law, but I have longed to hear white ministers say, "Follow this decree because integration is morally right and the Negro is your brother. " In the midst of blatant injustices inflicted upon the Negro, I have watched white churches stand on the sideline and merely mouth pious irrelevancies and sanctimonious trivialities. In the midst of a mighty struggle to rid our nation of racial and economic injustice, I have heard so many ministers say, "Those are social issues with which the gospel has no real concern," and I have watched so many churches commit themselves to a completely other-worldly religion which made a strange distinction between body and soul, the sacred and the secular.

So here we are moving toward the exit of the twentieth century with a religious community largely adjusted to the status quo, standing as a taillight behind other community agencies rather than a headlight leading men to higher levels of justice.

I have traveled the length and breadth of Alabama, Mississippi and all the other southern states. On sweltering summer days and crisp autumn mornings I have looked at her beautiful churches with their lofty spires pointing heavenward. I have beheld the impressive outlay of her massive religious education buildings. Over and over again I have found myself asking: "What kind of people worship here? Who is their God? Where were their voices when the lips of Governor Barnett dripped with words of interposition and nullification? Where were they when Governor Wallace gave the clarion call for defiance and hatred? Where were their voices of support when tired, bruised and weary Negro men and women decided to rise from the dark dungeons of complacency to the bright hills of creative protest?"

Yes, these questions are still in my mind. In deep disappointment, I have wept over the laxity of the church. But be assured that my tears have been tears of love. There can be no deep disappointment where there is not deep love. Yes, I love the church; I love her sacred walls. How could I do otherwise? I am in the rather unique position of being the son, the grandson and the great-grandson of preachers. Yes, I see the church as the body of Christ. But, oh! How we have blemished and scarred that body through social neglect and fear of being nonconformists.

There was a time when the church was very powerful. It was during that period when the early Christians rejoiced when they were deemed worthy to suffer for what they believed. In those days the church was not merely a thermometer that recorded the ideas and principles of popular opinion; it was a thermostat that transformed the mores of society. Wherever the early Christians entered a town the power structure got disturbed and immediately sought to convict them for being "disturbers of the peace" and "outside agitators." But they went on with the conviction that they were a "colony of heaven," and had to obey God rather than man. They were small in number but big in commitment. They were too God-intoxicated to be "astronomically intimidated." They brought an end to such ancient evils as infanticide and gladiatorial contest.

Things are different now. The contemporary church is often a weak, ineffectual voice with an uncertain sound. It is so often the arch-supporter of the status-quo. Far from being disturbed by the presence of the church, the power structure of the average community is consoled by the church's silent and often vocal sanction of things as they are.

But the judgment of God is upon the church as never before. If the church of today does not recapture the sacrificial spirit of the early church, it will lose its authentic ring, forfeit the loyalty of millions, and be dismissed as an irrelevant social club with no meaning for the twentieth century. I am meeting young people every day whose disappointment with the church has risen to outright disgust.

Maybe again, I have been too optimistic. Is organized religion too inextricably bound to the status quo to save our nation and the world? Maybe I must turn my faith to the inner spiritual church, the church within the church, as the true *ecclesia* and the hope of the world. But again I am thankful to God that some noble souls from the ranks of organized religion have broken loose from the paralyzing chains of conformity and joined us as active partners in the struggle for freedom. They have left their secure congregations and walked the streets of Albany, Georgia, with us. They have gone through the highways of the South on tortuous rides for freedom. Yes, they have gone to jail with us. Some have been kicked out of their churches, and lost support of their bishops and fellow ministers. But they have gone with the faith that right defeated is stronger than evil triumphant. These men have been the leaven in the lump of the race. Their witness has been the spiritual salt that has preserved the true meaning of the gospel in these troubled times. They have carved a tunnel of hope through the dark mountain of disappointment.

I hope the church as a whole will meet the challenge of this decisive hour. But even if the church does not come to the aid of justice, I have no despair about the future. I have no fear about the outcome of our struggle in Birmingham, even if our motives are presently misunderstood. We will reach the goal of freedom in Birmingham and all over the nation, because the goal of America is freedom. Abused and scorned though we may be, our destiny is tied up with the destiny of America. Before the Pilgrims landed at Plymouth we were here. Before the pen of Jefferson etched across the pages of history the majestic words of the Declaration of Independence, we were here. For more than two centuries our foreparents labored in this country without wages; they made cotton king; and they built the homes of their masters in the midst of brutal injustice and shameful humiliation—and yet out of a bottomless vitality they continued to thrive and develop. If the inexpressible cruelties of slavery could not stop us, the opposition we now face will surely fail. We will win our freedom because the sacred heritage of our nation and the eternal will of God are embodied in our echoing demands.

I must close now. But before closing I am impelled to mention one other point in your statement that troubled me profoundly You warmly commended the Birmingham police force for keeping "order" and "preventing violence. " I don't believe you would have so warmly commended the police force if you had seen its angry violent dogs literally biting six unarmed, nonviolent Negroes. I don't believe you would so quickly commend the policemen if you would observe their ugly and inhuman treatment of Negroes here in the city jail; if you would watch them push and curse old Negro women and young Negro girls; if you would see them slap and kick old Negro men and young boys; if you will observe them, as they did on two occasions, refuse to give us food because we wanted to sing our grace together. I'm sorry that I can't join you in your praise for the police department.

It is true that they have been rather disciplined in their public handling of the demonstrators. In this sense they have been rather publicly "nonviolent." But for what purpose? To preserve the evil system of segregation. Over the last few years I have consistently preached that nonviolence demands that the means we use must be as pure as the ends we seek. So I have tried to make it clear that it is wrong to use immoral means to attain moral. ends. But now I must affirm that it is just as wrong, or even more so, to use moral means to preserve immoral ends. Maybe Mr. Connor and his policeman have been rather publicly nonviolent, as Chief Pritchett was in Albany, Georgia, but they have used the moral means of nonviolence to maintain the immoral end of flagrant racial injustice. T. S. Eliot has said that there is no greater treason than to do the right deed for the wrong reason.

I wish you had commended the Negro sit-inners and demonstrators of Birmingham for their sublime courage, their willingness to suffer and their amazing discipline in the midst of the most inhuman provocation. One day the South will recognize its real heroes. They will be the James Merediths, courageously and with a majestic sense of purpose facing jeering and hostile mobs and the agonizing loneliness that characterizes the life of the pioneer. They will be old, oppressed, battered Negro women, symbolized in a seventy-two-year-old woman of Montgomery, Alabama, who rose up with a sense of dignity and with her people decided not to ride the segregated buses, and responded to one who inquired about her tiredness with ungrammatical profundity: "My feet is tired, but my soul is rested." They will be the young high school and college students, young ministers of the gospel and a host of their elders courageously and nonviolently sitting-in at lunch counters and willingly going to jail for conscience's sake. One day the South will know that when these disinherited children of God sat down at lunch counters they were in reality standing up for the best in the American dream and the most sacred values in our Judeo-Christian heritage, and thusly, carrying our whole nation back to those great wells of democracy which were dug deep by the Founding Fathers in the formulation of the Constitution and the Declaration of Independence. Never before have I written a letter this long (or should I say a book?). I'm afraid that it is much too long to take your precious time. I can assure you that it would have been much shorter if I had been writing from a comfortable desk, but what else is there to do when you are alone for days in the dull monotony of a narrow jail cell other than write long letters, think strange thoughts, and pray long prayers?

If I have said anything in this letter that is an overstatement of the truth and is indicative of an unreasonable impatience, I beg you to forgive me. If I have said anything in this letter that is an understatement of the truth and is indicative of my having a patience that makes me patient with anything less than brotherhood, I beg God to forgive me.

I hope this letter finds you strong in the faith. I also hope that circumstances will soon make it possible for me to meet each of you, not as an integrationist or a civil rights leader, but as a fellow clergyman and a Christian brother. Let us all hope that the dark clouds of racial prejudice will soon pass away and the deep fog of misunderstanding will be lifted from our fear-drenched communities and in some not too distant tomorrow the radiant stars of love and brotherhood will shine over our great nation with all of their scintillating beauty.

Yours for the cause of Peace and Brotherhood,
Martin Luther King, Jr.

April 16, 1963

REFERENCE
The Constitution of the United States

We the people of the United States, in Order to form a more perfect Union, establish justice, insure domestic Tranquility, provide for the common defence, promote the general Welfare, and secure the Blessings of Liberty to ourselves and our Posterity, do ordain and establish this Constitution for the United States of America.

Article I

Section 1. All legislative Powers herein granted shall be vested in a Congress of the United States, which shall consist of a Senate and House of Representatives.

Section 2. The House of Representatives shall be composed of Members chosen every second Year by the People of the several States, and the Electors in each State shall have the Qualifications requisite for Electors of the most numerous Branch of the State Legislature.

No Person shall be a Representative who shall not have attained to the Age of twenty five Years, and been seven Years a Citizen of the United States, and who shall not, when elected, be an Inhabitant of that State in which he shall be chosen.

Representatives and direct Taxes shall be apportioned among the several States which may be included within this Union, according to their respective Numbers, which shall be determined by adding to the whole Number of free Persons, including those bound to Service for a Term of Years, and excluding Indians not taxed, three fifths of all other Persons. The actual Enumeration shall be made within three Years after the first Meeting of the Congress of the United States, and within every subsequent Term of ten Years, in such Manner as they shall by Law direct. The Number of Representatives shall not exceed one for every thirty Thousand, but each State shall have at Least one Representative; and until such enumeration shall be made, the State of New Hampshire shall be entitled to chuse three, Massachusetts eight, Rhode-Island and Providence Plantations one, Connecticut five, New-York six, New Jersey four, Pennsylvania eight, Delaware one, Maryland six, Virginia ten, North Carolina five, South Carolina five, and Georgia three.

When vacancies happen in the Representation from any State, the Executive Authority thereof shall issue Writs of Election to fill such Vacancies. The House of Representatives shall chuse their Speaker and other Officers; and shall have the sole Power of Impeachment.

Section 3. The Senate of the United States shall be composed of two Senators from each State, chosen by the Legislature thereof, for six Years; and each Senator shall have one Vote.

Immediately after they shall be assembled in Consequence of the first Election, they shall be divided as equally as may be into three Classes. The Seats of the Senators of the first Class shall be vacated at the Expiration of the second Year, of the second Class at the Expiration of the fourth Year, and of the third Class at the Expiration of the sixth Year, so that one third may be chosen every second Year; and if Vacancies happen by Resignation, or otherwise, during the Recess of the Legislature of any State, the Executive thereof may make temporary Appointments until the next Meeting of the Legislature, which shall then fill such Vacancies.

No Person shall be a Senator who shall not have attained to the Age of thirty Years, and been nine Years a Citizen of the United States, and who shall not when elected, be an Inhabitant of that State for which he shall be chosen.

The Vice President of the United States shall be President of the Senate, but shall have no Vote, unless they be equally divided.

The Senate shall chuse their other, Officers, and also a President pro tempore, in the Absence of the Vice President or when he shall exercise the Office of President of the United States.

The Senate shall have the sole Power to try all Impeachments. When sitting for that Purpose, they shall be on Oath or Affirmation. When the President of the United States is tried, the Chief justice shall preside: And no Person shall be convicted without the Concurrence of two thirds of the Members present.

Judgment in Cases of Impeachment shall not extend further than to removal from Office, and disqualification to hold and enjoy any Office of honor, Trust or Profit under the United States: but the Party convicted shall nevertheless be liable and subject to Indictment Trial, Judgment and Punishment according to Law.

Section 4. The Times, Places and Manner of holding Elections for Senators and Representatives, shall be prescribed in each State by the Legislature thereof; but the Congress may at any time by Law make or alter such Regulations, except as to the Places of chusing Senators.

The Congress shall assemble at least once in every Year, and such Meeting shall be on the first Monday in December, unless they shall by Law appoint a different Day.

Section 5. Each House shall be the Judge of the Elections, Returns and Qualifications of its own Members, and a Majority of each shall constitute a Quorum to do Business; but a smaller Number may adjourn from day to day, and may be authorized to compel the Attendance of absent Members, in such Manner, and under such Penalties as each House may provide.

Each House may determine the Rules of its Proceedings, punish its Members for disorderly Behaviour, and, with the Concurrence of two thirds, expel a Member.

Each House shall keep a Journal of its Proceedings, and from time to time publish the same, excepting such Parts as may in their judgment require Secrecy; and the Yeas and Nays of the Members of either House on any question shall, at the Desire of one fifth of those Present, be entered on the Journal.

Neither House, during the Session of Congress, shall, without the Consent of the other, adjourn for more than three days, nor to any other Place than that in which the two Houses shall be sitting.

Section 6. The Senators and Representatives shall receive a Compensation for their Services, to be ascertained by Law, and paid out of the Treasury of the United States. They shall in all Cases, except Treason, Felony and Breach of the Peace, be privileged from Arrest during their Attendance at the Session of their respective Houses, and in going to and returning from the same; and for any Speech or Debate in either House, they shall not be questioned in any other Place.

No Senator or Representative shall, during the Time for which he was elected, be appointed to any civil Office under the Authority of the United States which shall have been created, or the Emoluments whereof shall have been encreased during such time; and no Person holding any Office under the United States, shall be a Member of either House during his Continuance in Office.

Section 7. All Bills for raising Revenue shall originate in the House of Representatives; but the Senate may propose or concur with Amendments as on other Bills.

Every Bill which shall have passed the House of Representatives and the Senate, shall, before it become a Law, be presented to the President of the United States; If he approve he shall sign it, but if not he shall return it with his Objections to that House in which it shall have originated, who shall enter the Objections at large on their Journal, and proceed to reconsider it. If after such Reconsideration two thirds of that House shall agree to pass the Bill, it shall be sent, together with the Objections, to the other House, by which it shall likewise be reconsidered, and if approved by two thirds of that House, it shall become a Law. But in all such Cases the Votes of both Houses shall be determined by Yeas and Nays, and the Names of the Persons voting for and against the Bill shall be entered on the journal of each House respectively. If any Bill shall not be returned by the President within ten Days (Sundays excepted) after it shall have been presented to him, the Same shall be a Law, in like Manner as if he had signed it, unless the Congress by their Adjournment prevent its Return, in which Case it shall not be a Law.

Every Order, Resolution, or Vote to which the Concurrence of the Senate and House of Representatives may be necessary (except on a question of Adjournment) shall be presented to the President of the United States; and before the Same shall take Effect shall be approved by him, or being disapproved by him, shall be repassed by two thirds of the Senate and House of Representatives, according to the Rules and Limitations prescribed in the Case of a Bill.

Section 8. The Congress shall have Power To lay and collect Taxes, Duties, Imposts and Excises, to pay the Debts and provide for the common Defence and general Welfare of the United States; but all Duties, Imposts and Excises shall be uniform throughout the United States;

To borrow Money on the credit of the United States;

To regulate Commerce with foreign Nations, and among the several States, and with the Indian Tribes;

To establish an uniform Rule of Naturalization, and uniform Laws on the subject of Bankruptcies throughout the United States;

To coin Money, regulate the Value thereof, and of foreign Coin, and fix the Standard of Weights and Measures;

To provide for the Punishment of counterfeiting the Securities and current Coin of the United States;

To establish Post Offices and post Roads;

To promote the Progress of Science and useful Arts, by securing for limited Times to Authors and Inventors the exclusive Right to their respective Writings and Discoveries;

To constitute Tribunals inferior to the supreme Court;

To define and punish Piracies and Felonies committed on the high Seas, and Offences against the Law of Nations;

To declare War, grant Letters of Marque and Reprisal, and make Rules concerning Captures on Land and Water;

To raise and support Armies, but no Appropriation of Money to that Use shall be for a longer Term than two Years;

To provide and maintain a Navy;

To make Rules for the Government and Regulation of the land and naval Forces;

To provide for calling forth the Militia to execute the Laws of the Union, suppress Insurrections and repel Invasions;

To provide for organizing, arming, and disciplining the Militia, and for governing such Part of them as may be employed in the Service of the United States, reserving to the States respectively, the Appointment of the Officers, and the Authority of training the Militia according to the discipline prescribed by Congress;

To exercise exclusive Legislation in all Cases whatsoever, over such District (not exceeding ten Miles square) as may, by Cession of particular States, and the Acceptance of Congress, become the Seat of the Government of the United States, and to exercise like Authority over all Places purchased by the Consent of the Legislature of the State in which the Same shall be, for the Erection of Forts, Magazines, Arsenals, dock-Yards, and other needful Buildings;—And

To make all Laws which shall be necessary and proper for carrying into Execution the foregoing Powers, and all other Powers vested by this Constitution in the Government of the United States, or in any Department or Officer thereof.

Section 9. The Migration or Importation of such Persons as any of the States now existing shall think proper to admit, shall not be prohibited by the Congress prior to the Year one thousand eight hundred and eight, but a Tax or duty may be imposed on such Importation, not exceeding ten dollars for each Person.

The Privilege of the Writ of Habeas Corpus shall not be suspended, unless when in Cases of Rebellion or Invasion the public: Safety may require it.

No Bill of Attainder or ex post facto Law shall be passed.

No Capitation, or other direct, Tax shall be laid, unless in Proportion to the Census or Enumeration herein before directed to be taken.

No Tax or Duty shall be laid on Articles exported from any State.

No Preference shall be given by any Regulation of Commerce or Revenue to the Ports of one State over those of another: nor shall Vessels bound to, or from, one State, be obliged to enter, clear, or pay Duties in another.

No Money shall be drawn from the Treasury, but in Consequence of Appropriations made by Law; and a regular Statement and Account of the Receipts and Expenditures of all public Money shall be published from time to time.

No Title of Nobility shall be granted by the United States: And no Person holding any Office of Profit or Trust under them, shall, without the Consent of the Congress, accept of any present, Emolument Office, or Title, of any kind whatever, from any King, Prince or foreign State.

Section 10. No State shall enter into any Treaty Alliance, or Confederation; grant Letters of Marque and Reprisal; coin Money; emit Bills of Credit; make any Thing but gold and silver Coin a Tender in Payment of Debts; pass any Bill of Attainder, ex post facto Law, or Law impairing the Obligation of Contracts, or grant any Title of Nobility.

No State shall, without the Consent of the Congress, lay any Imposts or Duties on Imports or Exports, except what may be absolutely necessary for executing it's inspection Laws: and the net Produce of all Duties and Imposts, laid by any State on Imports or Exports, shall be for the Use of the Treasury of the United States; and all such Laws shall be subject to the Revision and Controul of the Congress.

No State shall, without the Consent of Congress, lay any Duty of Tonnage, keep Troops, or Ships of War in time of Peace, enter into any Agreement or Compact with another State, or with a foreign Power, or engage in War, unless actually invaded, or in such imminent Danger as will not admit of delay.

Article II

Section 1. The executive Power shall be vested in a President of the United States of America. He shall hold his Office during the Term of four Years, and, together with the Vice President chosen for the same Term, be elected as follows:

Each State shall appoint in such Manner as the Legislature thereof may direct a Number of Electors, equal to the whole Number of Senators and Representatives to which the State may be entitled in the Congress: but no Senator or Representative, or Person holding an Office of Trust or Profit under the United States, shall be appointed an Elector.

The Electors shall meet in their respective States, and vote by Ballot for two Persons, of whom one at least shall not be an Inhabitant of the same State with themselves. And they shall make a List of all the Persons voted for, and of the Number of Votes for each; which List they shall sign and certify, and transmit sealed to the Seat of the Government of the United States, directed to the President of the Senate. The President of the Senate shall, in the Presence of the Senate and House of Representatives, open all the Certificates, and the Votes shall then be counted. The Person having the greatest Number of Votes shall be the President, if such Number be a Majority of the whole Number of Electors appointed; and if there be more than one who have such Majority and have an equal Number of Votes, then the House of Representatives shall immediately chuse by Ballot one of them for President; and if no person have a Majority then from the five highest on the List the said House shall in like Manner chuse the President. But in chusing the President the Votes shall be taken by States, the Representation from each State having one Vote; A quorum for this Purpose shall consist of a Member or Members from two thirds of the States, and a Majority of all the States shall be necessary to a Choice. In every Case, after the Choice of the President the Person having the greatest Number of Votes of the Electors shall be the Vice President. But if there should remain two or more who have equal Votes, the Senate shall chuse from them by Ballot the Vice President.

The Congress may determine the Time of chusing the Electors, and the Day on which they shall give their Votes; which Day shall be the same throughout the United States.

No Person except a natural born Citizen, or a Citizen of the United States, at the time of the Adoption of this Constitution, shall be eligible to the Office of President; neither shall any Person be eligible to that Office who shall not have attained to the Age of thirty five Years, and been fourteen Years a Resident within the United States.

In Case of the Removal of the President from Office, or of his Death, Resignation, or Inability to discharge the Powers and Duties of the said Office, the Same shall devolve on the Vice President and the Congress may by Law provide for the Case of Removal, Death, Resignation or Inability, both of the President and Vice President declaring what Officer shall then act as President, and such Officer shall act accordingly, until the Disability be removed, or a President shall be elected.

The President shall, at stated Times, receive for his Services, a Compensation, which shall neither be encreased nor diminished during the Period for which he shall have been elected, and he shall not receive within that Period any other Emolument from the United States, or any of them.

Before he enter on the Execution of his Office, he shall take the following Oath or Affirmation:— "I do solemnly swear (or affirm) that I will faithfully execute the Office of President of the United States, and will to the best of my Ability, preserve, protect and defend the Constitution of the United States."

Section 2. The President shall be Commander in Chief of the Army and Navy of the United States, and of the Militia of the several States, when called into the actual Service of the United States; he may require the Opinion, in writing, of the principal Officer in each of the executive Departments, upon any Subject relating to the Duties of their respective Offices, and he shall have Power to grant Reprieves and Pardons for Offences against the United States, except in Cases of Impeachment.

He shall have Power, by and with the Advice and Consent of the Senate, to make Treaties, provided two thirds of the Senators present concur; and he shall nominate, and by and with the Advice and Consent of the Senate, shall appoint Ambassadors, other public Ministers and Consuls, Judges of the supreme Court, and all other Officers of the United States, whose Appointments are not herein otherwise provided for, and which shall be established by Law: but the Congress may by Law vest the Appointment of such inferior Officers, as they think proper, in the President alone, in the Courts of Law, or in the Heads of Departments.

The President shall have Power to fill up all Vacancies that may happen during the Recess of the Senate, by granting Commissions which shall expire at the End of their next Session.

Section 3. He shall from time to time give to the Congress information of the State of the Union, and recommend to their Consideration such Measures as he shall judge necessary and expedient; he may, on extraordinary Occasions, convene both Houses, or either of them, and in Case of Disagreement between them, with Respect to the Time of Adjournment, he may adjourn them to such Time as he shall think proper; he shall receive Ambassadors and other public Ministers; he shall take Care that the Laws be faithfully executed, and shall Commission all the Officers of the United States.

Section 4. The President, Vice President and all civil Officers of the United States, shall be removed from Office on Impeachment for, and Conviction of, Treason, Bribery, or other high Crimes and Misdemeanors.

Article III

Section 1. The judicial Power of the United States, shall be vested in one supreme Court, and in such inferior Courts as the Congress may from time to time ordain and establish. The judges, both of the supreme and inferior Courts, shall hold their Offices during good Behaviour, and shall, at stated Times, receive for their Services, a Compensation, which shall not be diminished during their Continuance in Office.

Section 2. The judicial Power shall extend to all Cases, in Law and Equity, arising under this Constitution, the Laws of the United States, and Treaties made, or which shall be made, under their Authority;—to all Cases affecting Ambassadors, other public Ministers and Consuls;—to all Cases of admiralty and maritime jurisdiction;—to Controversies to which the United States shall be a Party;—to Controversies between two or more States;—between a State and Citizens of another State;—between Citizens of different States,—between Citizens of the same State claiming Lands under Grants of different States, and between a State, or the Citizens thereof, and foreign States, Citizens or Subjects.

In all Cases affecting Ambassadors, other public Ministers and Consuls, and those in which a State shall be Party, the supreme Court shall have original jurisdiction. In all the other cases before mentioned, the supreme Court shall have appellate jurisdiction, both as to Law and Fact with such Exceptions, and under such Regulations as the Congress shall make.

The Trial of all Crimes, except in Cases of Impeachment shall be by jury; and such Trial shall be held in the State where the said Crimes shall have been committed; but when not committed within any State, the Trial shall be at such Place or Places as the Congress may by Law have directed.

Section 3. Treason against the United States, shall consist only in levying War against them, or in adhering to their Enemies, giving them Aid and Comfort. No Person shall be convicted of Treason unless on the Testimony of two Witnesses to the same overt Act, or on Confession in open Court.

The Congress shall have Power to declare the Punishment of Treason, but no Attainder of Treason shall work Corruption of Blood, or Forfeiture except during the Life of the Person attainted.

Article IV

Section 1. Full Faith and Credit shall be given in each State to the Public Acts, Records, and judicial Proceedings of every other State. And the Congress may by general Laws prescribe the Manner in which such Acts, Records and Proceedings shall be proved, and the Effect thereof.

Section 2. The Citizens of each State shall be entitled to all Privileges and Immunities of Citizens in the Several States.

A Person charged in any State with Treason, Felony, or other Crime, who shall flee from justice, and be found in another State, shall on Demand of the executive Authority of the State from which he fled, be delivered up, to be removed to the State having jurisdiction of the Crime.

No Person held to Service or Labour in one State, under the Laws thereof, escaping into another, shall, in Consequence of any Law or Regulation therein, be discharged from such Service or Labour, but shall be delivered up on Claim of the Party to whom such Service or Labour may be due.

Section 3. New States may be admitted by the Congress into this Union; but no new States shall be formed or erected within the jurisdiction of any other State; nor any State be formed by the junction of two or more States, or Parts of States, without the Consent of the Legislatures of the States concerned as well as of the Congress.

The Congress shall have Power to dispose of and make all needful Rules and Regulations respecting the Territory or other Property belonging to the United States; and nothing in this Constitution shall be so construed as to Prejudice any Claims of the United States, or of any particular State.

Section 4. The United States shall guarantee to every State in this Union a Republican Form of Government and shall protect each of them against Invasion; and on Application of the Legislature, or of the Executive (when the Legislature cannot be convened) against domestic Violence.

Article V

The Congress, whenever two thirds of both Houses shall deem it necessary, shall propose Amendments to this Constitution, or, on the Application of the Legislatures of two thirds of the several States, shall call a Convention for proposing Amendments, which, in either Case, shall be valid to all Intents and Purposes, as Part of this Constitution, when ratified by the Legislatures of three fourths of the several States, or by Conventions in three fourths thereof, as the one or the other Mode of Ratification may be proposed by the Congress; Provided that no Amendment which may be made prior to the Year One thousand eight hundred and eight shall in any Manner affect the first and fourth Clauses in the Ninth Section of the first Article; and that no State, without its Consent shall be deprived of its equal Suffrage in the Senate.

Article VI

All Debts contracted and Engagements entered into, before the Adoption of this Constitution, shall be as valid against the United States under this Constitution, as under the Confederation.

This Constitution, and the Laws of the United States which shall be made in Pursuance thereof; and all Treaties made, or which shall be made, under the Authority of the United States, shall be the supreme Law of the Land; and the judges in every State shall be bound thereby, any Thing in the Constitution or Laws of any State to the Contrary notwithstanding.

The Senators and Representatives before mentioned, and the Members of the several State Legislatures, and all executive and judicial Officers, both of the United States and of the several States, shall be bound by Oath or Affirmation, to support this Constitution; but no religious Test shall ever be required as a Qualification to any Office or public Trust under the United States.

Article VII

The Ratification of the Conventions of nine States, shall be sufficient for the Establishment of this Constitution between the States so ratifying the Same.

Done in Convention by the Unanimous Consent of the States present the Seventeenth Day of September in the Year of our Lord one thousand seven hundred and Eighty seven and of the Independance of the United States of America the Twelfth. In witness whereof We have hereunto subscribed our Names,

Attest William Jackson Secretary

Go: Washington—Presidt.
and deputy from Virginia

Delaware

Geo: Read
Gunning Bedford junr
John Dickinson
Richard Bassett
Jaco: Broom

Maryland

James McHenry
Dan of St Thos. Jenifer
Danl Carroll

Virginia

John Blair—
James Madison Jr.

North Carolina

Wm. Blount
Richd. Dobbs Spaight.
Hu Williamson

South Carolina

J. Rutledge
Charles Cotesworth Pinckney
Charles Pinckney
Pierce Butler

Georgia

William Few
Abr Baldwin

New Hampshire
    John Langdon
    Nicholas Gilman

Massachusetts
    Nathaniel Gorham
    Rufus King

Connecticut
    Wm: Saml. Johnson
    Roger Sherman

New York
    Alexander Hamilton

New Jersey
    Wil: Livingston
    David Brearley
    Wm. Paterson.
    Jona: Dayton

Pennsylvania
    B Franklin
    Thomas Mifflin
    Robt Morris
    Geo. Clymer
    Thos. FtizSimons
    Jared Ingersoll
    James Wilson
    Gouv. Morris

ARTICLES in Addition to, and Amendment of, the Constitution of the United States of America, proposed by Congress, and ratified by the Legislatures of the several States, pursuant to the fifth Article of the original Constitution.

Article I

Congress shall make no law respecting an establishment of religion, or prohibiting the free exercise thereof; or abridging the freedom of speech, or of the press; or the right of the people peacably to assemble, and to petition the government for a redress of grievances.

Article II

A well regulated Militia, being necessary to the Security of a free State, the right of the people to keep and bear Arms, shall not be infringed.

Article III

No Soldier shall, in time of peace be quartered in any house, without the consent of the Owner, nor in time of war, but in a manner to be prescribed by law.

Article IV

The right of the people to be secure in their persons, houses, papers, and effects, against unreasonable searches and seizures, shall not be violated, and no Warrants shall issue, but upon probable cause, supported by Oath or affirmation, and particularly describing the place to be searched, and the persons or things to be seized.

Article V

No person shall be held to answer for a capital, or otherwise infamous crime, unless on a presentment or indictment of a Grand Jury, except in cases arising in the land or naval forces, or in the Militia, when in actual service in time of War or public danger; nor shall any person be subject for the same offence to be twice put in jeopardy of life or limb; nor shall be compelled in any criminal case to be a witness against himself, nor be deprived of life, liberty, or property, without due process of law; nor shall private property be taken for public use, without just compensation.

Article VI

In all criminal prosecutions, the accused shall enjoy the right to a speedy and public trial, by an impartial jury of the State and district wherein the crime shall have been committed, which district shall have been previously ascertained by law, and to be informed of the nature and cause of the accusation; to be confronted with the witnesses against him; to have compulsory process for obtaining witnesses in his favor, and to have the Assistance of Counsel for his defence.

Article VII

In Suits at common law where the value in controversy shall exceed twenty dollars, the right of trial by jury shall be preserved, and no fact tried by a jury, shall be otherwise re-examined in any Court of the United States, than according to the rules of the common law.

Article VIII

Excessive bail shall not be required, nor excessive fines imposed, nor cruel and unusual punishments inflicted.

Article IX

The enumeration in the Constitution, of certain rights, shall not be construed to deny or disparage others retained by the people.

Article X

The powers not delegated to the United States by the Constitution, nor prohibited by it to the States, are preserved to the States respectively, or to the people.

Articles I.–X. proposed to the states by Congress, September 25, 1789
Ratification completed, December 15, 1791
Ratification declared, March 1, 1792

Article XI

The Judicial power of the United States shall not be construed to extend to any suit in law or equity, commenced or prosecuted against one of the United States by Citizens of another State, or by Citizens or Subjects of any Foreign State.

Proposed to the states by Congress, March 4, 1794
Ratification completed, February 7, 1795
Ratification declared, January 8, 1798

Article XII

The Electors shall meet in their respective states, and vote by ballot for President and Vice-President one of whom, at least shall not be an inhabitant of the same state with themselves; they shall name in their ballots the person voted for as President and in distinct ballots the person voted for as Vice-President, and they shall make distinct lists of all persons voted for as President, and of all persons voted for as Vice-President, and of the number of votes for each, which lists they shall sign and certify; and transmit sealed to the seat of the government of the United States, directed to the President of the Senate;—The President of the Senate shall, in the presence of the Senate and House of Representatives, open all the certificates and the votes shall then be counted;—The person having the greatest number of votes for President, shall be the President, if such number be a majority of the whole number of Electors appointed; and if no person have such majority; then from the persons having the highest numbers not exceeding three on the list of those voted for as President, the House of Representatives shall choose immediately, by ballot, the President. But in choosing the President, the votes shall be taken by states, the representation from each state having one vote; a quorum for this purpose shall consist of a member or members from two-thirds of the states, and a majority of all the states shall be necessary to a choice. And if the House of Representatives shall not choose a President whenever the right of choice shall devolve upon them, before the fourth day of March next following, then the Vice-President shall act as President as in the case of the death or other constitutional disability of the President.—The person having the greatest number of votes as Vice-President shall be the Vice-President if such number be a majority of the whole number of Electors appointed, and if no person have a majority, then from the two highest numbers on the list the Senate shall choose the Vice-President; a quorum for the purpose shall consist of two-thirds of the whole number of Senators, and a majority of the whole number shall be necessary to a choice. But no person constitutionally ineligible to the office of President shall be eligible to that of Vice-President of the United States.

Proposed to the states by Congress, December 9, 1803
Ratification completed, June 15, 1804
Ratification declared, September 25, 1804

Article XIII

Section 1. Neither slavery nor involuntary servitude, except as a punishment for crime whereof the party shall have been duly convicted, shall exist within the United States, or any place subject to their jurisdiction.

Section 2. Congress shall have power to enforce this article by appropriate legislation.

Proposed to the states by Congress, January 31, 1865
Ratification completed, December 6, 1865
Ratification declared, December 18, 1865

Article XIV

Section 1. All persons born or naturalized in the United States, and subject to the jurisdiction thereof, are citizens of the United States and of the State wherein they reside. No State shall make or enforce any law which shall abridge the privileges or immunities of citizens of the United States; nor shall any State deprive any person of life, liberty; or property; without due process of law; nor deny to any person within its jurisdiction the equal protection of the laws.

Section 2. Representatives shall be apportioned among the several States according to their respective numbers, counting the whole number of persons in each State, excluding Indians not taxed. But when the right to vote at any election for the choice of electors for President and Vice President of the United States, Representatives in Congress, the Executive and Judicial officers of a State, or the members of the Legislature thereof, is denied to any of the male inhabitants of such State, being twenty-one years of age, and citizens of the United States, or in any way abridged, except for participation in rebellion, or other crime, the basis of representation therein shall be reduced in the proportion which the number of such male citizens shall bear to the whole number of male citizens twenty-one years of age in such State.

Section 3. No person shall be a Senator or Representative in Congress, or elector of President and Vice President, or hold any office, civil or military, under the United States, or under any State, who, having previously taken an oath, as a member of Congress, or as an officer of the United States, or as a member of any State legislature, or as an executive or judicial officer of any State, to support the Constitution of the United States, shall have engaged in insurrection or rebellion against the same, or given aid or comfort to the enemies thereof. But Congress may by a vote of two-thirds of each House, remove such disability.

Section 4. The validity of the public debt of the United States, authorized by law, including debts incurred for payment of pensions and bounties for services in suppressing insurrection or rebellion, shall not be questioned. But neither the United States nor any State shall assume or pay any debt or obligation incurred in aid of insurrection or rebellion against the United States, or any claim for the loss or emancipation of any slave; but all such debts, obligations and claims shall be held illegal and void.

Section 5. The Congress shall have power to enforce, by appropriate legislation, the provisions of this article.

Proposed to the states by Congress, June 13, 1866
Ratification completed, July 9, 1868
Ratification declared, July 28, 1868

Article XV

Section 1. The right of citizens of the United States to vote shall not be denied or abridged by the United States or by any State on account of race, color, or previous condition of servitude.

Section 2. The Congress shall have power to enforce this article by appropriate legislation.

Proposed to the states by Congress, February 26, 1869
Ratification completed, February 3, 1870
Ratification declared, March 30, 1870

Article XVI

The Congress shall have power to lay and collect taxes on incomes, from whatever source derived, without apportionment among the several States, and without regard to any census or enumeration.

Proposed to the states by Congress, July 12, 1909
Ratification completed, February 3, 1913
Ratification declared, February 25, 1913

Article XVII

The Senate of the United States shall be composed of two Senators from each State, elected by the people thereof, for six years; and each Senator shall have one vote. The electors in each State shall have the qualifications requisite for electors of the most numerous branch of the State legislatures.

When vacancies happen in the representation of any State in the Senate, the executive authority of such State shall issue writs of election to fill such vacancies: *Provided,* That the legislature of any State may empower the executive thereof to make temporary appointments until the people fill the vacancies by election as the legislature may direct.

This amendment shall not be so construed as to affect the election or term of any Senator chosen before it becomes valid as part of the Constitution.

Proposed to the states by Congress, May 13, 1912
Ratification completed, April 8, 1913
Ratification declared, May 31, 1913

Article XVIII

Section 1. After one year from the ratification of this article the manufacture, sale, or transportation of intoxicating liquors within, the importation thereof into, or the exportation thereof from the United States and all territory subject to the jurisdiction thereof for beverage purposes is hereby prohibited.

Section 2. The Congress and the several States shall have concurrent power to enforce this article by appropriate legislation.

Section 3. This article shall be inoperative unless it shall have been ratified as an amendment to the Constitution by the legislatures of the several States, as provided in the Constitution, within seven years from the date of the submission hereof to the States by the Congress.

Proposed to the states by Congress, December 18, 1917
Ratification completed, January 16, 1919
Ratification declared, January 29, 1919

Article XIX

The right of citizens of the United States to vote shall not be denied or abridged by the United States or by any State on account of sex.

Congress shall have power to enforce this article by appropriate legislation.

Proposed to the states by Congress, June 4, 1919
Ratification completed, August 18, 1920
Ratification declared, August 26, 1920

Article XX

Section 1. The terms of the President and Vice President shall end at noon on the 20th day of January and the terms of Senators and Representatives at noon on the 3d day of January, of the years in which such terms would have ended if this article had not been ratified; and the terms of their successors shall then begin.

Section 2. The Congress shall assemble at least once in every year, and such meeting shall begin at noon on the 3d day of January unless they shall by law appoint a different day.

Section 3. If; at the time fixed for the beginning of the term of the President the President elect shall have died, the Vice President elect shall become President. If a President shall not have been chosen before the time fixed for the beginning of his term, or if the President elect shall have failed to qualify then the Vice President elect shall act as President until a President shall have qualified; and the Congress may by law provide for the case wherein neither a President elect nor a Vice President elect shall have qualified, declaring who shall then act as President or the manner in which one who is to act shall be selected, and such person shall act accordingly until a President or Vice President shall have qualified.

Section 4. The Congress may by law provide for the case of the death of any of the persons from whom the House of Representatives may choose a President whenever the right of choice shall have devolved upon them, and for the case of the death of any of the persons from whom the Senate may choose a Vice President whenever the right of choice shall have devolved upon them.

Section 5. Sections 1 and 2 shall take effect on the 15th day of October following the ratification of this article.

Section 6. This article shall be inoperative unless it shall have been ratified as an amendment to the Constitution by the legislatures of three-fourths of the several States within seven years from the date of its submission.

Proposed to the states by Congress, March 2, 1932
Ratification completed, January 23, 1933
Ratification declared, February 6, 1933

Article XXI

Section 1. The eighteenth article of amendment to the Constitution of the United States is hereby repealed.

Section 2. The transportation or importation into any State, Territory or possession of the United States for delivery or use therein of intoxicating liquors, in violation of the laws thereof, is hereby prohibited.

Section 3. This article shall be inoperative unless it shall have been ratified as an amendment to the Constitution by conventions in the several States, as provided in the Constitution, within seven years from the date of the submission hereof to the States by the Congress.

Proposed to the states by Congress, February 20, 1933
Ratification completed, December 5, 1933
Ratification declared, December 5, 1933

Article XXII

Section 1. No person shall be elected to the office of the President more than twice, and no person who has held the office of President, or acted as President, for more than two years of a term to which some other person was elected President shall be elected to the office of the President more than once. But this Article shall not apply to any person holding the office of President when this Article was proposed by the Congress, and shall not prevent any person who may be holding the office of President, or acting as President, during the term within which this Article becomes operative from holding the office of President or acting as President during the remainder of such term.

Section 2. This article shall be inoperative unless it shall have been ratified as an amendment to the Constitution by the legislatures of three-fourths of the several States within seven years from the date of its submission to the States by the Congress.

Proposed to the states by Congress, March 21, 1947
Ratification completed, February 27, 1951
Ratification declared, March 1, 1951

Article XXIII

Section 1. The District constituting the seat of Government of the United States shall appoint in such manner as the Congress may direct:

A number of electors of President and Vice President equal to the whole number of Senators and Representatives in Congress to which the District would be entitled if it were a State, but in no event more than the least populous State; they shall be in addition to those appointed by the States, but they shall be considered, for the purposes of the election of President and Vice President, to be electors appointed by a State; and they shall meet in the District and perform such duties as provided by the twelfth article of amendment.

Section 2. The Congress shall have power to enforce this article by appropriate legislation.

Proposed to the states by Congress, March 21, 1947
Ratification completed, February 27, 1951
Ratification declared, March 1, 1951

Article XXIV

Section 1. The right of citizens of the United States to vote in any primary or other election for President or Vice President, for electors for President or Vice President or for Senator or Representative in Congress, shall not be denied or abridged by the United States or any State by reason of failure to pay any poll tax or other tax.

Section 2. The Congress shall have power to enforce this article by appropriate legislation.

Proposed to the states by Congress, August 27, 1962
Ratification completed, January 23, 1964
Ratification declared, February 4, 1964

Article XXV

Section 1. In case of the removal of the President from office or of his death or resignation, the Vice President shall become President.

Section 2. Whenever there is a vacancy in the office of the Vice President the President shall nominate a Vice President who shall take office upon confirmation by a majority vote of both Houses of Congress.

Section 3. Whenever the President transmits to the President pro tempore of the Senate and the Speaker of the House of Representatives his written declaration that he is unable to discharge the powers and duties of his office, and until he transmits to them a written declaration to the contrary, such powers and duties shall be discharged by the Vice President as Acting President.

Section 4. Whenever the Vice President and a majority of either the principal officers of the executive departments or of such other body as Congress may by law provide, transmit to the President pro tempore of the Senate and the Speaker of the House of Representatives their written declaration that the President is unable to discharge the powers and duties of his office, the Vice President shall immediately assume the powers and duties of the office as Acting President.

Thereafter, when the President transmits to the President pro tempore of the Senate and the Speaker of the House of Representatives his written declaration that no inability exists, he shall resume the powers and duties of his office unless the Vice President and a majority of either the principal officers of the executive department or of such other body as Congress may by law provide, transmit within four days to the President pro tempore of the Senate and the Speaker of the House of Representatives their written declaration that the President is unable to discharge the powers and duties of his office. Thereupon Congress shall decide the issue, assembling within forty-eight hours for that purpose if not in session. If the Congress, within twenty-one days after receipt of the latter written declaration, or, if Congress is not in session, within twenty-one days after Congress is required to assemble, determines by two-thirds vote of both Houses that the President is unable to discharge the powers and duties of his office, the Vice President shall continue to discharge the same as Acting President; otherwise, the President shall resume the powers and duties of his office.

Proposed to the states by Congress, July 6, 1965
Ratification completed, February 10, 1967
Ratification declared, February 23, 1967

Article XXVI

Section 1. The right of citizens of the United States, who are eighteen years of age or older, to vote shall not be denied or abridged by the United States or by any State on account of age.

Section 2. The Congress shall have power to enforce this article by appropriate legislation.

Proposed to the states by Congress, March 23, 1971
Ratification completed, July 2, 1971
Ratification declared, July 5, 1971

Article XXVII

No law, varying the compensation for the services of the Senators and Representatives, shall take effect, until an election of Representatives shall have intervened.

Proposed to the states by Congress, September 25, 1789
Ratification completed, May 7, 1992
Ratification declared, May 28, 1992

C. Religion and Military Ethics

Many people's moral beliefs are profoundly influenced by their religious beliefs. Indeed, for religious believers, there is no sharp distinction between religion and morality. Moral principles are, for the believer, understood as laws commanded by a Ruler of the Universe, valid and binding on us simply because this all-powerful Ruler commands us to abide by them.

This is not as obvious a connection as it might seem. Locating the source of morality in divine commands may threaten to undermine the goodness and glory of God by making morality seem wholly arbitrary, merely an expression of God's wholly inexplicable Will. Morality seems to have a constancy and universality that the very term "Will," including even Divine Will, lacks.

Also, if morality is merely a matter of Will and the exercise of omnipotent Power, would it not be possible, at least in principle, for the Author of the Moral Law to *change* that law, or, in any given instance, to override or suspend its normal provisions? Might God conceivably (as in the accompanying story of Abraham) command us to engage in acts that would transparently seem to be immoral (on other grounds)?

This possibility seems to imbue morality itself with a hopelessly paradoxical and arbitrary quality. Consequently, while some respected moral philosophers and theologians have advocated a *Divine Command* conception of moral principles as literal expressions of divine will, many other devout believers have thought that the moral validity of God's commands is due, not *per se* to God's will, but to God's intellect, to God's infallible apprehension of justice and goodness. Still others, like the great 19th-century Danish Christian theologian, Søren Kierkegaard, argue on the basis of the Abraham story specifically that *morality and ethics have nothing whatever to do with religion*, which is a profoundly personal or "existential" matter of the relationship of the believer to God, on a much "higher" level than the normal social relations of morality entail.

The Readings

Martin Cook engages this interesting and puzzling dilemma of the proper relationship between morality and religion, or more properly, between the individual's religious convictions and his or her public, moral obligations. Cook argues in part that religious belief contributes to morality by serving to form our moral character, and also by providing both an explanation of the origin of moral principles and a reason for obeying them. Albert Jonsen provides a detailed account of how moral reasoning and the resolution of complex moral dilemmas actually transpires for religious believers in several faith traditions

While military officers, like all citizens, may hold on to their personal beliefs, as officers in the U.S. military they are not entitled simply to operate on a wholly religious conception of morality without risking a serious and ongoing conflict with their Constitutional responsibilities and their individual commission and Oath of Office. By their oath they dedicate themselves to serve the U.S. Constitution, whose First Amendment requires that the government remain neutral regarding all religious beliefs.

What this means in practice is that the government and its military officers must be able to justify their decisions by appeal to moral principles that could be accepted by any reasonable person of good will, independent of his or her religious beliefs. While in military service, officers retain the right to the observance of their own religion. So do each of their subordinates, whose Constitutional rights would be violated by commands predicated on religious principles they do not accept. In all cases, religious conviction then may (but need not) serve as the motivation for acting in a morally responsible manner. An individual's faith, however, can never be found to provide the sole or only reasons that purport to justify an officer's individual conduct.

Religion and Morality: Exploring the Connections
Martin L. Cook

I. Historical Observations

It is obvious that there are connections between religious and moral dimensions of human life. Let us begin by noting just a few of the historical and psychological connections between them:

- In the Western family of religions (Judaism, Christianity and Islam), all members of the family revere the figure of Moses and view God's revelation of the divine law on Mt. Sinai as a central element of their common heritage.
- Many individuals look to their religious heritage for insight and guidance regarding the proper course of moral action, and the hope for divine reward and the fear of divine punishment provide motivation for their conduct.
- Many people seem to believe that faith in God is an essential grounding of morality: "If God does not exist, all things are permitted," one of Dostoevsky's characters says, illustrating this popular belief.
- At the level of popular belief, many in our society expect as a matter of course that religious leaders and institutions will live up to a higher standard of conduct than that of the general society—and are correspondingly more shocked and dismayed when those religious institutions fail to meet that expectation.

II. The Historical "Problem"

Despite the depth of the grounding of such beliefs in the popular mind, philosophy has found the close linking of religion and morality difficult almost from its origins. In Plato's dialogue *Euthyphro*, Socrates presses the question whether things are morally right simply because the gods say they are. Must it not be the opposite, he implies: that the gods, if they are good, would command only those things which are right and good, independently of their being divinely commanded? Implicit in Socrates' question is the suggestion that human beings have and use a rational and independent standard for assessing moral value. The gods' commands, if present at all, could at best serve to sanctify the already-known moral right; at worst, belief in the gods' commands might cause one to perform acts which violate rational morality because one believes the gods require them.

The theme of the independence of true morality from religious backing and grounding is picked up with a new energy at the time of the Enlightenment in Eighteenth Century Europe. The Enlightenment's ablest exponent and deepest thinker, Immanuel Kant, stressed the need for the autonomy of morality—independent and individually chosen motivation—as the precondition of any truly moral action. No matter how well one acts, Kant insisted, if the motivation for the action is deference to or fear of an external authority—even God—then one's actions are thereby deprived of genuine moral worth.

It remained to the Nineteenth Century for the Danish philosopher Soren Kierkegaard to draw out the full implications of the philosopher's account of morality. In *Fear and Trembling*, Kierkegaard explored in depth the story of Abraham as the exemplar par excellence of biblical faith. In particular, Kierkegaard was drawn to the account of Abraham's willingness to sacrifice Isaac. Rather than focusing on the "happy ending" of the story where God intervened to spare Isaac's life and to substitute an animal as sacrificial victim, Kierkegaard found Abraham to be the "Knight of Faith" en route to the sacrifice. If Abraham is the exemplar of faith, Kierkegaard reasoned, it was precisely in his obedience before he knew God's true intentions.

But if that is the case, it means the greatest obstacle to faith is the temptation to be ethical. According to the tradition of philosophical ethics, Kierkegaard argued, Abraham should have recognized in the divine command to sacrifice his son an immoral command and should have rejected it. Abraham's role as the exemplar of faith, by contrast, resulted precisely from his willingness to suspend the ethical—to act on distinctively religious motivations which supersede human moral standards—in the belief that God will (all appearances to the contrary) assure that his promises concerning Isaac will yet be fulfilled!

We are, at this juncture, left with an uncomfortable conclusion if we wish to view the connection between religion and morality positively. On the one hand, we have Plato and Kant speaking for most of our philosophical tradition, insisting that morality must appeal to rational and intelligible standards of human beings. On this analysis, religious motivations and sources of morality are, at best, potentially dangerous and misleading and, at worst, undermine the autonomy that is essential to true morality. On the other hand, in Kierkegaard we find the defender of the independence and priority of the religious over all rationally comprehensible moral systems. Abraham is the Knight of Faith precisely because he is willing and able to "suspend the ethical" in the blinding and transcendent submission of heart and will to the inscrutable divine purpose which, against all reason, he affirms to take precedence in his conduct.

Whichever horn of this dilemma we grasp, we are left with a difficult choice. In either direction, we encounter a sharp disjunction between rationally comprehensible morality and genuine religious understanding. Either religious language and thought provide a descant over the *cantus firmus* of rational morality (if we follow in the tradition of Plato and Kant), or it transcends rational morality and swamps it in the blinding light of religious submission of mind and will (if we prefer Kierkegaard's spirited defense of the independence of religion). Neither position provides much comfort for the popular belief in the deep and inherent connection of good morals with religion.

III. Religion and Rationality: Religious Syntheses

In Judaism, Christianity and Islam, the awareness of the difficulties we have just enumerated has been obvious from the first moments of their expansion into the world of Greek philosophical thinking. Each tradition has attempted to find a way of articulating at least the bulk of morality in such a way that it did not depend wholly, or perhaps even primarily, on strong claims of divine revelation.

Judaism developed the concept of the Noachide commandments—commandments which, because they were given to Noah and his immediate family, were binding on all Noah's descendants (*i.e.*, all humanity). This concept provided a means by which the rabbinical tradition could find a basis within the tradition to speak of common human morality. Medieval Jewish thinkers such as Maimonides, borrowing much from Aristotle, could elaborate these concepts into a version of common moral standards, intelligible to and knowable by all rational persons.

In Christianity, the brilliant synthesis of Aristotle and earlier Augustinian-Platonic Christian thought in the work of Thomas Aquinas performed a similar function. Aristotle (the philosopher for Thomas) had provided a thorough articulation of human morality as it could be understood by unaided reason—an understanding which revelation could supplement with important information about humanity's supernatural and spiritual ends, but not fundamentally subvert. Natural law, therefore, could provide a common moral knowledge for believer and non-believer alike, accessible to each by means of reason. God remained connected to this rational system because God created both the world of our experience and the reason by which we experience it. Nevertheless, we do not need to look directly to God's revealed will for the bulk of our moral information and can rely to a considerable degree on the human intellect, common to all humanity, for the basis of moral thought.

In Islam also, philosophical-theological thought moved in the direction of common rational bases for morality. Especially in the thought of Al-Ghazali (1058–1111 A.D.) and ibn-Rusd (Averroes, 1126–1198 A.D.), revealed law *(sharia)* could be understood only in the light of the rationally comprehensible moral standards available to careful and critical thinking.

It is an irony that all such attempts to bring together independent rational thought and religiously based morality become, in their later evolution, somewhat wooden systems of inherited thought. To use only the Christian example, Thomism by the Nineteenth Century had ceased to be a dynamic and open system of thought. Whereas Thomas had bravely attempted to come to terms with the newly rediscovered thought of Aristotle in the face of the church's opposition, Thomism itself had become the tradition of an inflexible church hierarchy and an ecclesiastical education system, both of which perpetuated themselves on the ground of their traditional status. Similar fates befell the Judaic and Islamic systems integrating philosophical and religious morality.

Still, such perhaps inevitable developments should not blind us to the universal impulses of the great philosophical thinkers of each of the Western religious traditions. For reasons internal to their most fundamental theological impulses, each of these traditions sought to find room within a religious framework to accommodate and acknowledge the worth of the best of secular and even pagan moral thinking.

To cite only these examples is, of course, to tell only half of the historical story. Each of the thinkers I have cited has been the object of severe attack and criticism from within his own religious community. Each has in his time been accused of showing too much accommodation to rationality and failing to emphasize sufficiently the strong role of revealed morality. Both in earlier times and at present there are movements we might broadly call "fundamentalist" which grow impatient with the hard intellectual work of harmonizing faith and reason. To cite only the Christian example once again, the very reading of Aristotle was prohibited at the time Thomas was writing his great synthesis, and Thomas' work remained officially condemned as too rational by the Church for a century until it was rehabilitated to become the official theology of Roman Catholicism.

Each of these traditions also has periodic resurgences of movements which look to scripture alone as the source of religious knowledge and deplore reason as "that great whore" (to use Martin Luther's phrase). Yet such movements are generally short-lived before the more thoughtful members of such communities look once again for a basis for linking up to rational morality. It is not long, for example, before Luther himself, having discarded reason in its scholastic guise, finds a need for "orders of creation" and "two kingdoms" which can guide political and moral thinking apart from direct use of biblical revelation.

To summarize: thoughtful religion always has, and in my judgment always will, end up advocating a variant of Natural Law thinking. It will inevitably evolve in the direction of acknowledging that most of the morality important from within the religious framework can be found and understood apart from a distinctively religious base. To that degree, Kant and Plato win the day; few religious thinkers, and few religious individuals, are willing to follow Kierkegaard in sharply separating the Knight of Faith from all ties to the morally comprehensible!

IV. The Contribution of Religion to Morality

We are now prepared to explore anew the contribution of religion to morality. We have established that strong claims that morality is and must be derived from distinctively religious sources is (1) potentially dangerous to the integrity of morality itself and (2) not a claim which is well-grounded in the best of religious thinking. On the other hand, this is a fact about the fundamental logic of morality and religion; it is obviously not a point about psychological, sociological, and personal connections individuals find between their religious beliefs and their moral lives.

One of the most astute treatments of this question is found in James M. Gustafson's small work, *Can Ethics Be Christian?* At the risk of oversimplifying Gustafson's subtle treatment, I wish to focus our attention on three areas he identifies where religion clearly has impact on the moral life of individuals. These are:

1) the "sort of person" one is;

2) reasons for being moral; and

3) the impact of religion on our interpretation of particular circumstances of action.

1. *The "sort of person" one is:* For much of post-Enlightenment ethics (until the recent beginnings of recovery of "virtue" ethics), ethical reflection and analysis have focused on the choice of particular actions. Yet it remains true that an equal or even more fundamental aspect of the moral life of individuals flows not from moments of crisis decision-making, but rather from the habits of character and dispositions they bring to action. Religious communities communicate powerful images of the preferred dispositions that should characterize their adherents. Some examples:

    - To have been raised from childhood thinking of compassion as the fundamental disposition toward the world and to revere the Bodhisattvas who forgo their own Nirvana in order to save all other sentient beings colors the style and personality of the Buddhist in a manner that long predates moments of decision.

    - To think of love as the fundamental moral requirement and to have that response illustrated from one's infancy in stories like the Good Samaritan and the images of the Crucifixion shape the Christian's fundamental personality and style in ways that, one hopes, surely shape action, and yet are not simple action-guiding rules.

    - To recite the Passover Haggadah is to be reminded, in the midst of the celebration of Israel's deliverance from Egypt, of the worth of all human lives, even those lost in Pharaoh's pursuing army, and deeply colors one's reverence for human lives, even those of the enemy.

    In terms of anthropologists, religious traditions hold up different "modal personalities" as embodying the ideals of a community. The songs religious communities sing, the parables and stories they tell, the "lives of the saints" recounted—each in their own way communicate in a powerful and yet pre-rational way the kinds of persons we should be in this community.

2. *Reasons for being moral:* Ethics as generally understood in philosophy devotes a great deal of time to the question of morally correct behavior. Yet such a focus often neglects a more fundamental issue, often neglected because it is wrongly assumed to have an obvious answer. That question is, "Why should I be moral?"

    It does little good to discuss the question of morally desirable action if one is not motivated to do the morally right thing in the first place—and not everyone is, of course. Religious ideas, symbols, and communities often serve to provide reasons why one should care about being moral.

    Religion is not unique in meeting this need, of course—any morally serious person has some answer to this question. Possible answers range from being true to one's own rational nature (Kant), because one aspires to human fulfillment (Aristotle), or because keeping one's contract with one's fellow citizens is necessary to prevent social chaos and warfare (Hobbes), to cite only a few examples. So I am by no means claiming that only religious motivations can meet this need. My claim is the more modest one; for people who are informed by religious ideas and practices, those religious matters commonly do provide their answer to the question, "Why be moral?"

    For the Western religions, gratitude to God for creating and sustaining the world often serves as a fundamental motivating factor for moral action. We should care about the welfare of the world and our fellow creatures because, for example, God has given us stewardship over the created order. Or we should help those in need, because God's fundamental stance toward us is one of undeserved favor. Or we are members of a community covenanted with God to keep God's commands.

Again, this level of moral thinking and discourse is considerably deeper and more fundamental than questions of the "What should I do?" variety. They go to the most fundamental levels of our interpretation of our place in the world and our responsibilities for it. Religion, since it provides that comprehensive and over-arching framework for many, is one of the ways individuals find meaning and motivation in their moral lives.

3. *Religion and the interpretation of the circumstances of action:* Another commonly neglected, but fundamental, aspect of moral thinking concerns interpretation of circumstances. Situations which raise moral difficulties and issues are not self-interpreting—a precondition for choosing to act is determining "what's going on," and that determination is itself value-laden.

To cite a deliberately extreme example, suppose we're observing the military destruction of a village of civilians. Depending on the religiously informed "glasses" we wear as we observe these events, we may spontaneously describe what's occurring as the divinely mandated destruction of infidels or the slaughter of innocent civilians—images informed, perhaps, by the biblical stories of the conquest of Canaan and the Slaughter of the Innocents, respectively.

Again, encountering a helpless and needy individual may equally be described as an encounter with a social parasite who should learn to tend to his own needs, or to the religiously informed imagination, as an encounter like that of the Good Samaritan.

In short, religious symbols, stories, and ideas fill the imagination of religiously sensitive individuals with analogies and metaphors which spontaneously inform their perceptions of the circumstances of action and, by means of that interpretation, with value-colored predispositions to approach moral choice in particular ways.

V. Conclusion

We have seen that religion and morality have complex interrelationships. At their best, both the religious and the philosophical traditions have found reasons to seek the normative content of much of morality in commonly available and rational standards. To that extent, the strong religious claim that morality is and must be founded on religious perception is a mistake.

On the other hand, we have also noted that religion has a profound impact, for those with serious religious perception, on the stance they take toward moral life. For the adherent, religion presents and sustains a vision of the kind of person he or she should be; it provides an interpretation of his or her place in the world that profoundly inspires reasons to be moral. And the images, symbols, and stories of the religious tradition condition pre-rationally our perceptions of "what's going on" — of the circumstances of action.

To be true to the best of their religious traditions and to provide the moral perspectives of greatest use to society, it is crucial that the religiously motivated be able to draw on and articulate those "natural law" *(i.e.,* commonly available to rational agents) aspects of their tradition which can enlighten and guide careful moral reflection in our ever more pluralistic society and military.

Practical Reasoning and Moral Casuistry
Albert R. Jonsen

Practical reasoning is a phrase used in Western moral philosophy to designate the intellectual process whereby an agent deliberates and decides about a particular course of action. Since moral decision and action is formulated in the light of some sort of general principles, applicable to all similarly situated agents, particular agents must determine how those general principles apply to the specific situation in which they will act. The logic of practical reasoning has been a topic of interest since Aristotle delineated his views in the *Nicomachean Ethics*, pointing out how deliberation and decision about practice differs from speculative or scientific reasoning (1994:III, iii). Modern philosophers have studied how reasons serve to explain, evaluate, and justify intentional decisions and have analyzed the forms of inference involved in statements containing such words as "ought," "should," etc. (Audi 1989; Gauthier 1963; Raz 1978). Since real moral agents decide and act in relation to specific, attainable ends and in the context of concrete circumstances, practical reasoning can be said to be about "cases," that is, "instances" (in Latin, *casus*), in which the particular agent's specific purposes and motives, as well as the extant circumstances of time, place, probability, and possibility, etc., can be described. Thus, the term II casuistry," although it has a more particular meaning to be noted below, may be used as somewhat synonymous with practical reasoning, and will so be used in this chapter about practical reasoning in religious ethics.

While diverse manners of practical reasoning appear in different traditions and cultures, this chapter will attend particularly to the forms of casuistic reasoning that have a prominent place in the ethical traditions of three historic religions: Judaism, Christianity, and Islam. This review of the casuistic traditions aims at a general account of casuistic reasoning, with recognition of some problems inherent in this form of practical reasoning.

Textual Sources for Casuistry

Judaism finds the source of divine revelation in the five books of the Law given by Yahweh to Moses: Genesis, Exodus, Leviticus, Numbers, and Deuteronomy, called collectively the Torah. Christianity accepts the same five books, together with another 34 books of the Hebrew scriptures, called collectively the Old Testament, and the four gospels and 23 other writings, called the New Testament. Muhammed designated Jews and Christians as "People of the Book," meaning that they had received a divine revelation contained in the written words of an inspired text, the Bible. Islam also has its Book, the Qur'ān, in which Allāh conveys to humankind a vision of the meaning of life and commands about how to live it.

Thus, in each of these historic faiths, a written text incorporates the fundamentals of belief, not in abstract terms, but in the specific communications of the Lord, Creator, and Redeemer. The form of that communication, in each of the holy books, is varied: stories about divine creation and providence, poetry about divine mercy and justice, and extensive rules about worship and about the behavior of believers in every aspect of their lives.

The faithful, who are guided by the sacred texts, also live in times and societies not exactly like those in which the divine words were spoken to the inspired scribes. Thus, while the commands are clearly in universal form and are intended to bind the faithful through all time, the differences in cultural, social, and linguistic settings of the faithful through history, require additional interpretation, beyond the casuistry found within the text itself. Further, there is the theological question about inspired scripture itself: Does it represent the literal words of the divine source or does the human intermediary necessarily introduce the human elements of fallibility or cultural relativity and, if so, to what extent? Faced with these questions, each of the three faiths, through their long history, developed institutions for interpretation and commentary on the originating texts. Over many centuries, theological, moral, and legal commentary accumulated into elaborate systems of scholarly theory and method and of rules and opinions taught to the faithful.

Jewish Casuistry

Judaism reveres Torah as the primary source of revelation. These books contain not only the Decalogue but also, according to Jewish tradition, another 613 commandments applicable to Jewish life. A scholarly class, the rabbis, reflected on every word of those books to reveal their deepest meaning and their relevance for observant Jews. Schools of interpretation, with differing emphases but a common purpose of elucidating the sacred text, grew up in Palestine, Babylon, and the Diaspora; the vast literary collection of their reflections constitute Talmud. The Talmudic literature, incorporating the commentary of a multitude of rabbis over many centuries, becomes the major secondary source of Jewish Law. Codes, composed for the most part during the Middle Ages, systematize the multifarious reflections of the rabbis. In general, the sum of Torah, Talmud, and Mishnaic commentaries and Codes make up Halakhah, the law of Jewish life, in its ritual and moral dimensions. However. all these sources are themselves written works, in need of continual interpretation for current times and problems. Rabbinic activity includes a dynamic process: responses to particular questions posed by the observant Jew facing a situation in which some aspect of the tradition seems challenged by previously unfamiliar circumstances. These responses, called Teshuva, (often given the Latin name *Responsa*), are directed to the immediate question but reflect the entire tradition and become part of the tradition (Freehof 1955).

The responding rabbi (or rabbinical council) first examines the texts of Torah and of Talmudic and Mishnaic commentary and the Cones for the most relevant guidance, notes any contradictions or obscurity in these sources, then attempts to reconcile these difficulties in order to find a way toward a resolution appropriate to the immediate case. In the most orthodox view, rabbinic resolution is more than advice; it binds the conscience of the questioner. These responses, which are given in the course of daily life, are often private but are sometimes recorded and collected. particularly if they issue from a rabbi of renowned piety and scholarship. The responses then enter the stream of interpretation of the Law and are used by subsequent rabbis as sources of wisdom. Among the vast number of known Teshuva (estimated at around 250,000 dating from several centuries before the Common Era), a recent collection is most poignant and revealing of rabbinic method and wisdom: responses issued during the Holocaust to Jewish questioners under the most extreme circumstances. One eloquent and heartrending response, issued by the Chief Rabbi in the Warsaw ghetto, considers whether an infant's crying might be stifled as Nazi stormtroopers hunted for hidden Jews, even if the stifling might smother the child. The rabbi answers in a long, scholarly review of the tradition about endangering life, which is almost uncompromising about protecting life, yet he comes to the conclusion that, although the child might not be killed, its life might be endangered to save the hidden Jews, for its own life would certainly be extinguished if the parents were discovered. This casuistry manifests how a tradition with the highest respect for the preservation of life might find grounds for an exception (Kirschner 1985).

Christian Casuistry

Christianity accepts the same first five books of the Hebrew Bible. At the same time, the moral imperatives of Jesus announced in the gospels and the teachings of the first disciples reported in the rest of the New Testament, provide extensive teachings about the moral life, many of which, in Christian eyes, surpassed the Jewish Law in rigor and in sublimity. Early Christians endeavored to put into practice some of the "hard teachings" of Jesus, such as the imperative to leave father and mother to follow him, to turn the other cheek and put up the sword when attacked, and to refrain from marriage, and were forced to interpret these commands within the demands of daily life. In the early church, distinctions between counsels of perfection, as enunciated in the words of Jesus, and moral imperatives, stated in the Decalogue, marked the beginnings of casuistry. Clement, Bishop of Alexandria, for example, wrote extensive treatises to advise his Christian flock about how to live in the pagan surroundings of that great metropolis. When the church was officially recognized within the empire, the strict pacifism that appeared to be the teaching of Jesus had to accommodate the needs of civil defense. Christians served as soldiers and could not be simply told to put up the sword. St. Augustine, formulating the theology of church and state, initiated a casuistry of "just war," an idea already adumbrated in Roman authors, such as Cicero, in which the conditions of legitimate self-defense of person and nation allowed Christians to shed blood. A large casuistry developed around these questions (Kelsay and Johnson 1991).

A practice of confession of sins by a believer to a priest appeared in the late middle ages. Practiced in different ways in different churches, the church in the British Isles, particularly in Ireland, imposed on the faithful a particularly demanding regimen. Moral offenses were catalogued in detail and penances for each specified in books entitled Penitentials. Despite the detail, priests had to consider the grounds for excuse or mitigation and so a rudimentary casuistry developed that was carried by monks from Ireland onto the Continent. Toward the end of the eleventh century, there appeared a movement to organize church law and practice, which had developed largely out of the decrees of local bishops and regional counsels. Collections of these "canons," or rules, were made, inspired by the desire to order and reconcile them. As canonists, the lawyers of the church, performed these tasks, they encountered the many cases that had given rise, over time, to the decrees they were reconciling and, in so doing, formulated certain rules for the interpretation of cases. In 1215 a major church council, Lateran IV, promoted the practice of personal, private confession to a central place in Christian life, requiring all Christians to confess to a priest at least once a year. In the wake of this decree, the need arose to educate clergy throughout Europe about how to judge the seriousness of the sins confessed to them and to make discretionary decisions about penance and absolution. Much more sophisticated Penitentials were produced, often by the theological scholars of the new universities, in which the nature of virtues and vices was analyzed and the circumstances that rendered them more or less serious, that is, as mortal or venial sins, were explained, usually under the general heading of the Ten Commandments. These explanations were illustrated by "for examples," that is, "cases." These cases were sometimes reports of actual ones familiar to confessors or were fictitious ones used for didactic purposes.

The work of canonists and theologians drew on several sources. They not only referenced the precepts of the Bible, the comments of the Fathers, and the decrees of councils. They also utilized the rational techniques of scholastic thought, inspired by the Aristotelian renaissance in the thirteenth century. Also, a theory of natural law, inspired by Roman law, and Ciceronian and Stoic philosophy, provided a conceptual framework for ethics which was not dependant on revelation. Christian casuistry, unlike Talmudic or Islamic, could proceed with wide ranging exploration of rational ethics and only peripheral references to revealed sources.

In the sixteenth century, casuistry emerged as a special branch of moral theology and a multitude of books, analyzing every conceivable moral act, appeared. The authors of these books were frequently members of the newly founded Jesuit order, whose interest in the education of young Catholics and in the ministry of the confessional made them the casuists *par excellence*. In 1656 the mathematical genius Blaise Pascal published *The Provincial Letters*, a scathing criticism of Jesuit casuistry in which he claimed that the techniques of rational analysis had been carried to extremes, justifying by clever reasoning the most outrageous violations of Christian morals and submerging the gospel message under sophistry. His criticism gave casuistry a bad reputation for centuries, and ascribed to the word "casuistry" an almost entirely pejorative meaning—a cynical, sophistic, deceptive distortion of moral rules in order to avoid obeying them through specious rationalization (Pascal 1967; Jonsen 1993). Despite Pascal's criticism and the temptation to abuse, a sound, serious practice of casuistry continued within the Catholic and Anglican, and to some extent, Lutheran and Calvinist, churches (Jonsen and Toulmin 1988; Mahoney 1987; Keenan and Shannon 1995).

During and after the Reformation, casuistry also came in for severe criticism from the Protestant reformers. As Protestantism developed, the formal techniques of casuistry withered, although all moral judgment required some sort of practical reasoning. In Anglicanism alone, a vigorous scholarly casuistry prevailed, closely resembling Roman Catholic casuistry, although notably more liberated from hierarchical doctrine (Kirk 1999).

Islamic Casuistry

Islam holds the Qur'ān to have been dictated to Mohammed by Allāh and to contain the substance of divine prescriptions for life. In addition, the Hadīth, or traditions ascribing certain words and practices to the Prophet, and to his first companions, was also held in high esteem as a guide to the moral life. Islam, within the first century of its existence, became sovereign over a wide region of the Near East and North Africa. Its military and political dominance prevailed over Persian, Syrian, Byzantine, and Hellenistic states where substantial systems of law and moral custom already existed. Islamic rulers desired to bring those legal and customary systems into some conformity with the divine law expressed in Qur'ān. Early in Islamic history, then, judges (*quadis*) were appointed not simply to decide cases but to reinterpret them in light of Qur'ān and Hadīth, and out of these judicial activities arose Sharī'a, the law of Islam which contained both legal norms for social life and ethical norms for personal behavior, whreh were not sharply distinguished: all life for the believer, as well as all rules for the state, were included in Sharī'a. The task of interpreting Sharī'a, the jurisprudence and ethics of Islam, is called *Ficq* (Schacht 1964; Hourani 1971).

The daily life of the faithful encounters difficulties with the familiar norms of Sharī'a and those who wish to fulfill the law are encouraged to seek the opinion of scholars of Ficq. These scholars, *mufti*, devote themselves to the study of Qur'ān, Hadīth, and the traditions of interpretation. Organized in many schools of thought over centuries, these divided into traditionalists and rationalists: the former adhere closely to text and traditions, the latter allow the use of logic and rational methods, derived, as in medieval Christianity, from Aristotelian thought, to guide interpretation. Unlike Christian thinkers, Islamic scholars, even the rationalists, did not develop a formal theory of natural law ethics, although they left a large place for customary moral practices in their interpretation of Sharī'a, The mufti, either as individual scholars or in schools, consider cases submitted to them, formulate their opinion, *Ra 'y*, and issue advice, called *Fatwā*, about behavior that would most closely conform to Sharī'a, The format of a Fatwā resembles the format of the Jewish Teshuva: reference to relevant scriptural texts, examination of scholarly opinion, rational efforts to interpret and reconcile opinions, allusion to analogies and formulation of advice. Fatwā can address political or personal problems. The consensus or agreement of the scholars, in a particular place or of a particular school or in all of the lands of Islam, provides an important criterion for the reliability of opinion. In the latter centuries of the Ottoman empire a bureaucracy was established, headed by a Grand Mufti of Istanbul, devoted entirely to the preparation of Fatwā for the guidance of government and of individual behavior.

Among the earliest and most fundamental moves of Islamic casuistry is the distinction of all moral law into five categories: obligatory, recommended, permitted, disapproved, and forbidden (there is debate about whether any acts are neutral). Working within these categories, mufti can draw careful distinctions about how stringently laws bind the believer. Thus, the ominous words of Qur'ān, "Fight in the cause of Allāh those who fight you . . . and slay them wherever you catch them" (S. 2, 190) can be interpreted as permissive, not obligatory, and it can be recommended that women, children, old men, and other non-combatants can be left unmolested. The words "those who fight you" can be interpreted to exclude these parties but if they, although non-combatants, give aid to the enemies of Allāh, it is permitted, perhaps obligatory, to execute them. It is disapproved to slay any conquered person who may be of use to Islam and it is forbidden to slay those who convert to Islam. Principles of mercy and moderation temper literal fulfillment of the law in the light of circumstances (Khadduri 1955).

Casuistic Method

The casuistries of these three faiths, while different in content and inspiration, are remarkably similar in their fundamental methodology. Crucial to each form of casuistry is the move between revealed text or universal ethical principle and the particular decision. The revealed text itself is rarely specific enough to meet the perplexity of the presented case and the texts themselves often seem to the uninitiated rather peripheral to the problem. For example, a text of the Book of Leviticus (19:16), "thou shalt not stand by the blood of your neighbor," serves as the starting point for a rich Talmudic casuistry about the duty to heal the sick. These words are certainly not an explicit command to heal, but one of the greatest Talmudists, Moses Maimonides (also a physician), reads them to say that anyone capable of rescuing another from drowning, from marauders, or from attacking beasts, who fails to do so, transgresses this command (Malmonldes, *Hilkhot* 1.14). Maimonides clearly goes beyond the words of the text to what he considers its meaning. Modern Jewish bioethicists still reflect on cases of withholding technological life-support in the light of this text and its interpretations over time. Drawing any text close to an actual case demands both reverence for the text as well as intellectual ingenuity. It is this intellectual ingenuity that forms the heart of any casuistry.

However, intellectual ingenuity may also challenge reverence for the text or the principle. The human intellect is capable not only of reasoning but also of rationalization: thus, scholars may utilize the techniques of reasoning, such as definition of terms, distinction and division of concepts and the subtle, often fallacious, steps of logic to reach any conclusion dictated by their preferences. Religious casuistry attempts to reign in the unfettered use of rational techniques by embedding the reasoning within the tradition of scholarship. Many casuists have considered similar cases and frequently come up with similar conclusions, sometimes even by different routes of reasoning. Their opinions are to be respected and the divergent opinions that may also appear are to be carefully analyzed, in order to discern the grounds for the differences and to reconcile them, insofar as possible. If reconciliation is not possible and a genuine difference of defensible opinion remains, the questioner is free to act in accord with various opinions. Here the reputation of the competing casuists and the reasonableness of the various opinions must be weighed. In each casuistic tradition, rules for this weighing of opinion are devised.

Reference to the tradition of scholarly casuists involves reference to the cases that those scholarly predecessors and contemporaries have decided. The cases are, as noted above, "similar" to the instant case and their conclusions are "similar." Again, the casuistries of the three faiths utilize reasoning by analogy as an indispensable technique. Reasoning by analogy involves identification of a case that seems similar to the present case and then carefully distinguishing the precise ways in which the present case resembles the former one, in expectation of finding some feature that allows the casuist to claim that this previously decided case is similar enough to the present case that the prior decision, or consensus of decisions, should be taken as the key to the resolution of the present case. It is, however, only the key, for the circumstances of the present case are likely to differ from the former one, or the circumstances of the former one might he too sketchily described to allow direct comparison. Thus, the importance of the circumstances must be carefully specified and their bearing on diverse resolutions evaluated.

These are central activities of casuistry: use of analogical reasoning and evaluation of circumstances. They too are open to abuse, since the choice of appropriate analogies and the description of circumstances may be very subjective. Here, the recourse to tradition is of less value, since at this point in the analysis, the casuist is on his own. He has been able to check his subjectivity by placing himself within the long conversation of his casuistic colleagues, but only up to the point where he must choose the appropriate analogy and weigh the current circumstances. Here the "prudence" or "practical wisdom" of the casuist, his ability to consider fairly and comprehensively the circumstances of this case, in light of relevant princlples, becomes central. It is a virtue or talent of the experienced casuist. Casuistry, then, is both conservative and creative: it places cases within a tradition and then moves the tradition ahead by the decision in the present case. The primary check on unfettered ingenuity now becomes the response of the contemporary casuistic community, and the broader community of believers, to the new resolution of the new case.

Religious casuistry, then, is the practical reasoning of communities of faith. In the background always stand the scripture and traditional beliefs of the community. In the foreground stand the scholarly communities who interpret that scriptural and traditional background and teach it to the faithful. They use common rational methods of interpretation and, in addition, refer that interpretation to the preceding tradition of scholarship. They not only theorize but also respond to actual cases in which the faithful may face perplexity in acting according to their faith in current circumstance. They utilize to a high degree analogical reasoning, working from case to case, in order to reach a resolution that is both traditional and novel.

Bibliography

Aristotle. 1994: *The Nicomachean Ethics*, ed. H. Rackham. Cambridge, MA: Harvard University Press.

Audi, R. 1989: *Practical Reasoning*. London: Routledge.

Freehof, S. B. 1955: *The Responsa Literature*. Philadelphia, PA: Jewish Publication Society of America.

Gauthier, D. P. 1963: *Practical Reasoning: The Structure and Foundations of Prudential and Moral Arguments and their Exemplification in Discourse*. Oxford: Clarendon Press.

Hourani, G. F. 1971: *Islamic Rationalism: The Ethics of 'Abd al-Jabbar*. Oxford: Clarendon Press.

Jonsen, A. R. 1993: "Casuistical." *Common Knowledge* 2 (2). 48–66.

Jonsen, A. R. and Toulmin, S. E. 1988: *The Abuse of Casuistry: A History of Moral Reasoning*. Berkeley: University of California Press.

Keenan, J. F. and Shannon, T. A. (eds.) 1995: *The Context of Casuistry*. Washington, DC. Georgetown University Press.

Kelsay, J. and Johnson, J. T. (eds.) 1991: *Just War and Jihad: Historical and Theoretical Perspectives on War and Peace in Western and Islamic Traditions*. New York: Greenwood Press.

Khadduri, M. 1955: *War and Peace in the Law of Islam*. Baltimore. MD: Johns Hopkins University Press.

Kirk, K. E. 1999: *Conscience and Its Problems: An Introduction to Casuistry*. ed. D. H. Smith, Louisville, KY: John Knox/Westminster Press.

Kirschner, R. 1985: *Rabbinic Responsa of the Holocaust Era*. New York: Schocken Books.

Mahoney, J. 1987: *The Making of Moral Theology: A Study of the Roman Catholic Tradition*. Oxford: Oxford University Press.

Pascal, B. 1967: *The Provincial Letters*, trans. A. J. Krailsheimer. London: Penguin Books.

Raz, J. 1978: *Practical Reasoning*. New York: Oxford University Press.

Schacht, J. 1964: *An Introduction to Islamic Law*. Oxford: Clarendon Press.

Abraham's Obedience Tested
Genesis 22:1–12

Later on God tested Abraham's faith and obedience. "Abraham!" God called.

"Yes," he replied. "Here I am."

"Take your son, your only son—yes, Isaac, whom you love so much—and go to the land of Moriah. Sacrifice him there as a burnt offering on one of the mountains, which I will point out to you."

The next morning Abraham got up early. He saddled his donkey and took two of his servants with him, along with his son Isaac. Then he chopped wood to build a fire for a burnt offering and set out for the place where God had told him to go. On the third day of the journey, Abraham saw the place in the distance. "Stay here with the donkey," Abraham told the young men. "The boy and I will travel a little while farther. We will worship there, and then we will come right back."

Abraham placed the wood for the burnt offering on Isaac's shoulders, while he himself carried the knife and the fire. As the two of them went on together, Isaac said, "Father?"

"Yes, my son," Abraham replied.

"We have the wood and the fire," said the boy, "but where is the lamb for the sacrifice."

"God will provide a lamb, my son," Abraham answered. And they both went on together.

When they arrived at the place where God had told Abraham to go, he built an altar and placed the wood on it. Then he tied Isaac up and laid him on the altar over the wood. And Abraham took the knife and lifted it up to kill his son as a sacrifice to the Lord. At that moment the angel of the Lord shouted to him from heaven, "Abraham, Abraham."

"Yes," he answered. "I'm listening."

"Lay down the knife," the angel said. "Do not hurt the boy in any way, for now I know that you truly fear God. You have not withheld even your beloved son from me."

Part III
Traditions of Moral Reasoning in Western Culture

Part III

Part II explained how a military officer's oath embodies unwavering commitments to a distinctive set of moral principles, purposes and priorities. The Preamble to the U.S. Constitution declares that this Constitution was ordained "to establish justice . . . promote the general Welfare, and secure the Blessing of Liberty." Thus, the oath to serve the Constitution is simultaneously a commitment to act in the service of liberty and justice, and in behalf of the general Welfare. The Constitution, however, does not create the principles of justice, nor does it define liberty, or the common good. It presupposes, instead, that the performance of the government and its agents is to be evaluated by principles of justice, and by concepts of liberty and the common good that are prior to and independent of the laws created in and by the Constitution.

That is, our actions, and our government's actions (and even the Constitution itself) are subject to evaluation from the perspective of what the original "Founding Fathers" and the long succession of American political philosophers from Madison and Jefferson to Martin Luther King, Jr. describe as "higher laws" or "self-evident truths." Well, then: *just exactly what are* these "self-evident truths," higher laws, or bedrock moral principles?

In the preceding units on religion and morality, and on relativism, we have discovered that most normal individuals seem to have a fairly good sense of what is expected of them, and how they ought to behave toward one another. That is, they have a fairly robust sense of morality, and an innate ability to reason about moral dilemmas. Thus, most people recognize a moral principle of symmetry or reciprocity, as captured in teachings like the "Golden Rule," that we ought not to do things to other moral agents that we would not want done to ourselves. Thomas Jefferson likewise merely confirmed and modified slightly the recognition of the English political philosopher, John Locke, that every human being has certain inalienable "rights," including the right to life, liberty, and the pursuit of happiness.

Reference to ancient and venerable teachings like the Golden Rule, and to famous historical figures like Locke and Jefferson, however, suggests that there is a long and ongoing "conversation" down through the ages of which we are inheritors. That conversation concerns these very questions: how should we live? What traits of individual character should we cultivate, and what forms of behavior ought we to avoid? What are our obligations to ourselves and others? How might we best flourish as individuals and in society? What is the meaning of often-cited ideals such as human rights, justice, duty, liberty, or "the common good?" And how might we ensure that these worthy ideals (such as liberty and rights) are respected, or that these worthy goals (such as justice and the pursuit of happiness) are attained in our society?

This is the subject matter of moral philosophy and moral reasoning, to which we now briefly turn. Some people refer to the historical figures and their writings that we will consider in this unit as constituting a collection of "moral theories." Utilitarianism is often portrayed as one such theory, for example. Thus, its great advocate, the 19th-century English philosopher, John Stuart Mill, is understood as having propounded a "theory" about how to attain the greatest good for the greatest number. To this is contrasted a very different and opposing "theory," advocated by philosophers like the great German Enlightenment figure, Immanuel Kant, according to which we are to do our duty, regardless of the consequences. Meanwhile, the ancient Greek philosopher, Aristotle, is portrayed as regarding the essence of morality as the cultivation of the right "excellences" or habits of the heart, without regard either for duty or the common good.

These resulting "theories" are then set against one another as if they were competing options for the allegiance of students. As a consequence, moral philosophy down through the ages seems to its perplexed students as little more than a disjointed collection of contrasting editorial opinions, with little agreement or coherence. Kant, whose own difficult and demanding writing style has frustrated generations of students, nevertheless worried greatly about the negative impact of this philosophical wrangling on subsequent generations. He cautioned that philosophy, with its endless and seemingly irresolvable conflicts and contradictions might inspire what he termed *"misology,"* the *hatred of reason* itself.

This resulting anti-intellectualism would be tragic, Kant maintained, especially in the case of morality. For, unlike animals (who are presumably driven to action solely on the basis of uncontrollable and inscrutable desires), Kant believed the very quality that makes human beings unique and worthy of dignity and respect is their ability to reason about practical (as well as theoretical) dilemmas, and to set aside their individual desires on occasion in order to act, in a responsible manner, in behalf of rational and reasonable ends. That ability to deliberate and to choose responsibly, rather than the ability just to "do as I please," is what constitutes *freedom*.

If Kant is correct about this, then there is something very important at stake in attempting to study this ongoing philosophical conversation about living a moral life. *Doing so helps us to cultivate our freedom*—our ability to deliberate about complex cases, and thus to govern our own lives and behavior.

The contrasting "theories" (if that is even what they are)[1] do not serve primarily to tell us "right" from "wrong." In fact, most moral philosophy presumes that, as educated adults, we already know "right" from "wrong." We no more require a moral theory to tell us this than we require a theory of gravitation in order to walk! Instead, the theories or perspectives offered about morality help us understand *why* certain kinds of behavior, such as wanton killing or indiscriminate lying, are wrong. They also provide insight and encouragement into the positive inclinations we have to treat others with respect, and to do good things for ourselves and others when we are able, by explaining *why* these kinds of actions are good, and *how* they can lead to a useful and satisfying life.

Moral "theories" thus function in the practical world much like scientific theories in the theoretical disciplines. A theory of gravity doesn't teach us how to walk, for example, but explains (among many other things) why and how walking is possible, how it works. More importantly, contrasting theories of gravity (the classical and relativistic) help to explain why, under unusual circumstances, physical objects behave in unexpected ways, and help us better anticipate and respond appropriately in these complex situations. Likewise, contrasting traditions of moral reasoning help to explain, and sometimes aid in resolving conflicts and help us to determine what would be best under difficult and complex circumstances. In all cases, *intellectual understanding offers the promise of increasing our freedom* (or what moral philosophers term our *agency*, our ability to act as we ultimately, rationally, wish to act).

This would be especially important for present and future military officers to understand, inasmuch as we fund, equip, train and deploy military forces primarily to protect this precious commodity of freedom and the right of self-determination. Thus, the study of moral philosophy turns out to be of great value for military officers. By developing or improving our understanding of basic ethical concepts and the underlying structures of moral reasoning, we become better equipped to deliberate intelligently and resolve the diverse and often novel moral dilemmas we may confront.

Of course, when military leaders explain their decisions to their subordinates and superiors it is rarely useful to cite Aristotle, Mill or Kant as authorities! The point of studying these excerpts is surely not to enable young officers to "show off" by quoting a complex text or invoking one or another famous historical authority. It is also not possible, nor is it expected, that students will gain a mastery of the complicated points of view of each of these important figures merely by reading a short, representative sample of their work. Expertise in philosophy is no more the goal of this exercise than is "showing off!"

---

[1] A theory is usually a very specific, and highly coherent body of testable or falsifiable propositions concerning a specific subject—as in the theory of special relativity, quantum electrodynamics, or Stephen Jay Gould's theory of punctuated equilibrium as an account of evolution in biology. The moral philosophies we will study are not characterizable in this fashion: they are neither rigorous nor especially precise in their application, and thus it would be extremely difficult to disprove or "falsify" them. It is somewhat more humble, as well as more accurate, to view the several philosophical approaches to moral reasoning presented here as precisely that: approaches, traditions of analysis and application, provisional attempts at a general or universal formulation of underlying principles. Thus, we prefer to use the term "moral traditions," or "traditions of moral reasoning," for the views of Mill, Aristotle, Kant and others. It sounds less pretentious.

The point of studying these venerable and thoroughly-examined traditions of moral reasoning is rather to improve our overall ability to act wisely, judiciously, and well. By studying moral philosophy, a leader becomes better equipped to make sensible decisions, issue proper orders under duress, and to explain those decisions to others, including both those serving under their command, and also up the chain of command within which they themselves operate. Thoughtful, informed, well-read and well-educated individuals make better decisions, handle conflicts with more finesse, and so may avoid making the kinds of elementary blunders that can frustrate an important mission, ruin an honorable career, and often do immeasurably great harm to others. So it is, as John Paul Jones famously observed, that the qualifications for making an outstanding military officer *include a good liberal education*, together with "the nicest sense of personal honor." The study of moral philosophy makes a substantial contribution to both.

A. Utilitarianism and the Greatest Good

Utilitarianism is a formal name for a familiar, and quite common kind of practical reasoning, in which we first decide *how* to act, and then subsequently *justify* our choices and actions, by evaluating which of the feasible alternatives is likely to result in the greatest good, or best possible outcome in comparison with all the other options. Inasmuch as "the good" is a difficult notion to determine, and open to competing interpretations, utilitarian moral reasoning often ends up focusing instead on avoiding, or at least minimizing harm, such as suffering, pain, or loss of life. Thus, a simple formulation of utilitarian principles would be that we should always choose to act so as to avoid, or at least minimize, the amount of suffering or harm (or loss of life) our actions will cause. This fundamental principle discerning right from wrong actions is termed "The Principle of Utility." A much more familiar, if somewhat poorly defined and elusive formulation of this fundamental principle, is that one should always act so as to bring about "the greatest good for the greatest number."

Utilitarian reasoning is described explicitly in the moral philosophies of eighteenth-century English and Scottish philosophers, such as Francis Hutchinson, Adam Smith (the father of modern economic theory), and David Hume. It was popularized and advocated as a distinctive "moral theory," however, by Jeremy Bentham (1748–1832), a controversial legal scholar and social reformer in the late eighteenth and early nineteenth-century England. In his *Introduction to the Principles of Morals and Legislation* (1789), for example, Bentham argued that utilitarian reasoning should supplant religious beliefs, natural law (see Chapter IV: D, below), or custom and tradition as a guide to the formulation of appropriate laws and public policies. Bentham's utilitarianism, in particular, was put forward as the most rational and humane method of determining reasonable and appropriate forms of punishment to inflict upon lawbreakers. Against both time-honored practices and received prejudices, Bentham insisted on a single test of any law: does it, considering all its manifold effects on the happiness of the members of the community, do more net good than any alternative?

Appearing as it did in the same year as the French Revolution, Bentham's work engendered fear and hostility, and "Benthamites" (as his followers and advocates were known) were thought by their contemporaries to be radical, atheistic, unconventional, and largely untrustworthy agitators, bent on overturning the established social order. Bentham himself was a colorful and flamboyant figure who did little to assuage the fears of such critics. Under his influence, for example, his followers founded the University of London (University College), thereby making higher education available to the general populace without regard to social rank or religious belief, a thinly-veiled insult to the authority of the ancient and well-established religious colleges in Oxford and Cambridge. In a final, amusing act of eccentricity and of blatant disregard for the dictates of custom, good taste, and what he dismissed as "mere convention," moreover, Bentham ordered that, following his death, his body should be embalmed and placed prominently in a glass cabinet in the foyer of his university, so that the good-natured wrangling of the faculty would never occur in his absence! Thus, Mr. Bentham continues to greet students, faculty, and visitors to University College in London to this very day.

Bentham was a harsh critic of the conception, central to the French and American revolutions, of "natural rights." He regarded the concept of "rights" as illusory and nonsensical. Instead, he maintained, *morality should be like a science, and its judgments should proceed on the basis of logic and evidence*. Moral decision-making for Bentham involved a kind of "calculus," which he termed "hedonic calculus." Bentham identified seven key variables in this calculus of "pleasure and pain," including components such as duration, intensity, purity, and *extent of distribution* of the pleasure or happiness to be produced by a given action or policy. His utilitarian calculus demanded the simultaneous maximization of all seven variables, each measuring a key component the quantity of pleasure and pain produced by a given action, or social policy, in order to determine which among the available alternative actions or policies under consideration would serve to maximize the resulting amount of pleasure over pain.

That last variable is especially significant. Utilitarianism has been criticized ever since Bentham's time as, at least in principle, sanctioning immoral practices such as slavery, on the grounds that such a system might conceivably result in the greatest possible good for the society as a whole. Thus utilitarianism could conceivably approve of injustice. Bentham, however, explicitly disallowed such a perverse application of his "theory" by demanding that everyone affected by a policy must be shown to derive at least some benefit from it. As the great American philosopher, John Rawls, later demonstrated (Chapter V: B, below), this important constraint on the distribution of utilitarian benefits and burdens is a cornerstone of the concept of justice.

Utilitarianism's most celebrated advocate, however, was Jeremy Bentham's godson, the renowned nineteenth-century English philosopher, John Stuart Mill (1806–1873). Mill is widely regarded as the greatest and most influential philosopher in the English language during the nineteenth century. He had a remarkably rigorous education, supervised by Bentham and by Mill's father, James Mill, a close friend and dedicated "Benthamite." Mill's difficult, austere, and somewhat lonely adult life, his short but extremely happy marriage to Harriet Taylor, and the flourishing of his later work in social and political philosophy as a tribute to her following her death in 1858, are movingly portrayed in an elegantly written biography by Nicholas Capaldi (*John Stuart Mill: A Biography*: Cambridge University Press, 2003). These later writings range over a broad array of topics, from the unjustifiable subjugation of women, to the justification of military intervention. Mill's greatest and perhaps most influential work, *On Liberty* (1859), will form the cornerstone of our studies in Chapter V: A, below, on the concepts of liberty and human rights.

The Readings

We begin our consideration of utilitarian moral reasoning with a selection from Mill's classic, and widely-read pamphlet, *Utilitarianism* (1861). In this selection, the author defends the basic concept of utilitarian reasoning from criticisms and objections raised by its many 19th-century critics, who saw the doctrine of utilitarianism as an attack on traditional morality, and mistakenly thought it to recommend hedonism, atheism, and selfishness as the basis for moral action.

In response, Mill's defense of the concept of considering the balance of benefit over harm as a basis for moral reasoning is elegant, eloquent, and extremely clever. To those who object that utilitarian reasoning ignores the role of religion in morality, for example, Mill responds with a rather startling (and somewhat unorthodox) theological claim: that since God desires above all else the happiness of his creatures, God might well be thought to be the ultimate utilitarian. Indeed, he boldly asserts, the "Golden Rule" itself ought to be understood as a succinct formulation of utilitarian doctrine.

More plausibly, Mill moves in this reading selection to divorce utilitarian reasoning from the mechanical and somewhat simplistic theories about pleasure and pain advocated by his predecessors, and denounced as "hedonism" by utilitarianism's critics. Actions and policies differ not only in the quantity of pleasure and pain produced, Mill suggests, but also in the caliber or quality of their "goodness." To Aristotle's famous question—whether it would be better (all things considered) to be Socrates dissatisfied, than to be a pig, satisfied—Mill responds that the pig is not the final authority on the matter, as it is incapable of judging both sides of the question. Utilitarianism's desire to maximize the quantity of "goodness" in the world must, he maintains, reflect the judgment and the collective wisdom of human experience as to the sorts of actions that bring genuine, long-lasting, and satisfying pleasure and happiness.

Finally, in this reading, Mill addresses a hotly-disputed concern about the so-called Principle of Utility. The fundamental principle itself, according to Mill, defines an action as "right" or morally praiseworthy if it results in happiness, or "wrong" if it results in the opposite: unhappiness or misery.

Are we to apply this principle, individually, to each and every choice we face? Would this even be feasible? Mill retorts that the Principle of Utility is to be regarded like the *Nautical Almanac*, that is, as a framework of reference. The Principle of Utility, on Mill's account, is not a "first-order principle" that we use as a decision-guiding procedure on a case-by-case basis. Rather, it is a "second-order principle" that we consult when in doubt. The Principle of Utility provides the fundamental ground of moral justification to which we might refer, that is, when our normally reliable moral rules or moral intuitions (which are themselves to be understood as having been derived from that Principle) are confused or in conflict with one another. Thus, given that we usually know right from wrong, the Principle of Utility explains *why* an action or policy is right or wrong, and so serves to provide both a justification for living the moral life, and a means of resolving conflict, or addressing doubt or uncertainty when these arise.

Utilitarianism is an important perspective on moral reasoning, and, as West Point philosopher Louis Pojman indicates in his reading selection, it continues to receive further refinements to this day. It is, and has always been an important feature of decision-making in government and the military, as well as in the field of economics, in which public officials and military leaders are faced with implementing policies or making decisions in which the well-being (and even the life or death) of large numbers of individuals is at stake. Presumably, in such circumstances, the public official or military leader is obligated to minimize harm (e.g., overall suffering or loss of life) to the greatest extent possible, or to distribute benefits as widely and impartially as possible. Thus, U.S. President Harry S. Truman resorted to utilitarian criteria in defending his controversial decision to use nuclear weapons against Japan during World War II, on the grounds that doing so would, in the end, minimize the total number of deaths and the extent of destruction on both sides by avoiding the necessity of a prolonged, conventional armed invasion of the Japanese islands. Although Truman subsequently claimed that he never lost a night's sleep over this decision, the results of his very complex utilitarian calculations in this instance continue to provoke controversy to the present day.

Some of the variations of utilitarian reasoning, and problems in its application, are also discussed at some length in Pojman's article. In particular, Pojman warns (as does Mill himself) against a common fallacy in attempting to understand and apply utilitarianism, namely, that it is a doctrine of selfishness. In contrast to utilitarianism, which is strictly impartial regarding the distribution of happiness, *egoism* is the very different moral doctrine that we should make all moral decisions solely on the basis of our own individual happiness (or, by extension, the happiness only of those who matter to us).

Egoism is not difficult to understand psychologically, and many (including Socrates' friend Glaucon) have argued that, morality aside, this is how people *actually do* make moral choices. Egoism is, however, a very difficult doctrine to define or to justify, and it is certainly not to be equated with utilitarian reasoning. Finally, a dilemma regarding the fundamental principle of utilitarian reasoning, that the happiness of the greatest number is the most important single consideration, is challenged in the fictional story, "The Ones who Walk Away from Omelas."

Utilitarianism*
John Stuart Mill

Chapter 1: General Remarks

Little progress has been made towards deciding the controversy concerning *the criterion of right and wrong*. From the dawn of philosophy the question concerning the foundation of morality, has

- ·been regarded as the main problem in speculative thought,
- ·occupied the most gifted intellects, and
- ·divided them into sects and schools, vigorously warring against one another.

And after more than two thousand years the same discussions continue! Philosophers still line up under the same opposing battle-flags, and neither thinkers nor people in general seem to be any nearer to being unanimous on the subject than when young Socrates listened to old Protagoras and asserted the theory of utilitarianism against the popular morality of the so-called 'sophist'.

But it wouldn't be hard to show that whatever steadiness or consistency mankind's moral beliefs *have* achieved has been mainly due to the silent influence of a standard that hasn't been ·consciously· recognised. In the absence of an acknowledged first principle, ethics has been not so much a •guide to men in forming their moral views as a •consecration of the views they actually have; but men's views—both for and against—are greatly influenced by what *effects on their happiness* they suppose things to have; and so the principle of utility—or, as Bentham eventually called it, 'the greatest happiness principle'—has had a large share in forming the moral doctrines even of those who most scornfully reject its authority. And *every* school of thought admits that the influence of actions on happiness is a very significant and even predominant consideration in many of the details of morals, however unwilling they may be to allow the production of happiness as the fundamental principle of morality and the source of moral obligation.

Chapter 2: What Utilitarianism Is

The doctrine that the basis of morals is utility, or the greatest happiness principle, holds that actions are right in proportion as they tend to promote happiness, wrong in proportion as they tend to produce the reverse of happiness. By 'happiness' is meant pleasure and the absence of pain; by 'unhappiness' is meant pain and the lack of pleasure. To give a clear view of the moral standard set up by the theory, much more needs to be said, especially about what things the doctrine includes

---

* Copyright © Jonathan Bennett. Last amended: April 2008 [Brackets] enclose editorial explanations. Small ·dots· enclose material that has been added, but can be read as though it were part of the original text. Occasional •bullets, and also indenting of passages that are not quotations, are meant as aids to grasping the structure of a sentence or a thought. Every four-point ellipsis . . . . indicates the omission of a brief passage that seems to present more difficulty than it is worth. Further editing by Lawrence Lengbeyer.

in the ideas of pain and pleasure, and to what extent it leaves this as an open question. But these supplementary explanations don't affect the theory of life on which this theory of morality is based—namely the thesis that

> pleasure and freedom from pain are the only things that are *desirable as ends*, and that
>
> everything that is *desirable at all* is so either •for the pleasure inherent in it or •as means to the promotion of pleasure and the prevention of pain.

Now, such a theory of life arouses utter dislike in many minds, including some that are among the most admirable in feeling and purpose. The view that life has (as they express it) no higher end—no better and nobler object of desire and pursuit—than *pleasure* they describe as utterly mean and grovelling, a doctrine worthy only of pigs. The followers of Epicurus were contemptuously compared with pigs, very early on, and modern holders of the utilitarian doctrine are occasionally subjected to equally polite comparisons by its German, French, and English opponents.

•Higher and Lower Pleasure•

When attacked in this way, the Epicureans have always answered that it is not they but their accusers who represent human nature in a degrading light, because the accusation implies that human beings are capable only of pleasures that pigs are also capable of. If this were true, there'd be no defence against the charge, but then it wouldn't be a *charge*; for if the sources of pleasure were precisely the same for humans as for pigs, the rule of life that is good enough for *them* would be good enough for *us*. The comparison of the Epicurean life to that of beasts is felt as degrading precisely because a beast's pleasures do *not* satisfy a human's conceptions of happiness. Human beings have •higher faculties than the animal appetites, and once they become conscious of •them they don't regard anything as happiness that doesn't include •their gratification. Admittedly the Epicureans were far from faultless in drawing out the consequences of the utilitarian principle; to do this at all adequately one must include—which they didn't—many Stoic and some Christian elements. But every Epicurean theory of life that we know of assigns to the •pleasures of the intellect, of the feelings and imagination and of the moral sentiments a much higher value as pleasures than to •those of mere sensation. But it must be admitted that when utilitarian writers have said that mental pleasures are better than bodily ones they have mainly based this on mental pleasures being more permanent, safer, less costly and so on—i.e. from their circumstantial advantages rather than from their intrinsic nature. And on all these points utilitarians have fully proved their case; but they *could*, quite consistently with their basic principle, have taken the other route—occupying the higher ground, as we might say. It is quite compatible with the principle of utility to recognize that some kinds of pleasure are more desirable and more valuable than others. In estimating ·the value of· anything else, we take into account •quality as well as •quantity; it would be absurd if the value of pleasures were supposed to depend on •quantity alone.

'What do you mean by "difference of quality in pleasures"? What, according to you, makes one pleasure •more valuable than another, merely *as a pleasure*, if not its being •greater in amount?' There is only one possible answer to this.

> Pleasure P1 is more desirable than pleasure P2 if: all or almost all people who have had experience of both give a decided preference to P1, irrespective of any feeling that they *ought* to prefer it.

If those who are competently acquainted with both these pleasures place P1 so far above P2 that •they prefer it even when they know that a greater amount of discontent will come with it, and •wouldn't give it up in exchange for any quantity of P2 that they are capable of having, we are justified in ascribing to P1 a superiority in *quality* that so greatly outweighs *quantity* as to make quantity comparatively negligible.

Now, it is an unquestionable fact that the way of life that employs the higher faculties is strongly preferred ·to the way of life that caters only to the lower ones· by people who are equally acquainted with both and equally capable of appreciating and enjoying both. Few human creatures would agree to be changed into any of the lower animals in return for a promise of the fullest allowance of animal pleasures;

- no intelligent human being would consent to be a fool,
- no educated person would prefer to be an ignoramus,
- no person of feeling and conscience would rather be selfish and base,

even if they were convinced that the fool, the dunce or the rascal is better satisfied with his life than they are with theirs. . . . If they ever think they *would*, it is only in cases of unhappiness so extreme that to escape from it they would exchange their situation for almost any other, however undesirable they may think the other to be. Someone with higher faculties •requires more to make him happy, •is probably capable of more acute suffering, and •is certainly vulnerable to suffering at more points, than someone of an inferior type; but in spite of these drawbacks he can't ever really wish to sink into what he feels to be a lower grade of existence. Explain this unwillingness how you please! We may attribute it to

- *pride*, a name that is given indiscriminately to some of the most and to some of the least admirable feelings of which human beings are capable;
- the *love of liberty and personal independence* (for the Stoics, that was one of the most effective means for getting people to value the higher pleasures); or
- the *love of power*, or the *love of excitement*, both of which really do play a part in it.

But the most appropriate label is *a sense of dignity*. All human beings have this sense in one form or another, and how strongly a person has it is roughly proportional to how well endowed he is with the higher faculties. In those who have a strong sense of dignity, their dignity is so essential to their happiness that they couldn't want, for more than a moment, anything that conflicts with it.

Anyone who thinks that this preference takes place at a sacrifice of happiness—anyone who denies that the superior being is, other things being anywhere near equal, happier than the inferior one—is confusing two very different ideas, those of *happiness* and of *contentment*. It is true of course that the being whose capacities of enjoyment are low has the greatest chance of having them fully satisfied ·and thus of being contented·; and a highly endowed being will always feel that any happiness that he can look for, given how the world is, is imperfect. But he can learn to bear its imperfections, if they are at all bearable; and they won't make him envy the person who isn't conscious of the imperfections only because he has no sense of the good that those imperfections are imperfections *of*— ·for example, the person who isn't bothered by the poor quality of the conducting because he doesn't enjoy music anyway·. It is better to be a human being dissatisfied than a pig satisfied; better to be Socrates dissatisfied than a fool satisfied. And if the fool or the pig think otherwise, that is because they know only their own side of the question. The other party to the comparison knows both sides.

'But many people who are capable of the higher pleasures do sometimes, under the influence of temptation, give preference to the lower ones.' Yes, but this is quite compatible with their fully appreciating the intrinsic superiority of the higher. Men's infirmity of character often leads them to choose the *nearer* good over the *more valuable* one; and they do this just as much when •it's a choice between two bodily pleasures as when •it is between a bodily pleasure and a mental one. They pursue sensual pleasures at the expense of their health, though they are perfectly aware that health is the greater good, ·doing this because the sensual pleasures are *nearer*·.

'Many people who begin with youthful enthusiasm for everything noble, as they grow old sink into laziness and selfishness.' Yes, this is a very common charge; but I don't think that those who undergo it voluntarily choose the lower kinds of pleasures in preference to the higher. I believe that before they devote themselves exclusively to the lower pleasures they have already become incapable of the higher ones. In most people a capacity for the nobler feelings is a very tender plant that is easily killed, not only by hostile influences but by mere lack of nourishment; and in the majority of young persons it quickly dies away if their jobs and their social lives aren't favourable to keeping that higher capacity in use. Men lose their high aspirations as they lose their intellectual tastes, because they don't have time or opportunity for indulging them; and they addict themselves to lower pleasures not because they deliberately prefer them but because they are either •the only pleasures they can get or the only pleasures they can still enjoy. It may be questioned whether anyone who has remained equally capable of both kinds of pleasure has ever knowingly and calmly preferred the lower kind; though throughout the centuries many people have broken down in an ineffectual attempt to have both at once.

I don't see that there can be any appeal against this verdict of the only competent judges! On a question as to which is the better worth having of two pleasures, or which of two ways of life is the more agreeable to the feelings (apart from its moral attributes and from its consequences), the judgment of those who are qualified by knowledge of both must be admitted as final—or, if they differ among themselves, the judgment of the majority among them. And we can be encouraged to accept this judgment concerning the *quality* of pleasures by the fact that there is no other tribunal to appeal to even on the question of *quantity*. What means do we have for deciding which is the more acute of two pains, or the more intense of two pleasurable sensations, other than the collective opinion of those who are familiar with both? ·Moving back now from quantity to quality·: there are different kinds of pain and different kinds of pleasure, and every pain is different from every pleasure. What can decide whether a particular ·kind of· pleasure is worth purchasing at the cost of a particular ·kind of· pain, if not the feelings and judgment of those who are experienced ·in both kinds·? When, therefore, those feelings and judgments declare the pleasures derived from the higher faculties to be preferable in kind, apart from the question of intensity, to those that can be enjoyed by animals that don't have the higher faculties, their opinion on this subject too should be respected.

I have dwelt on this point because you need to understand it if you are to have a perfectly sound conception of utility or happiness, considered as the governing rule of human conduct. But you could rationally accept the utilitarian standard without having grasped ·that people who enjoy the higher pleasures are happier than those who don't·. That's because the utilitarian standard is not •the agent's own greatest happiness but •the greatest amount of happiness altogether; and even if it can be doubted whether a noble character is always happier because of its nobleness, there can be no doubt that such a character makes other people happier, and that the world in general gains immensely from its existence. So utilitarianism would achieve its end only through the general cultivation of nobleness of character, even if

> each individual got benefit only from the nobleness of others, with his own nobleness serving to *reduce* his own happiness.

But mere *statement* of this last supposition brings out its absurdity so clearly that there is no need for me to argue against it.

·Happiness as an Aim·

According to the greatest happiness principle as I have explained it, the ultimate end . . . . , for the sake of which all other things are desirable (whether we are considering our own good or that of other people) is an existence as free as possible from pain and as rich as possible in enjoyments. This means rich in •quantity and in •quality; the test of •quality, and the rule for measuring it against •quantity, being the preferences of those who are best equipped to make the comparison—equipped, that is, by the range of their experience and by their habits of self-consciousness and self-observation. If the greatest happiness of all is (as the utilitarian opinion says it is) •the end of human action, is must also be •the standard of morality; which can therefore be defined as:

> the rules and precepts for human conduct such that: the observance of them would provide the best possible guarantee of an existence such as has been described—for all mankind and, so far as the nature of things allows, for the whole sentient creation.

...

·Self-Sacrifice·

Only while the world is in a very imperfect state can it happen that anyone's best chance of serving the happiness of others is through the absolute sacrifice of his own happiness; but while the world is in that imperfect state, I fully admit that the readiness to make such a sacrifice is the highest virtue that can be found in man. . . .

... The utilitarian morality *does* recognize that human beings can sacrifice their own greatest good for the good of others; it merely refuses to admit that the sacrifice is *itself* a good. It regards as *wasted* any sacrifice that doesn't increase, or tend to increase, the sum total of happiness. The only self-renunciation that it applauds is devotion to the happiness, or to some of the means to happiness, of others. ...

I must again repeat something that the opponents of utilitarianism are seldom fair enough to admit, namely that the happiness that forms the utilitarian standard of what is right in conduct is not •the agent's own happiness but •that of all concerned. As between his own happiness and that of others, utilitarianism requires him to be as strictly impartial as a disinterested and benevolent spectator. [Here and everywhere Mill uses 'disinterested' in its still-correct meaning = 'not self-interested' = 'not swayed by any consideration of how the outcome might affect one's own welfare'.] In the golden rule of Jesus of Nazareth we read the complete spirit of the ethics of utility. To

> do as you would be done by, and to
> love your neighbour as yourself

constitute the ideal perfection of utilitarian morality. As the ·practical· way to get as close as possible to this ideal, ·the ethics of· utility would command two things. **(1)** First, laws and social arrangements should place the happiness (or what for practical purposes we may call the *interest*) of every individual as much as possible in harmony with the interest of the whole. **(2)** Education and opinion, which have such a vast power over human character, should use that power to establish in the mind of every individual an unbreakable link between •his own happiness and •the good of the whole; especially between •his own happiness and •the kinds of conduct (whether *doing* or *allowing*) that are conducive to universal happiness. If **(2)** is done properly, it will tend to have two results: **(2a)** The individual won't be able to *conceive* the possibility of being personally happy while acting in ways opposed to the general good. **(2b)** In each individual a direct impulse to promote the general good will be one of the habitual motives of action, and the feelings connected with it will fill a large and prominent place in his sentient existence. This is the true character of the utilitarian morality. If those who attack utilitarianism see it as being like *this*, I don't know •what good features of some other moralities they could possibly say that utilitarianism lacks, •what more beautiful or more elevated developments of human nature any other ethical systems can be supposed to encourage, or •what motivations for action that aren't available to the utilitarian those other systems rely on for giving effect to their mandates.

·Setting the Standard too High?·

The objectors to utilitarianism can't be accused of always representing it in a •discreditable light. On the contrary, objectors who have anything like a correct idea of its disinterested character sometimes find fault with utilitarianism's standard as being •too high for humanity. To require people always to act from the •motive of promoting the general interests of society—that is demanding too much, they say. But this is to mistake the very meaning of a standard of morals, and confuse the •rule of action with the •motive for acting. It is the business of ethics to tell us what are our duties, or by what test we can know them; but no system of ethics requires that our only motive in everything we do shall be a feeling of duty; on the contrary, ninety-nine hundredths of all our actions are done from other motives, and *rightly* so if the •rule of duty doesn't condemn them. It is especially unfair to utilitarianism to object to it on the basis of this particular misunderstanding, because utilitarian moralists have gone beyond almost everyone in asserting that the motive has nothing to do with the *morality of the action* though it has much to do with the *worth of the agent*. He who saves a fellow creature from drowning does what is morally right, whether his motive is duty or the hope of being paid for his trouble; he who betrays a friend who trusts him is guilty of a crime, even if his aim is to serve another friend to whom he is under greater obligations.

Let us now look at actions that are done from the motive of duty, in direct obedience to ·the utilitarian· principle: it is a misunderstanding of the utilitarian way of thinking to conceive it as implying that people should fix their minds on anything as wide as *the world* or *society in general*. The great majority of good actions are intended not for •the benefit of the world but for *parts* of the good of the world, namely •the benefit of individuals. And on these occasions the thoughts of the most virtuous man need not go beyond the particular persons concerned, except to the extent that he has to assure himself that in benefiting those individuals he isn't violating the rights (i.e. the legitimate and authorised expectations) of anyone else. According to the utilitarian ethics the object of virtue is to multiply happiness; for any person (except one in a thousand) it is only on exceptional occasions that he has it in his power to do this on an extended scale, i.e. to be a public benefactor; and it is only on these occasions that he is called upon to consider public utility; in every other case he needs to attend only to private utility, the interest or happiness of some few persons. The only people who need to concern themselves regularly about so large an object as *society in general* are those ·few· whose actions have an influence that extends that far. ·Thoughts about the *general* welfare do have a place in everyone's moral thinking· in the case of *refrainings*—things that people hold off from doing, for moral reasons, though the consequences in the particular case might be beneficial. The thought in these cases is like this: 'If I acted in that way, my action would belong to a class of actions which, if practised generally, would be generally harmful, and for that reason I ought not to perform it.' It would be unworthy of an intelligent agent not to be consciously aware of such considerations. But the amount of regard for the public interest implied in this kind of thought is no greater than is demanded by every system of morals, for they all demand that one refrain from anything that would obviously be pernicious to society; ·so there is no basis here for a criticism of utilitarianism in particular·.

·Is Utilitarianism Chilly?·

The same considerations dispose of another reproach against the doctrine of utility, based on a still grosser misunderstanding of the purpose of a standard of morality and of the very meanings of the words 'right' and 'wrong'. It is often said that utilitarianism •makes men cold and unsympathising; that it •chills their moral feelings towards individuals; that it •makes them attend only to

- the dry and hard consideration of the consequences of actions, leaving out of their moral estimate
- the ·personal· qualities from which those actions emanate.

If this means that they don't allow their judgment about the rightness or wrongness of an action to be influenced by their opinion of the qualities of the person who does it, this is a complaint not against •utilitarianism but against •having any standard of morality at all; for certainly no known ethical standard declares that an action is good or bad because it is done by a good or a bad man, still less because it is done by a lovable, brave or benevolent man, or by an unfriendly, cowardly or unsympathetic one. These considerations ·of personal virtue· are relevant to how we estimate *persons*, not *actions*; and the utilitarian theory in no way conflicts with the fact that there are other things that interest us in persons besides the rightness and wrongness of their actions. The Stoics, indeed, with the paradoxical misuse of language which was part of their system and by which they tried to raise themselves to a level at which their only concern was with *virtue*, were fond of saying that he who has virtue has everything; that it is the virtuous man, and *only* the virtuous man, who is rich, is beautiful, is a king. But the utilitarian doctrine doesn't make any such claim on behalf of the virtuous man. Utilitarians are well aware that there are other desirable possessions and qualities besides virtue, and are perfectly willing to allow to all of them their full worth. They are also aware that •a right action doesn't necessarily indicate a virtuous character, and that •actions that are blamable often come from ·personal· qualities that deserve praise. When this shows up in any particular case, it modifies utilitarians' estimation not of the act but of the agent. They do hold that in the long run the best proof of a good character is good actions; and they firmly refuse to consider any mental disposition as good if its predominant tendency is to produce bad conduct. This, which I freely grant, makes utilitarians unpopular with many people; but this is an unpopularity that they must share with everyone who takes seriously the distinction between right and wrong; and the criticism is not one that a conscientious utilitarian need be anxious to fend off.

If the objection means only this:

> Many utilitarians look on the morality of actions, as measured by the utilitarian standard, in too exclusive a manner, and don't put enough emphasis on the other beauties of character that go towards making a human being lovable or admirable,

this may be admitted. Utilitarians who have cultivated their moral feelings but not their sympathies or their artistic perceptions do fall into this mistake; and so do all other moralists under the same conditions. What can be said in excuse of other moralists is equally available for utilitarians, namely that if one is to go wrong about this, it is better to go wrong on that side, ·rather than caring about lovability etc. and ignoring the morality of actions·. As a matter of fact, utilitarians are in this respect like the adherents of other systems: there is every imaginable degree of rigidity and of laxity in how they apply their standard ·of right and wrong·: some are puritanically rigorous, while others are as forgiving as any sinner or sentimentalist could wish! But on the whole, a doctrine that highlights the interest that mankind have in the repression and prevention of conduct that violates the moral law is likely to do as good a job as any other in turning the force of public opinion again such violations. It is true that the question 'What *does* violate the moral law?' is one on which those who recognise different standards of morality are likely now and then to differ. But ·*that* isn't a point against utilitarianism·; difference of opinion on moral questions wasn't first introduced into the world by utilitarianism! And that doctrine does supply a tangible and intelligible way—if not always an easy one—of deciding such differences.

·Utilitarianism as 'Godless'·

It may be worthwhile to comment on a few more of the common misunderstandings of utilitarian ethics, even those that are so obvious and gross that it might appear impossible for any fair and intelligent person to fall into them. ·It might *appear* impossible but unfortunately it *isn't*·: the crudest misunderstandings of ethical doctrines are continually met with in the deliberate writings of persons with the greatest claims both to high principle and to philosophy. That is because people—even very able ones—often take little trouble to understand the likely influence of any opinion against which they have a prejudice, and are unaware of this deliberate ignorance as a defect. We quite often hear the doctrine of utility denounced as a godless doctrine. If this mere *assumption* needs to be replied to at all, we may say that the question depends on what idea we have formed of the moral character of the Deity. If it is true that God desires the happiness of his creatures above all else, and that this was his purpose in creating them, then utilitarianism, far from being a godless doctrine, is the most deeply religious of them all. If the accusation is that utilitarianism doesn't recognise the *revealed* will of God as the supreme law of morals, I answer that a utilitarian who believes in the perfect goodness and wisdom of God has to believe that whatever God has thought fit to reveal on the subject of morals must fulfil the requirements of utility in a supreme degree. Others besides utilitarians have held this:

> The Christian revelation was intended (and is fitted) to bring into the hearts and minds of mankind a spirit that will enable them to find *for themselves* what is right, and incline them to do right when they have found it; rather than to tell them—except in a very general way—what it is. And we need a doctrine of ethics, carefully followed out, to know what the will of God is.

We needn't discuss here whether this is right; because whatever aid religion—either natural or revealed—can provide to ethical investigation is as open to the utilitarian moralist as to any other. He is as entitled to cite it as God's testimony to the usefulness or hurtfulness of a course of action as others are to cite it as pointing to a transcendental law that has no connection with usefulness or happiness.

·Expediency·

Again, utilitarianism is often slapped down as an immoral doctrine by giving it the name 'Expediency', and taking advantage of the common use of that term to contrast it with 'Principle'. But when 'expedient' is opposed to 'right', it usually means *what is expedient for the particular interest of the agent himself*, as when a high official sacrifices the interests of his country in order to keep himself in place. When it means anything better than this, it means *what is expedient for some immediate temporary purpose, while violating a rule whose observance is much* more *expedient*. The 'expedient' in *this* sense, instead of being the same thing as the •useful, is a branch of the •hurtful. For example, telling a lie would often be *expedient* for escaping some temporary difficulty or getting something that would be immediately useful to ourselves or others. But **(1)** the principal support of all present social well-being is people's ability to trust one another's assertions, and the lack of that trust does more than anything else to keep back civilisation, virtue, everything on which human happiness on the largest scale depends. Therefore (2) the development in ourselves of a sensitive feeling about truthfulness is one of the most useful things that our conduct can encourage, and the weakening of that feeling is one of the most harmful. Finally, (3) any deviation from truth—even an unintentional one—does something towards weakening the trustworthiness of human assertion. For these reasons we feel that (4) to obtain an immediate advantage by violating such an overwhelmingly expedient rule is *not expedient*, and that someone who acts in that way does his bit towards depriving mankind of the good, and inflicting on them the harm, involved in the greater or less reliance that they can place in each other's word, thus acting as though he were one of mankind's worst enemies. Yet all moralists agree that even this rule ·about telling the truth·, sacred as it is, admits of possible exceptions. The chief one is the case where the withholding of some fact from someone would save an individual (especially someone other than oneself) from great and undeserved harm, and the only way of withholding it is to lie about it. (Examples: keeping information ·about the whereabouts of a weapon· from a malefactor ·who would use it·, or keeping bad news from a person who is dangerously ill.) But in order that this exception ·to the truth-telling rule· doesn't extend itself beyond the need for it, and has the least possible effect of weakening reliance on truth-telling, it ought to be ·openly and explicitly· *recognised*, and if possible its limits should be defined; and if the principle of utility is good for anything, it must be good for weighing these conflicting utilities against one another, and marking out the region within which one or the other dominates.

·Time to Calculate?·

Again, defenders of utility often find themselves challenged to reply to such objections as this: 'Before acting, one doesn't have *time* to calculate and weigh the effects on the general happiness of any line of conduct.' This is just like saying: 'Before acting, one doesn't have time on each occasion to read through the Old and New Testaments; so it is impossible for us to guide our conduct by Christianity.' The answer to the objection is that *there has been plenty of time*, namely, the whole past duration of the human species. During all that time, mankind have been learning by •experience what sorts of consequences actions are apt to have, this being something on which all the morality of life depends, as well as all the prudence [= 'decisions about what will further one's own interests']. The objectors talk as if the start of this course of •experience had been put off until now, so that when some man feels tempted to meddle with the property or life of someone else he has to start at that moment considering *for the first time* whether murder and theft are harmful to human happiness! Even if that were how things stand, I don't think he would find the question very puzzling. . . .

If mankind were agreed in considering utility to be the test of morality, they would *of course*—it would be merely fanciful to deny it—reach some agreement about what is useful, and would arrange for their notions about this to be taught to the young and enforced by law and opinion. Any ethical standard whatever can easily be 'shown' to work badly if we suppose •*universal idiocy* to be conjoined with it! But on any hypothesis short of •that, mankind must by this time have acquired positive beliefs as to the effects of some actions on their happiness; and the beliefs that have thus come down ·to us from the experience of mankind· are *the rules of morality* for the people in general—and for the philosopher until he succeeds in finding something better. I admit, or rather I strongly assert, that

- philosophers might easily find something better, even now, on many subjects; that
- the accepted code of ethics is not God-given; and that
- mankind have still much to learn about how various kinds of action affect the general happiness.

The corollaries from the principle of utility, like the rules of every practical art, can be improved indefinitely, and while the human mind is progressing they are constantly improving.

But to consider the intermediate rules of morality as unprovable is one thing; to pass over them entirely, trying to test each individual action directly by the first principle, is another. It is a strange notion that having a •first principle is inconsistent with having •secondary ones as well. When you tell a traveller the location of the place he wants to get to, you aren't forbidding him to use landmarks and direction-posts along the way! The proposition that *happiness is the end and aim of morality* doesn't mean that no road ought to be laid down to that goal, or that people going to it shouldn't be advised to take one direction rather than another. Men really ought to stop talking a kind of *nonsense* on this subject - nonsense that they wouldn't utter or listen to with regard to any other practically important matter. Nobody argues that the art of navigation is not based on astronomy because sailors can't wait to calculate the Nautical Almanack. Because they are rational creatures, sailors go to sea with the calculations already done; and *all* rational creatures go out on the sea of life with their minds made up on the common questions of *right and wrong*, as well as on many of the much harder questions of *wise and foolish*. And we can presume that they will continue to do so long as foresight continues to be a human quality. Whatever we adopt as the fundamental principle of morality, we need subordinate principles through which to apply it; the absolute need for them is a feature of *all* ·moral· systems, so it doesn't support any argument against any one system in particular. To argue solemnly in a manner that presupposes this:

> No such secondary principles can be had; and mankind never did and never will draw any general conclusions from the experience of human life

is as totally absurd, I think, as anything that has been advanced in philosophical controversy.

·Bad Faith·

The remainder of the standard arguments against utilitarianism mostly consist in blaming it for •the common infirmities of human nature and •the general difficulties that trouble conscientious persons when they are shaping their course through life. We are told that a utilitarian will be apt to make his own particular case an exception to moral rules; and that when he is tempted ·to do something wrong· he will see more utility in doing it than in not doing it. But is utility the only morality that can provide us with excuses for evil doing, and means of cheating our own conscience? ·Of course not!· Such excuses are provided in abundance by •all doctrines that recognise the existence of conflicting considerations as a fact in morals; and this is recognized by every doctrine that any sane person has believed. It is the fault not •of any creed but •of the complicated nature of human affairs that rules of conduct can't be formulated so that they require no exceptions, and hardly any kind of action can safely be stated to be either always obligatory or always condemnable.

Every ethical creed softens the rigidity of its laws by giving the morally responsible agent some •freedom to adapt his behaviour to special features of his circumstances; and under every creed, at the •opening thus made, self-deception and dishonest reasoning get in. *Every* moral system allows for clear cases of conflicting obligation. These are real difficulties, knotty points both in the •theory of ethics and in the •practical personal matter of living conscientiously. In practice they are overcome, more or less successfully, according to the person's intellect and virtue; but it can't be claimed that having an ultimate standard to which conflicting rights and duties can be referred will make one *less* qualified to deal with them! If utility is the basic source of moral obligations, utility can be invoked to decide between obligations whose demands are incompatible. The ·utility· standard may be hard to apply, but it is better than having no standard. In other systems, the moral laws all claim independent authority, so that there's no common umpire entitled to settle conflicts between them; when one of them is claimed to have precedence over another, the basis for this is little better than sophistry, allowing free scope for personal desires and preferences (unless the conflict is resolved by the unadmitted influence of considerations of utility). It is only in these cases of conflict between secondary principles that there is any need to appeal to first principles. In every case of moral obligation some secondary principle is involved; and if there is only one, someone who recognizes that principle can seldom be in any real doubt as to which one it is.

Utilitarianism
Louis P. Pojman

... [A]s a moral philosophy, utilitarianism begins with the work of Scottish philosophers Frances Hutcheson (1694–1746), David Hume (1711–1776), and Adam Smith (1723–1790); and comes into its classical stage in the writings of English social reformers Jeremy Bentham (1748–1832) and John Stuart Mill (1806–1873). They were the nonreligious ancestors of the 20th-century secular humanists, optimistic about human nature and our ability to solve our problems without recourse to providential grace. Engaged in a struggle for legal as well as moral reform, they were impatient with the rule-bound character of law and morality in 18th and 19th-century Great Britain and tried to make the law serve human needs and interests.

Bentham's concerns were mostly practical rather than theoretical. He worked for a thorough reform of what he regarded as an irrational and outmoded legal system. He might well have paraphrased Jesus, making his motto "Morality and Law were made for man, not man for Morality and Law." What good was adherence to outworn deontological rules that served no useful purpose, that only kept the poor from enjoying a better life, and that supported punitive codes that served only to satisfy sadistic lust for vengeance?

The changes the utilitarians proposed were not done in the name of justice, for—they believed—even justice must serve the human good. The poor were to be helped, women were to be liberated, and criminals were to be rehabilitated if possible, not in the name of justice, but because doing so could bring about more utility: ameliorate suffering and promote more pleasure or happiness.

The utilitarian view of punishment is a case in point. Whereas deontologists believe in retribution—that all the guilty should be punished in proportion to the gravity of their crime—the utilitarians' motto is "Don't cry over spilt milk!" They believe that the guilty should be punished only if the punishment would serve some deterrent (or preventive) purpose. Rather than punish John in exact proportion to the heinousness of his deed, we ought to find the right punishment that will serve as the optimum deterrent.

The proper amount of punishment to be inflicted upon the offender is the amount that will do the most good (or the least harm) to all those who will be affected by it. The measure of harm inflicted on the criminal, John, should be preferable to the harm avoided by setting that particular penalty rather than one slightly lower. If punishing John will do no good (because John is not likely to commit the crime again and no one will be deterred by the punishment), then John should go free.

It is the *threat* of punishment that is the important thing. Every *act* of punishment is an admission of the failure of the threat; if the threat were successful, there would be no punishment to justify. Of course, utilitarians believe that, given human failing, punishment is vitally necessary as a deterrent, so that the guilty will seldom if ever be allowed to go free.

There are two main features of utilitarianism: the *consequentialist principle* (or its teleological aspect) and the *utility principle* (or its hedonic aspect). The consequentialist principle states that the rightness or wrongness of an act is determined by the goodness or badness of the results that flow from it. It is the end, not the means, that counts; the end justifies the means. The utility principle states that the only thing that is good in Itself is some specific type of state (e.g., pleasure, happiness, welfare). Hedonistic utilitarianism views pleasure as the sole good and pain as the only evil. To quote Bentham, the first one to systematize classical utilitarianism, "Nature has placed mankind under the governance of two sovereign masters, pain and pleasure. It is for them alone to point out what we ought to do, as well as what we shall do."[1] An act is right if it either brings about more pleasure than pain or prevents pain, and an act is wrong if it either brings about more pain than pleasure or prevents pleasure from occurring.

Bentham invented a scheme for measuring pleasure and pain that he called the *hedonic calculus*: The quantitative score for any pleasure or pain experience is obtained by summing the seven aspects of a pleasurable or painful experience: its intensity, duration, certainty, nearness, fruitfulness, purity, and extent. Adding up the amounts of pleasure and pain for each possible act and then comparing the scores would enable us to decide which act to perform. With regard to our example of deciding between giving the dying man's money to the Yankees or to the East African famine victims, we would add up the likely pleasures to all involved, for all seven qualities. If we find that giving the money to the famine victims will cause at least 3 million *hedons* (units of happiness) but that giving the money to the Yankees will cause less than 1,000 hedons, we would have an obligation to give the money to the famine victims.

There is something appealing about Bentham's utilitarianism. It is simple in that there is only one principle to apply: Maximize pleasure and minimize suffering. It is commonsensical in that we think that morality really is about ameliorating suffering and promoting benevolence. It is scientific: Simply make quantitative measurements and apply the principle impartially, giving no special treatment to ourselves or to anyone else because of race, gender, or religion.

However, Bentham's philosophy may be too simplistic in one way and too complicated in another. It may be too simplistic in that there are other values than pleasure (as we saw in the previous chapter), and it seems too complicated in its artificial hedonic calculus. The calculus is encumbered with too many variables and has problems assigning scores to the variables. For instance, what score do we give a cool drink on a hot day or a warm shower on a cool day? How do we compare a 5-year-old's delight over a new toy with a 50-year-old's delight with a new lover? Can we take your second car from you and give it to Beggar Bob, who does not own a car and would enjoy it more than you? And if it's simply the overall benefits of pleasure that we are measuring, then if Jack or Jill would be "happier" in the Pleasure Machine or the Happiness Machine or on drugs than in the "real world," would we not have an obligation to ensure that these conditions obtain? Because of such considerations, Bentham's version of utilitarianism was, even in his own day, referred to as the "pig-philosophy," since a pig enjoying his life would constitute a higher moral state than a slightly dissatisfied Socrates.

It was to meet these sorts of objections and save utilitarianism from the charge of being a pig-philosophy that Bentham's brilliant successor, John Stuart Mill, sought to distinguish happiness from mere sensual pleasure. His version of utilitarianism—*eudaimonistic* (from the Greek *eudaimonia*, meaning "happiness") utilitarianism—defines happiness in terms of certain types of higher-order pleasures or satisfactions, such as intellectual, aesthetic, and social enjoyments, as well as in terms of minimal suffering. That is, there are two types of pleasures: the lower, or elementary (e.g., eating, drinking, sexuality, resting, and sensuous titillation), and the higher (e.g., intellectuality, creativity, and spirituality). Though the lower pleasures are more intensely gratifying, they also lead to pain when overindulged in. The spiritual, or achieved, pleasures tend to be more protracted, continuous, and gradual.

Mill argues that the higher, or more refined, pleasures are superior to the lower ones: "A being of higher faculties requires more to make him happy, is capable probably of more acute suffering, and certainly accessible to it at more points, than one of an inferior type," but still he is qualitatively better off than the person without these higher faculties. "It is better to be a human being dissatisfied than

---

[1] Jeremy Bentham, *An Introduction to the Principles of Morals and Legislation* (1789), Ch 1; reprinted in Louis Pojman, ed., *Ethical Theory* (Wadsworth, 1989) pp. 111–114.

a pig satisfied; better to be Socrates dissatisfied than a fool satisfied."[2] Humans are the kind of creatures who require more to be truly happy. They want the lower pleasures, but they also want deep friendship, intellectual ability, culture, ability to create and appreciate art, knowledge, and wisdom.

But, one may object, how do we know that it really is better to have these higher pleasures? Here Mill imagines a panel of experts, and says that, of those who have had wide experience of pleasures of both kinds almost all give a decided preference to the higher type. Since Mill was an *empiricist*—one who believed that all knowledge and justified belief was based in experience—he had no recourse but to rely on the composite consensus of human history. By this view, people who experience both rock music and classical music will, if they appreciate both, prefer Bach and Beethoven to the Rolling Stones or the Dancing Demons. That is, we generally move up from appreciating simple things (e.g., nursery rhymes) to more complex and intricate things (e.g., poetry that requires great talent) rather than the other way around.

Mill has been criticized for not giving a better reply—for being an elitist and for unduly favoring the intellectual over the sensual. But he has a point. Don't we generally agree, if we have experienced both the lower and the higher types of pleasure, that even though a full life would include both, a life with only the former is inadequate for human beings? Isn't it better to be Socrates dissatisfied than the pig satisfied—and better still to be Socrates satisfied?

The point is not merely that humans would not be satisfied with what satisfies a pig, but that somehow the quality of these pleasures is *better*. But what does it mean to speak of better pleasure? Is Mill assuming some non-hedonic notion of intrinsic value to make this distinction—that is, that knowledge, intelligence, freedom, friendship, love, health, and so forth are good things in their own right? Or is Mill simply saying that the lives of humans are generally such that they will be happier with more developed, refined, spiritual values? Which thesis would you be inclined to defend?

The formula he comes up with, finally, is the following:

> Happiness . . . [is] not a life of rapture; but moments of such, in all existence made up of few and transitory pains, many and various pleasures, with a decided predominance of the active over the passive, and having as the foundation of the whole, not to expect more from life than it is capable of bestowing.[3]

It does seem that intellectual activity, autonomous choice, and other non-hedonic qualities supplement the notion of pleasure.

Two Types of Utilitarianism

There are two classical types of utilitarianism: *act-* and *rule-utilitarianism*. In applying the principle of utility, act-utilitarians, such as Bentham, say that ideally we ought to apply the principle to all of the alternatives open to us at any given moment. We may define act-utilitarianism in this way:

> *Act-utilitarianism:* An act is right if and only if it results in as much good as any available alternative.

Of course, we cannot do the necessary calculations to determine which act is the correct one in each case, for often we must act spontaneously and quickly. So rules of thumb (e.g., "In general don't lie," and "Generally keep your promises") are of practical importance. However, the right act is still that alternative that results in the most utility.

The obvious criticism of act-utility is that it seems to fly in the face of fundamental intuitions about minimally correct behavior. Consider Richard Brandt's criticism of act-utilitarianism:

> It implies that if you have employed a boy to mow your lawn and he has finished the job and asks for his pay, you should pay him what you promised only if you cannot find a better use for your money. It implies that when you bring home your monthly paycheck you should use it to support your family and yourself only if it cannot be

---

[2] John Stuart Mill, *Utilitarianism* (1863), Ch II; reprinted in Pojman, ed., *Ethical Theory,* p. 165.
[3] Mill, *Utilitarianism,* in Pojman, ed., *Ethical Theory,* pp, 166-167.

used more effectively to supply the needs of others. It implies that if your father is ill and has no prospect of good in his life, and maintaining him is a drain on the energy and enjoyments of others, then, if you can end his life without provoking any public scandal or setting a bad example, it is your positive duty to take matters into your own hands and bring his life to a close.[4]

Rule-utilitarians like Brandt attempt to offer a more credible version of the theory. They state that an act is right if it conforms to, a valid rule within a system of rules that, if followed, will result in the best possible state of affairs (or the least bad state of affairs, if it is a question of all the alternatives being bad). We may define rule-utilitarianism this way:

> *Rule-utilitarianism:* An act is right if and only if it is required by a rule that is itself a member of a set of rules whose acceptance would lead to greater utility for society than any available alternative.

An oft-debated question in ethics is whether rule-utilitarianism is a consistent version of utilitarianism. Briefly, the argument that rule-utilitarianism is an inconsistent version that must either become a deontological system or transform itself into act-utilitarianism goes like this: Imagine that following the set of general rules of a rule-utilitarian system yields 100 hedons (positive utility units). We could always find a case where breaking the general rule would result in additional hedons without decreasing the sum of the whole. So, for example, we could imagine a situation in which breaking the general rule "Never lie" in order to spare someone's feelings would create more utility (e.g., 102 hedons) than keeping the rule would. It would seem that we could always improve on any version of rule-utilitarianism by breaking the set of rules whenever we judge that by so doing we could produce even more utility than by following the set.

One way of resolving the difference between act- and rule-utilitarians is to appeal to the notion of *levels of rules.* For the sophisticated utilitarian there will be three levels of rules to guide actions. On the lowest level is a set of utility-maximizing rules of thumb that should always be followed unless there is a conflict between them, in which case a second-order set of conflict-resolving rules should be consulted. At the top of the hierarchy is the *remainder rule* of act utilitarianism: When no other rule applies, simply do what your best judgment deems to be the act that will maximize utility.

An illustration of this might be the following: Two of our lower-order rules might be "Keep your promises" and "Help those in need when you are not seriously inconvenienced in doing so." Suppose you promised to meet your teacher at 3 P.M. in his office. On your way there you come upon an accident victim stranded by the wayside who desperately needs help. It doesn't take you long to decide to break the appointment with your teacher, for it seems obvious in this case that the rule to help others overrides the rule to keep promises. We might say that there is a second order rule prescribing that the first-order rule of helping people in need when you are not seriously inconvenienced in doing so overrides the rule to keep promises. However, there may be some situation where no obvious rule of thumb applies. Say you have $50 you don't really need now. How should you use this money? Put it into your savings account? Give it to your favorite charity? Use it to throw a party? Here and only here, on the third level, the general act-utility principle applies without any other primary rule; that is, do what in your best judgment will do the most good.

It is a subject of keen debate whether John Stuart Mill was a rule- or an act-utilitarian. He doesn't seem to have noticed the difference, and there seem to be aspects of both theories in his work. Philosophers like J. J. C. Smart and Kai Nielsen hold views that are clearer examples of act-utilitarianism. Nielsen attacks what he calls *moral conservatism,* which is any

> normative ethical theory which maintains that there is a privileged moral principle or cluster of moral principles, prescribing determinate actions, with which it would always be wrong not to act in accordance no matter what the consequences.

---

[4]Richard Brandt, "Towards a Credible Form of Utilitarianism," in H. Castaneda and G. Naknikian, eds., *Morality and the Language of Conduct* (Wayne State University Press, 1963), pp. 109–110.

For Nielsen, no rules are sacrosanct; but differing situations call forth different actions, and potentially any rule could be overridden (though in fact we may need to treat some as absolutes for the good of society).

Nielsen's argument in favor of utilitarianism makes strong use of the notion of *negative responsibility*: We are responsible not only for the consequences of our actions, but also for the consequences of our nonactions. Suppose that you are the driver of a trolley car and suddenly discover that your brakes have failed. You are just about to run over five workmen on the track ahead of you. However, if you act quickly, you can turn the trolley onto a side track where only one man is working. What should you do? One who makes a strong distinction between active and passive evil (*allowing* versus *doing* evil) would argue that you should do nothing and merely allow the trolley to kill the five men, but one who denies that this is an absolute distinction would prescribe that you do something positive in order to minimize evil. Negative responsibility means that you are going to be responsible for someone's death in either case. Doing the right thing, the utilitarian urges, means minimizing the amount of evil. So you should actively cause the one death in order to save the other five lives.[5]

Critics of utilitarianism contend either that negative responsibility is not a strict duty or that it can be worked into other systems besides utilitarianism.

---

[5]Kai Nielsen, "Against Moral Conservatism," *Ethics* (1972), reprinted in Pojman, ed., *Ethical Theory*. Nielsen no longer holds the position espoused in this article; J. J. C. Smart, "Outlines of a System of Utilitarian Ethics" in J. J. C. Smart and Bernard Williams, *Utilitarianism: For and Against* (Cambridge University Press, 1973).

The Ones Who Walk Away from Omelas
*Ursula K. Le Guin**

With a clamor of bells that set the swallows soaring, the Festival of Summer came to the city. Omelas, bright-towered by the sea. The rigging of the boats in harbor sparkled with flags. In the streets between houses with red roofs and painted walls, between old moss-grown gardens and under avenues of trees, past great parks and public buildings, processions moved. Some were decorous: old people in long stiff robes of mauve and grey, grave master workmen, quiet, merry women carrying their babies and chatting as they walked. In other streets the music beat faster, a shimmering of gong and tambourine, and the people went dancing, the procession was a dance. Children dodged in and out, their high calls rising like the swallows' crossing flights over the music and the singing. All the processions wound towards the north side of the city, where on the great water-meadow called the Green Fields boys and girls, naked in the bright air, with mud-stained feet and ankles and long, lithe arms, exercised their restive horses before the race. The horses wore no gear at all but a halter without bit. Their manes were braided with streamers of silver, gold, and green. They flared their nostrils and pranced and boasted to one another; they were vastly excited, the horse being the only animal who has adopted our ceremonies as his own. Far off to the north and west the mountains stood up half encircling Omelas on her bay. The air of morning was so clear that the snow still crowning the Eighteen Peaks burned with white-gold fire across the miles of sunlit air, under the dark blue of the sky. There was just enough wind to make the banners that marked the racecourse snap and flutter now and then. In the silence of the broad green meadows one could hear the music winding through the city streets, farther and nearer and ever approaching, a cheerful faint sweetness of the air that from time to time trembled and gathered together and broke out into the great joyous clanging of the bells.

Joyous! How is one to tell about joy? How describe the citizens of Omelas?

They were not simple folk, you see, though they were happy. But we do not say the words of cheer much any more. All smiles have become archaic. Given a description such as this one tends to make certain assumptions. Given a description such as this one tends to look next for the King, mounted on a splendid stallion and surrounded by his noble knights, or perhaps in a golden litter borne by great-muscled slaves. But there was no king. They did not use swords, or keep slaves. They were not barbarians. I do not know the rules and laws of their society, but I suspect that they were singularly few. As they did without monarchy and slavery, so they also got on without the stock exchange, the advertisement, the secret police, and the bomb. Yet I repeat that these were not simple folk, not dulcet shepherds, noble savages, bland utopians. They were not less complex than us. The trouble is that we have a bad habit, encouraged by pedants and sophisticates, of considering happiness as something rather stupid. Only pain is intellectual, only evil interesting. This is the treason of the artist: a refusal to admit the banality of evil and the terrible boredom of pain. If you can't lick 'em, join 'em. If it hurts, repeat it. But to praise despair is to condemn delight, to embrace violence is to lose hold of everything

---

*"The Ones Who Walk Away from Omelas" by Ursula K. Le Guin. Copyright © 1973 by Ursula K. Le Guin first appeared in *New Dimensions 3*; reprinted by permission of the author and the author's agent, Virginia Kidd.

else. We have almost lost hold; we can no longer describe a happy man, nor make any celebration of joy. How can I tell you about the people of Omelas? They were not naive and happy children—though their children were, in fact, happy. They were mature, intelligent, passionate adults whose lives were not wretched. O miracle! But I wish I could describe it better. I wish I could convince you. Omelas sounds in my words like a city in a fairy tale, long ago and far away, once upon a time. Perhaps it would be best if you imagined it as your own fancy bids, assuming it will rise to the occasion, for certainly I cannot suit you all. For instance, how about technology? I think that there would be no cars or helicopters in and above the streets; this follows from the fact that the people of Omelas are happy people. Happiness is based on a just discrimination of what is necessary, what is neither necessary nor destructive, and what is destructive. In the middle category, however—that of the unnecessary but undestructive, that of comfort, luxury, exuberance, etc.—they could perfectly well have central heating, subway trains, washing machines, and all kinds of marvelous devices not yet invented here, floating light-sources, fuelless power, a cure for the common cold. Or they could have none of that: it doesn't matter. As you like it. I incline to think that people from towns up and down the coast have been coming in to Omelas during the last days before the Festival on very fast little trains and double-decked trams and that the train station of Omelas is actually the handsomest building in town, though plainer than the magnificent Farmers' Market. But even granted trains, I fear that Omelas so far strikes some of you as goody-goody. Smiles, bells, parades, horses, bleh. If so, please add an orgy. If an orgy would help, don't hesitate. Let us not, however, have temples from which issue beautiful nude priests and priestesses already half in ecstasy and ready to copulate with any man or woman, lover or stranger, who desires union with the deep godhead of the blood, although that was my first idea. But really it would be better not to have any temples in Omelas—at least, not manned temples. Religion yes, clergy no. Surely the beautiful nudes can just wander about, offering themselves like divine soufflés to the hunger of the needy and the rapture of the flesh. Let them join the processions. Let tambourines be struck above the copulations, and the glory of desire be proclaimed upon the gongs, and (a not unimportant point) let the offspring of these delightful rituals be beloved and looked after by all. One thing I know there is none of in Omelas is guilt. But what else should there be? I thought at first there were no drugs, but that is puritanical. For those who like it, the faint insistent sweetness of *drooz* may perfume the ways of the city, *drooz*, which first brings a great lightness and brilliance to the mind and limbs, and then after some hours a dreamy languor, and wonderful visions at last of the very arcana and inmost secrets of the Universe, as well as exciting the pleasure of sex beyond all belief; and it is not habit-forming. For more modest tastes I think there ought to be beer. What else, what else belongs in the joyous city? The sense of victory, surely, the celebration of courage. But as we did without clergy; let us do without soldiers. The joy built upon successful slaughter is not the right kind of joy; it will not do; it is fearful and it is trivial. A boundless and generous contentment, a magnanimous triumph felt not against some outer enemy but in communion with the finest and fairest in the souls of all men everywhere and the splendor of the world's summer: this is what swells the hearts of the people of Omelas, and the victory they celebrate is that of life. I really don't think many of them need to take *drooz*.

Most of the processions have reached the Green Fields by now. A marvelous smell of cooking goes forth from the red and blue tents of the provisioners. The faces of small children are amiably sticky; in the benign grey beard of a man a couple of crumbs of rich pastry are entangled. The youths and girls have mounted their horses and are beginning to group around the starting line of the course. An old woman, small, fat, and laughing, is passing out flowers from a basket, and tall young men wear her flowers in their shining hair. A child of nine or ten sits at the edge of the crowd, alone, playing on a wooden flute. People pause to listen, and they smile, but they do not speak to him, for he never ceases playing and never sees them, his dark eyes wholly rapt in the sweet, thin magic of the tune.

He finishes, and slowly lowers his hands holding the wooden flute.

As if that little private silence were the signal, all at once a trumpet sounds from the pavillion near the starting line: imperious, melancholy, piercing. The horses rear on their slender legs, and some of them neigh in answer. Sober-faced, the young riders stroke the horses' necks and soothe them, whispering, "Quiet, quiet, there my beauty, my hope. . . ." They begin to form in rank along the starting line. The crowds along the racecourse are like a field of grass and flowers in the wind. The Festival of Summer has begun.

Do you believe? Do you accept the festival, the city, the joy? No? Then let me describe one more thing.

In a basement under one of the beautiful public buildings of Omelas, or perhaps in the cellar of one of its spacious private homes, there is a room. It has one locked door, and no window. A little light seeps in dustily between cracks in the boards, secondhand from a cobwebbed window somewhere across the cellar. In one corner of the little room a couple of mops, with stiff, clotted, foul-smelling heads, stand near a rusty bucket. The floor is dirt, a little damp to the touch, as cellar dirt usually is. The room is about three paces long and two wide: a mere broom closet or disused tool room. In the room a child is sitting. It could be a boy or a girl. It looks about six, but actually is nearly ten. It is feeble-minded. Perhaps it was born defective, or perhaps it has become imbecile through fear, malnutrition, and neglect. It picks its nose and occasionally fumbles vaguely with its toes or genitals, as it sits hunched in the corner farthest from the bucket and the two mops. It is afraid of the mops. It finds them horrible. It shuts its eyes, but it knows the mops are still standing there; and the door is locked; and nobody will come. The door is always locked; and nobody ever comes, except that sometimes— the child has no understanding of time or interval—sometimes the door rattles terribly and opens, and a person, or several people, are there. One of them may come in and kick the child to make it stand up. The others never come close, but peer in at it with frightened, disgusted eyes. The food bowl and the water jug are hastily filled, the door is locked, the eyes disappear. The people at the door never say anything, but the child, who has not always lived in the tool room, and can remember sunlight and its mother's voice, sometimes speaks. "I will be good," it says. "Please let me out. I will be good!" They never answer. The child used to scream for help at night, and cry a good deal, but now it only makes a kind of whining, "eh-haa, eh-haa," and it speaks less and less often. It is so thin there are no calves to its legs; its belly protrudes; it lives on a half-bowl of corn meal and grease a day. It is naked. Its buttocks and thighs are a mass of festered sores, as it sits in its own excrement continually.

They all know it is there, all the people of Omelas. Some of them have come to see it, others are content merely to know it is there. They all know that it has to be there. Some of them understand why, and some do not, but they all understand that their happiness, the beauty of their city, the tenderness of their friendships, the health of their children, the wisdom of their scholars, the skill of their makers, even the abundance of their harvest and the kindly weathers of their skies, depend wholly on this child's abominable misery.

This is usually explained when they are between eight and twelve, whenever they seem capable of understanding; and most of those who come to see the child are young people, though often enough an adult comes, or comes back, to see the child. No matter how well the matter has been explained to them, these young spectators are always shocked and sickened at the sight. They feel disgust, which they had thought themselves superior to. They feel anger, outrage, impotence, despite all the explanations. They would like to do something for the child. But there is nothing they can do. If the child were brought up into the sunlight out of that vile place, if it were cleaned and fed and comforted, that would be a good thing, indeed; but if it were done, in that day and hour all the prosperity and beauty and delight of Omelas would wither and be destroyed. Those are the terms. To exchange all the goodness and grace of every life in Omelas for that single, small improvement: to throw away the happiness of thousands for the chance of the happiness of one: that would be to let guilt within the walls indeed.

The terms are strict and absolute; there may not even be a kind word spoken to the child.

Often the young people go home in tears, or in a tearless rage, when they have seen the child and faced this terrible paradox. They may brood over it for weeks or years. But as time goes on they begin to realize that even if the child could be released, it would not get much good of its freedom: a little vague pleasure of warmth and food, no doubt, but little more. It is too degraded and imbecile to know any real joy. It has been afraid too long ever to be free of fear. Its habits are too uncouth for it to respond to humane treatment. Indeed, after so long it would probably be wretched without walls about it to protect it, and darkness for its eyes, and its own excrement to sit in. Their tears at the bitter injustice dry when they begin to perceive the terrible justice of reality and to accept it. Yet it is their tears and anger, the trying of their generosity and the acceptance of their helplessness, which are perhaps the true source of the splendor of their lives. Theirs is no vapid, irresponsible happiness. They know that they, like the child, are not free. They know compassion. It is the existence of the child, and their knowledge of its existence, that makes possible the nobility of their architecture, the poignancy of their music, the profundity of their science. It is because of the child that they are so gentle with children. They know that if the wretched one were not there snivelling in the dark, the other one, the flute-player, could make no joyful music as the young riders line up in their beauty for the race in the sunlight of the first morning of summer.

Now do you believe in them? Are they not more credible? But there is one more thing to tell, and this is quite incredible.

At times one of the adolescent girls or boys who go to see the child does not go home to weep or rage, does not, in fact, go home at all. Sometimes also a man or woman much older falls silent for a day or two, and then leaves home. These people go out into the street, and walk down the street alone. They keep walking, and walk straight out of the city of Omelas, through the beautiful gates. They keep walking across the farmlands of Omelas. Each one goes alone, youth or girl, man or woman. Night falls; the traveler must pass down village streets, between the houses with yellow-lit windows, and on out into the darkness of the fields. Each alone, they go west or north, toward the mountains. They go on. They leave Omelas, they walk ahead into the darkness, and they do not come back. The place they go towards is a place even less imaginable to most of us than the city of happiness. I cannot describe it at all. It is possible that it does not exist. But they seem to know where they are going, the ones who walk away from Omelas.

B. Kantian Ethics and the Basis of Duty

The German Enlightenment philosopher, Immanuel Kant (1724–1804), is regarded as among the greatest and most influential of Western philosophers, and also undeniably as one of the most difficult to read and understand. The difficulty in reading the dense and complex structure of Kant's formal writings (as opposed to his posthumously-published lecture notes for his college students, or essays for a broader public audience) stems in part from the fact that this solitary, disciplined figure—who lived his entire life as a bachelor university professor in a tiny, remote town on the eastern edge of the Baltic Sea—insisted on devoting his life to understanding and (where necessary) re-conceptualizing the foundations of scientific knowledge and practical, moral reasoning.

The two were not, and in fact are not, as unrelated as they seem to us today. Kant's earliest essays, and the prefaces to his major (so-called "Critical") works all reveal that he regarded moral reasoning as on par with mathematical reasoning. Both are essential expressions of the nature of Reason or rationality itself. Right off the bat, it is hard to imagine an attitude more at odds with contemporary thinking about the historically-conditioned, cultural grounding of morality and the limitations on rationality characteristic of intellectual fashions in the twenty-first century. We now tend to think of the natural sciences as utterly distinct from practical, moral reasoning, and have even come to entertain occasional doubts concerning the impartiality and authenticity of scientific reasoning itself.

For Kant, by contrast, our abilities to delve into the workings of nature, solve mathematical equations, and discern with utmost certainty our moral duties—including our social and political responsibilities to establish justice, equality, and bring about a cosmopolitan rule of law that would make war itself unnecessary—were of a piece, constituting essential elements in the life of Reason. Both the achievements of reason in science and mathematics as well as its promise in establishing an international political order that would one day be altogether free of war, together conferred a special dignity on humanity itself. Two things there are, he once remarked, that inspire our awe: "the starry heavens above, and *the Moral Law within*." It is this attitude of quiet confidence in the power and possibilities of reason that characterize the meaning of the phrase, "Enlightenment" in western history, and Kant is justifiably regarded as the modern European Enlightenment's chief spokesman and exemplar.

Kant was inspired in part by the historical example of the great Polish astronomer, Copernicus (Mikolai Kopernika), whose own birthplace, Torún, lay only about 200 kilometers south of Kant's home town of Königsberg, in what was then known as "East Prussia." Copernicus had almost single-handedly brought about a fundamental revolution in the way we understood our place in the universe—a conceptual shift in perspective that simplified considerably the mathematical models that astronomers had used for centuries to record and predict the motions of celestial bodies. The revolutionary "heliocentric" hypothesis of Copernicus, however, also required that human beings give up the notion of our occupying a "privileged place" as the "unmoved center" of that universe—and surrender, along with this, the most basic observations and intuitions of common sense, that we "stand still" while the sun, moon, and planets revolve around us.

Kant held, however, that the dignity of humanity did not derive from this false belief in our *privileged location in* the universe, but instead derived from our *privileged perspective on* the universe. This perspective is the perspective of Reason. It is through the power of Reason alone that otherwise insignificant creatures like ourselves can successfully probe the secrets of nature, and, of even greater importance, can devise reasonable rules or principles that should govern our individual and collective lives, and then proceed (and this is the key notion) to *choose as individuals to subject ourselves voluntarily* to the rules and laws that Reason has thus legislated. It is this fundamental insight regarding the nature of Reason and morality that we find enshrined in the reading selection below, from Kant's brief, dense, seminal work, *Groundwork of the Metaphysics of Morals* (1783). It emerges, in particular, in the so-called "Third Form" of the "Categorical Imperative of Reason" (that Kant terms "the Kingdom of Ends") from which, he claims, the preceding two (and much better known) forms of that moral principle are derived, the "Formula of Universal Law," and the "Formula of Humanity as an End in Itself."

Thus, like Copernicus, Kant believed that a fundamental, revolutionary reorientation of the manner in which we understood ourselves as rational beings was urgently required in order to explain how we are apparently able both to gain knowledge of the external world, and to effect or "legislate" our rational moral will within it. Copernicus had been preoccupied with the inconsistencies in what was known as Ptolemaic astronomy, the conceptual scheme that still forms the cornerstone of modern celestial navigation to this very day. The inconsistencies that plagued Kant, by comparison, were to explain how reason could prove itself so monumentally successful in gaining reliable knowledge of the natural world when practiced within the disciplines of the natural sciences and mathematics, and yet find itself so hopelessly mired in confusion and contradiction when that very same Reason turned to other, perfectly "reasonable" questions such as the origin of the cosmos, the nature and existence of God, and the perennial debates concerning, for example, whether human nature is free or is, like the physical world, determined solely by laws of cause and effect (determinism).

In the Preface of his great work, *The Critique of Pure Reason* (1781), Kant expressed grave concern that if he could not resolve these apparent inconsistencies, subsequent generations of students might grow discouraged with Reason itself, and come to doubt the certainty of both mathematics and the Moral Law. Hence, he believed, only a careful, foundational study of Reason as it applied itself to these different areas could bring about what he termed a "Copernican revolution" in our understanding of Reason itself, and of Reason's place (and its limitations) in the scheme of human life and human affairs.

The revolutionary perspective that Kant advocated was this: that human reason is not passive, but highly active in developing our understanding and judgment about our world. With respect to what he termed "pure reason" (the kind of disciplined thought characteristic of mathematics and the natural sciences) our reason actively organizes the complex welter of sense impressions each of us receives into a coherent, organized picture or scheme of events. Because Reason works the same way in each of us, the result is an organized "picture" or operational appearance of nature to the human mind. Because this picture or perspective is a collective product or "synthesis" of reason and experience, we who are rational beings will thus always be able to agree in common on what each of us perceives as the basic, organizing principles of that natural world.

These basic organizing principles themselves are, in turn, what we know as the "laws of nature" that constituted the crowning achievements of Newtonian or classical physics—a subject which Kant taught, along with mathematics, astronomy, philosophy, logic, and arguably the first college-level course in the western world on the topic we now call "anthropology." The common or generic features of pure reason itself, considered independently, are the elements of thought that give rise to arithmetic, algebra, logic, and Euclidean geometry. These features in turn help us understand why these areas of "pure thought" apart from experience are so successful when "applied" to the world of normal experience, and are almost magically able to "model" or "interpret" it unambiguously. It is not, as Aristotle believed, that these forms or patterns are "in" natural things; rather, it is the case that these mathematical forms or patterns are modes of rational thought that we, as rational beings, superimpose upon our perceptions of the natural world, thereby bringing order out of chaos. This remarkable consensus and certainty, however, fail to materialize when we attempt to turn our reason loose on problems that transcend, or lie beyond, the realm of normal experience (as when we argue, for example, about the existence of God).

What has all this to do with morality? Reason, working in the practical realm of forming intentions and strategies for achieving our normal wants and desires (Kant calls these strategies "maxims" of behavior) operates in much the same manner when deliberating about our resulting practical choices and responsibilities as it does in the natural sciences. That is, *Reason superimposes its form, order, and discipline* upon the welter of disorganized and frequently conflicting desires that each of us entertains. The resulting impartial ordering of desire, issuing in commands or obligations, is what we experience as morality, as (in Kant's phrase) the imperatives of duty.

As living creatures, we all experience a wide range of varying and often conflicting wants and desires—for food, power, comfort, sexual satisfaction, or security. If we merely responded to these desires and acted to attain them, we would be merely driven by them, and our behavior would be accurately described as having been *causally determined*, almost mechanically, by our desires. Kant in fact thought, rightly or wrongly, that animals *are* largely driven by their desires in this fashion, and accordingly, cannot be said to be free or "autonomous." We recall that Socrates' friend, Glaucon, back in the beginning of this book, expressed his belief that, when all is said and done, human beings themselves are (or would be) little better than selfish, driven animals in this respect when freed from all accountability and social constraints.

What most of us experience in attempting to live our moral lives, however, is not just the compulsion of desire, but something more like a *conflict between* these individual desires, on one hand, and something else, on the other: not merely a sense of *what each of us wants to do*, but also a sense of *what any one of us ought to do*. That is, we feel a conflict between desire and what Kant calls "our duty." Much of the time, fortunately, our normal, healthy desires do not conflict with the normal duties our Reason lays upon us, for example, not to kill wantonly, or to lie, steal, cheat, break promises, or behave in other immoral ways. But sometimes our desires, like those of the shepherd, Gyges, do clearly conflict with what we acknowledge to be our duty. If we then choose to forego our individual, immediate desires and conform to the requirements of duty or obligation, then we have chosen to act morally.

That choice, however, still may not be entirely free: that is, we may choose to do our duty out of fear of getting caught or punished if we should choose to do otherwise, or we may reasonably expect others to notice our "selfless" and noble acts, and hold us in high esteem. As the Gyges story illustrates, however, we are in deep trouble if our individual behavior is solely dependent upon such external constraints, for these can sometimes "disappear" or cease to function as constraints on our behavior. For Kant, we could hardly be described as "acting freely" or "autonomously" if *only* the external fear of punishment or hope of reward is motivating our actions.

The Readings

Thus, Kant advances in the first portion of our reading selection what seems to most readers a very puzzling point. Granted, we all presumably agree that actions which are "contrary to duty"—that is, actions in behalf of our individual wants and desires that are in direct conflict with the normal constraints that impartial Reason places upon our individual actions—are, by definition, immoral (and so Kant begins by stating he won't even bother with evaluating these kinds of actions). But what of actions, such as those above, that "conform to duty," in which we seem to act *in accordance with duty* out of fear of punishment, or hope of recognition or reward?

The answer Kant offers is, "that depends upon our reason, or motivation, for acting." If, like Gyges, we can only be counted upon to "do our duty" when others are watching, and when the normal social systems of accountability are in place, then we (once again like Gyges) might be tempted to "go wild" and act solely on the basis of our selfish desires whenever those external forms of accountability disappeared. Perhaps, Kant speculates, this is what the human condition is, but even if so, we can hardly be described as having acted in a morally worthy or praiseworthy manner, simply because we happen to do our duty.

Instead, he claims, an action has moral worth only if it is done not merely in accordance with duty, "but for the sake of duty." This puzzling phrase, however, amounts to a very straightforward claim: we are acting in a morally worthy fashion, and possess what Kant terms "a morally good will," only if we are the kind of individuals who can be counted upon to do what we know we must and ought to do, even when there are no external forms of incentive or accountability in place. This, then, is Kant's answer to Glaucon's timeless challenge: morality is not necessarily an illusion, or "merely" a social construct. Instead, the morally worthy individual (if such a person exists) does what he or she *ought* to do, whether he feels like it or not, or whether anyone else notices, cares, rewards, praises, or blames or punishes her or not.

Put in this way, this strange-sounding demand is not quite as odd as it seems. We might conclude, correctly, that Kant is demanding that *all of us act* as we expect military officers routinely to act: to do our duty, no matter what. But Kant is, in addition, *demanding that all of us, including military officers, act freely, of our own accord*—autonomously—when we honor and obey our obligations, or when we place obligation before desire. Morality is not to be understood as some alien social construction, imposed upon us against our will. We do not simply do our duty because we *have to*, we are to do our duty because *we choose to*. Reason alone makes it clear to us both what our universal moral obligations are (e.g., not to lie, cheat, kill or steal), and Reason also demands that we should always place our universal obligations ahead of our individual desires whenever the two conflict. Because human beings experience this conflict, and individually have the power to choose between duty and desire, it is consequently self-evident (what Kant himself terms a "fact of Reason") that human nature is free and not causally determined by external circumstances.

In this astonishing fashion, Kant has "solved" the age-old quarrel about freedom and determinism, by equating freedom with Reason, with our ability to deliberate, judge wisely and impartially, and govern our resulting actions ourselves, apart from any external constraints or inducements. If Kant is correct, and if we are truly "free" in any meaningful sense, then it follows that we should not require external inducements or constraints to force us to behave morally. Instead, we should "desire" in a rational sense to behave this way out of respect for ourselves and others like us.

Likewise, Kant's first two formulations of this rational "imperative" of duty suddenly seem almost straightforward. When considering our actions, we should act only on those individual impulses or "maxims" that we could simultaneously generalize as universal laws. In effect, we should only act in ways that we could also imagine ourselves giving our consent for others to act. More significantly, we should not unilaterally grant ourselves exceptions to reasonable and rational rules that we acknowledge as binding on all, and that we fully expect to apply to everyone else. This is the so-called "first form" of Kant's "Categorical Imperative," the "Formula of Universal Law."

When considering other moral agents like ourselves, we should never treat them in ways that make them solely a means to attaining our own ends. This is the "second form" of the Categorical Imperative, the "Formula of Humanity as an End in Itself." Put in this way, it calls attention to the reason and freedom essential to each moral agent. It suggests that each of us has a duty to avoid acting in ways that interfere with or preempt the ability of others to use their reason, or to exercise their own freedom to choose, as these together form the very basis of our moral and social lives, the source of our dignity and autonomy.

Thus, in an example Kant cites twice in the following reading selection, making false promises, or lying, is wrong both because I cannot will such practice consistently as a universal law (in fact, lying succeeds only if we assume that everyone else is telling the truth[1]), and because the lie in this case confounds my friend's reason, and destroys his ability to evaluate my request and freely give or withhold his consent to lend me money. I myself would not wish to be deceived in this fashion; indeed, no rational being would wish to be the object of a scheme that violated their freedom or "informed consent" by deceiving and withholding the truth from them, thereby destroying their own abilities to deliberate and choose. That is what makes such behavior morally wrong.

Particularly when considering this "second formulation" of the categorical imperative, I discover that I am not merely enjoined to refrain from bad or immoral actions that conflict with the demands of reason and violate the moral freedom of others. I am also commanded to do good, to "act beneficently" in the world, to undertake actions that could and should have the character of universal law for all to follow, and which serve to enable the freedom and dignity of every member of the moral community. For example, if I came across a trapped passenger in a burning automobile, I am not merely enjoined not to injure such a person further, but I must act positively to save her if I can (although what specific actions are demanded is left unclear). Many have thought this implication of Kant's theory to apply to the problem of humanitarian military intervention: it would be what Kant

---

[1] Mathematicians and quantitative economists engaged in what is now called "game theory" label this dilemma "the free-rider problem." It is significant that Kant first called attention to its logical inconsistency, in that unilaterally making oneself an exception to the operant institutional constraints is ultimately incompatible with the continuance of the very institution upon which "free-riding" parasitically depends.

terms an "imperfect duty" or obligation not to stand by idly while others suffer harm, though it is unclear just who is obligated, and what they are obligated specifically to do in response.

To many of Kant's contemporaries (and to many readers since), these injunctions, and especially the "second form" of the Categorical Imperative, seemed like needlessly complicated re-statements of the ancient and venerable "Golden Rule." Certainly there is much to which we might draw comparison in these two conceptions of morality. Kant, however, was visibly annoyed with such comparisons, and thought himself to be proposing something much more extensive. We see the point of his frustrations when we turn to the final form of the Categorical Imperative, the "Kingdom of Ends."

If we join the insights of the first two variations of this fundamental moral insight together, we see that Kant envisions morality as a social arrangement, almost as a kind of social contract. In everything that we do, he urges, we should remind ourselves that each of us, and all of us, are "lawgivers," participating in making or in legislating the common and universal constraints on our individual behavior necessary (as philosopher Onora O'Neill explains) in order to make it possible for a diverse multitude of free and rational beings to live together in reasonable peace, security, and harmony. We make the laws, the Moral Laws of Reason, and then we each agree to live under the laws we thus make (that is the "first form" of the Categorical Imperative, restated). Those Moral Laws, in turn, serve to guarantee to each of us the maximum degree of individual freedom of action, respect, and dignity possible, compatible with simultaneously granting an equal degree of freedom and respect to everyone else (that is the "second form" of the Categorical Imperative, restated). These insights are thus drawn together in the concluding injunction: act as if each of us were both a legislator and a citizen in the moral community, the "Kingdom of Ends." This is the democratic order of law, finally, that our Reason imposes upon all of us, giving rise to our individual sense of moral obligation.[2]

So, to return to the analysis of Reason in science and morality with which we began: we now see that Reason functions in both science and morality in a "legislative capacity." The underlying nature of Reason is juridical: that is, reason "makes laws," and so organizes, or orders, its world. This organization of experience in the natural world results in the "laws of nature" that we study in the physical sciences. This juridical ordering of desire and practical experience in our social world results in Duty, the Laws of Morality.

This is a stunning achievement, worthy of our hard work and careful study. We customarily think of Law and Obligation as the antithesis of freedom, as constraints upon the ability to do as we please. Kant demonstrates, however, that this view is mistaken. Inasmuch as these laws or constraints are rationally necessary as the foundation of our life together, and are the collective result of our common deliberation together as rational beings about the nature of that social foundation—and inasmuch as each of us furthermore gives our consent to live according to the laws we have made—these laws cannot be opposed to our individual freedom. Rather, *the Moral Law is itself the very expression of that freedom*: moral principles are, in Kant's phrase, "the Laws of Freedom."

* * * * *

Nothing remains today of the tiny village of Königsberg, or of the small university to which a very industrious, disciplined, an exceptionally gifted professor and Rector once brought world-wide fame and recognition. Every last building was leveled to dust in the destructive world wars of the twentieth century, leaving not a trace of the birthplace of the scholar whose consuming passion had been to bring the blessings of peace, under the rule of international law, to the war-ravaged world of his own century. The town of Königsberg was completely rebuilt under Soviet occupation following World War II, and is now known by its equivalent Russian name, "Kalinengrad," the City of the King. It serves to this day, in one of history's odd ironies, as the headquarters and home port of the once-powerful Soviet nuclear submarine fleet.

---

[2]This particular way of presenting Kant's project can be found in more detail in G. R. Lucas, "Moral Order and the Constraints of Agency: Toward a New Metaphysics of Morals," *New Essays in Metaphysics*, ed. Robert C. Neville (Albany, NY: State University of New York Press, 1987), 117–139.

from Groundwork of the Metaphysic of Morals
*Immanuel Kant**

Chapter I

[The Good Will.]

It is impossible to conceive anything at all in the world, or even out of it, which can be taken as good without qualification, except a *good will*. Intelligence, wit, judgement, and any other *talents of the mind* we may care to name, or courage, resolution, and constancy of purpose, as qualities of *temperament*, are without doubt good and desirable in many respects; but they can also be extremely bad and hurtful when the will is not good which has to make use of these gifts of nature, and which for this reason has the term *'character'* applied to its peculiar quality. It is exactly the same with *gifts of fortune*. Power, wealth, honour, even health and that complete well-being and contentment with one's state which goes by the name of *'happiness,'* produce boldness, and as a consequence often over-boldness as well, unless a good will is present by which their influence on the mind—and so too the whole principle of action—may be corrected and adjusted to universal ends; not to mention that a rational and impartial spectator can never feel approval in contemplating the uninterrupted prosperity of a being graced by no touch of a pure and good will, and that consequently a good will seems to constitute the indispensable condition of our very worthiness to be happy.

Some qualities are even helpful to this good will itself and can make its task very much easier. They have none the less no inner unconditioned worth, but rather presuppose a good will which sets a limit to the esteem in which they are rightly held and does not permit us to regard them as absolutely good. Moderation in affections and passions, self-control, and sober reflexion are not only good in many respects: they may even seem to constitute part of the *inner* worth of a person. Yet they are far from being properly described as good without qualification (however unconditionally they have been commended by the ancients). For without the principles of a good will they may become exceedingly bad; and the very coolness of a scoundrel makes him, not merely more dangerous, but also immediately more abominable in our eyes than we should have taken him to be without it.

[The Good Will and Its Results.]

A good will is not good because of what it effects or accomplishes—because of its fitness for attaining some proposed end: it is good through its willing alone—that is, good in itself. Considered in itself it is to be esteemed beyond comparison as far higher than anything it could ever bring about merely in order to favour some inclination or, if you like, the sum total of inclinations. Even if, by some special

---

*These excerpts were prepared and edited by Lawrence Lengbeyer, Assistant Professor of Philosophy, U.S. Naval Academy.

disfavour of destiny or by the niggardly endowment of step-motherly nature, this will is entirely lacking in power to carry out its intentions; if by its utmost effort it still accomplishes nothing, and only good will is left (not admittedly, as a mere wish, but as the straining of every means so far as they are in our control); even then it would still shine like a jewel for its own sake as something which has its full value in itself. Its usefulness or fruitlessness can neither add to, nor subtract from, this value.

[The Function of Reason.]
Yet in this Idea of the absolute value of a mere will, all useful results being left out of account in its assessment, there is something so strange that . . . there must arise the suspicion that perhaps . . . we may have misunderstood the purpose of nature in attaching reason to our will as its governor. We will therefore submit our Idea to an examination from this point of view.

In the natural constitution of an organic being—that is, of one contrived for the purpose of life—let us take it as a principle that in it no organ is to be found for any end unless it is also the most appropriate to that end and the best fitted for it. Suppose now that for a being possessed of reason and a will the real purpose of nature were his *preservation*, his *welfare*, or in a word his *happiness*. In that case nature would have hit on a very bad arrangement by choosing reason in the creature to carry out this purpose. For all the actions he has to perform with this end in view, and the whole rule of his behaviour, would have been mapped out for him far more accurately by instinct; and the end in question could have been maintained far more surely by instinct than it ever can be by reason.

. . . In actual fact too we find that the more a cultivated reason concerns itself with the aim of enjoying life and happiness, the further does man get away from true contentment. This is why there arises in many . . . a certain degree of *misology*—that is, a hatred of reason; for when they balance all the advantage they draw, . . . they discover that they have in fact only brought more trouble on their heads than they have gained in the way of happiness. On this account they come to envy rather than to despise, the more common run of men, who are closer to the guidance of mere natural instinct, and who do not allow their reason to have much influence on their conduct. . . .

[S]ince reason is not sufficiently serviceable for guiding the will safely as regards its objects and the satisfaction of all our needs (which it in part even multiplies)—a purpose for which an implanted natural instinct would have led us much more surely; and since none the less reason has been imparted to us as a practical power—that is, as one which is to have influence on the *will*; its true function must be to produce a *will* which is *good*, not as *means* to some further end, but in *itself* and for this function reason was absolutely necessary in a world where nature, in distributing her aptitudes, has everywhere else gone to work in a purposive manner. Such a will need not on this account be the sole and complete good, but it must be the highest good and the condition of all the rest, even of all our demands for happiness. In that case . . . the cultivation of reason . . . can even reduce happiness to less than zero without nature proceeding contrary to its purpose.

[The Good Will and Duty.]
We have now to elucidate the concept of a will estimable in itself and good apart from any further end. This concept, which is already present in a sound natural understanding and requires not so much to be taught as merely to be clarified, always holds the highest place in estimating the total worth of our actions and constitutes the condition of all the rest. We will therefore take up the concept of *duty*, which includes that of a good will, exposed, however, to certain subjective limitations and obstacles. These, so far from hiding a good will or disguising it, rather bring it out by contrast and make it shine forth more brightly.

[The Motive of Duty.]
[A human action is morally good, not because it is done from immediate inclination—still less because it is done from self-interest—but because it is done for the sake of duty.] For example, it certainly accords with duty that a grocer should not overcharge his inexperienced customer; and where there is much competition a sensible shopkeeper refrains from so doing and keeps to a fixed and general price for everybody so that a child can buy from him just as well as anyone else. Thus people are served *honestly*; but this is not nearly enough to justify us in believing that the shopkeeper has acted

in this way from duty or from principles of fair dealing; his interests required him to do so. We cannot assume him to have in addition an immediate inclination towards his customers, leading him, as it were out of love, to give no man preference over another in the matter of price. Thus the action was done neither from duty nor from immediate inclination, but solely from purposes of self-interest.

On the other hand, to preserve one's life is a duty, and besides this every one has also an immediate inclination to do so. But on account of this the often anxious precautions taken by the greater part of mankind for this purpose have no inner worth, and the maxim of their action is without moral content. They do protect their *lives in conformity with duty*, but not *from the motive of duty*. When on the contrary, disappointments and hopeless misery have quite taken away the taste for life; when a wretched man ... longs for death and still preserves his life without loving it—not from inclination or fear but from duty; then indeed his maxim has a moral content.

To help others where one can is a duty, and besides this there are many spirits of so sympathetic a temper that, without any further motive of vanity or self-interest, they find an inner pleasure in spreading happiness around them and can take delight in the contentment of others as their own work. Yet I maintain that in such a case an action of this kind, however right and however amiable it may be, has still no genuinely moral worth. It stands on the same footing as other inclinations—for example, the inclination for honour, which if fortunate enough to hit on something beneficial and right and consequently honourable, deserves praise and encouragement, but not esteem; for its maxim lacks moral content, namely, the performance of such actions, not from inclination, but ... *from duty*. Suppose then that the mind of this friend of man were over-clouded by sorrows of his own, ... but that he still, ... though no longer stirred by the need of others, ... does the action ... for the sake of duty alone; then for the first time his action has its genuine moral worth. Still further: if nature had implanted little sympathy in this or that man's heart; if (being in other respects an honest fellow) he were cold in temperament and indifferent to the sufferings of others—perhaps because, being endowed with the special gift of patience and robust endurance in his own sufferings, he assumed the like in others or even demanded it; if such a man (who would in truth not be the worst product of nature) were not exactly fashioned by her to be a philanthropist, would he not still find in himself a source from which he might draw a worth far higher than any that a good-natured temperament can have? Assuredly he would. It is precisely in this that the worth of character begins to show—a moral worth and beyond all comparison the highest—namely, that he does good, not from inclination, but from duty.

To assure one's own happiness is a duty (at least indirectly); for discontent with one's state ... might easily become a great *temptation to the transgression of duty*. But here also, apart from regard to duty, all men have already of themselves the strongest and deepest inclination towards happiness, because precisely in this Idea of 'happiness' all inclinations are combined into a sum total. ... [H]owever, ... a single inclination which is determinate as to what it promises and as to the time of its satisfaction may outweigh a wavering Idea; and ... a man, for example, a sufferer from gout may choose to enjoy what he fancies and put up with ... [the costs to his] health. But in this case also, when the universal inclination towards happiness has failed to determine his will, when good health, at least for him, has not entered into his calculations as so necessary, what remains over, here as in other cases, is a law—the law of furthering his happiness, not from inclination, but from duty; and in this for the first time his conduct has a real moral worth.

It is doubtless in this sense that we should understand too the passages from Scripture in which we are commanded to love our neighbour and even our enemy. For love out of inclination cannot be commanded; but kindness done from duty—although no inclination impels us, and even although natural and unconquerable disinclination stands in our way—is *practical*, and not *pathological*, love, residing in the will and not in the propensions of feeling, in principles of action and not of melting compassion; and it is this practical love alone which can be an object of command.

[The Formal Principle of Duty.]
Our second proposition is this: An action done from duty has its moral worth, *not in the purpose* to be attained by it, but in the maxim in accordance with which it is decided upon; it depends therefore, not on the realization of the object of the action, but solely on the *principle of volition* in accordance with which, irrespective of all objects of the faculty of desire, the action has been performed.

[Reverence for the Law.]

Our third proposition, as an inference from the two preceding I would express thus: *Duty is the necessity to act out of reverence for the law*. For . . . the effect of my proposed action I can have an inclination, but *never reverence*, precisely because it is merely the effect, and not the activity, of a will. Similarly for inclination as such; whether my own or that of another, I cannot have reverence: I can at most in the first case approve, and in the second case sometimes even love—that is, regard it as favourable to my own advantage. Only something . . . which does not serve my inclination, but outweighs it or at least leaves it entirely out of account in my choice—and therefore only bare law for its own sake, can be an object of reverence and therewith a command. Now an action done from duty has to set aside altogether the influence of inclination, and along with inclination every object of the will; so there is nothing left able to determine the will except objectively the *law* and subjectively *pure reverence* for this practical law, and therefore the maxim[1] of obeying this law even to the detriment of all my inclinations.

Thus the moral worth of an action does not depend on the result expected from it. . . . For all these results (agreeable states and even the promotion of happiness in others) could have been brought about by other causes as well, and consequently their production did not require the will of a rational being, in which, however, the highest and unconditioned good can alone be found. Therefore nothing but the *idea of the law* in itself, *which admittedly is present only in a rational being*—so far as it, and not an expected result, is the ground determining the will—can constitute that pre-eminent good which we call moral, a good which is already present in the person acting on this idea and has not to be awaited merely from the result.[2]

[The Categorical Imperative.]

But what kind of law can this be the thought of which, even without regard to the results expected from it, has to determine the will if this is to be called good absolutely and without qualification? Since I have robbed the will of every inducement that might arise for it as a consequence of obeying any particular law, nothing is left but the conformity of actions to universal law as such, and this alone must serve the will as its principle. That is to say, I ought never to act except in such a way *that I can also will that my maxim should become a universal law*. Here bare conformity to universal law as such (without having as its base any law prescribing particular actions) is what serves the will as its principle. . . .

Take this question, for example. May I not, when I am hard pressed, make a promise with the intention of not keeping it? Here I readily distinguish the two senses which the question can have—Is it prudent, or is it right, to make a false promise? The first no doubt can often be the case. . . . To tell the truth for the sake of duty is something entirely different from doing so out of concern for inconvenient results; for in the first case the concept of the action already contains in itself a law for me, while in the second case I have first of all to look around elsewhere in order to see what effects may be bound up with it for me. When I deviate from the principle of duty, this is quite certainly bad; but if I desert my prudential maxim, this can often be greatly to my advantage, though it is admittedly safer to stick to it. Suppose I seek, however, to learn in the quickest way and yet unerringly how to solve the problem 'Does a lying promise accord with duty?' I have then to ask myself 'Should I really be content that my maxim (the maxim of getting out of a difficulty by a false promise) should hold as a universal law (one valid both for myself and others)? And could I really say to myself that every one may make a

---

[1] A *maxim* is the subjective principle of a volition: an objective principle (that is, one which would also serve subjectively as a practical principle for all rational beings if reason had full control over the faculty of desire) is a practical *law*.

[2] What I recognize immediately as law for me, I recognize with reverence, which means merely consciousness of the *subordination* of my will to a law without the mediation of external influences on my senses. Immediate determination of the will by the law and consciousness of this determination is called *'reverence.'* Reverence is properly awareness of a value which demolishes my self-love. The *object* of reverence is the *law* alone—that law which we impose on *ourselves* but yet as necessary in itself. Considered as a law, we are subject to it without any consultation of self-love; considered as self-imposed it is a consequence of our will. All reverence for a person is properly only reverence for the law (of honesty and so on) of which that person gives us an example. Because we regard the development of our talents as a duty, we see too in, a man of talent a sort of *example of the law* (the law of becoming like him by practice), and this is what constitutes our reverence for him.

false promise if he finds himself in a difficulty from which he can extricate himself in no other way?' I then become aware at once that I can indeed will to lie, but I can by no means will a universal law of lying: for by such a law there could properly be no promises at all, since it would be futile to profess a will for future action to others who would not believe my profession or who, if they did so overhastily, would pay me back in like coin; and consequently my maxim, as soon as it was made a universal law, would be bound to annul itself.

Thus I need no far-reaching ingenuity to find out what I have to do in order to possess a good will. . . . I ask myself only 'Can you also will that your maxim should become a universal law?' Where you cannot, it is to be rejected, and that not because of a prospective loss to you or even to others, but because it cannot fit as a principle into a possible enactment of universal law. For such an enactment reason compels my immediate reverence. . . . [R]everence is the assessment of a worth which far outweighs all the worth of what is commended by inclination, and the necessity for me to act out of pure reverence for the practical law is what constitutes duty to which every other motive must give way because it is the condition of a will good *in itself* whose value is above all else.

[Ordinary Practical Reason.]
In studying the moral knowledge of ordinary human reason we have now arrived at its first principle. This principle it admittedly does not conceive thus abstractly in its universal form; but it does always have it actually before its eyes and does use it as a norm of judgement. . . . [I]n consequence there is no need of science or philosophy for knowing what man has to do in order to be honest and good, and indeed to be wise and virtuous. . . . Might it not then be more advisable in moral questions to abide by the judgement of ordinary reason and, at the most, to bring in philosophy only in order to set forth the system of morals more fully and intelligibly and to present its rules in a form more convenient for use?

[The Need for Philosophy.]
Innocence is a splendid thing, only it has the misfortune not to keep very well and to be easily misled. . . . Man feels in himself a powerful counterweight to all the commands of duty presented to him by reason as so worthy of esteem—the counterweight of his needs and inclinations, whose total satisfaction he grasps under the name of 'happiness'. . . . From this there arises . . . a disposition to quibble with these strict laws of duty, to throw doubt on their validity or at least on their purity and strictness, and to make them, where possible, more adapted to our wishes and inclinations; that is, to pervert their very foundations and destroy their whole dignity. . . .

In this way the *common reason of mankind* is impelled, . . . on practical grounds themselves, to . . . take a step into the field of *practical philosophy*. It there seeks to acquire . . . precise instruction about the source of its own principle, and about the correct function of this principle in comparison with maxims based on need and inclination, in order that it may escape from the embarrassment of antagonistic claims and may avoid the risk of losing all genuine moral principles because of the ambiguity into which it easily falls.

Chapter II

. . .

[Imperatives in General.]
Everything in nature works in accordance with laws. Only a rational being has the power to act *in accordance with his idea* of laws—that is, in accordance with principles—and only so has he a *will*. Since *reason* is required in order to derive actions from laws, the will is nothing but practical reason. If reason infallibly determines the will [as in a perfectly rational angel], then in a being of this kind the actions which are recognized [by reason] to be objectively necessary are also subjectively necessary—that is to say, the will . . . [can] choose *only that* which reasons independently of inclination recognizes to be . . . good. But if [as in humans] . . . the will is exposed also to subjective . . . impulsions . . . which do not always harmonize with the [dictates of] . . . reason . . .; then . . . the determining of such a will in accordance with objective laws [given by reason] is . . . [experienced as a constraint,] a command (of reason), and the formula of this command is called an *Imperative*.

[Classification of Imperatives.]

All *imperatives* command either *hypothetically* or *categorically*. Hypothetical imperatives declare a possible action to be practically necessary as a means to the attainment of something else that one wills.... A categorical imperative ... represent[s] an action as objectively necessary in itself apart from its relation to a further end.

...

[A categorical imperative] is concerned, not with the matter of the action and its presumed results, but with its form and with the principle from which it follows; and what is essentially good in the action consists in the mental disposition, let the consequences be what they may. This imperative may be called the imperative of *morality*.

...

[How Are Imperatives Possible?]

...

[T]he categorical imperative alone purports to be a practical *law* . . .; for an action necessary merely in order to achieve ... [some selected] purpose ... [is not necessary], and we can always escape from the precept if we abandon the purpose; whereas an unconditioned [i.e., categorical] command does not leave it open to the will to do the opposite at its discretion and therefore alone carries with it the necessity which we demand from a law.

...

[The Formula of Universal Law.]

...

When I conceive a *hypothetical* imperative in general, I do not know beforehand what it will contain—until its condition is given. But if I conceive a *categorical* imperative, I know at once what it contains. For since besides the law this imperative contains only the necessity that our maxim[3] should conform to this law, while the law, as we have seen, contains no condition to limit it, there remains nothing over to which the maxim has to conform except the universality of a law as such; and it is this conformity alone that the imperative properly asserts to be necessary.

[T]here is therefore only a single categorical imperative and it is this: '*Act only on that maxim through which you can at the same time will that it should become a universal law.*' ...

[T]he universal imperative of duty may also run as follows: '*Act as if the maxim of your action were to become through your will a universal law of nature.*'

[Illustrations.]

We will now enumerate a few duties, following their customary division into duties towards self and duties towards others and into perfect and imperfect duties.[4]

1. A man feels sick of life as the result of a series of misfortunes that has mounted to the point of despair, but he is still so far in possession of his reason as to ask himself whether taking his own life may not be contrary to his duty to himself. He now applies the test 'Can the maxim of my action really become a universal law of nature?' His maxim is 'From self-love I make it my principle to shorten my life if its continuance threatens more evil than it promises pleasure.' The only further question to ask is whether this principle of self-love can become a universal law of nature. It is then seen at once that a system of nature by whose law the very same feeling whose function (*Bestimmung*) is to stimulate the furtherance of life should actually destroy life would

---

[3] A *maxim* is a subjective principle of action and must be distinguished from an *objective principle*—namely, a practical law. The former contains a principle on which the subject *acts*. A law, on the other hand, is an objective principle valid for every rational being; and it is a principle on which he *ought to act*—that is, an imperative.

[4] I understand here by perfect duty one which allows no exception in the interests of inclination.

contradict itself and consequently could not subsist as a system of nature. Hence this maxim cannot possibly hold as a universal law of nature and is therefore entirely opposed to the supreme principle of all duty.

2. Another finds himself driven to borrowing money because of need. He well knows that he will not be able to pay it back; but he sees too that he will get no loan unless he gives a firm promise to pay it back within a fixed time. He is inclined to make such a promise; but he has still enough conscience to ask 'Is it not unlawful and contrary to duty to get out of difficulties in this way?' Supposing, however, he did resolve to do so, the maxim of his action would run thus: 'whenever I believe myself short of money, I will borrow money and promise to pay it back, though I know that this will never be done'. Now this principle of self-love or personal advantage is perhaps quite compatible with my own entire future welfare; only there remains the question 'Is it right?' I therefore transform the demand of self-love into a universal law and frame my question thus: 'How would things stand if my maxim became a universal law?' I then see straight away that this maxim can never rank as a universal law of nature and be self-consistent, but must necessarily contradict itself. For the universality of a law that every one believing himself to be in need can make any promise he pleases with the intention not to keep it would make promising, and the very purpose of promising, itself impossible, since no one would believe he was being promised anything, but would laugh at utterances of this kind as empty shams.

3. A third finds in himself a talent whose cultivation would make him a useful man for all sorts of purposes. But he sees himself in comfortable circumstances, and he prefers to give himself up to pleasure rather than to bother about increasing and improving his fortunate natural aptitudes. Yet he asks himself further 'Does my maxim of neglecting my natural gifts, besides agreeing in itself with my tendency to indulgence, agree also with what is called duty?' He then sees that a system of nature could indeed always subsist under such a universal law, although (like the South Sea Islanders) every man should let his talents rust and should be bent on devoting his life solely to idleness, indulgence, procreation, and, in a word, to enjoyment. Only he cannot possibly *will* that this should become a universal law of nature or should be implanted in us as such a law by a natural instinct. For as a rational being he necessarily wills that all his powers should be developed since they serve him, and are given him, for all sorts of possible ends.

4. Yet a fourth is himself flourishing, but he sees others who have to struggle with great hardships (and whom he could easily help); and he thinks 'What does it matter to me? Let every one be as happy as Heaven wills or as he can make himself; I won't deprive him of anything; I won't even envy him; only I have no wish to contribute anything to his wellbeing or to his support in distress!' Now admittedly if such an attitude were a universal law of nature, mankind could get on perfectly well—better no doubt than if everybody prates about sympathy and goodwill, and even takes pains, on occasion, to practice them, but on the other hand cheats where he can, traffics in human rights, or violates them in other ways. But although it is possible that a universal law of nature could subsist in harmony with this maxim, yet it is impossible to *will* that such a principle should hold everywhere as a law of nature. For a will which decided in this way would be in conflict with itself, since many a situation might arise in which the man needed love and sympathy from others, and in which, by such a law of nature sprung from his own will, he would rob himself of all hope of the help he wants for himself.

[The Canon of Moral Judgment.]

These are some of the many actual duties—or at least of what we take to be such—whose derivation from the single principle cited above leaps to the eye. We must *be able to will* that a maxim of our action should become a universal law—this is the general canon for all moral judgement of action. Some actions are so constituted that their maxim cannot even be *conceived* as a universal law of nature without contradiction, let alone be *willed* as what *ought* to become one. In the case of others we do not find this inner impossibility, but it is still impossible to *will* that their maxim should be raised to the universality of a law of nature, because such a will would contradict itself.

. . . [W]henever we transgress a duty, . . . we in fact do not will that our maxim should become a universal law— . . . but rather that its opposite should remain a law universally: we only take the liberty of making an *exception* to it for ourselves (or even just for this once) to the advantage of our inclination. . . . [W]e in fact recognize the validity of the categorical imperative and . . . merely permit ourselves a few exceptions which are, . . . we pretend . . . forced upon us.

. . .

[The Formula of the End in Itself.]

The will is conceived as a power of determining oneself to action in *accordance with the idea of certain laws*. And such a power can be found only in rational beings. Now what serves the will as a subjective ground of its self-determination is an *end*; and this, if it is given by reason alone, must be equally valid for all rational beings. . . . Ends that a rational being adopts arbitrarily as *effects* of his action (material ends) are in every case only relative; for it is solely their relation to . . . the subject's power of appetition which gives them their value. Hence this value can provide no universal principles, no principles valid and necessary for all rational beings and also for every volition—that is, no practical laws. Consequently all these relative ends can be the ground only of hypothetical imperatives.

Suppose, however, there were something *whose existence* has *in itself* an absolute value, something which *as an end in itself* could be a ground of determinate laws; then in it, and in it alone, would there be the ground of a possible categorical imperative—that is, of a practical law.

Now I say that man, and in general every rational being, *exists* as an end in himself, *not merely as a means* for arbitrary use by this or that will: he must in all his actions, whether they are directed to himself or to other rational beings, always be viewed *at the same time as an end*. All the objects of inclination have only a conditioned value; for if there were not these inclinations and the needs grounded on them, their object would be valueless. Inclinations themselves, as sources of needs, are so far from having an absolute value to make them desirable for their own sake that it must rather be the universal wish of every rational being to be wholly free from them. Thus the value of all objects that can be *produced* by our action is always conditioned. Beings whose existence depends, not on our will, but on nature, have none the less, if they are non-rational beings, only a relative value as means and are consequently called *things*. Rational beings, on the other hand, are called *persons* because their nature already marks them out as ends in themselves—that is, as something which ought not to be used merely as a means—and consequently imposes to that extent a limit on all arbitrary treatment of them (and is an object of reverence). Persons, therefore, are not merely subjective ends whose existence as an object of our actions has a value *for us*: they are *objective ends*—that is, things whose existence is in itself an end, and indeed an end such that in its place we can put no other end to which they should serve *simply* as means; for unless this is so, nothing at all of *absolute* value would be found anywhere.

. . . *Rational nature exists as an end in itself*. This is the way in which a man necessarily conceives his own existence: it is therefore so far a *subjective* principle of human actions. But it is also the way in which every other rational being conceives his existence on the same rational ground which is valid also for me; hence it is at the same time an objective principle, from which, as a supreme practical ground, it must be possible to derive all laws for the will. The practical imperative will therefore be as follows: *Act in such a way that you always treat humanity, whether in your own person or in the person of any other, never simply as a means, but always at the same time as an end*. We will now consider whether this can be carried out in practice.

[Illustrations.]

Let us keep to our previous examples.

First, as regards the concept of necessary duty to oneself the man who contemplates suicide will ask 'Can my action be compatible with the Idea of humanity *as an end in itself?*' If he does away with himself in order to escape from a painful situation, he is making use of a person merely as a *means* to maintain a tolerable state of affairs till the end of his life. But man is not a thing—not something to be used *merely* as a means: he must always in all his actions be regarded as an end in himself. Hence I cannot dispose of man in my person by maiming, spoiling, or killing.

. . . Secondly, so far as necessary or strict duty to others is concerned, the man who has a mind to make a false promise to others will see at once that he is intending to make use of another man *merely as a means* to an end he does not share. For the man whom I seek to use for my own purposes by such a promise cannot possibly agree with my way of behaving to him, and so cannot himself share the end of the action. This incompatibility with the principle of duty to others leaps to the eye more obviously when we bring in examples of attempts on the freedom and property of others. For then it is manifest that a violator of the rights of man intends to use the person of others merely as a means without taking into consideration that, as rational beings, they ought always at the same time to be rated as ends—that is, only as beings who must themselves be able to share in the end of the very same action.[5]

Thirdly, in regard to contingent (meritorious) duty to oneself, it is not enough that an action should refrain from conflicting with humanity in our own person as an end in itself: it must also *harmonize with this end*. Now there are in humanity capacities for greater perfection which form part of nature's purpose for humanity in our person. To neglect these can admittedly be compatible with the *maintenance* of humanity as an end in itself but not with the *promotion* of this end.

Fourthly, as regards meritorious duties to others, the natural end which all men seek is their own happiness. Now humanity could no doubt subsist if everybody contributed nothing to the happiness of others but at the same time refrained from deliberately impairing their happiness. This is, however, merely to agree negatively and not positively with *humanity as an end in itself* unless every one endeavours also, so far as in him lies, to further the ends of others. For the ends of a subject who is an end in himself must if this conception is to have its *full* effect in me, be also, as far as possible, *my* ends.

[The Formula of Autonomy.]

This principle of humanity, and in general of every rational agent, *as an end in itself* (a principle which is the supreme limiting condition of every man's freedom of action) is not borrowed from experience; firstly, because it is universal, applying as it does to all rational beings as such, and no experience is adequate to determine universality; secondly, because in it humanity is conceived, not as an end of man (subjectively)—that is, as an object which, as a matter of fact, happens to be made an end—but as an objective end—one which, be our ends what they may, must, as a law, constitute the supreme limiting condition of all subjective ends and so must spring from pure reason. That is to say, the ground for every enactment of practical law lies *objectively in the rule* and in the form of universality which (according to our first principle) makes the rule capable of being a law (and indeed a law of nature); *subjectively*, however, it lies in the *end*; but (according to our second principle) the subject of all ends is to be found in every rational being as an end in himself. From this there now follows our third practical principle for the will—as the supreme condition of the will's conformity with universal practical reason—namely, the Idea *of the will of every rational being as a will which makes universal law*.

By this principle all maxims are repudiated which cannot accord with the will's own enactment of universal law. The will is therefore not merely subject to the law, but is so subject that it must be considered as also *making the law* for itself and precisely on this account as first of all subject to the law (of which it can regard itself as the author).

. . .

[The Formula of the Kingdom of Ends.]

The concept of every rational being as one who must regard himself as making universal law by all the maxims of his will, and must seek to judge himself and his actions from this point of view, leads to a closely connected and very fruitful concept—namely, that of a *kingdom of ends*.

I understand by a '*kingdom*' a systematic union of different rational beings under common laws. Now since laws determine ends as regards their universal validity, we shall be able—if we abstract

---

[5]Let no one think that here the trivial '*quod tibi non vis fieri, etc.*' [ie. 'Don't do to others what you don't want done to yourself'] can serve as a standard or principle. For it is merely derivative from our principle, although subject to various qualifications: it cannot be a universal law since it contains the ground neither of duties to oneself nor of duties of kindness to others (for many a man would readily agree that others should not help him if only he could be dispensed from affording help to them), nor finally of strict duties towards others; for on this basis the criminal would be able to dispute with the judges who punish him, and so on.

from the personal differences between rational beings, and also from all the content of their private ends—to conceive a whole of all ends in systematic conjunction (a whole both of rational beings as ends in themselves and also of the personal ends which each may set before himself); that is, we shall be able to conceive a kingdom of ends which is possible in accordance with the above principles.

For rational beings all stand under the *law* that each of them should treat himself and all others, *never merely as a means*, but always *at the same time as an end in himself*. But by so doing there arises a systematic union of rational beings under common objective laws—that is, a kingdom. Since these laws are directed precisely to the relation of such beings to one another as ends and means, this kingdom can be called a kingdom of ends (which is admittedly only an Ideal).

A rational being belongs to the kingdom of ends as a member, when, although he makes its universal laws, he is also himself subject to these laws. He belongs to it as its *head*, when as the maker of laws he is himself subject to the will of no other.

A rational being must always regard himself as making laws in a kingdom of ends which is possible through freedom of the will—whether it be as member or as head. The position of the latter he can maintain, not in virtue of the maxim of his will alone, but only if he is a completely independent being, without needs and with an unlimited power adequate to his will.

Thus morality consists in the relation of all action to the making of laws whereby alone a kingdom of ends is possible. This making of laws must be found in every rational being himself and must be able to spring from his will. The principle of his will is therefore never to perform an action except on a maxim such as can also be a universal law, and consequently such *that the will can regard itself as at the same time making universal law by means of its maxim*. Where maxims are not already by their very nature in harmony with this objective principle of rational beings as makers of universal law, the necessity of acting on this principle is practical necessitation—that is, *duty*. Duty does not apply to the head in a kingdom of ends, but it does apply to every member and to all members in equal measure.

The practical necessity of acting on this principle—that is, duty—is in no way based on feelings, impulses, and inclinations, but only on the relation of rational beings to one another, a relation in which the will of a rational being must always be regarded as *making universal law*, because otherwise he could not be conceived as an *end in himself*. Reason thus relates every maxim of the will, considered as making universal law, to every other will and also to every action towards oneself: it does so, not because of any further motive or future advantage, but from the Idea of the *dignity* of a rational being who obeys no law other than that which he at the same time enacts himself.

[The Dignity of Virtue.]

In the kingdom of ends everything has either a *price* or a *dignity*. If it has a price, something else can be put in its place as an *equivalent*; if it is exalted above all price and so admits of no equivalent, then it has a dignity.

What is relative to universal human inclinations and needs has a *market price*; what, even without presupposing a need, accords with a certain taste—that is, with satisfaction in the mere purposeless play of our mental powers—has a *fancy price (Affektionspreis)*; but that which constitutes the sole condition under which anything can be an end in itself has not merely a relative value—that is, a price—but has an intrinsic value—that is, *dignity*.

Now morality is the only condition under which a rational being can be an end in himself; for only through this is it possible to be a lawmaking member in a kingdom of ends. Therefore morality, and humanity so far as it is capable of morality, is the only thing which has dignity. Skill and diligence in work have a market price; wit, lively imagination, and humour have a fancy price; but fidelity to promises and kindness based on principle (not on instinct) have an intrinsic worth. In default of these, nature and art alike contain nothing to put in their place; for their worth consists, not in the effects which result from them, not in the advantage or profit they produce, but in the attitudes of mind—that is, in the maxims of the will—which are ready in this way to manifest themselves in action even if they are not favoured by success. Such actions too need no recommendation from any subjective disposition or taste in order to meet with immediate favour and approval; they need no immediate propensity or feeling for themselves; they exhibit the will which performs them as an object of immediate reverence; nor is anything other than reason required to *impose* them upon the will, not to *coax* them from the will—which last would anyhow be a contradiction in the case of duties. This assessment reveals as dignity the value of such a mental attitude and puts it infinitely above all price, with which it cannot be brought into reckoning or comparison without, as it were, a profanation of its sanctity.

What is it then that entitles a morally good attitude of mind—or virtue—to make claims so high? It is nothing less than the *share* which it affords to a rational being in the making of universal law, and which therefore fits him to be a member in a possible kingdom of ends. For this he was already marked out in virtue of his own proper nature as an end in himself and consequently as a maker of laws in the kingdom of ends—as free in respect of all laws of nature, obeying only those laws which he makes himself and in virtue of which his maxims can have their part in the making of universal law (to which he at the same time subjects himself). For nothing can have a value other than that determined for it by the law. But the law-making which determines all value must for this reason have a dignity—that is, an unconditioned and incomparable worth—for the appreciation of which, as necessarily given by a rational being, the word *'reverence'* is the only becoming expression. *Autonomy* is therefore the ground of the dignity of human nature and of every rational nature.

. . .

A Simplified Account of Kant's Ethics
Onora O'Neill

Kant's moral theory has acquired the reputation of being forbiddingly difficult to understand and, once understood, excessively demanding in its requirements. I don't believe that this reputation has been wholly earned, and I am going to try to undermine it. . . . I shall try to reduce some of the difficulties. . . . Finally, I shall compare Kantian and utilitarian approaches and assess their strengths and weaknesses.

The main method by which I propose to avoid some of the difficulties of Kant's moral theory is by explaining only one part of the theory. This does not seem to me to be an irresponsible approach in this case. One of the things that makes Kant's moral theory hard to understand is that he gives a number of different versions of the principle that he calls the Supreme Principle of Morality, and these different versions don't look at all like one another. They also don't look at all like the utilitarians' Greatest Happiness Principle. But the Kantian principle is supposed to play a similar role in arguments about what to do.

Kant calls his Supreme Principle the *Categorical Imperative;* its various versions also have sonorous names. One is called the Formula of Universal Law; another is the Formula of the Kingdom of Ends. The one on which I shall concentrate is known as the *Formula of the End-in-Itself.* To understand why Kant thinks that these picturesquely named principles are equivalent to one another takes quite a lot of close and detailed analysis of Kant's philosophy. I shall avoid this and concentrate on showing the implications of this version of the Categorical Imperative.

The Formula of the End in Itself

Kant states the Formula of the End in Itself as follows:

> Act in such a way that you always treat humanity, whether in your own person or in the person of any other, never simply as a means but always at the same time as an end.

To understand this we need to know what it is to treat a person as a means or as an end. According to Kant, each of our acts reflects one or more *maxims.* The maxim of the act is the principle on which one sees oneself as acting. A maxim expresses a person's policy, or if he or she has no settled policy, the principle underlying the particular intention or decision on which he or she acts. Thus, a person who decides "This year I'll give 10 percent of my income to famine relief" has as a maxim the principle of tithing his or her income for famine relief. In practice, the difference between intentions and maxims is of little importance, for given any intention, we can formulate the corresponding maxim by deleting references to particular times, places, and persons. In what follows I shall take the terms 'maxim' and 'intention' as equivalent.

Whenever we act intentionally, we have at least one maxim and can, if we reflect, state what it is. (There is of course room for self-deception here—"I'm only keeping the wolf from the door" we may claim as we wolf down enough to keep ourselves overweight, or, more to the point, enough to feed someone else who hasn't enough food.)

When we want to work out whether an act we propose to do is right or wrong, according to Kant, we should look at our maxims and not at how much misery or happiness the act is likely to produce, and whether it does better at increasing happiness than other available acts. We just have to check that the act we have in mind will not use anyone as a mere means, and, if possible, that it will treat other persons as ends in themselves.

Using Persons as Mere Means

To use someone as a *mere means* is to involve them in a scheme of action to *which they could not in principle consent.* Kant does not say that there is anything wrong about using someone as a means. Evidently we have to do so in any cooperative scheme of action. If I cash a check I use the teller as a means, without whom I could not lay my hands on the cash; the teller in turn uses me as a means to earn his or her living. But in this case, each party consents to her or his part in the transaction. Kant would say that though they use one another as means, they do not use one another as mere means. Each person assumes that the other has maxims of his or her own and is not just a thing or a prop to be manipulated.

But there are other situations where one person uses another in a way to which the other could not in principle consent. For example, one person may make a promise to another with every intention of breaking it. If the promise is accepted, then the person to whom it was given must be ignorant of what the promisor's intention (maxim) really is. If one knew that the promisor did not intend to do what he or she was promising, one would, after all, not accept or rely on the promise. It would be as though there had been no promise made. Successful false promising depends on deceiving the person to whom the promise is made about what one's real maxim is. And since the person who is deceived doesn't know that real maxim, he or she can't in principle consent to his or her part in the proposed scheme of action. The person who is deceived is, as it were, a prop or a tool—a mere means—in the false promisor's scheme. A person who promises falsely treats the acceptor of the promise as a prop or a thing and not as a person. In Kant's view, it is this that makes false promising wrong.

One standard way of using others as mere means is by deceiving them. By getting someone involved in a business scheme or a criminal activity on false pretenses, or by giving a misleading account of what one is about, or by making a false promise or a fraudulent contract, one involves another in something to which he or she in principle cannot consent, since the scheme requires that he or she doesn't know what is going on. Another standard way of using others as mere means is by coercing them. If a rich or powerful person threatens a debtor with bankruptcy unless he or she joins in some scheme, then the creditor's intention is to coerce; and the debtor, if coerced, cannot consent to his or her part in the creditor's scheme. To make the example more specific: If a moneylender in an Indian village threatens not to renew a vital loan unless he is given the debtor's land, then he uses the debtor as a mere means. He coerces the debtor, who cannot truly consent to this "offer he can't refuse." (Of course the outward form of such transactions may look like ordinary commercial dealings, but we know very well that some offers and demands couched in that form are coercive.)

In Kant's view, acts that are done on maxims that require deception or coercion of others, and so cannot have the consent of those others (for consent precludes both deception and coercion), are wrong. When we act on such maxims, we treat others as mere means, as things rather than as ends in themselves. If we act on such maxims, our acts are not only wrong but unjust: such acts wrong the particular others who are deceived or coerced.

Treating Persons as Ends in Themselves

Duties of justice are, in Kant's view (as in many others'), the most important of our duties. When we fail in these duties, we have used some other or others as mere means. But there are also cases where, though we do not use others as mere means, still we fail to use them as ends in themselves in the fullest possible way. To treat someone as an end in him or herself requires in the first place that one not use him or her as mere means, that one respect each as a rational person with his or her own maxims. But beyond that, one may also seek to foster others' plans and maxims by sharing some of their ends. To act beneficently is to seek others' happiness, therefore to intend to achieve some of the things that those others aim at with their maxims. If I want to make others happy, I will adopt maxims that not merely do not manipulate them but that foster some of their plans and activities. Beneficent acts try to achieve what others want. However, we cannot seek everything that others want; their wants are too numerous and diverse, and, of course, sometimes incompatible. It follows that beneficence has to be selective.

There is then quite a sharp distinction between the requirements of justice and of beneficence in Kantian ethics. Justice requires that we act on no maxims that use others as mere means. Beneficence requires that we act on some maxims that foster others' ends, though it is a matter for judgment and discretion which of their ends we foster. Some maxims no doubt ought not to be fostered because it would be unjust to do so. Kantians are not committed to working interminably through a list of happiness-producing and misery-reducing acts; but there are some acts whose obligatoriness utilitarians may need to debate as they try to compare total outcomes of different choices, to which Kantians are stringently bound. Kantians will claim that they have done nothing wrong if none of their acts is unjust, and that their duty is complete if in addition their life plans have in the circumstances been reasonably beneficent.

In making sure that they meet all the demands of justice, Kantians do not try to compare all available acts and see which has the best effects. They consider only the proposals for action that occur to them and check that these proposals use no other as mere means. If they do not, the act is permissible; if omitting the act would use another as mere means, the act is obligatory. Kant's theory has less scope than utilitarianism. Kantians do not claim to discover whether acts whose maxims they don't know fully are just. They may be reluctant to judge others' acts or policies that cannot be regarded as the maxim of any person or institution. They cannot rank acts in order of merit. Yet, the theory offers more precision than utilitarianism when data are scarce. One can usually tell whether one's act would use others as mere means, even when its impact on human happiness is thoroughly obscure.

The Limits of Kantian Ethics: Intentions and Results

Kantian ethics differs from utilitarian ethics both in its scope and in the precision with which it guides action. Every action, whether of a person or of an agency, can be assessed by utilitarian methods, provided only that information is available about all the consequences of the act. The theory has unlimited scope, but, owing to lack of data, often lacks precision. Kantian ethics has a more restricted scope. Since it assesses actions by looking at the maxims of agents, it can only assess intentional acts. This means that it is most at home in assessing individuals' acts; but it can be extended to assess acts of agencies that (like corporations and governments and student unions) have decision-making procedures. It can do nothing to assess patterns of action that reflect no intention or policy, hence it cannot assess the acts of groups lacking decision-making procedures, such as the student movement, the women's movement, or the consumer movement.

It may seem a great limitation of Kantian ethics that it concentrates on intentions to the neglect of results. It might seem that all conscientious Kantians have to do is to make sure that they never intend to use others as mere means, and that they sometimes intend to foster other's ends. And, as we all know, good intentions sometimes lead to bad results and correspondingly, bad intentions sometimes do no harm, or even produce good. If Hardin is right, the good intentions of those who feed the starving lead to dreadful results in the long run. If some traditional arguments in favor of capitalism are right, the greed and selfishness of the profit motive have produced unparalleled prosperity for many.

But such discrepancies between intentions and results are the exception and not the rule. For we cannot just *claim* that our intentions are good and do what we will. Our intentions reflect what we expect the immediate results of our action to be. Nobody credits the "intentions" of a couple who practice neither celibacy nor contraception but still insist "we never meant to have (more) children." Conception is likely (and known to be likely) in such cases. Where people's expressed intentions ignore the normal and predictable results of what they do, we infer that (if they are not amazingly ignorant) their words do not express their true intentions. The Formula of the End in Itself applies to the intentions on which one acts—not to some prettified version that one may avow. Provided this intention—the agent's real intention—uses no other as mere means, he or she does nothing unjust. If some of his or her intentions foster others' ends, then he or she is sometimes beneficent. It is therefore possible for people to test their proposals by Kantian arguments even when they lack the comprehensive causal knowledge that utilitarianism requires. Conscientious Kantians can work out whether they will be doing wrong by some act even though it blurs the implications of the theory. If we peer through the blur, we see that the utilitarian view is that lives may indeed be sacrificed for the sake of a greater good even when the persons are not willing. There is nothing wrong with using another as a mere means provided that the end for which the person is so used is a happier result than could have been achieved any other way, taking into account the misery the means have caused. In utilitarian thought, persons are not ends in themselves. Their special moral status derives from their being means to the production of happiness. Human life has therefore a high though derivative value, and one life may be taken for the sake of greater happiness in other lives, or for ending of misery in that life. Nor is there any deep difference between ending a life for the sake of others' happiness by not helping (e.g., by triaging) and doing so by harming. Because the distinction between justice and beneficence is not sharply made within utilitarianism, it is not possible to say that triaging is a matter of not benefiting, while other interventions are a matter of injustice.

Utilitarian moral theory has then a rather paradoxical view of the value of human life. Living, conscious humans are (along with other sentient beings) necessary for the existence of everything utilitarians value. But it is not their being alive but the state of their consciousness that is of value. Hence, the best results may require certain lives to be lost—by whatever means—for the sake of the total happiness and absence of misery that can be produced.

Kant and Respect for Persons

Kantians reach different conclusions about human life. Human life is valuable because humans (and conceivably other beings, e.g., angels or apes) are the bearers of rational life. Humans are able to choose and to plan. This capacity and its exercise are of such value that they ought not to be sacrificed for anything of lesser value. Therefore, no one rational or autonomous creature should be treated as mere means for the enjoyment or even the happiness of another. We may in Kant's view justifiably even nobly risk or sacrifice our lives for others. For in doing so we follow our own maxim and nobody uses us as mere means. But no others may use either our lives or our bodies for a scheme that they have either coerced or deceived us into joining. For in doing so they would fail to treat us as rational beings; they would use us as mere means and not as ends in ourselves.

It is conceivable that a society of Kantians, all of whom took pains to use no other as mere means, would end up with less happiness or with fewer persons alive than would some societies of complying utilitarians. For since the Kantians would be strictly bound only to justice, they might without wrongdoing be quite selective in their beneficence and fail to maximize either survival rates or happiness, or even to achieve as much of either as a strenuous group of utilitarians, who they know that their foresight is limited and that they may cause some harm or fail to cause some benefit. But they will not cause harms that they can foresee without this being reflected in their intentions.

Utilitarianism and Respect for Life

From the differing implications that Kantian and utilitarian moral theories have for our actions towards those who do or may suffer famine, we can discover two sharply contrasting views of the value of human life. Utilitarians value happiness and the absence or reduction of misery. As a utilitarian one ought (if conscientious) to devote one's life to achieving the best possible balance of happiness over misery. If one's life plan remains in doubt, this will be because the means to this end are often unclear. But whenever the causal tendency of acts is clear, utilitarians will be able to discern the acts they should successively do in order to improve the world's balance of happiness over unhappiness.

This task is not one for the faint-hearted. First, it is dauntingly long, indeed interminable. Second, it may at times require the sacrifice of happiness, and even of lives, for the sake of a greater happiness. Such sacrifice may be morally required not only when the person whose happiness or even whose life is at stake volunteers to make the sacrifice. It may be necessary to sacrifice some lives for the sake of others. As our control over the means of ending and preserving human life has increased, analogous dilemmas have arisen in many areas for utilitarians. Should life be preserved at the cost of pain when modern medicine makes this possible? Should life be preserved without hope of consciousness? Should triage policies, because they may maximize the number of survivors, be used to determine who should be left to starve? Should population growth be fostered wherever it will increase the total of human happiness—or on some views so long as average happiness is not reduced? All these questions can be fitted into utilitarian frameworks and answered if we have the relevant information. And sometimes the answer will be that human happiness demands the sacrifice of lives, including the sacrifice of unwilling lives. Further, for most utilitarians, it makes no difference if the unwilling sacrifices involve acts of injustice to those whose lives are to be lost. It might, for example, prove necessary for maximal happiness that some persons have their allotted rations, or their hard-earned income, diverted for others' benefit. Or it might turn out that some generations must sacrifice comforts or liberties and even lives to rear "the fabric of felicity" for their successors. Utilitarians do not deny these possibilities, though the imprecision of our knowledge of consequences often somehow make the right calculations. On the other hand, nobody will have been made an instrument of others' survival or happiness in the society of complying Kantians.

On a Supposed Right to Lie because of Philanthropic Concerns*

Immanuel Kant

In an article bearing the title "On Political Reactions" by Benjamin Constant [(1767–1830), the renowned French statesman and writer] there is contained on p. 123 the following passage:

> "The moral principle stating that it is a duty to tell the truth would make any society impossible if that principle were taken singly and unconditionally. We have proof of this in the very direct consequences which a German philosopher [Kant] has drawn from this principle. This philosopher goes as far as to assert that it would be a crime to tell a lie to a murderer who asked whether our friend who is being pursued by the murderer had taken refuge in our house."

[Constant] on p. 124 refutes this [moral] principle in the following way:

> "To tell the truth is a duty, but is a duty only with regard to one who has a right to the truth. But no one has a right to a truth that harms others."

The first fallacy here lies in the statement, "To tell the truth is a duty, but is a duty only with regard to one who has a right to the truth."

Now, the first question is whether a man (in cases where he cannot avoid answering Yea or Nay) has the warrant (right) to be untruthful. The second question is whether he is not actually bound to be untruthful in a certain statement which he is unjustly compelled to make in order to prevent a threatening misdeed against himself or someone else.

Truthfulness in statements that cannot be avoided is the formal duty of man to everyone, however great the disadvantage that may arise therefrom for him or for any other. And even though by telling an untruth I do no wrong to him who unjustly compels me to make a statement, yet by this lie I do wrong to duty in general. That is, as far as in me lies I bring it about that statements (declarations) in general find no credence, and hence also that all rights based on contracts become void and lose their force, and this is a wrong done to mankind in general.

Hence a lie defined merely as an intentionally untruthful declaration to another man does not require the additional condition that it must do harm to another. For a lie always harms another; if not some other human being, then it nevertheless does harm to humanity in general, inasmuch as it vitiates the very source of right.

[T]his well-intentioned lie can become punishable in accordance with civil law; and that which avoids liability to punishment only by accident can also be condemned as wrong. For example, if by telling a lie you have in fact hindered someone who was even now planning a murder, then you are legally responsible for all the consequences that might result therefrom. But if you have adhered strictly to the truth, then public justice cannot lay a hand on you, whatever the unforeseen consequence might be. It is indeed possible that after you have honestly answered Yes to the murderer's question as to

---

* http://philosophy.ucsd.edu/faculty/rarneson/Courses/KANTsupposedRightToLie.pdf (3-29-10). This essay appeared in September of 1799 in Berlinische Blaetter (*Berlin Press*), published by Biester.

to the truth, then public justice cannot lay a hand on you, whatever the unforeseen consequence might be. It is indeed possible that after you have honestly answered Yes to the murderer's question as to whether the intended victim is in the house, the latter went out unobserved and thus eluded the murderer, so that the deed would not have come about. However, if you told a lie and said that the intended victim was not in the house, and he has actually (though unbeknownst to you) gone out, with the result that by so doing he has been met by the murderer and thus the deed has been perpetrated, then in this case you may be justly accused as having caused his death. For if you had told the truth as best you knew it, then the murderer might perhaps have been caught by neighbors who came running while he was searching the house for his intended victim, and thus the deed might have been prevented. Therefore, whoever tells a lie, regardless of how good his intentions may be, must answer for the consequences resulting therefrom even before a civil tribunal and must pay the penalty for them, regardless of how unforeseen those consequences may be. This is because truthfulness is a duty that must be regarded as the basis of all duties founded on contract, and the laws of such duties would be rendered uncertain and useless if even the slightest exception to them were admitted.

To be truthful (honest) in all declarations is, therefore, a sacred and unconditionally commanding law of reason that admits of no expediency whatsoever.

Monsieur Constant remarks thoughtfully and correctly with regard to the decrying of such principles that are so strict as to be alleged to [be] impracticable and therefore to be rejected. He says on page 125 thus: "A principle acknowledged as true must hence never be abandoned, however obviously there seems to be danger involved in it." And yet the good man himself abandoned the unconditional principle of truthfulness on account of the danger which that principle posed for society.

[Constant] confuses the action whereby someone does *harm* to another by telling the truth when its avowal cannot be avoided with the action whereby someone does *wrong* to another. For every man has not only a right but even the strictest duty to be truthful in statements that are unavoidable, whether this truthfulness does harm to himself or to others. Therefore he does not *himself* by this [truthfulness] actually harm the one who suffers because of it; rather, this harm is caused by accident. For he is not at all free to choose in such a case, inasmuch as truthfulness (if he must speak [i.e. must answer Yea or Nay]) is an unconditional duty. The "German philosopher" [Kant] will, therefore, not take as his principle the proposition "To tell the truth is a duty, but is a duty only with regard to the man who has a right to the truth." He will not do so, first, because truth is not a possession the right to which can be granted to one person but refused to another. But, secondly, he will not do so mainly because the duty of truthfulness makes no distinction between persons to whom one has this duty and to whom one can be excused from this duty; it is, rather, an unconditional duty which holds in all circumstances.

Right must never be adapted to politics; rather, politics must always be adapted to right.

[W]rongdoing would occur if I made the duty of truthfulness, which is wholly unconditional and which constitutes the supreme juridical condition in assertions, into a conditional duty subordinate to other considerations. And although by telling a certain lie I in fact [might] not wrong anyone, I nevertheless violate the principle of right in regard to all unavoidabl[e] statements generally. This is much worse than committing an injustice against some individual person.

The man who is asked whether or not he intends to speak truthfully in the statement that he is now to make and who does not receive the very question with indignation as regards the suspicion thereby expressed that he might be a liar, but who instead asks permission to think first about possible exceptions—that man is already a liar. This is because he shows that he does not acknowledge truthfulness as in itself a duty but reserves for himself exceptions from a rule which by its very nature does not admit of any exceptions, inasmuch as to admit of such would be self-contradictory.

This is because such exceptions would destroy the universality on account of which alone they bear the name of principles.

C. Aristotle and the Ethics of Virtue and Character

In the Principle of Utility and in the Categorical Imperative, we encounter fundamental statements of two significant and different conceptions outlining the basic structure of the moral life. The first calls attention primarily to the results or outcomes of our actions, and judges actions (or policies) morally right or wrong on their tendency to promote the public good, or to alleviate or prevent widespread human suffering. The other focuses upon "duty," that is, upon the obligations that each of us must recognize and, if we are morally worthy, to which we must then deliberately and voluntarily cause our individual behavior to conform, regardless of the consequences. Often these two conceptions are portrayed by moral philosophers as representing *competing* principles, each complete and self-contained (even though diametrically opposed in their fundamental conception of what determines right and wrong behavior), designed by their respective founders or promoters to be used as algorithms or decision procedures in guiding our every action.

While this is a convenient summary, it is not a very accurate or helpful portrayal of what each conception demands, or of the extent to which each calls attention to complementary features of a moral life. While both Kant and Mill discussed individual actions and behavior, both were more concerned to understand the basic foundations of morality as an institution, rather than to construct an iron-clad decision procedure or formula that individuals could mechanically apply to each and every moral choice they faced. The underlying assumption is that moral reasoning improves if we understand more adequately what morality consists in, what it is about. Simply providing an unreflective procedure for memorization and application would hardly have served that purpose. This, again, is why John Paul Jones rightly noted that effective military leadership requires a sound liberal education.

This is most evident when we turn to the question of character and virtue. Mill's conception of utilitarianism, for example, makes considerable demands upon those who adopt it. It is not only (as in the case of President Truman and the atomic bomb) that the calculations are difficult to carry out, and subject to controversy. Utilitarianism, as Mill observes, also demands that in evaluating the sum total of benefit or harm to result from our actions, we remain strictly impartial. Our own interests and welfare, or those of our family and friends, are to count no more or less than those of total strangers. That is an extraordinary demand to lay upon moral agents. It is unlikely, in the example we are considering, that President Truman thought himself bound by it, or that he felt obligated to extend any more impartial concern to the welfare of Japanese enemy soldiers and non-combatants than they themselves had extended to enemy soldiers and citizens (or to American and British prisoners of war) in the various Pacific-rim nations that they occupied.

We might say that this extraordinary demand for impartiality places heavy demands upon *the character* of the moral agent, or sets forth impartiality as *a "virtue" or excellence of character* to be cultivated. Likewise, utilitarianism could hardly succeed as an approach to moral reasoning if we could not presuppose the basic benevolence or good intentions of most moral agents. Someone who cares not a whit for the happiness or welfare of others, or who is too intellectually dense or psychologically insensitive to conceive of what a generally desirable result might be, can hardly be put forth as a moral exemplar in this conception. As Mill himself remarks, rather testily, "there is no difficulty in proving any ethical standard whatever to work ill if we suppose universal idiocy to be conjoined with it!"

The same anomalies emerge when we turn to Kant's "theory." Recall the feature of Kant's conception that strikes many as so odd: namely, that an action derives its moral worth, not from its form, but from its underlying intention, and that an action is morally worthy not simply because it outwardly conforms to the requirements of duty, but only if, in fact, it is also done "for the sake of duty." That is likewise to say, in Kant's conception as well, that *the moral agent must have an exemplary character*, one which recognizes the rational demands of duty upon him even when there are no external incentives or constraints to compel, constrain, or otherwise shape his behavior. A person, such as our ideal of the military officer, who can be counted on to do what she ought to do, whether she can be made to or not, whether she derives anything personally of benefit or not, or whether she faces reward or punishment or not, must be *a person of outstanding character*. So uncompromising and stringent is this demand for virtuous character that Kant has the good sense, in the *Groundwork*, to wonder whether any actual human being has ever fully and completely lived up to it.

It is, accordingly, ridiculous to do as generations of moral philosophers have done: to present these views or conceptions of the foundations of moral life as if they were "rival decision theories," offering diametrically opposed formulae to be unthinkingly applied in specific situations. Kant and Mill, and their philosophical audiences, want what we all want: a full and complete understanding of the institutions of morality, one sufficient to explain what role moral obligations play in our lives, and why we should take the demands of morality itself seriously. And, if the conceptions of morality and of moral reasoning they both put forth end up making extraordinary demands on the character and what we might call the "moral courage" of each of us, we might also wish to know what these demands are, and whether, though difficult, they are reasonable to aim at or strive for as goals to be exemplified in our lives. Kant, in particular, spent a good deal of time in his university lectures on ethics, as well as in one-half of a later, lengthy volume on "the metaphysics of morals" attempting to grapple with this problem, what he termed "the doctrine of virtue."

In comparing the two conceptions of morality that we have studied thus far, we might cautiously observe that one approach tends to favor the outcome or impact of actions or policies as the principal criterion by which our actions or policies are to be assessed, while the other focuses on the form or structure of actions and policies of two types:

1. those we are *strictly prohibited* from undertaking (Kant termed these the "perfect" or negative duties of justice); and
2. those that we would be, not simply permitted, but actually enjoined as obligations, to undertake (Kant termed these the "imperfect" or positive *duties of virtue*).

All this suggests that there is a third perspective on the moral situation, that both of our preceding "theories" or approaches address indirectly: namely, *the quality or character of the individual moral agents* engaging in these acts and living with their resultant outcomes. We might, that is, wonder not only:

1. *what outcomes or results we desire*, and
2. *what kinds of actions we may be enjoined or prohibited from undertaking* to achieve those results, but also, finally,
3. *what are the habits or qualities of character we might be best advised to cultivate within ourselves* in order to equip us to live the kinds of lives the other two conceptions recommend: namely, a life of duty, and of concern for and service to the public good.

This third approach is the domain of what is sometimes labeled "virtue ethics," and one of its principal and most famous exponents is the ancient Greek philosopher and student of Plato, Aristotle (384–322 B.C.)

Greek ethical theory is marked by a focus on the question: What is the good for man? Aristotle's answer is: *eudaimonia*, a term often loosely translated as *happiness* (a matter of *feeling* well), but perhaps more accurately rendered as *human flourishing* (a matter of *being* well, or of *living* well). Our *eudaimonia*, Aristotle argues, is a life of excellence, of aspiring to, and attaining, excellence in what we are or are meant to be "by our nature," as a navigator of a ship may strive to excel in the practice or activity of navigating her ship. What we human beings most essentially are by nature, Aristotle observes, is rational, social animals. Our perfection is therefore constituted by habits of thought, feeling, desire and choice in accordance with Reason. This is hardly a conclusion with which either Kant or Mill would disagree!

On its face, this is at once an obvious, and also a paradoxical approach to take. On the one hand, we all want to live well, and to "be excellent." But, given our profound differences in situation, historical context, and individual abilities, what can this mean? If we ask about a specific practice or profession—such as that of being a physician, a fire-fighter, or a military officer—we may likely be able to agree on a list of qualities or characteristics that define "excellence" with respect to that particular practice. So we might wish the doctor to be knowledgeable about and skilled in the practice of medicine, and also wish that individual to be compassionate, caring, truthful, respectful, and so forth. Surely the fire fighter must be brave, physically fit, and knowledgeable about the causes of, and responses to, fire that are likely to rescue its victims and minimize its harm.

Likewise, we might recognize that an "excellent" military officer would possess a number of requisite characteristics: honor, courage, integrity, physical fitness, for example, and that the officer would be steadfast and unwavering in his devotion to duty (the characteristic that Socrates cited). Like the doctor or the fire-fighter, we might also wish the military officer to be proficient in the knowledge and skills of her profession. It would avail the officer little to be a person of integrity if he or she proceeds to unleash destructive power on the wrong people, at the wrong time, or in the wrong place, or incompetently (in the wrong way). Note that, with respect to each of these practices, the list of "excellences," or what we might call "virtues of the profession" is a diverse list: some of the items refer to skill and intelligence, others seem to encompass careful and proficient practice (what we might call "prudence" or good judgment), and still others are transparently moral qualities, such as honor, truthfulness, integrity, and so forth.

Once we recognize that it makes sense to raise such questions concerning specific and well-defined human practices, such as the various professions, we might then generalize the question, as Aristotle did, to apply to the decidedly more ambiguous and less well defined "practice" of "being a human being," or of living a well-rounded and fully human life. Are there (we might now ask) "virtues," a list of characteristics of excellence, that pertain to being an excellent human being, or to pursuing and living a good human life, *per se*? Presumably, if there were such a list, attempting to habituate ourselves in these ways, and to imbue ourselves with these proper habits of heart and mind, would lead to our flourishing, or to our "happiness," both individually and collectively.

The Readings

Unlike the utilitarian and Kantian conceptions of morality, Aristotle offers nothing like a fundamental criterion or principle for resolving conflicts, let alone for identifying what specific acts we should perform in particular circumstances, or what policies we might wish to enact (or prohibit) in our society. As our reading selections from his great work, the *Nichomachean Ethics* indicate, his sole attempt at generalization in this respect is not itself a "formula" or algorithm—although later interpreters down through the ages often mistook it as such, and labeled it "Aristotle's Golden Mean" (a phrase that Aristotle himself never uses).

What Aristotle actually proposed was the sketch of a grand conception of the moral life, within which this practical advice or rule of thumb of "seeking the mean in all things" was included: namely, that a specific virtue may often, but not always, be understood as the "mean" or mid-point between two extremes, one of which constitutes a deficiency, and the other extreme an excess, respectively, of some given quality or desirable characteristic of behavior. His account of the virtue of courage in these reading selections constitutes an excellent and very well-known example of how this prudential advice is to be understood. So, courage is described as a "mean" or mid-point between the vices of cowardice one the on hand (a deficiency), and recklessness or foolhardiness (an excess) on the other. (The lack, incidentally, of precise and unambiguous labels for these virtues and vices is another hallmark of this approach to moral reasoning.)

Hitting the "bull's eye," whether in marksmanship or in morality, requires patience and practice. We are likely often to miss the mark. But if we are to miss the mark, Aristotle also advises in this instance, it is often, but not always better to err on the side of an excess rather than a deficiency. Presumably, in the example of courage, if one cannot achieve perfect courage itself, one would prefer to be found a foolhardy hot-head than to be exposed as a sniveling coward.

These sketches of "the mean" do not present a foolproof method for moral reasoning, nor were they ever meant to do so. So, for example, Aristotle himself suggests some instances in which this "method" of seeking the mean would seem absurd: there is no "mean" or mid-point for behaviors like adultery, or robbery, or murder. Yet even where it seems to apply, the concept is not a "doctrine" or a formal principle in anything like the sense intended by Mill's Principle of Utility, or Kant's Categorical Imperative, and is surely not reliable as a foolproof guide to action in specific instances. In the American civil war, for example, it is well known that the Confederate general Robert E. Lee was neither himself a coward, nor would he have been likely to tolerate cowardice in his ranks. That said, in marked contrast to Aristotle's advice about preferring an excess to a deficiency, it is likely that Lee, in retrospect would have preferred caution and cowardice at Gettysburg to the kind of spirited hot-headedness demonstrated by the brash young Jeb Stuart, whose undisciplined and disorganized forays with the Confederate cavalry likely cost Lee that fateful battle, and thus the war.

This, too, is a characteristic of so-called "virtue theory" as an approach to ethics. Even less than the other conceptions we have considered, Aristotle's conception hardly qualifies as a "theory." It is useful as a way of thinking about how individuals and societies might instill or habituate desirable traits character in their children, pupils, professional apprentices, or citizens through moral education, prudent legislation, and wise example. It is of little use, however, in deciding what to do in certain vexed or confusing circumstances. Nor can we even predict with certainty what an individual possessed of the requisite virtues is likely to do when confronted with a specific moral dilemma.

Instead, we find that Aristotle's method is first to inquire into what persons and cultures actually do and believe about virtues, and then weave these observations into a grand conception of the good life at which all persons of reason and good will might sensibly aim throughout the course of their lives. Aristotle offers along the way, as in these cases or examples, what we might take as wise, sound, or prudent advice about cultivating the kinds of balanced, moderate dispositions that aid in attaining happiness and living a good life. Moral virtues are dispositions of feeling and action regulated by reason. Ultimately, the standard for what are the proper feelings and actions is the judgments of the good man. For Aristotle, "the good man is the norm and measure of each class of things." This seems at once true and wise, on one hand, and hopelessly tautological, on the other, since it is precisely the nature of goodness (and the precise pathway to it) about which we would like to inquire more specifically.

If, rather than finding fault or probing for inconsistencies, we realize that Aristotle begins at the point at which the other "rival theories" terminate—that is, with an inquiry into moral character—then we realize we have in these readings encompassed all aspects of what might best be termed "the moral situation."[1] That is,

1. we have discussed the kind of expectations we have of morality, its *"outcome" or results*;
2. we have discussed the law-like restraints, *the "duties" governing or specifying the forms of action* necessary to achieve those results; and
3. we now conclude by considering *the traits of character of the individuals* who would find themselves ready and able to undertake these actions, live according to these rules, and be reasonably content with their results (which, Kant warned, were often not what we deserve nor always what we might reasonably hope for).

Aristotle's more detailed analysis of the virtue of courage is selected here for its obvious pertinence for warriors. Likewise, friendship, or what better be called comradeship, is also of special salience for warriors, perhaps more so than for any other profession. Aristotle's analysis of *philia* (here translated as "friendship") is specially apt since he has in mind a kind of bonding that need not (but may) involve much close personal mutual knowledge. Such bonding can extend beyond the members of one's squad or platoon to members of one's company, battalion, brigade, division, and even more widely. Aristotle says that, in the end, it is friendship that "seems to hold states together." The sense in which friendship holds combatants together during times of unimaginable stress, fear, and destruction is the topic of the companion piece by the late psychologist (and World War II combat veteran), J. Glenn Gray, excerpted from his classic study, *The Warriors: Reflections on Men in Battle*.

---

[1]This concept, particularly as its relates virtue ethics to Kant's moral philosophy, is described in more detail in G. R. Lucas, "Agency after Virtue," *International Philosophical Quarterly*, 28, no. 3 (September, 1988) 293–311.

The Moral Virtues
*Aristotle**

W.T. Jones, a historian of philosophy, conveniently summarizes Aristotle's views in the following table:

Activity	Vice (excess)	Virtue (mean)	Vice (deficit)
Facing death	Too much fear (i.e., cowardice)	Right amount of fear (i.e., courage)	Too little fear (i.e., foolhardiness)
Bodily actions (eating, drinking, sex, etc.)	Profligacy	Temperance	No name for this state, but it may be called "insensitivity"
Giving money	Prodigality	Liberality	Illiberality
Large-scale giving	Vulgarity	Magnificence	Meanness
Claiming honors	Vanity	Pride	Humility
Social intercourse	Obsequiousness	Friendliness	Sulkiness
According honors	Injustice	Justice	Injustice
Retribution for wrongdoing	Injustice	Justice	Injustice

Source: W. T. Jones, *The Classical Mind* (New York: Harcourt, Brace, & World, 1952, 1969), p. 268.

No doubt people will say, "To call happiness the highest good is a truism. We want a more distinct account of what it is." We might arrive at this if we could grasp what is meant by the "function" of a human being. If we take a flautist or a sculptor or any craftsman—in fact any class of men at all who have some special job or profession—we find that his special talent and excellence comes out in that job, and this is his function. The same thing will be true of man simply as man—that is of course if

---

*From *The Ethics of Aristotle*. Translated by J.A.K. Thomson. Reprinted with permission of Routledge, Ltd. UK.

"man" does have a function. But is it likely that joiners and shoemakers have certain functions or specialized activities, while man as such has none but has been left by Nature a functionless being? Seeing that eye and hand and foot and every one of our members has some obvious function, must we not believe that in like manner a human being has a function over and above these particular functions? Then what exactly is it? The mere act of living is not peculiar to man—we find it even in the vegetable kingdom—and what we are looking for is something peculiar to him. We must therefore exclude from our definition the life that manifests itself in mere nurture and growth. A step higher should come the life that is confined to experiencing sensations. But that we see is shared by horses, cows, and the brute creation as a whole. We are left, then, with a life concerning which we can make two statements. First, it belongs to the rational part of man. Secondly, it finds expression in actions. The rational part may be either active or passive: passive in so far as it follows the dictates of reason, active in so far as it possesses and exercises the power of reasoning. A similar distinction can be drawn within the rational life; that is to say, the reasonable element in it may be active or passive. Let us take it that what we are concerned with here is the reasoning power in action, for it will be generally allowed that when we speak of "reasoning" we really mean *exercising* our reasoning faculties. (This seems the more correct use of the word.) Now let us assume for the moment the truth of the following propositions. (a) The function of a man is the exercise of his non-corporeal faculties or "soul" in accordance with, or at least not divorced from, a rational principle. (b) The function of an individual and of a *good* individual in the same class—a harp player, for example, and a good harp player, and so through the classes—is generically the same, except that we must add superiority in accomplishment to the function, the function of the harp player being merely to play on the harp, while the function of the good harp player is to play on it well. (c) The function of man is a certain form of life, namely an activity of the soul exercised in combination with a rational principle or reasonable ground of action. (d) The function of a good man is to exert such activity well. (e) A function is performed well when performed in accordance with the excellence proper to it.—If these assumptions are granted, we conclude that the good for man is "an activity of soul in accordance with goodness" or (on the supposition that there may be more than one form of goodness) "in accordance with the best and most complete form of goodness."

. . . Let us begin, then, with this proposition. Excellence of whatever kind affects that of which it is the excellence in two ways. (1) It produces a good state in it. (2) It enables it to perform its function well. Take eyesight. The goodness of your eye is not only that which makes your eye good, it is also that which makes it function well. Or take the case of a horse. The goodness of a horse makes him a good horse, but it also makes him good at running, carrying a rider, and facing the enemy. Our proposition, then, seems to be true, and it enables us to say that virtue in a man will be the disposition which (a) makes him a good man, (b) enables him to perform his function well. We have already touched on this point, but more light will be thrown upon it if we consider what is the specific nature of virtue.

Every form, then, of applied knowledge, when it performs its function well, looks to the mean and works to the standard set by that. It is because people feel this that they apply the *cliché*, "You couldn't add anything to it or take anything from it" to an artistic masterpiece, the implication being that too much and too little alike destroy perfection, while the mean preserves it. Now if this be so, and if it be true, as we say, that good craftsmen work to the standard of the mean, then, since goodness like Nature is more exact and of a higher character than any art, it follows that goodness is the quality that hits the mean. By "goodness" I mean goodness of moral character, since it is moral goodness that deals with feelings and actions, and it is in them that we find excess, deficiency, and a mean. It is possible, for example, to experience fear, boldness, desire, anger, pity, and pleasures and pains generally, too much or too little or to the right amount. If we feel them too much or too little, we are wrong. But to have these feelings at the right times on the right occasions towards the right people for the right motive and in the right way is to have them in the right measure, that is, somewhere between the extremes; and this is what characterizes goodness. The same may be said of the mean and extremes in actions. Now it is in the field of actions and feelings that goodness operates; in them we find excess, deficiency, and, between them, the mean, the first two being wrong, the mean right and praised as such. Goodness, then, is a mean condition in the sense that it aims at and hits the mean.

Consider, too, that it is possible to go wrong in more ways than one. (In Pythagorean terminology evil is a form of the Unlimited, good of the Limited.) But there is only one way of being right. That is why going wrong is easy, and going right is difficult; it is easy to miss the bull's-eye and difficult to hit it. Here, then, is another explanation of why the too much and the too little are connected with evil and the mean with good. As the poet says,

> The Good are good simply, while the bad are evil in every sort of way.

We may now define virtue as a disposition of the soul in which, when it has to choose among actions and feelings, it observes the mean relative to us, this being determined by such a rule or principle as would take shape in the mind of a man of sense or practical wisdom. We call it a mean condition as lying between two forms of badness, one being excess and the other deficiency; and also for this reason, that, whereas badness either falls short of or exceeds the right measure in feelings and actions, virtue discovers the mean and deliberately chooses it. Thus, looked at from the point of view of its essence as embodied in its definition, virtue no doubt is a mean; judged by the standard of what is right and best, it is an extreme.

Aristotle enters a caution. Though we have said that virtue observes the mean in actions and passions, we do not say this of all acts and all feelings. Some are essentially evil and, when these are involved, our rule of applying the mean cannot be brought into operation.[1]

But choice of a mean is not possible in every action or every feeling. The very names of some have an immediate connotation of evil. Such are malice, shamelessness, envy among feelings, and among actions adultery, theft, murder. All these and more like them have a bad name as being evil in themselves; it is not merely the excess or deficiency of them that we censure. In their case, then, it is impossible to act rightly; whatever we do is wrong. Nor do circumstances make any difference in the rightness or wrongness of them. When a man commits adultery there is no point in asking whether it is with the right woman or at the right time or in the right way, for to do anything like that is simply wrong. It would amount to claiming that there is a mean and excess and defect in unjust or cowardly or intemperate actions. If such a thing were possible, we should find ourselves with a mean quantity of excess, a mean of deficiency, an excess of access and a deficiency of deficiency. But just as in temperance and justice there can be no mean or excess or deficiency, because the mean in a sense is an extreme, so there can be no mean or excess or deficiency in those vicious actions—however done, they are wrong. Putting the matter into general language, we may say that there is no mean in the extremes, and no extreme in the mean, to be observed by anybody.

After the definition comes its application to the particular virtues. In these it is always possible to discover a mean—at which the virtue aims—between an excess and a deficiency. Here Aristotle found that a table or diagram of the virtues between their corresponding vices would be useful, and we are to imagine him referring to this in the course of his lectures.

But a generalization of this kind is not enough; we must show that our definition fits particular cases. When we are discussing actions particular statements come nearer the heart of the matter, though general statements cover a wider field. The reason is that human behaviour consists in the performance of particular acts, and our theories must be brought in to harmony with them.

You see here a diagram of the virtues. Let us take our particular instances from that. In the section confined to the feelings inspired by danger you will observe that the mean state is "courage." Of those who go to extremes in one direction or the other the man who shows an excess of fearlessness has no name to describe him, the man who exceeds in confidence or daring is called "rash" or "foolhardy," the man who shows an excess of fear and a deficiency of confidence is called a "coward." In the pleasures and pains—though not all pleasures and pains, especially pains—the virtue which observes the mean is "temperance," the excess is the vice of "intemperance." Persons defective in the power to enjoy pleasures are a somewhat rare class, and so have not had a name assigned to them: suppose we call them "unimpressionable." Coming to the giving and acquiring of money, we find that the mean is "liberality," the excess "prodigality," the deficiency "meanness." But here we meet a complication. The

---

[1] The italicized interpolations in this selection are the translator's.

prodigal man and the mean man exceed and fall short in opposite ways. The prodigal exceeds in giving and falls short in getting money, whereas the mean man exceeds in getting and falls short in giving it away. Of course this is but a summary account of the matter—a bare outline. But it meets our immediate requirements. Later on these types of character will be more accurately delineated.

But there are other dispositions which declare themselves in the way they deal with money. One is "lordliness" or "magnificence," which differs from liberality in that the lordly man deals in large sums, the liberal man is small. Magnificence is the mean state here, the excess is "bad taste" or "vulgarity," the defect is "shabbiness." These are not the same as the excess and defect on either side of liberality. How they differ is a point which will be discussed later. In the matter of honour the mean is "proper pride," the excess "vanity," the defect "poor-spiritedness." And just as liberality differs, as I said, from magnificence in being concerned with small sums of money, so there is a state related to proper pride in the same way being concerned with small honours, while pride is concerned with great. For it is possible to aspire to small honours in the right way, or to a greater or less extent than is right. The man who has this aspiration to excess is called "ambitious;" if he does not cherish it enough, he is "unambitious;" but the man who has it to the right extent—that is, strikes the mean—has no special designation. This is true also of corresponding dispositions with one exception, that of the ambitious man, which is called "ambitiousness." This will explain why each of the extreme characters stakes out a claim in the middle region. Indeed we ourselves call the character between the extremes sometimes "ambitious" and sometimes "unambitious." That is proved by our sometimes praising a man for being ambitious and sometimes for being unambitious. The reason will appear later. In the meantime let us continue our discussion of the remaining virtues and vices, following the method already laid down.

Let us next take anger. Here too we find excess, deficiency, and the mean. Hardly one of the states of mind involved has a special name; but, since we call the man who attains the mean in this sphere "gentle," we may call his disposition "gentleness." Of the extremes the man who is angry over-much may be called "irascible," and his vice "irascibility;" while the man who reacts too feebly to anger may be called "poor-spirited" and his disposition "poor-spiritedness."

. . . As regards veracity, the character who aims at the mean may be called "truthful" and what he aims at "truthfulness." Pretending, when it goes too far, is "boastfulness" and the man who shows it is a "boaster" or "braggart." If it takes the form of understatement, the pretence is called "irony" and the man who shows it "ironical." In agreeableness in social amusement the man who hits the mean is "witty" and what characterizes him is "wittiness." The excess is "buffoonery" and the man who exhibits that is a "buffoon." The opposite of the buffoon is the "boor" and his characteristic is "boorishness." In the other sphere of the agreeable—the general business of life—the person who is agreeable in the right way is "friendly" and his disposition "friendliness." The man who makes himself too agreeable, supposing him to have no ulterior object, is "obsequious;" if he has such an object, he is a "flatterer." The man who is deficient in this quality and takes every opportunity of making himself disagreeable may be called "peevish" or "sulky" or "surly."

> *But it is not only in settled dispositions that a mean may be observed in passing states of emotion.*

Even when feelings and emotional states are involved one notes that mean conditions exist. And here also, it would be agreed, we may find one man observing the mean and another going beyond it, for instance, the "shamefaced" man, who is put out of countenance by anything. Or a man may fall short here of the due mean. Thus any one who is deficient in a sense of shame, or has none at all, is called "shameless." The man who avoids both extremes is "modest," and him we praise. For, while modesty is not a form of goodness, it is praised; it and the modest man. Then there is "righteous indignation." This is felt by any one who strikes the mean between "envy" and "malice," by which last word I mean a pleased feeling at the misfortunes of other people. These are emotions concerned with the pains and pleasures we feel at the fortunes of our neighbours. The man who feels righteous indignation is pained by undeserved good fortune; but the envious man goes beyond that and is pained at anybody's success. The malicious man, on the other hand, is so far from being pained by the misfortunes of another that he is actually tickled by them.

However, a fitting opportunity of discussing these matters will present itself in another place. And after that we shall treat of justice. In that connexion we shall have to distinguish between the various kinds of justice—for the word is used in more senses than one—and show in what way each of them is a mean.

> *But after all, proceeds Aristotle, the true determinant of the mean is not the geometer's rod but the guiding principle in the good man's soul. The diagram of the virtues and vices, then, is just an arrangement and, as Aristotle goes on to show, an unimportant one at that.*

Thus there are three dispositions, two of them taking a vicious form (one in the direction of excess, the other of defect) and one a good form, namely, the observance of the mean. They are all opposed to one another, though not all in the same way. The extreme states are opposed both to the mean and one another, and the mean is opposed to both extremes. For just as the equal is greater compared with the less, and less compared with the greater, so the mean states (whether in feelings or actions) are in excess if compared with the deficient, and deficient if compared with the excessive, states. Thus a brave man appears rash when set beside a coward, and cowardly when set beside a rash man; a temperate man appears intemperate beside a man of dull sensibilities, and dull if contrasted with an intemperate man. This is the reason why each extreme character tries to push the mean nearer the other. The coward calls the brave man rash, the rash man calls him a coward. And so in the other cases. But, while all the dispositions are opposed to one another in this way, the greatest degree of opposition is that which is found between the two extremes. For they are separated by a greater interval from one another than from the mean, as the great is more widely removed from the small, and the small from the great, than either from the equal. It may be added that sometimes an extreme bears a certain resemblance to a mean. For example, rashness resembles courage, and prodigality resembles liberality. But between the extremes there is always the maximum dissimilarity. Now opposites are by definition things as far removed as possible from one another. Hence the farther apart things are, the more opposite they will be. Sometimes it is the deficiency, in other instances it is the excess, that is more directly opposed to the mean. Thus cowardice, a deficiency, is more opposed to courage than is rashness, an excess. And it is not insensibility, the deficiency, that is more opposed to temperance but intemperance, the excess. This arises from one or other of two causes. One lies in the nature of the thing itself and may be explained as follows. When one extreme is nearer to the mean and resembles it more, it is not that extreme but the other which we tend to oppose to the mean. For instance, since rashness is held to be nearer and liker to courage than is cowardice, it is cowardice which we tend to oppose to courage on the principle that the extremes which are remoter from the mean strike us as more opposite to it. The other cause lies in ourselves. It is the things to which we are naturally inclined that appear to us more opposed to the mean. For example, we have a natural inclination to pleasure, which makes us prone to fall into intemperance. Accordingly we tend to describe as opposite to the mean those things towards which we have an instinctive inclination. For this reason intemperance, the excess, is more opposed to temperance than is insensibility to pleasure, the deficiency.

I have said enough to show that moral excellence is a mean, and I have shown in what sense it is so. It is, namely, a mean between two forms of badness, one of excess and the other of defect, and is so described because it aims at hitting the mean point in feelings and in actions. This makes virtue hard of achievement, because finding the middle point is never easy. It is not everybody, for instance, who can find the centre of a circle—that calls for a geometrician. Thus, too, it is easy to fly into a passion—anybody can do that—but to be angry with the right person and to the right extent and at the right time and with the right object and in the right way—that is not easy, and it is not everyone who can do it. This is equally true of giving or spending money. Hence we infer that to do these things properly is rare, laudable, and fine.

> *Aristotle now suggests some rules for our guidance.*

In view of this we shall find it useful when aiming at the mean to observe these rules. (1) *Keep away from that extreme which is the more opposed to the mean.* It is Calypso's advice:

> *Swing around the ship clear of this surf and surge.*

For one of the extremes is always a more dangerous error than the other; and—since it is hard to hit the bull's-eye—we must take the next best course and choose the least of the evils. And it will be easiest for us to do this if we follow the rule I have suggested. (2) *Note the errors into which we personally are most liable to fall.* (Each of us has his natural bias in one direction or another.) We shall find out what ours are by noting what gives us pleasure and pain. After that we must drag ourselves in the opposite direction. For our best way of reaching the middle is by giving a wide berth to our darling sin. It is the method used by a carpenter when he is straightening a warped board. (3) *Always be particularly on your guard against pleasure and pleasant things.* When Pleasure is at the bar the jury is not impartial. So it will be best for us if we feel towards her as the Trojan elders felt towards Helen, and regularly apply their words to her. If we are for packing her off, as they were with Helen, we shall be the less likely to go wrong.

To sum up. These are the rules by observation of which we have the best chance of hitting the mean. But of course difficulties spring up, especially when we are confronted with an exceptional case. For example, it is not easy to say precisely what is the right way to be angry and with whom and on what grounds and for how long. In fact, we are inconsistent on this point, sometimes praising people who are deficient in the capacity for anger and calling them "gentle," sometimes praising the choleric and calling them "stout fellows." To be sure we are not hard on a man who goes off the straight path in the direction of too much or too little, if he goes off only a little way. We reserve our censure for the man who swerves widely from the course, because then we are bound to notice it. Yet it is not easy to find a formula by which we may determine how far and up to what point a man may go wrong before he incurs blame. But this difficulty of definition is inherent in every object of perception; such questions of degree are bound up with the circumstances of the individual case, where our only criterion is the perception.

So much, then, has become clear. In all our conduct it is the mean state that is to be praised. But one should lean sometimes in the direction of the more, sometimes in that of the less, because that is the readiest way of attaining to goodness and the mean.

Habit and Virtue
Aristotle*

Virtues of Character in General

How a virtue of character is acquired

Virtue, then, is of two sorts, virtue of thought and virtue of character. Virtue of thought arises and grows mostly from teaching, and hence needs experience and time. Virtue of character [i.e. of *ethos*] results from habit [*ethos*]; hence its name "ethical," slightly varied from *"ethos."*

Virtue comes about, not by a process of nature, but by habituation

Hence it is also clear that none of the virtues of character arises in us naturally.

(1) What is natural cannot be changed by habituation

For if something is by nature [in one condition], habituation cannot bring it into another condition. A stone, e.g., by nature moves downwards, and habituation could not make it move upwards, not even if you threw it up ten thousand times to habituate it; nor could habituation make fire move downwards, or bring anything that is by nature in one condition into another condition.

Thus the virtues arise in us neither by nature nor against nature, but we are by nature able to acquire them, and reach our complete perfection through habit.

(2) Natural capacities are not acquired by habituation

Further, if something arises in us by nature, we first have the capacity for it, and later display the activity. This is clear in the case of the senses; for we did not acquire them by frequent seeing or hearing, but already had them when we exercised them, and did not get them by exercising them.

Virtues, by contrast, we acquire, just as we acquire crafts, by having previously activated them. For we learn a craft by producing the same product that we must produce when we have learned it, becoming builders, e.g., by building and harpists by playing the harp; so also, then, we become just by doing just actions, temperate by doing temperate actions, brave by doing brave actions.

(3) Legislators concentrate on habituation

What goes on in cities is evidence for this also. For the legislator makes the citizens good by habituating them, and this is the wish of every legislator; if he fails to do it well he misses his goal. [The right] habituation is what makes the difference between a good political system and a bad one.

---

*"Habit and Virtue" from *Nicomachean Ethics* by Aristotle. Translated by Terence Irwin (Indianapolis, IN: Hackett, 1985), pp. 33–40.

(4) Virtue and vice are formed by good and bad actions

Further, just as in the case of a craft, the sources and means that develop each virtue also ruin it. For playing the harp makes both good and bad harpists, and it is analogous in the case of builders and all the rest; for building well makes good builders, building badly, bad ones. If it were not so, no teacher would be needed, but everyone would be born a good or a bad craftsman.

It is the same, then, with the virtues. For actions in dealings with [other] human beings make some people just, some unjust; actions in terrifying situations and the acquired habit of fear or confidence make some brave and others cowardly. The same is true of situations involving appetites and anger; for one or another sort of conduct in these situations makes some people temperate and gentle, others intemperate and irascible.

Conclusion: The importance of habituation

To sum up, then, in a single account: A state [of character] arises from [the repetition of] similar activities. Hence we must display the right activities, since differences in these imply corresponding differences in the states. It is not unimportant, then, to acquire one sort of habit or another, right from our youth; rather, it is very important, indeed all-important.

What Is the Right Sort of Habituation?

This is an appropriate question, for the aim of ethical theory is practical

Our present inquiry does not aim, as our others do, at study; for the purpose of our examination is not to know what virtue is, but to become good, since otherwise the inquiry would be of no benefit to us. Hence we must examine the right way to act, since, as we have said, the actions also control the character of the states we acquire.

First, then, actions should express correct reason. That is a common [belief], and let us assume it; later we will say what correct reason is and how it is related to the other virtues.

But let us take it as agreed in advance that every account of the actions we must do has to be stated in outline, not exactly. As we also said at the start, the type of accounts we demand should reflect the subject-matter; and questions about actions and expediency, like questions about health, have no fixed [and invariable answers].

And when our general account is so inexact, the account of particular eases is all the more inexact. For these fall under no craft or profession, and the agents themselves must consider in each case what the opportune action is, as doctors and navigators do.

The account we offer, then, in our present inquiry is of this inexact sort; still, we must try to offer help.

The right sort of habituation must avoid excess and deficiency

First, then, we should observe that these sorts of states naturally tend to be ruined by excess and deficiency. We see this happen with strength and health, which we mention because we must use what is evident as a witness to what is not. For both excessive and deficient exercises ruin strength; and likewise, too much or too little eating or drinking ruins health, while the proportionate amount produces, increases and preserves it.

The same is true, then, of temperance, bravery and the other virtues. For if, e.g., someone avoids and is afraid of everything, standing firm against nothing, he becomes cowardly, but if he is afraid of nothing at all and goes to face everything he becomes rash. Similarly, if he gratifies himself with every pleasure and refrains from none, he becomes intemperate, but if he avoids them all, as boors do, he becomes some sort of insensible person. Temperance and bravery, then, are ruined by excess and deficiency but preserved by the mean.

The same actions, then, are the sources and causes both of the emergence and growth of virtues and of their ruin; but further, the activities of the virtues will be found in these same actions. For this is also true of more evident cases, e.g. strength, which arises from eating a lot and from withstanding much hard labour, and it is the strong person who is most able to do these very things. It is the same with the virtues. Refraining from pleasures makes us become temperate, and when we have become temperate we are most able to refrain from pleasures. And it is similar with bravery; habituation in disdaining what is fearful and in standing firm against it makes us become brave, and when we have become brave we shall be most able to stand firm.

Pleasure and pain are important in habituation

But [actions are not enough]; we must take as a sign of someone's state his pleasure or pain in consequence of his action. For if someone who abstains from bodily pleasures enjoys the abstinence itself, then he is temperate, but if he is grieved by it, he is intemperate. Again, if he stands firm against terrifying situations and enjoys it, or at least does not find it painful, then he is brave, and if he finds it painful, he is cowardly.

[Pleasures and pains are appropriately taken as signs] because virtue of character is concerned with pleasures and pains.

Virtue is concerned with pleasure and pain

(1) For it is pleasure that causes us to do base actions, and pain that causes us to abstain from fine ones. Hence we need to have had the appropriate upbringing—right from early youth, as Plato says—to make us find enjoyment or pain in the right things; for this is the correct education.

(2) Further, virtues are concerned with actions and feelings; but every feeling and every action implies pleasure or pain; hence, for this reason too, virtue is concerned with pleasures and pains.

(3) Corrective treatment [for vicious actions] also indicates [the relevance of pleasure and pain], since it uses pleasures and pains; it uses them because such correction is a form of medical treatment, and medical treatment naturally operates through contraries.

(4) Further, as we said earlier, every state of soul is naturally related to and concerned with whatever naturally makes it better or worse; and pleasures and pains make people worse, from pursuing and avoiding the wrong ones, at the wrong time, in the wrong ways, or whatever other distinctions of that sort are needed in an account.

These [bad effects of pleasure and pain] are the reason why people actually define the virtues as ways of being unaffected and undisturbed [by pleasures and pains]. They are wrong, however, because they speak [of being unaffected] unconditionally, not of being unaffected in the right or wrong way, at the right or wrong time, and the added specifications.

We assume, then, that virtue is the sort of state [with the appropriate specifications] that does the best actions concerned with pleasures and pains, and that vice is the contrary. The following points will also make it evident that virtue and vice are concerned with the same things.

(5) There are three objects of choice—fine, expedient and pleasant—and three objects of avoidance—their contraries, shameful, harmful and painful. About all these, then, the good person is correct and the bad person is in error, and especially about pleasure. For pleasure is shared with animals, and implied by every object of choice, since what is fine and what is expedient appear pleasant as well.

(6) Further, since pleasure grows up with all of us from infancy on, it is hard to rub out this feeling that is dyed into our lives; and we estimate actions as well [as feelings], some of us more, some less, by pleasure and pain. Hence, our whole inquiry must be about these, since good or bad enjoyment or pain is very important for our actions.

(7) Moreover, it is harder to fight pleasure than to fight emotion, [though that is hard enough], as Heraclitus says. Now both craft and virtue are concerned in every case with what is harder, since a good result is even better when it is harder. Hence, for this reason also, the whole inquiry, for virtue and political science alike, must consider pleasures and pains; for if we use these well, we shall be good, and if badly, bad.

In short, virtue is concerned with pleasures and pains; the actions that are its sources also increase it or, if they are done differently, ruin it; and its activity is concerned with the same actions that are its sources.

But our claims about habituation raise a puzzle: How can we become good without being good already?

However, someone might raise this puzzle: "What do you mean by saying that to become just we must first do just actions and to become temperate we must first do temperate actions? For if we do what is grammatical or musical, we must already be grammarians or musicians. In the same way, then, if we do what is just or temperate, we must already be just or temperate."

First reply: Conformity versus understanding

But surely this is not so even with the crafts, for it is possible to produce something grammatical by chance or by following someone else's instructions. To be a grammarian, then, we must both produce something grammatical and produce it in the way in which the grammarian produces it, i.e. expressing grammatical knowledge that is in us.

Second reply: Crafts versus virtues

Moreover, in any case what is true of crafts is not true of virtues. For the products of a craft determine by their own character whether they have been produced well; and so it suffices that they are in the right state when they have been produced. But for actions expressing virtue to be done temperately or justly [and hence well] it does not suffice that they are themselves in the right state. Rather, the agent must also be in the right state when he does them. First, he must know [that he is doing virtuous actions]; second, he must decide on them, and decide on them for themselves; and, third, he must also do them from a firm and unchanging state.

As conditions for having a craft these three do not count, except for the knowing itself. As a condition for having a virtue, however, the knowing counts for nothing, or rather for only a little, whereas the other two conditions are very important, indeed all-important. And these other two conditions are achieved by the frequent doing of just and temperate actions.

Hence actions are called just or temperate when they are the sort that a just or temperate person would do. But the just and temperate person is not the one who [merely] does these actions, but the one who also does them in the way in which just or temperate people do them.

It is right, then, to say that a person comes to be just from doing just actions and temperate from doing temperate actions; for no one has even a prospect of becoming good from failing to do them.

Virtue requires habituation, and therefore requires practice, not just theory

The many, however, do not do these actions but take refuge in arguments, thinking that they are doing philosophy, and that this is the way to become excellent people. In this they are like a sick person who listens attentively to the doctor, but acts on none of his instructions. Such a course of treatment will not improve the state of his body; any more than will the many's way of doing philosophy improve the state of their souls.

Courage
*Aristotle**

Book III—Moral Virtue
Courage

Courage is concerned with the feelings of fear and confidence—strictly speaking, with the fear of death in battle.

6. That it is a mean with regard to feelings of fear and confidence has already been made evident[1]; and plainly the things we fear are fearful things, and these are, to speak without qualification, evils; for which reason people even define fear as expectation of evil. Now we fear all evils, e.g. disgrace, poverty, disease, friendlessness, death, but the brave man is not thought to be concerned with all; for to fear some things is even right and noble, and it is base not to fear them—e.g. disgrace; he who fears this is good and modest and he who does not is shameless. He is, however, by some people called brave, by a transference of the word to a new meaning; for he has in him something which is like the brave man, since the brave man also is a fearless person. Poverty and disease we perhaps ought not to fear, nor in general the things that do not proceed from vice and are not due to a man himself. But not even the man who is fearless of these is brave. Yet we apply the word to him also in virtue of a similarity; for some who in the dangers of war are cowards are liberal and are confident in face of the loss of money. Nor is a man a coward if he fears insult to his wife and children or envy or anything of the kind; nor brave if he is confident when he is about to be flogged. With what sort of fearful things, then, is the brave man concerned? Surely with the greatest; for no one is more likely than he to stand his ground against what is awe-inspiring. Now death is the most fearful of all things; for it is the end, and nothing is thought to be any longer either good or bad for the dead. But the brave man would not seem to be concerned even with death in all circumstances, e.g. at sea or in disease. In what circumstances, then? Surely in the noblest. Now such deaths are those in battle; for these take place in the greatest and noblest danger. And these are correspondingly honoured in city-states and at the courts of monarchs. Properly, then, he will be called brave who is fearless in face of a noble death, and of all emergencies that involve death; and the emergencies of war are in the highest degree of this kind. Yet at sea also, and in disease, the brave man is fearless, but not in the same way as the seamen; for he has given up hope of safety, and is disliking the thought of death in this shape, while they are hopeful because of their experience. At the same time, we show courage in situations where there is the opportunity of showing prowess or where death is noble; but in these forms of death neither of these conditions is fulfilled.

---

*Excerpts from "The Nicomachean Ethics" are reprinted from *Introduction to Aristotle* by Richard McKeon. Copyright © 1973 Zahava McKeon.

[1] 1107^a33–$^b4$.

The motive of courage is the sense of honour: characteristics of the opposite vices, cowardice and rashness

7. What is fearful is not the same for all men; but we say there are things fearful even beyond human strength. These, then, are fearful to every one—at least to every sensible man; but the fearful things that are not beyond human strength differ in magnitude and degree, and so too do the things that inspire confidence. Now the brave man is as dauntless as man may be. Therefore, while he will fear even the things that are not beyond human strength, he will face them as he ought and as the rule directs, for honour's sake; for this is the end of virtue. But it is possible to fear these more, or less, and again to fear things that are not fearful as if they were. Of the faults that are committed, one consists in fearing what we should not, another in fearing as we should not, another in fearing when we should not, and so on; and so too with respect to the things that inspire confidence. The man, then, who faces and who fears the right things and from the right motive, in the right way and at the right time, and who feels confidence under the corresponding conditions, is brave; for the brave man feels and acts according to the merits of the case and in whatever way the rule directs. Now the end of every activity is conformity to the corresponding state of character. This is true, therefore, of the brave man as well as of others. But courage is noble. Therefore the end also is noble; for each thing is defined by its end. Therefore it is for a noble end that the brave man endures and acts as courage directs.

Of those who go to excess he who exceeds in fearlessness has no name (we have said previously that many states of character have no names),[2] but he would be a sort of madman or insensitive to pain if he feared nothing, neither earthquakes nor the waves, as they say the Celts do not; while the man who exceeds in confidence about what really is fearful is rash. The rash man, however, is also thought to be boastful and only a pretender to courage; at all events, as the brave man is with regard to what is fearful, so the rash man wishes to *appear*; and so he imitates him in situations where he can. Hence also most of them are a mixture of rashness and cowardice; for, while in these situations they display confidence, they do not hold their ground against what is really fearful. The man who exceeds in fear is a coward; for he fears both what he ought not and as he ought not, and all the similar characterizations attach to him. He is lacking also in confidence; but he is more conspicuous for his excess of fear in painful situations. The coward, then, is a despairing sort of person; for he fears everything. The brave man, on the other hand, has the opposite disposition; for confidence is the mark of a hopeful disposition. The coward, the rash man, and the brave man, then, are concerned with the same objects but are differently disposed towards them; for the first two exceed and fall short, while the third holds the middle, which is the right, position; and rash men are precipitate, and wish for dangers beforehand but draw back when they are in them, while brave men are excited in the moment of action, but collected beforehand.

As we have said, then, courage is a mean with respect to things that inspire confidence or fear, in the circumstances that have been stated[3]; and it chooses or endures things because it is noble to do so, or because it is base not to do so.[4] But to die to escape from poverty or love or anything painful is not the mark of a brave man, but rather of a coward; for it is softness to fly from what is troublesome, and such a man endures death not because it is noble but to fly from evil.

Five kinds of courage improperly so called

8. Courage, then, is something of this sort, but the name is also applied to five other kinds. First comes the courage of the citizen-soldier; for this is most like true courage. Citizen-soldiers seem to face dangers because of the penalties imposed by the laws and the reproaches they would otherwise incur, and because of the honours they win by such action; and therefore those peoples seem to be bravest among whom cowards are held in dishonour and brave men in honour. This is the kind of courage that Homer depicts, *e.g.* in Diomede and in Hector:

---

[2] 1107^b2, cf. 1107^b29, 1108^a5.
[3] Ch.6.
[4] $1115^b11–24$.

> First will Polydamas be to heap reproach on me then[5]; For Hector one day mid the
> Trojans shall utter his vaunting harangue:
> 'Afraid was Tydeides, and fled from my face.'[6]

This kind of courage is most like to that which we described earlier,[7] because it is due to virtue; for it is due to shame and to desire of a noble object (i.e. honour) and avoidance of disgrace, which is ignoble. One might rank in the same class even those who are compelled by their rulers; but they are inferior, inasmuch as they do what they do not from shame but from fear, and to avoid not what is disgraceful but what is painful; for their masters compel them, as Hector[8] does:

> But if I shall spy any dastard that cowers far from the fight,
> Vainly will such an one hope to escape from the dogs.

And those who give them their posts, and beat them if they retreat, do the same, and so do those who draw them up with trenches or something of the sort behind them; all of these apply compulsion. But one ought to be brave not under compulsion but because it is noble to be so.

Experience with regard to particular facts is also thought to be courage; this is indeed the reason why Socrates thought courage was knowledge. Other people exhibit this quality in other dangers, and professional soldiers exhibit it in the dangers of war; for there seem to be many empty alarms in war, of which these have had the most comprehensive experience; therefore they seem brave, because the others do not know the nature of the facts. Again, their experience makes them most capable in attack and in defence, since they can use their arms and have the kind that are likely to be best both for attack and for defence; therefore they fight like armed men against unarmed or like trained athletes against amateurs; for in such contests too it is not the bravest men that fight best, but those who are strongest and have their bodies in the best condition. Professional soldiers turn cowards, however, when the danger puts too great a strain on them and they are inferior in numbers and equipment; for they are the first to fly, while citizen-forces die at their posts, as in fact happened at the temple of Hermes.[9] For to the latter flight is disgraceful and death is preferable to safety on those terms; while the former from the very beginning faced the danger on the assumption that they were stronger, and when they know the facts they fly, fearing death more than disgrace; but the brave man is not that sort of person.

Passion also is sometimes reckoned as courage; those who act from passion, like wild beasts rushing at those who have wounded them, are thought to be brave, because brave men also are passionate, for passion above all things is eager to rush on danger, and hence Homer's 'put strength into his passion'[10] and 'aroused their spirit and passion'[11] and 'hard he breathed panting'[12] and 'his blood boiled.'[13] For all such expressions seem to indicate the stirring and onset of passion. Now brave men act for honour's sake, but passion aids them; while wild beasts act under the influence of pain; for they attack because they have been wounded or because they are afraid, since if they are in a forest they do not come near one. Thus they are not brave because, driven by pain and passion, they rush on danger without foreseeing any of the perils, since at that rate even asses would be brave when they are hungry; for blows will not drive them from their food; and lust also makes adulterers do many daring things. Those creatures are not brave, then, which are driven on to danger by pain or passion. The

---

[5]Il. xxii 100.

[6]Il. viii. 148, 149.

[7]Chs. 6, 7.

[8]Aristotle's quotation is more like Il. ii. 391–3, where Agamemnon speaks, than xv. 348–51, where Hector speaks.

[9]The reference is to a battle at Coronea in the Sacred War, c. 353 B.C., in which the Phocians defeated the citizens of Coronea and some Boeotian regulars.

[10]This is a conflation of Ii. xi. 11 or xiv. 151 and xvi. 529.

[11]Cf. Il. v. 470, xv. 232, 594.

[12]Cf. Od. xxiv. 318f.

[13]The phrase does not occur in Homer; it is found in Theocr xx. 15.

'courage' that is due to passion seems to be the most natural, and to be courage if choice and motive be added.

Men, then, as well as beasts, suffer pain when they are angry; and are pleased when they exact their revenge; those who fight for these reasons, however, are pugnacious but not brave; for they do not act for honour's sake nor as the rule directs, but from strength of feeling; they have, however, something akin to courage.

Nor are sanguine people brave; for they are confident in danger only because they have conquered often and against many foes. Yet they closely resemble brave men, because both are confident; but brave men are confident for the reasons stated earlier, while these are so because they think they are the strongest and can suffer nothing. (Drunken men also behave in this way; they become sanguine.) When their adventures do not succeed, however, they run away; but it was[14] the mark of a brave man to face things that are, and seem, terrible for a man, because it is noble to do so and disgraceful not to do so. Hence also it is thought the mark of a braver man to be fearless and undisturbed in sudden alarms than to be so in those that are foreseen; for it must have proceeded more from a state of character, because less from preparation; acts that are foreseen may be chosen by calculation and rule, but sudden actions must be in accordance with one's state of character.

People who are ignorant of the danger also appear brave, and they are not far removed from those of a sanguine temper, but are inferior inasmuch as they have no self-reliance while these have. Hence also the sanguine hold their ground for a time; but those who have been deceived about the facts fly if they know or suspect that these are different from what they supposed, as happened to the Argives when they fell in with the Spartans and took them for Sicyonians.[15]

Relation of courage to pain and pleasure

9. We have, then, described the character both of brave men and of those who are thought to be brave.

Though courage is concerned with confidence and fear, it is not concerned with both alike, but more with the things that inspire fear; for he who is undisturbed in face of these and bears himself as he should towards these is more truly brave than the man who does so towards the things that inspire confidence. It is for facing what is painful, then, as has been said[16] that men are called brave. Hence also courage involves pain, and is justly praised; for it is harder to face what is painful than to abstain from what is pleasant. Yet the end which courage sets before itself would seem to be pleasant, but to be concealed by the attending circumstances, as happens also in athletic contests; for the end at which boxers aim is pleasant—the crown and the honours—but the blows they take are distressing to flesh and blood, and painful, and so is their whole exertion; and because the blows and the exertions are many the end, which is but small, appears to have nothing pleasant in it. And so, if the case of courage is similar, death and wounds will be painful to the brave man and against his will, but he will face them because it is noble to do so or because it is base not to do so. And the more he is possessed of virtue in its entirety and the happier he is, the more he will be pained at the thought of death; for life is best worth living for such a man, and he is knowingly losing the greatest goods, and this is painful. But he is none the less brave, and perhaps all the more so, because he chooses noble deeds of war at that cost. It is not the case, then, with all the virtues that the exercise of them is pleasant except in so far as it attains its end. But it is quite possible that the best soldiers may be not men of this sort but those who are less brave but have no other good; for these are ready to face danger, and they sell their life for trifling gains.

So much, then, for courage; it is not difficult to grasp its nature in outline, at any rate, from what has been said.

---

[14] 1115^{b}11–24.

[15] At the Long Walls of Corinth, 392 B.C. Cf Xen. *Hell.* iv 4.10.

[16] 1115^{b}7–13.

Friendship
Aristotle

Book VIII

Chapter 1. It will be natural to discuss friendship next, for friendship is a kind of virtue or implies virtue. It is also indispensable to life. For without friends no one would choose to live, even though he possessed every other good. It even seems that people who are rich and hold official and powerful positions have the greatest need of friends; for what is the good of this sort of prosperity without some opportunity for generosity, which is never so freely or so admirably displayed as toward friends? Or how can prosperity be preserved in safety and security without friends? The greater a person's importance, the more liable it is to disaster. And in poverty and other misfortunes our friends are our only refuge. Again, when we are young, friends are a help to us, in saving us from error, and when we grow old, in taking care of us and doing the things for us we are too feeble to do for ourselves. When we are all in the prime of life, they prompt us to noble actions, as the line runs, "Two going together," for two people are better than one both in thought and in action.

Friendship or love seems the natural instinct of a parent toward a child, and of a child toward a parent, not only among men but among birds and animals generally. . . .

Again, it seems that friendship is the bond which holds states together, and that lawmakers set more store by it than by justice; for harmony is something like friendship, and it is harmony that they especially try to promote, and discord that they try to expel, as the enemy of the state. When people are friends there is no need of justice between them; but when they are just, they yet need friendship too. Indeed justice, in its supreme form, assumes the character of friendship.

Nor is friendship indispensable only; it is also noble. We praise those who love their friends, and to have many friends is thought to be a fine thing. Some people hold that to be a friend is the same thing as to be a good man.

The subject of friendship gives room for a good many differences of opinion. Some define it as a sort of likeness, and say people are friends because they are like each other. Hence the sayings, "Like seeks like," "Birds of a feather," and so on. Others, on the contrary, say "Two of a trade never agree." So philosophical thinkers indulge in more profound physical speculations on the subject. . . . Heraclitus declares that "contending things drew together," that "harmony most beautiful is formed of discords," and that "all things are by strife engendered." Others, among whom is Empedocles, take the opposite view and insist that "like desires like," . . .

Chapter 2. It is possible, I think to shed light on the subject of friendship, by determining what is lovable or an object of love. For plainly not everything is loved, but only that which is lovable, which is what is good or pleasant or useful. A thing too is useful if it is a means of gaining something good or pleasant. If so, it follows that it is the good and the pleasant that are lovable because they are ends.

We may ask, then, do we love what is good in itself, or what is good for us? For there is sometimes a difference between them. The same question may be asked in regard to what is pleasant. It is said that everyone loves what is good for himself, and that, while the good is lovable in an absolute sense, it is what is good for each individual that is lovable in his eyes. It may even be said that a man loves not what is good for him but what seems good. But this will make no difference; for in that case, what is lovable will be what seems lovable.

Now there are three motives for love. We do not, it must be noted, apply the term "love" to our feeling for lifeless things. The reason is (1) that they are incapable of returning our affection, and (2) that we do not wish their good; for it would, of course, be ridiculous to wish good to the wine. If we wish it at all, it is only in the sense of wishing the wine to keep well, so that we may enjoy it ourselves. But everyone knows that we ought to wish our friend's good for his sake. If we wish people good in this sense, we call it good will, unless our good wishes are returned; reciprocal good will we call friendship.

We must add too that the good will must not be unknown. A person often wishes well to people whom he has not seen, but whom he supposes to be good or useful; and it is possible that one of these persons may entertain the same feelings toward him. Such people, then, it is clear, wish well to one another, but they cannot properly be called friends, so long as their feeling is unknown to each other. If they are to be friends, they must feel good will to each other and wish each other's good for one of the motives aforesaid, and each of them must know that the other wishes him well.

Now as the reasons for friendship differ in kind, so accordingly do the corresponding kinds of affection and friendship. The kinds of friendship therefore are three, being equal in number to the things which are lovable or the objects of friendship, for every such object may arouse a reciprocal affection between two persons.

People who love each other wish each other's good up to the point on which their love is fixed. Accordingly, those who love each other for reasons of utility do not love each other for themselves, but only as far as they get some benefit from one another. So with those who love for pleasure's sake. They are fond of witty people, not for their character, but because they are pleasant to them. People then who love for utility's sake are moved to affection by what is good for themselves, and people who love for pleasure, by what is pleasant to themselves. They love a person not for what he is in himself, but only for being useful or pleasant to them. Such friendships then are friendships incidentally only; for the person loved is not loved for being what he is, but merely for being a source of some good or pleasure. Such friendships accordingly are easily dissolved, if the parties do not continue always the same; for they cease loving once they cease to be pleasant or useful to each other.

Now utility is not a permanent quality; it varies at different times. Hence when the reason for the friendship disappears, the friendship itself is dissolved, since it depended on that reason. Friendship of this kind seems to arise especially among old people, for in old age we look for profit rather than pleasure, and also among those in the prime of life or youth who have an eye to their own interest. Friends of this kind do not generally live together; for sometimes they are not even congenial. Nor do they want such companionship, except when they are of use to one another, since the pleasure they give each other goes no further than the hopes they entertain of getting benefit from it. Among these friendships we may count the friendship which exists between host and guest.

The friendship of the young is based apparently on pleasure; for they live by emotion and are inclined to pursue most the pleasure of the moment. But as their age increases, their pleasures alter with it. They are therefore quick at making friendships and quick at abandoning them; for their friendships shift with the object that pleases them, and their pleasure is liable to sudden change. Young people are amorous too, amorousness being generally a matter of emotion and pleasure. Hence they fall in love and soon afterwards fall out of love, passing from one condition to another many times in a single day. But amorous people wish to spend their days and lives together, since thus they attain the object of their friendship.

Perfect friendship is the friendship of people who are good and alike in virtue; for they are alike in wishing each other's good, inasmuch as they are good and good in themselves. Those who wish the good of their friends for their friend's sake are in the truest sense friends, since their friendship is the consequence of their own character, and not an accident. Their friendship therefore lasts as long as their goodness, and goodness is a permanent quality. So each of them is good in an absolute sense, and good in relation to his friend. For good men are not only good in an absolute sense, but helpful to each other. They are pleasant too; for the good are pleasant in an absolute sense, and pleasant to one another. For everybody finds pleasure in actions proper to him and in others like him, and all good people act alike or nearly alike.

Such a friendship is naturally permanent, for it unites in itself all the right conditions of friendship. For the aim of all friendship is good or pleasure, either absolute or relative to the person who feels the affection; and it is founded on a certain similarity. In the friendship of good men all the conditions just described are realized in the friends themselves; other friendships bear only a resemblance to the perfect friendship. That which is good in an absolute sense is pleasant also in an absolute sense. They are too the most lovable objects of affection, and for this reason love and friendship in this highest and best sense are found most among such men.

Friendships of this kind are likely to be rare; for such people are few. Such friendships require time and familiarity too; for, as the adage puts it, men cannot know one another until they have eaten salt together; nor can they admit one another to friendship, or be friends at all, until each has been proved lovable and trustworthy by the other. People who are quick to treat one another as friends wish to be friends but are not so really, unless they are lovable and know each other to be so; for the wish to be friends may arise in a minute, but not friendship.

Chapter 4. . . . For pleasure or profit then it is possible that even bad men may be friends to one another, and good people to bad, and one who is neither good nor bad to any sort of person; but clearly none but the good can be friends for the friends' own sake, since bad people do not delight in one another unless to gain something thereby.

. . . It is impossible to be friends with a great number of people in the perfect sense of friendship as it is to be in love with a great number of people at once. . . . It is not easy for a great number of people to give intense pleasure to the same person at the same time, or, I may say, to seem even good to him at all. Friendship too involves experience and familiarity, which are very difficult. But it is possible to find a great number of acquaintances who are simply useful or pleasant or agreeable; for people of this kind are numerous and their services do not take much time.

. . . We have said that a good man is at once pleasant and useful. But such a man does not become the friend of one superior to him in rank, unless he is himself superior to that person in goodness. Otherwise there is no equality, such as does occur when his superiority in virtue is proportionate to his inferiority in some other respect. Friendships of this kind, however, are exceedingly rare.

. . . There is another kind of friendship that is based on inequality, such as the friendship of a father for his son, or of any elder person for a younger, or of a ruler for a subject. These friendships are of different sorts.

In all friendships that involve the principle of inequality, the love also should be proportional; the better or the more useful party, or whoever may be the superior, should receive more love than he gives. For when the love is proportional to the merit, a sort of equality or other justice is established; and this equality seems to be a condition of friendship.

. . . The good, being constant themselves, remain unchanged in relation to one another, and neither ask others to do wrong nor do it themselves. They may even be said to prevent it; for good people do no wrong nor allow their friends to do it. But in wicked friends there is no stability; for they do not remain the same themselves for long. And if they become friends, it is only for a short time, and for the satisfaction they take in each other's wickedness. . . .

Book IX

Chapter 3. Another question that presents a problem is whether we ought or ought not to break off friendships with people whose character is no longer what it once was. If the motive of the friendship was utility or pleasure, then when the utility or the pleasure comes to an end, there is nothing unreasonable in breaking off the friendship. For it was the utility or the pleasure that we loved, and when they have ceased to exist, it is only reasonable that our love should come to an end too.

. . . But suppose we take a person into our friendship, believing him to be a good man, and he turns out and is recognized as a rascal, is it still our duty to love him? Love, it would seem, is now an impossibility, because . . . it is not our duty to love the wicked, or to make ourselves like bad men. We have said already that like loves like. Is it right then in such circumstances to break off the friendship at once? Or, perhaps, if not in all cases, at least where the vice is incurable? If there is any possibility of reforming the friend who has gone wrong, we should indeed come to the help of his character even more than of his property, since character is a better thing than property and enters more closely into friendship. Still a person who breaks off a friendship under these circumstances is not thought to be acting at all unreasonably. He was not a friend of the person as that person is now; therefore, if his friend has altered and it is impossible to reclaim him, he lets him go.

Again, suppose A stays as he was but B becomes better and vastly superior to A in virtue. Ought B then to treat A still as a friend? It is, I think, impossible. The case becomes clearest when the distance is wide between the two friends, as happens with childhood friendships, when one of two friends remains a child in mind and the other is a fully developed man. How can they be friends, when they sympathize with each other neither in their ideas nor in their pleasures and pains? There will be no personal understanding between them.

. . . Is it right then, when two friends cease to be sympathetic, for one to treat the other exactly as if he had never been his friend? Surely we must not entirely forget the old intimacy, but even as we think we should oblige friends rather than strangers, so for old friends we should show some consideration for the sake of the past friendship, provided that the break in the friendship was not caused by some extraordinary wickedness. . . .

The Warriors: Reflections on Men in Battle
J. Glenn Gray[*]

Love: War's Ally and Foe

What is it that men are concerned to preserve and to care for in battle? The most obvious answer is self-preservation, taking care of their own lives. This is true in a different sense from the common biological teaching of self-preservation as a basic instinct that men share with other animals. It is also true in a different sense from the egoistic psychology that traces all motivation to self-interest. He who has seen men throw away their lives in battle when caught up in communal passion or expose themselves recklessly and carelessly to mortal danger will be cured forever of such easy interpretations of human motivation. Nothing is clearer than that men can act contrary to the alleged basic instinct of self-preservation and against all motives of self-interest and egoism. Were it not so, the history of warfare in our civilization would be completely different from what it has been.

Nevertheless, self-preservation is a dependable and pervasive feature of human existence in a deeper sense than egoistic theories suppose. The philosopher Spinoza called it the striving to persevere in our own being, and the phrase is exact. Though striving to persevere in our own being is not absolute (for men may deliberately choose suicide) and not merely biological selfishness (since men are capable of dying for others), it is a power that lies both in and beyond the conscious, rational life. Many a soldier has been surprised to discover the desire to continue in being as a final hold and support, after superhuman exertion and mental strain had robbed him of conscious will, and any religious faith he may have possessed had ceased to be meaningful. The literature of war is full of the accounts of armies, beaten and bled, starved and weakened, yet tenaciously staying alive and rescuing a remnant of their strength and numbers. The account of the ancient Greek in the Anabasis may be taken as a classic example of this survival power in soldiers. Hardly a major war since Xenophon's time has been without similar feats of endurance, though few have had a chronicler such as he.

. . . Now friendship has often enough been defined in our tradition as that relationship between human beings in which each dispassionately seeks the welfare of the other. Friendship is thus thought to be the most unselfish form of love, since in the pure state it devotes itself without reserve to the interests of the other. Accordingly, many societies have exalted friendship as the noblest of all relationships, and even the founder of Christianity, to whom another form of love took precedence, is declared to have said: "Greater love has no man than this, that a man lay down his life for his friends."

---

[*]*The Warriors: Reflections on Men in Battle* (New York: Harcourt, Brace & Co. 1959) Copyright © 1959 J. Glenn Gray, 1987 Ursula A. Gray.

What meaning has friendship for warriors? How can a young man endure battle when the fear of death is doubled, when not only his own life but that of his friend is at stake? Is the quality of this relationship heightened or reduced by the dread strain of war? Before trying to answer these questions, I must first attempt to make clear a basic difference between friends and comrades. Only those men or women can be friends, I believe, who possess an intellectual and emotional affinity for each other. They must be predetermined for each other, as it were, and then must discover each other, something that happens rarely enough in peace or war.

Though many men never have a friend, and even the most fortunate of us can have few, comradeship is fortunately within reach of the vast majority. Suffering and danger cannot create friendship, but they make all the difference in comradeship. Men who have lived through hard and dangerous experiences together are frequently deceived about their relationship. Comrades love one another like brothers, and under the influence of shared experience commonly vow to remain true friends for the rest of their lives. But when other experiences intervene and common memories dim, they gradually become strangers. At veterans' conventions they can usually regain the old feelings only with the aid of alcoholic stimulation. The false heartiness and sentimentality of such encounters are oppressive and pathetic. Men who once knew genuine closeness to each other through hazardous experience have lost one another forever. And since most men rarely attain anything closer to friendship than this, the loss of comradeship cannot be taken lightly. When veterans try to feel for their old buddies what they felt in battle and fail, they frequently cherish somewhere in their secret memories the unsentimental original passion.

The essential difference between comradeship and friendship consists, it seems to me, in a heightened awareness of the self in friendship and in the suppression of self-awareness in comradeship. Friends do not seek to lose their identity, as comrades and erotic lovers do. On the contrary, friends find themselves in each other and thereby gain greater self-knowledge and self-possession. They discover in their own breasts, as a consequence of their friendship, hitherto unknown potentialities for joy and understanding. This fact does not make friendship a higher form of selfishness, as some misguided people have thought, for we do not seek such advantages in friendship for ourselves. Our concern, insofar as we are genuine friends, is for the friend. That we ourselves also benefit so greatly reveals one of the hidden laws of human affinity. While comrade-ship wants to break down the walls of self, friendship seeks to expand these walls and keep them intact. The one relationship is ecstatic, the other is wholly individual. Most of us are not capable of meeting the demands on self that friendship brings, whereas comradeship is in most respects an easing of these demands. Comrades are content to be what they are and to rest in their emotional bliss. Friends must always explore and probe each other, in the attempt to make each one complete through drawing out the secrets of another's being. Yet each recognizes that the inner fountain of the other is inexhaustible. Friends are not satiable, as comrades so often are when danger is past.

"That a man lay down his life for his friends" is indeed a hard saying and testifies to a supreme act of fortitude. Friends live for each other and possess no desire whatsoever for self-sacrifice. When a man dies for his friend, he does it deliberately and not in an ecstasy of emotion. Dying for one's comrades, on the other hand, is a phenomenon occurring in every war, which can hardly be thought of as an act of superhuman courage. The impulse to self-sacrifice is an intrinsic element in the association of organized men in pursuit of a dangerous and difficult goal.

For friends, however, dying is terribly hard, even for each other; both have so much to lose. The natural fear of dying is not so hard for them to overcome. What is hard is the loss or diminution of companionship through death. Friends know—I am tempted to say, only friends know—what they are giving up through self-sacrifice. It is said, to be sure, that they can communicate with one another even beyond death, but the loss is nevertheless cruel and final. Too often at moments of greatest need, when one's friend is dead, communication is broken off and one's dialogue becomes monologue. Friends can hardly escape the recognition of death as unmitigated evil and the most formidable opponent of their highest value.

War and battle create for this love both a peculiar kind of security and a kind of exposure, which other forms of love seldom know. The security arises from the insulation that friendship affords against the hatreds and the hopelessness that combat often brings. Even though one friend may be in safety at home, the friend who fights knows that somewhere the other is participating in his life.

Through letters he can communicate his deepest feelings and his explorations of the evil experiences through which he is passing. Even when letters are cut off, friends can communicate in their memories of each other, each explaining in imagination to the other and having the assurance of being understood. There is joy in having a person who understands completely and whom you understand. It insulates the soldier's heart without closing his mind to the experiences he is undergoing.

... Friends ... can thus endure much of war's horror without losing the zest for life. More than that, they can discover meaning in experiences of the most gruesome sort which others do not see. Friendship opens up the world to us by insulating us against passions that narrow our sympathies. It gives us an assurance that we belong in the world and helps to prevent the sense of strangeness and lostness that afflicts sensitive people in an atmosphere of hatred and destruction. When we have a friend, we do not feel so much accidents of creation, impotent and foredoomed. The assurance of friendship has been enough to help soldiers over many dreadful things without harm to their integrity.

But friendship makes life doubly dear, and war is always a harvest of death. Hence friends are exposed to an anxiety even greater than that of other lovers.

... In every slain man on the battlefield, one can recognize a possible friend of someone. His fate makes all too clear the horrible arbitrariness of the violence to which my friend is exposed. Therefore, in love as friendship we have the most dependable enemy of war. The possible peaks of intensity and earlier maturity which war may bring to friendship are as nothing compared with the threats of loss it holds.

... The companionship of a lost friend is not replaceable.

... Love as friendship must subsist haphazardly and as best it may in the midst of war. Its true domain is peace, only peace.

D. The Tradition of Natural Law

Having considered the "moral situation" from all possible perspectives—the moral agent, the form of her actions, and their outcome or results—there would appear to be little more to review. A slightly different perspective on the nature of morality, however, suggests that there is at least one further topic to consider on the road to a thorough introduction to this subject, even if this particular perspective or tradition, that of "natural law," is no longer widely taught or understood outside of parochial schools and some Roman Catholic institutions of higher education.

This, in and of itself, is unfortunate, in that many of America's most cherished founding principles derive from this tradition of moral reflection, and these are not the property or the exclusive province of any single religious tradition or denomination. Indeed, the teachings of "natural law" suggest, quite to the contrary, that moral resources are democratically available to any and all with eyes to see and ears to hear, and can be understood without recourse either to the dogmas of religious faith, or to the pretensions of elaborate philosophical "theories."

When Thomas Jefferson, following his reading of the English philosopher, John Locke, writes in the *Declaration of Independence* (1776) that we hold certain truths to be "self-evident," and claims that the Creator has endowed every human being with certain "inalienable rights," he is drawing explicitly on a tradition of moral reflection dating back centuries, well before the Roman empire, at least to the golden era of the Roman Republic. Jefferson's point (as was Martin Luther King's, explicitly, two centuries later) in drawing on this *natural law tradition*, is that there are straightforward moral "truths" or principles that any reasonably competent adult can plainly discern for themselves, upon careful reflection. Moreover, it does not require initiation into a specific community of faith, or exhaustive tutoring at the hands of philosophical "experts," in order for individuals to discern these truths. We derive them solely through "the clear light of reason," reflecting on human experience, unaided by either the doctrines of the clergy or the theories of the sophists.

Aristotle can be blamed in part for promulgating this provocative heresy. We notice, if we read carefully, that morality in Aristotle's framework is a matter of observation, life-long training, and habituation. It is not something that, strictly speaking, can be taught in a classroom (even though Aristotle discussed it with his own students, and obviously wrote about it). This corresponds to our widespread contemporary prejudice that morality, and basic moral principles, are "caught" rather than "taught"—learned early on as we incorporate and internalize our experiences of equality and inequality, justice and injustice, respect and disrespect, civility and incivility.[1]

Later Stoic philosophers, like the great Roman orator and statesman, Cicero, came to believe on the basis of their understanding of Aristotle and Plato in particular, that the thread of Reason—the "Logos"—was woven into the fabric of the universe, holding the cosmos together in a grand order of Law that was plainly evident for rational beings like ourselves to apprehend, should we ever care to

---

[1]Psychologists would add that this rational moral development is possible insofar as human beings have the capacity for both sympathy (feeling concern for the well-being of others), and even more importantly, of empathy (the ability to imagine, and to take into account, the standpoint or perspective of others). While we possess these qualities to different degrees (some of us may be a bit dense or obtuse in being unable to see ourselves as others see us, for example), a complete absence of both is a serious pathology, rendering such a person a danger to society, not simply a practitioner of some alternative form of morality.

do so. Cicero, whose own political and philosophical writings were required reading in every classroom in America well into the nineteenth century, summarized this perspective eloquently when he wrote:

> "True law is *right reason in agreement with nature*; it is of universal application, unchanging and everlasting; it summons to duty by its commands, and averts from wrongdoing by its prohibitions . . . We cannot be freed from its obligations by Senate or People, and we need not look outside ourselves for an expounder or interpreter of it . . . There will not be different laws at Rome and at Athens, or different laws now and in the future, but *one eternal and unchangeable law will be valid for all nations and all times*, and there will be *one master and ruler, that is God, over us all, for he is the author of this law, its promulgator, and its enforcing judge."* (*Republic,* Book III; emphasis added)

A century later, the great Christian theologian and apostle, Paul of Tarsus, who was well versed in Stoic philosophy, defended his ministry to the wider population of the Hellenistic world by arguing, in his profound letter to the Christian congregation in Rome, that all of these "Gentiles" were already acquainted with the basic provisions of God's moral law. They acquired this knowledge, not through divine revelation, or through special election as a "chosen people," but by virtue of being (as we all presumably are, in Paul's view, as in Jefferson's, and King's) created by God. Thus Paul claimed:

> When Gentiles, who have not the Law [of Moses], nonetheless *do by nature* what that Law requires, they are a law unto themselves, even though they do not have the Law. They show that *what the Law requires is written on their hearts* . . . (Romans 2: 14–15; emphasis added)

We shall have occasion to revisit Stoic philosophy in particular, and also to study the impact of the "natural law tradition" on our conception of "natural rights"—as well as its specific historical contributions to understanding the criteria necessary for justifying the use of military force, and for placing certain moral restraints on the behavior of combatants in wartime—as we reflect in conclusion on the Stoic example of Vice Admiral James B. Stockdale, in the final chapter of this book. Suffice it to say, at present, that this tradition, relatively constant over two millennia, offers the hypothesis that the universe, or Nature, is a rationally-ordered Whole or system, whose governing principles are those of Reason. Human beings, whom Aristotle described as "rational animals," have within themselves the capacity for discerning these Laws in Nature. In this clear light of reason, we discern intuitively how each of us is expected to live, and how each of us is obligated to treat one another with dignity and respect. We find these insights captured in the otherwise distinct and diverse lamentations of Hebrew slaves during the Second Dynasty in Egypt, of Athenian commoners resisting the Tyrants, of Spartacus and his follow bondsmen in the Roman empire, echoing down through the pages of history in every society, including our own, in which oppression and injustice have been allowed to flourish. This echoing sentiment is captured in Kant's insight that we are not placed on this earth to serve merely as means to the morally unscrupulous ends of others.

These are the concepts woven throughout America's founding documents, including the Declaration and the Constitution, as well as the basis of this nation's historical commitment to individual liberty and human rights, even when found in the breach, rather than in the observance. All of these founding ideals, which continue to inspire (or to indict) our civic behavior at present, are all but unintelligible without reference to the natural law tradition.

The Readings

St. Paul's claim is taken up as the starting point for our first selection on "ethics and natural law" from the *Summa Theologica* of St. Thomas Aquinas (1224/5–74). This great medieval scholar and theologian is widely regarded as the preeminent figure in the history of the natural law tradition, and accordingly, we consider briefly his views on the normative theological order of Eternal Law (the Logos), Divine or "Revealed" Law (the Jewish Torah and Christian incarnation), Natural law (of which St. Paul speaks), and Human law, each with its respective sphere of influence. Of particular importance (especially, again, for Martin Luther King, Jr., who completed a doctoral dissertation at Boston

University on Thomas and the natural law tradition) human legislation may add to, but may never contravene, Natural or Divinely-revealed law. Instead, these various orders of laws exist in a hierarchy of authority, with the lower orders subordinate to, and in harmony with, the higher.

Thomas's conception blends the Stoic-Hellenistic conception of the all-pervasive Logos with a uniquely-Aristotelian conception of purpose. In this sacred tradition of natural law, integral to Catholic moral theology since the time of St. Thomas, one of the principal responsibilities of interpreters is therefore to inquire into the purposes, or "natural inclinations" of specific beings or actions in attempting to discern what sorts of "natural laws" govern their behavior. A full understanding of this purposive, or "teleological," order of nature is deemed essential to discerning our specific responsibilities as human moral agents, and, in particular, for understanding the kinds of behaviors that are permitted or required as "conforming to natural purpose," while distinguishing these from others which are prohibited or condemned as "contrary to nature."

This Catholic tradition represents, on the one hand, a remarkable synthesis and a profound intellectual achievement. Among its principal findings, as outlined in the reading selection by natural law theologian C.E. Harris, for example, are the Principle of Forfeiture and the Doctrine of Double Effect. The former suggests with deceptive simplicity that a moral agent who violates or threatens to violate a specific natural right of another automatically forfeits that right himself.

The Doctrine of Double Effect, central to moral debates in medical and military ethics, distinguishes between the *intended effects* of an action, and the *unintended, but merely foreseen effects*. The doctrine itself intends to prohibit moral agents from pursuing morally justifiable ends through the use of morally questionable means: specifically, if an evil result is deliberately intended as the primary means (and not merely foreseen as the unintentional side-effect) of attaining an otherwise-appropriate end, such a means is morally prohibited. So, for example, a doctor may not deliberately kill a healthy patient and harvest his organs, even for the otherwise morally worthy purpose of saving, through organ transplants, the lives of several other patients who are dying of heart or kidney disease. Likewise, a military commander may not, in order to win an otherwise justifiable war, order the deliberate killing of non-combatants, although the collateral deaths of non-combatants as an accidental consequence of an otherwise-legitimate military engagement against enemy combatants is not deemed blameworthy. Such deaths are unintended, and not the legitimate part of any military action, even though they are, in principle, foreseen as likely possibilities during wartime.

To some critics of natural law, these findings do not seem as straightforward as they appear. Indeed, with the doctrine of double effect, it may sometimes appear as if an intellectual "sleight of hand" is being pursued to justify almost any means to a desired end, rather than serving as a constraint upon morally acceptable behavior, as Thomas originally intended. With the principle of forfeiture, it might seem obvious, for example, that a brutal serial killer has long since forfeited his own right to life, and so the State would be justified in putting him to death as punishment. It does not automatically follow, however, from the forfeiture of that individual's *right to life* that the State is automatically entitled, let alone obligated, to proceed actually to *extinguish that life* itself. Even if it could be argued that the State has this right, as a result of the principle of forfeiture, it still does not follow that the State is obligated to exercise it, or that it would be wise for it to do so. It was for this reason that utilitarians like Jeremy Bentham, in particular, criticized the prevailing influence of the natural law tradition of jurisprudence. We need to ask instead, he argued, whether engaging in capital punishment (for example) would deter or encourage crimes of this sort, or otherwise result in discernable, quantifiable benefits for society as a whole.

The most problematic areas of the sacred tradition, however, come not from the rational basis of natural law *per se*, but from the historically-attendant Aristotelian assumptions about purpose and "teleology" in nature. Even though such thinking has long since been purged from the sciences of physics and biology as pointless and ungrounded speculation, these considerations still play a powerful role in the Catholic or sacred tradition of natural law. Thus, Harris self-confidently describes certain natural inclinations and the "biological values" that follow from them, such as the tendency of all animals to engage in sexual intercourse for the sake of procreation. From this, he concludes, we discern through natural law that we have an obligation to produce and rear children, and therefore, even further, that practices such as artificial contraception or homosexuality would be wrong.

All that may be so, but it is unclear whether these conclusions follow from the "natural" evidence, and that point is crucial for evaluating the validity of this kind of reasoning. It sometimes appears, for example, that we are to take our cue from the practices encountered in nature, as being "natural." But when we leave the armchair or the theological seminary for the nature preserve, we discover that homosexuality is widely practiced, especially among primates, and that there are other practices that occur routinely in nature, such as killing and eating one's offspring, that call into question the status of procreation as the sole purpose of sexuality. In any case, no reasonable person would wish to advocate the latter behavior as a moral practice, whether or not it is "natural" or even serves some "natural purpose." Such discoveries cast into severe doubt our ability to discern any normative biological values in nature that we might, with reasonable certainty, translate into valid moral principles. There are some "natural" practices that we might find morally commendable, and many others (as the biologist Thomas Huxley observed) that we would properly find morally repugnant. In neither case does their occurrence "in nature" seem to have anything to do with their moral status: the latter is, rather, conferred or withheld by us, as moral agents.

Perhaps the dilemma is not with "natural law" *per se*, but with the longstanding conflation, within an otherwise-venerable historical tradition, of elements of rational moral reflection (the true core of the Stoic tradition) together with a great deal of outmoded and discredited ancillary beliefs concerning "Nature" and about what constitutes "natural" behavior. The latter, while part of the sacred tradition, has nothing essentially to do with "natural law." Indeed, if we take Paul and Thomas at their word (just as their more orthodox religious critics feared might happen!), we recognize that once Reason has been admitted as the ground of judgment in the restricted sphere of natural law, it is no longer necessary to make any explicit reference to the speculative origins, background, or context of that law, whether in nature or confessional theology, in order to avail ourselves of its moral guidance.

Our final reading provides an account, therefore, of what some would call "the new natural law tradition," in which pre-theoretic rational intuitions, applied to historical cases or to hypothetical thought experiments, are used exclusively to tease out general moral principles by induction, wholly without reference religious faith, or to speculative theories about God, nature, purpose, or "natural inclinations." The Doctrine of Double Effect emerges in these discussions, not from sacred texts or scholastic authorities, but simply from considerations of odd and somewhat amusing scenarios involving various strategies that might be permitted, or prohibited, in attempting to save victims of shark attacks, disease, or runaway trolley cars.

from Summa Theologica, Ethics and Natural Law
*St. Thomas Aquinas**

Aquinas (1225–1274 A.D.) here discusses eternal law, natural law, human law, and divine law and argues for their specific realms of application.

What is the Essence of Law?

... [L]aw ... is nothing else than an ordinance of reason for the common good, made by him who has care of the community, and promulgated [announced].

The natural law is promulgated by the very fact that God instilled it into man's mind so as to be known by him naturally.

Is There an Eternal Law?

... [T]he whole community of the universe is governed by Divine Reason. . . . [T]his kind of law must be called eternal.

Is There in Us a Natural Law?

A gloss on Rom. 2:14 ("When the Gentiles, who have not the law, do by nature those things that are of the law") comments as follows: "Although they have no written law, yet they have the natural law, whereby each one knows, and is conscious of, what is good and what is evil."

... [S]ince all things subject to Divine providence are ruled and measured by the eternal law, ... it is evident that all things partake somewhat of the eternal law, insofar as, namely, from its being imprinted on them, they derive their respective inclinations to their proper acts and ends. Now among all others, the rational creature is subject to Divine providence in the most excellent way, insofar as it partakes of a share of providence, by being provident both for itself and for others. Wherefore it has a share of the Eternal Reason, whereby it has a natural inclination to its proper act and end; and this participation of the eternal law in the rational creature is called the natural law. . . . [T]he light of natural reason, whereby we discern what is good and what is evil, which is the function of the natural law, is nothing else than an imprint on us of the Divine light. It is therefore evident that the natural law is nothing else than the rational creature's participation [in] the eternal law.

Is There a Human Law?

Augustine distinguishes two kinds of law: the one eternal, the other temporal, which he calls human.

... [J]ust as ... from naturally known indemonstrable principles we draw the conclusions of the various sciences ... by the efforts of reason, so too it is from the precepts of the natural law ... that the

---

*Edited excerpt from St. Thomas Aquinas, *The Summa Theologica of St. Thomas Aquinas,* trans. Fathers of the English Dominican Province, vol. 8 2nd ed (London: Burns, Oates & Washbourne Ltd, 1927), pp. 8, 10-16, 32, 36-39, 43-44, 45-48, 52.

human reason needs to proceed to the more particular determination of certain matters. These particular determinations, devised by human reason, are called human laws. ... Wherefore Tully says in his *Rhetoric* that "justice has its source in nature; thence certain things came into custom by reason of their utility; afterwards these things which emanated from nature and were approved by custom were sanctioned by fear and reverence for the law."

[But, someone might object, is not natural law sufficient for the ordering of all human affairs, without the need for human law? No.] The human reason cannot have a full participation [in] the ... Divine Reason, but according to its own mode, and imperfectly. Consequently, ... [just as in the sciences] there is in us the knowledge of certain general principles, but not proper knowledge of each single [specific] truth ... contained in the Divine Wisdom, so too ... [in matters of decisionmaking and action] man has a natural participation of the eternal law according to certain general principles but not as regards the particular determinations of individual cases which are ... contained in the eternal law. Hence the need for human reason to proceed further to particular legal sanctions.

[Why, Given the Natural Law and Human Law,] Was There Any Need for a Divine Law?

Besides the natural and the human law, it was necessary for the directing of human conduct to have a Divine law [set out in the Bible]. And this for four reasons. First, ... if man were ordained to no other end than that which is proportionate to his natural faculty, there would be no need for man to have any further direction ... besides the natural law and human law which is derived from it. But since man is ordained to an end of eternal happiness ... , ... therefore it was necessary that, besides the natural and the human law, man should be directed to his end by a law given by God.

Secondly, because, on account of the uncertainty of human judgment, ... different people form different judgments on human acts, whence also different and contrary laws result. In order, therefore, that man may know without any doubt what he ought to do and what he ought to avoid, it was necessary for man to be directed in his proper acts by a law given by God, for it is certain that such a law cannot err.

Thirdly, because ... man is not competent to judge of interior [mental processes], that are hidden, but only of exterior acts which appear; and yet for the perfection of virtue it is necessary for man to conduct himself aright in both kinds of acts. Consequently, human law could not sufficiently curb and direct interior acts, and it was necessary for this purpose that a Divine law should supervene.

Fourthly, because, as Augustine says, human law cannot punish or forbid all evil deeds, since, while aiming at doing away with all evils, it would do away with many good things and would hinder the advance of the common good, which is necessary for human intercourse. In order, therefore, that no evil might remain unforbidden and unpunished, it was necessary for the Divine law to supervene, whereby all sins are forbidden.

Is Every [Human] Law Derived from the Eternal Law?

Human law has the nature of law insofar as it partakes of right reason; and it is clear that, in this respect, it is derived from the eternal law. But insofar as it deviates from reason, it is called an unjust law, and has the nature, not of law but of violence.

What Is Subject to the Eternal Law?

... God imprints on the whole of nature the principles of its proper actions. ... And thus all actions and movements of the whole of nature are subject to the eternal law. Consequently, irrational creatures are subject to the eternal law, through being moved by Divine providence; but not, as rational creatures are, through understanding. ...

There are two ways in which a thing is subject to the eternal law ... : first, ... by way of knowledge; secondly, by way of action and passion, i.e., ... by way of an inward motive principle. [(It is) ... in this second way [that] irrational creatures are subject to the eternal law. ... [)] [By contrast,] each rational creature has some knowledge of the eternal law, ... [and] it also has a natural inclination to that which is in harmony with the eternal law. . . .

Both ways, however, are imperfect, and to a certain extent destroyed, in the wicked; because in them the natural inclination to virtue is corrupted by vicious habits, and, moreover, the natural knowledge of good is darkened by passions and habits of sin. . . .

. . . Nevertheless, in no man does the . . . flesh dominate so far as to destroy the whole good of his nature; . . . there remains . . . the inclination to act in accordance with the eternal law.

Does the Natural Law Contain Several Precepts, or One Only?

. . . *[G]ood is that which all things seek after*. Hence this is the first precept of law, that *good is to be done . . . , and evil is to be avoided*. All other precepts of the natural law are based upon this. . . .

. . . [A]ll those things to which man has a natural inclination are naturally apprehended by reason as being good, and consequently as objects of pursuit, and their contraries as evil, and objects of avoidance. Wherefore according to the order of natural inclinations, is the order of the precepts of the natural law: . . . [I]n man there is first of all an inclination to good in accordance with the nature which he has in common with all substances, . . . the preservation of its own being; and by reason of this inclination, whatever is a means of preserving human life, and of warding off its obstacles, belongs to the natural law. Secondly, there is in man an inclination . . . according to that nature which he has in common with other animals; and in virtue of this inclination, those things . . . belong to the natural law "which nature has taught to all animals," such as sexual intercourse, education of offspring, and so forth. Thirdly, there is in man an inclination to good according to . . . reason, which . . . is [special] to him; thus man has a natural inclination to know the truth about God, and to live in society; and . . . whatever pertains to this inclination belongs to the natural law—for instance, to shun ignorance, to avoid offending those among whom one has to live, and other such things. . . .

Are All Acts of Virtue Prescribed by the Natural Law?

[S]ince the rational soul is the proper form of man, there is in every man a natural inclination to act according to reason, and this is to act according to virtue. Consequently, considered thus, all acts of virtue are prescribed by the natural law, since each one's reason naturally dictates to him to act virtuously. But . . . many things are done virtuously to which nature does not incline at first, but which, through the inquiry of reason, have been found by men to be conducive to well-living.

. . .

By human nature we may mean either that which is [distinctive of] man—and in this sense all sins, as being against reason, are also against nature. . .—or we may mean that nature which is common to man and other animals, and in this sense certain special sins are said to be against nature: thus, contrary to [hetero]sexual intercourse, which is natural to all animals, is unisexual lust, which has received the special name of the unnatural crime.

Is the Natural Law the Same in All Men?

. . . [I]n speculative [theoretical, scientific] matters, truth is the same for all men both as to [general] principles and as to [specific] conclusions, although the truth is not known to all as regards the conclusions but only as regards the principles which are called common notions. But in matters of action, truth or [rightness] is not the same for all as to matters of detail but only as to the general principles, and where . . . [it] is the same . . . in matters of detail, it is not equally known to all.

. . . [So], as regards the general principles, . . . truth or [rightness] is the same for all and is equally known by all. As to the proper [specific] conclusions of the speculative reason, the truth is the same for all but is not equally known to all; thus it is true for all that the three angles of a triangle are together equal to two right angles, although it is not known to all. But as to the proper [specific] conclusions of the practical reason [in matters of decisionmaking and action], neither is the truth or [rightness] the same for all, nor, where it is the same, is it equally known by all. Thus it is right and true for all to act according to reason, and from this [general] principle, it follows as a proper [specific] conclusion that goods entrusted to another should be restored to their owner. Now this is true for the majority of cases, but it may happen in a particular case that it would be injurious, and therefore unreasonable, to restore goods held in trust, for instance if they are claimed for the purpose of fighting against one's country. And this principle will be found to fail the more . . . as we descend further into detail, e.g., if one were to say that goods held in trust should be restored with such and such a guarantee or in such and such a way, because the greater the number of conditions added, the greater the number of ways in which the principle may fail, so that it be not right to restore or not to restore.

Consequently, we must say that the natural law as to general principles is the same for all, both as to [rightness] and as to knowledge. But as to certain matters of detail which are conclusions . . . of those general principles, it is the same for all in the majority of cases both as to [rightness] and as to knowledge, and yet, in some few cases, it may fail [to be] . . . , since in some the reason is perverted by passion or evil habit or an evil disposition of nature; thus, formerly, theft, although it is expressly contrary to the natural law, was not considered wrong among the Germans, as Julius Caesar relates.

Can the Law of Nature Be Abolished from the Heart of Man?

. . . [T]here belong to the natural law, first, certain most general precepts that are known to all; and secondly, certain secondary and more detailed precepts which are . . . conclusions following closely from first principles. As to those general principles, the natural law . . . can nowise be blotted out from men's hearts. But it is blotted out in the case of particular action insofar as reason is hindered from applying the general principle to a particular point of practice on account of [desire] or some other passion. . . . But as to the . . . secondary precepts, the natural law can be blotted out from the human heart either by evil persuasions, . . . or by vicious customs and corrupt habits, as among some men theft and even unnatural vices, as the Apostle states, were not esteemed sinful.

from The Ethics of Natural Law
C.E. Harris*

The term *natural law* can be misleading because it inevitably brings to mind some kind of ethical legalism—the belief that hard-and-fast guidelines cover every possible detail of conduct. This characterization, however, is unfair to the natural-law tradition. The greatest proponent of natural law, Thomas Aquinas (1224–1274), believed that the basic outlines of proper human behavior are relatively clear. But he also taught that the closer we come to particular moral judgments, the more prone we are to error and the more room we make for differences of opinion. Some contemporary natural-law theorists even believe that natural law has a historical dimension, so that what is right in one epoch may not be right in another. Whether or not the view is accepted, the lively discussions of ethical issues in the Roman Catholic Church, where natural-law thinking is especially prominent, show that natural-law theorists by no means believe that all ethical problems have already been solved. The word *law* merely refers to the prescriptive character of the rules that should govern human behavior.

The natural-law theorist does, however, believe in an objective standard for morality. Moral truth exists just as scientific truth exists. The natural-law theorist cannot be a radical ethical relativist or an ethical skeptic; rather, he is committed to some form of moral realism. He generally believes we know the basic outlines of this standard, but this belief does not mean we have interpreted the implications of this standard correctly in every case. In ethics, as in science, human beings continually search for truth. The belief in objective truth should be no more stifling of human freedom and creativity in ethics than it is in science.

Human Nature and Natural Inclinations

What is the standard of truth in natural-law ethics? As an approximation we can say that the standard is human nature. People should do whatever promotes the fulfillment of human nature. How then do we determine what human nature is? Let us consider some analogous situations that illustrate the difficulty in describing human nature. It is often useful to describe something's nature in terms of its function—that is, the purpose it serves. For example, we can describe the nature of a pencil in terms of its function of enabling humans to make marks on paper. A "good" pencil is one that performs this function well—without smudging or scratching or breaking, for example. Similarly, if an automobile's function is to provide transportation, a good automobile is one that provides comfortable and reliable transportation. The function of a tomato plant is to produce tomatoes, and a good tomato plant is one that produces an abundance of tomatoes of high quality.

We can also determine the function of human beings if we confine a person to one particular social role. The function of a farmer is to grow food, and a good farmer produces food efficiently and with proper care for the animals and the land for which he is responsible. By similar reasoning we can say that a good father is one who attends diligently to the welfare of his children. But now let us take

---

*Excerpts from "The Ethics of Natural Law" by C.E. Harris are from *Applying Moral Theories.* Copyright © 1996 Wadsworth Publishing Company.

human beings out of their social roles and ask simply, "What is the function of a human being?" Here we see the problem faced by those who attempt to base ethics on human nature. Generally speaking, the more complex the animal, the more varied is its behavior and presumably the less clearly defined is its function. The freedom of action possessed by human beings makes it plausible to argue, as some philosophers have, that human beings are characterized precisely by the fact that they have no set nature or function. How can we make sense out of natural law in the face of these problems?

Fortunately we can take another, more promising, approach to discovering what human nature is like. One way to determine the characteristics of a thing is to observe its behavior. In chemistry we learn about the nature of iron by observing how it reacts with other elements. Perhaps we can find out what human nature is like by ascertaining those "natural inclinations," as Aquinas puts it, that human beings have in common. To put it another way, perhaps we can discover what human nature is by identifying those goals that human beings generally tend to seek. These values would presumably reflect the structure of our human nature, which natural law directs us to follow. Therefore, we shall propose the following statement as the moral standard of natural law:

> MS: Those actions are right that promote the values specified by the natural inclinations of human beings.

How do we find out what these natural inclinations are? We might first consult psychologists, sociologists, or anthropologists. Some contemporary natural-law theorists use studies from the social sciences to defend their conclusions. However, the natural-law tradition developed before the rise of the social sciences, and a more informal method of observation was used to discover the basic human inclinations. Most natural-law theorists would maintain that these observations are still valid. We can divide the values specified by natural human inclinations into two basic groups: (1) biological values, which are strongly linked with our bodies and which we share with other animals, and (2) characteristically human values, which are closely connected with our more specifically human aspects. (We will not call this second group uniquely human values because some of the inclinations that point to these values, such as the tendency to live in societies, are not unique to human beings.) We can summarize the values and the natural inclinations that point to them as follows:

1. Biological Values

a. **Life.** From the natural inclinations that we and all other animals have to preserve our own existence, we can infer that life is good, that we have an obligation to promote our own health, and that we have the right of self-defense. Negatively, this inclination implies that murder and suicide are wrong.

b. **Procreation.** From the natural inclination that we and all animals have to engage in sexual intercourse and to rear offspring, we can infer that procreation is a value and that we have an obligation to produce and rear children. Negatively, this inclination implies that such practices as sterilization, homosexuality and artificial contraception are wrong.

2. Characteristically Human Values

a. **Knowledge.** From the natural tendency we have to know, including the tendency to seek knowledge of God, we can infer that knowledge is a value and that we have an obligation to pursue knowledge of the world and of God. Negatively this inclination implies that the stifling of intellectual curiosity and the pursuit of knowledge, including the pursuit of the knowledge of God, is wrong. It also implies that a lack of religion is wrong.

b. **Sociability.** From the natural tendency we have to form bonds of affection and love with other human beings and to form groups or societies, we can infer that friendship and love are good and that the state, as an outgrowth of this tendency to form societies, is a natural institution and therefore good. We thus have an obligation to pursue close relationships with other human beings and to submit to the legitimate authority of the state. We can also infer that war can be justified under certain conditions if it is necessary to defend the state. Negatively, this inclination implies that activities that interfere with proper human relationships, such as spreading slander and lies, are wrong. Actions that destroy the power of the state are also wrong, so natural law offers a basis for an argument against revolution and treason, except when the state is radically unjust.

These natural inclinations are reflections of human nature, and the pursuit of the goods they specify is the way to individual fulfillment. Aquinas himself makes it clear that his enumeration of basic values, which closely parallels our account, is incomplete; other natural-law theorists have expanded the list to include such things as play and aesthetic experience. However, the list given here has had the greatest historical influence, and we shall assume it is basically complete.

The more important issue raised by this list is the potential for conflict between the various values. What should we do when our need to defend ourselves requires that we kill someone else? What should we do when sterilization is necessary to prevent a life-threatening pregnancy? What should be done when contraception seems necessary in order to limit family size so that families can properly educate the children they already have? In each of these examples, one aspect of natural law seems to conflict with another, and the question arises whether these values have a hierarchy on which a decision can be based. The answer to this question brings into focus one of the most important and controversial aspects of natural-law moral philosophy, namely its moral absolutism.

Moral Absolutism and the Qualifying Principle

Moral Absolutism

Suppose you are on a military convoy from the United States to England during World War II. Your ship is attacked and sunk. Your life raft is carrying twenty-four persons, although it was designed to carry only twenty. You have good reason to believe that the raft will sink unless four people are eliminated, and four people on board have been so seriously injured in the catastrophe that they are probably going to die anyway. Because no one volunteers to jump overboard, you, as the ranking officer on the boat decide to have the four pushed overboard. Are you morally justified in doing so? Many of us would say that under the circumstances you are, but natural-law theorists would say that you are not justified, even if everyone on the raft would die otherwise.

Consider another wartime example. Suppose you know that some prisoners have information that will save a large number of lives. The only way to obtain the information is to threaten to kill the prisoners, but you know that they will not reveal what they know unless your threat is absolutely serious. To show them how serious you are, you have another prisoner, who has done nothing to deserve death, shot before their eyes. As a result of your action, the information is revealed and many lives are saved. Is this action justified? Many people would say that under these extreme circumstances it is justified, but natural-law theorists would say that it is not.

. . .

These examples point out one of the most significant aspects of natural-law theory, namely its absolutism.

Moral absolutism can refer either to the belief that some objective standard of moral truth exists independently of us (what we have referred to as moral realism) or that certain actions are right or wrong regardless of their consequences. Natural law is an absolutist moral theory in both senses, but the second meaning of absolutism is highlighted by the preceding illustrations. Natural-law theorists believe that *none of the values specified by natural inclinations may be directly violated.* Innocent people may not be killed for any reason, even if other innocent people can thereby be saved. The procreative function that is a part of our biological nature may not be violated by such practices as contraception and sterilization, even if these practices are necessary to preserve other values, such as a child's education or even the mother's life. Similarly, homosexuality violates the value of procreation and is prohibited, even if it is the only kind of sex a person can enjoy.

Natural-law theorists believe that basic values specified by natural inclinations cannot be violated because *basic values cannot be measured or compared;* that is, basic values cannot be quantified or measured by some common unit, so they cannot be traded off against one another. For example, we cannot divide the good of knowledge into units of value and the good of procreation into units of value so that the two can be compared on a common scale. Nor can the good of a single life be compared with the good of a number of lives; thus we cannot say that a single life may be sacrificed to preserve a number of other lives. This idea is sometimes called the doctrine of the "absolute value" or "infinite value" of a human life, suggesting that a human life cannot be weighed against anything else, including another human life. Natural-law theorists also make this point by saying that basic values are incommensurable. Because we cannot measure values, we cannot calculate which consequences of an action are more important. Therefore, consequences cannot be used to determine the moral status of actions.

There is another characteristic of natural law that is non-consequentialist, even though may not rule out consideration of consequences. Natural-law theorists insist that *moral judgments must include evaluation of the intentions of the person performing the action*. The intention of an action is what a person wants to accomplish or "has in mind," as we say, in performing the action. For example, a person can give money to charity because she wants a good reputation in the community. The consequences of the action are good, but the intention of the person is not morally praiseworthy. Some moral philosophers distinguish between a moral evaluation of the action and a moral evaluation of the intention of the person performing the action. Using this distinction, we can say that the action of giving money to charity was praiseworthy, but that the person giving the money was not to be commended, because the intention of the person was not praiseworthy.

Qualifying Principles

Because values are incommensurable and may not ever be directly violated, we may find ourselves in a moral dilemma in which any action we could perform violates some value and hence is immoral. For example, self-defense may sometimes require that we override the natural inclination of another human being to self-preservation. If we do nothing, we allow ourselves to be killed; if we defend ourselves, we kill someone else. To avoid the paralysis of action that would result from such moral dilemmas, natural-law theorists have developed two principles that are crucial in making moral judgments: the principle of forfeiture and the principle of double effect.

According to the *principle of forfeiture*, a person who threatens the life of an innocent person forfeits his or her own right to life. (An innocent person is one who has not threatened anyone's life.) Suppose you are a pioneer tilling his land one morning when two men approach you and express an intent to kill you and your family in order to take your land. Is it morally permissible for you to defend yourself, even to the point of killing them? Natural-law theorists answer the question in the affirmative. Even though you might have to kill your would-be assailants, they have forfeited their innocence by unjustifiably threatening your life. Therefore, they have forfeited their claim to have their natural inclination to self-preservation respected. We can make this point by distinguishing between killing and murder. *Killing* is taking the life of a non-innocent person, whereas *murder* is taking the life of an innocent person. When you take the life of a person who is attempting to kill you, you are killing him but you are not committing murder.

The principle of forfeiture can be used to justify not only acts of individual self-defense, but also war and capital punishment. A defensive war may be justified under certain conditions, even though it involves killing other people, because the aggressors have forfeited their right to life. Similarly, murderers may justly be put to death because they have forfeited their right to life by killing others.

According to the *principle of double effect*, it is morally permissible to perform an action that has two effects, one good and the other bad, if the following conditions are met.

1. The act, considered in itself and apart from its consequences, is good, or at least morally permissible. We would not need to analyze an act of murder from the standpoint of the other three conditions of the principle of double effect, because murder is bad in itself and apart from its consequences.
2. The bad effect cannot be avoided if the good effect is to be achieved. The moral significance of this criterion lies in the belief that if an alternative method that does not produce the bad effect is available and is not used, we must assume that the bad effect was intended. This criterion illustrates the important place that consideration of intent has

in natural law. An action with improper intent is morally unacceptable even if it does not otherwise violate natural law. There is, however, another test that must be passed before we can say that an action is unintended. It is embodied in the next criterion.

3. The bad effect is not the means of producing the good effect, but only a side effect. If the bad effect is a means of achieving the good effect, the bad effect must be intended along with the good effect to which it is a means, so the action is morally impermissible. This criterion also illustrates the importance of intention in natural law.

4. The criterion of proportionality is satisfied, in that the good effect and the bad effect are more or less equally balanced in importance. If the bad effect of an action is of far greater significance than any good effect, the action should not be done, even if the other criteria are met. The criterion of proportionality represents a consideration of consequences, but consequences may be considered only if the other criteria are met.

If these four criteria are met, the violation of a fundamental value may be considered as indirect rather than direct. Although we may still be said to *bring about an evil,* we cannot be said to do *an* evil, according to natural law. The best way to explain the principle of double effect is by examples, so let us consider several applications.

In the first example, a pregnant woman who has tuberculosis wants to take a drug that will cure her disease, but the drug also has the effect of aborting the fetus. Is taking the drug morally permissible? The principle of double effect justifies taking the drug in this case because all four of its conditions are met.

First, the act of taking the drug to cure a disease is itself morally permissible. In fact, considered in itself and apart from its consequences, it is morally obligatory for the mother to take the drug, for she is obligated to do what she can to preserve her own life.

Second, if we assume that the drug is the only one that will cure the disease and that the mother cannot put off taking the drug until after the baby is born, then the bad effect is unavoidable. By this criterion, then, the death of the child is not intended. We must clarify here what natural-law theorists mean. The bad effect is certainly foreseen; the woman knows the drug will produce an abortion. But an effect may be foreseen without also being intended, that is, without being the goal of the action. If another drug were available that would cure her tuberculosis without causing the abortion, presumably the woman would take it. Otherwise it would be difficult to argue that she did not intend to have an abortion.

Third, the bad effect is not the means of achieving the good effect. An abortion is not a necessary step in curing a person of tuberculosis; rather, it just happens that the only drug that will cure the woman also causes an abortion. The abortion is an unfortunate and unintended side effect, due to the particular nature of the drug.

Fourth, a proportionally serious reason exists for causing the abortion. The death of the fetus is at least balanced by the saving of the mother's life. If the bad effect were serious (as in this case), but the good effect were relatively insignificant, the action would not be justifiable by the principle of double effect, even if the other conditions were met.

The criterion of proportionality is an exception to the earlier statement that values are incommensurable and that human lives cannot be weighed against one another. We have seen that it may also be considered an exception to the claim that consequences are not considered in moral evaluation. Here, consequences do play a part in natural-law reasoning. But note that consequences can be considered only when the other three conditions have been met. It is therefore more accurate to say that in natural-law theory the consideration of consequences occupies some place in moral evaluation but that it is of secondary importance.

Two other examples will further illustrate how the principle of double effect functions. Suppose I want to turn on a light so that I can read a book on ethics, but I know that throwing the switch on the wall that turns on the light will result in the electrocution of a workman on the floor below. Is it morally permissible to throw the switch?

First, turning on a light to read a book on ethics is in itself a permissible—even praiseworthy—action.

Second, the bad effect is unavoidable if the good effect is to be achieved. If there were another light that could be turned on and I deliberately failed to use it knowing the consequence is the death of the workman, then I could not argue that I did not intend to kill the workman.

Third, the bad effect (killing the workman) is not a means to reading philosophy, but rather only an unfortunate and unintended side effect. Killing someone is not ordinarily a consequence of turning on a light.

But the fourth condition of the principle of double effect is not satisfied. The killing of a human being is not outweighed by the value of reading a book on ethics. Therefore, turning on the light is not justified by the principle of double effect.

Consider another example.[1] In the process of attempting to deliver a fetus, a physician discovers that it is hydrocephalic. The fetus's large cranium makes normal vaginal delivery impossible; both the woman and the fetus would die in the attempt. Neither the mother nor the fetus would survive a Cesarean section, so the only way to save the mother's life is to crush the skull of the fetus (craniotomy), thus rendering a vaginal delivery of the stillborn fetus possible. Would the craniotomy be justifiable by the principle of double effect?

First the act of attempting to save the mother's life is morally permissible, even commendable.

Second, there is no way to save the mother's life except by killing the fetus. The bad effect cannot be avoided if the good effect is to be achieved.

Third, the bad effect can only be seen as the means of producing the good effect. It makes no sense to talk about crushing the head of the fetus without also killing it. Since the death of the fetus must be considered the means of achieving the good effect, the third criterion is not satisfied. Because each of the four criteria must be met in order for the action to be permissible, we already know that the craniotomy is impermissible. However, for the sake of completeness, we shall consider the fourth criterion.

Fourth, since both the fetus and the mother will die if the craniotomy is not performed, the criterion of proportionality is satisfied. But since the third criterion is not met, the craniotomy may not be performed and the fetus and mother must both die.

Natural-law theorists admit that this is a tragic case, and various attempts have been made to justify the craniotomy on other grounds. For example, some natural-law theorists argue that the principle of forfeiture can be invoked, since the fetus should be considered an aggressor on the life of the mother. Even though the fetus is innocent of any conscious motive to harm its mother, the actual effect of its growth is to threaten the life of its mother. Natural-law theorists sometimes say that the fetus, having no malicious intent is *subjectively innocent* but not objectively innocent because it does threaten the mother's life. Whether this argument justifies the craniotomy will be left to the reader to decide.

Checklist for Applying Natural-Law Ethics

1. Determine whether an action is in accord with the four fundamental values specified by human inclinations, or whether there is some apparent violation of one or more of these values.

2. If there is an apparent violation, determine whether the qualifying principle of forfeiture applies to the action.

3. If there is an apparent violation, determine whether the qualifying principle of double effect applies to the action. In order for the principle of double effect to apply, all four of the following conditions must be met:

    a. The act itself must be good, or at least morally permissible.

    b. The bad effect of the action must be unavoidable if the good effect is to be achieved.

    c. The bad effect is not the means of producing the good effect, but only a side effect.

    d. The criterion of proportionality must be satisfied, in that the good effect must be at least as morally desirable as the bad effect is morally undesirable.

4. Make a final decision on the morality of the action.
    a. If the action is in accord with the fundamental values, or if it is not in accord but is excused by one of the qualifying principles, the action is morally permissible.
    b. If the action is permissible and the alternative to the action is a violation of a fundamental value, the action is morally obligatory.
    c. If the action is a violation of a fundamental value and the qualifying principles do not apply, the action is morally impermissible.

Concept Summary

The basic idea of natural law is that a person should promote those values that are the object of our fundamental human inclinations or tendencies. The realization of these values in a person's life will lead to a fulfillment of his or her human nature. As analyzed by natural-law theorists, these values include the biological values of life and procreation and the characteristically human values of knowledge and sociability.

Because natural law stipulates that no fundamental values may be directly violated, the question arises as to what action should be taken when situations seem to force a person to violate one of the values regardless of what is done. The qualifying principles of forfeiture and double effect are designed to remedy this problem. According to the principle of forfeiture, a person who threatens the life of an innocent person forfeits his or her own right to life. The principle of double effect provides four criteria that must be met for an action could be considered only an indirect violation of a fundamental value.

Natural Law and the Principle of Double Effect: Six Hypothetical Cases

George R. Lucas, Jr.

Introduction to Natural Law and Case-Based Reasoning (Moral Casuistry)

Moral argument and analysis frequently takes place in the context of what might best be termed "thought experiments." If two persons were marooned on a desert island, or adrift in a lifeboat, for example, would it be morally permissible (we might ask) for one to kill the other and consume him for food in order to survive? Would our moral intuitions, or our specific answer, be different if we varied the situation so that, instead, one of the two persons dies first of natural causes, and the remains are then cannibalized by the survivor? What is it, if anything, that seems to make a moral difference between these two, otherwise similar, situations?

The method of the "thought experiment" is quite familiar in other contexts, such as physics, in which scientists like George Gamov or Albert Einstein constructed elaborate and famous scenarios (such as the Einstein-Poldasky-Rosen paradox) outside the normal limits of actual experiment in order to test the coherence and validity of scientific intuitions about quantum theory. Charles Darwin likewise sketched a number of hypothetical or speculative scenarios in *The Origin of Species* (1859) in order to examine his intuitions regarding the process of biological evolution, while the early modern French philosopher and mathematician, Rene Descartes, described a number of famous and imaginative thought experiments in his writings, such as the "Wax Experiment" at the end of the second of his *Meditations on First Philosophy* (1641), and the conditional hypothesis of the "evil genius" that might be responsible for distorting our normal sense perceptions in Meditation I.

Until quite recently, however, the method of the hypothetical thought experiment was largely limited to the realm of the natural sciences and so-called "natural philosophy." It was unusual to find moral philosophers using this approach as a means of discovery of new insights. Instead, examples or hypothetical cases were, more often than not, used to illustrate moral principles already introduced. Immanuel Kant, for example, treats readers of his brief but densely-argued "Groundwork of the Metaphysics of Morals" (1783) to four "cases" illustrating the application of categorical imperative reasoning to proposed practices involving suicide, lying, and duties to engage in morally virtuous actions. The *examples are not used to derive the procedure*; instead, the procedure already derived and outlined earlier by Kant is then used in these examples as the basis for analyzing the hypothetical cases.

Until very recently, however, it was much rarer to find moral philosophers engaged in an imaginary "voyage of discovery" using thought experiments to develop new moral theories or modes of analysis. With rare exceptions (such as J. L. Austin's "three ways of spilling ink") there is little of this methodology employed in the work of renowned moral philosophers in the nineteenth and twentieth centuries, from John Stuart Mill, Henry Sidgwick, and G.E. Moore to Roderick Firth and John Rawls.

Reliance on real-life case studies as the basis for moral analysis came increasingly into play as philosophers turned their attention to moral dilemmas arising in the practice of medicine or in otherwise routine business practice in the 1960s and 1970s. Then, quite abruptly during this period, the moral "thought experiment" as an analytical tool of discovery was introduced primarily through the pioneering work of philosophers like Judith Anscombe and the late Bernard Williams of Cambridge University, and Judith Jarvis Thompson at the Massachusetts Institute of Technology. Their cases, including "Jim and the Indians," the "Trolley Problem," the "Woman and the Violinist," and numerous others have taken on a kind of mythological character within professional philosophy.

However it caught on, the development and use of such fictitious or hypothetical cases or "thought experiments" as a method of discovering and testing moral intuitions is now quite routine and almost second nature among moral philosophers. Inasmuch as the case analysis proceeds on the basis of shared rational intuitions, rather than upon any explicit moral theory, we might label this "natural law reasoning," or "secular natural law." This would distinguish this approach from the more familiar, religiously-grounded tradition of natural law reflection practiced by Catholic moral philosophers and scholastic theologians in the Medieval era. In contrast, the "new" natural law reasoning does not appear to require any prior commitment of faith, nor does it attempt to draw any larger metaphysical or theological conclusions from our ability to reason collaboratively about perplexing moral dilemmas.

Natural law reasoning in this secular mode is not without its drawbacks. The method of the "thought experiment" in moral philosophy sometimes strikes those newly introduced to it, for example, as exaggerated, strange, and sometimes even bizarre. "Why," the novice might ask, "ought we to concern ourselves with the ridiculously implausible situation of one person throwing a second, overweight pedestrian off a bridge in order to stop a runaway trolley car from killing five other innocent people? How absurd! What nonsense! How could anyone learn anything useful from such a ridiculous scenario?" Philosophers trying to discuss such cases are lucky if their audience doesn't consider throwing *them* off a nearby bridge!

Actually, the cases themselves are just as frequently perceived as delightful "brain-teasers," and are often, by themselves, quite useful in generating good discussion and insights. In my own experience, it is far more the stylized, somewhat pompous, jargon-laden, tendentious and pseudo-technical way that philosophers proceed to *present and discuss* such cases that alienates students and the public. Thus, philosophers themselves occasionally, through their pretentious commentary, ruin the invitation to thoughtful reflection that their creative imaginations have helped to stimulate. In any case, readers should not be discouraged by the unusual and sometimes fantastic details of individual cases, but see each for what they are: an attempt to isolate or fix a range of relevant variables or parameters until the specific question at issue is brought precisely into focus.

A Classical Example of This Method in Action

Though it was admittedly rare, and even more rarely well done, the contemporary method does have venerable antecedents. A famous example from ancient times is found in Plato's *Republic*, in which Socrates's friend, Glaucon, presents a hypothetical or "mythological" case, the "Ring of Gyges." Glaucon first describes the myth as his fellow dinner companions would have known it: a fantastic story of a shepherd, an earthquake, a journey into the bowels of the earth, and a chance discovery by that shepherd, Gyges, of a magical ring that makes him invisible.

What Glaucon then does with this remarkably vivid myth is to offer it as a case study that seems to support an argument or a moral position that he himself is advocating. Glaucon's position, or hypothesis, about morality is that "justice" (or what we might better translate as "righteousness," or individual moral "goodness" or rectitude) is really little more than a kind of social convention. People, he argues, are led to do what seems to us as "morally right" as the result of a kind of complicated risk analysis. Each of us would, if we could, do whatever we pleased that was in our individual and narrowly-defined self-interest, *if only we could get away with our actions*. Each of us fears, however, what might become of us if everyone else were to act likewise, *and so we hold each other strictly accountable* for the results of individual actions. The result, Glaucon triumphantly concludes, is "justice:" an implicit agreement to limit the sphere of action or power of choice for each individual in order to secure our collective order and security. The resulting equilibrium and order are what we mistakenly perceive as individuals "choosing" to act morally. That is, what we think of as "morality," or good individual character, is really the result of enlightened self-interest, a kind of amoral *realpolitik*.

So far, so good: this is an old argument, even though we hear endless variations of it put forth all the time as a novel modern insight. But what is the proof or "evidence" that Glaucon adduces for this claim? The answer is: the story, or "case study," of the shepherd, Gyges.

Note than when the magic ring frees the shepherd from the normal restraints of public accountability by making him (and his actions) invisible, Gyges proceeds to act ruthlessly, and certainly without the slightest regard for the constraints of morality. He seduces the king's wife, kills the king, and assumes the dictatorial power of a tyrant. Thus Glaucon believes his case against "justice"—or against the broader institutions of morality as representing anything other than a self-interested, crass social arrangement of mutual convenience—is demonstrated by the case or the "thought experiment" he introduces.

Another feature of the use of hypothetical case studies is our ability to limit, or to vary and manipulate, what might be thought of as the "boundary conditions" of the case in order to fine-tune or focus the moral argument on specific and salient features in question. Thus does Glaucon proceed to vary the original myth: what if, he asks, there were now two such magic rings, and we gave one to someone we believed to be morally corrupt, and the other to someone who was generally thought to be morally righteous? Would we be able to discern any difference in their behavior? Glaucon clearly thinks the answer negative, and takes this variation of the case as further proof that "justice" is an outward social convention, rather than (as Socrates will continue to maintain) a feature of an individual's moral character. Less we remain unconvinced, Glaucon introduces a third variant of the case, dispensing altogether with the magic ring(s), and focusing instead upon the social nature of morality directly. What if a morally upright person, and another who was debased and immoral (as measured by our intuitive understanding of these terms) went through their entire lives with everyone else mistakenly thinking the bad man good, and the good man bad, honoring the former and reviling the latter? Who would be the happier? Who would have lived the richer, more satisfying life?

Natural Law and the "Light of Reason"

The glaring defect of Glaucon's examples is that they "beg the question," by presuming in their very construction the issue that Glaucon hopes to prove by invoking them as evidence. By contrast, modern cases, such as those diagrammed below, intend to present (insofar as humanly possible) value-neutral situations in which the circumstances of the case help "tease out" features or implications of what might be called our "pre-theoretic" intuitions. The method is thus intuitive and inductive, and the goal is to see whether normal human reason, unaided by interpretation or theory, can make sense of the cases, sort out the variables, and develop some general guiding principles from these specific instances.

If we consent to call this "natural law" reasoning, it is important to clarify yet again that we are proposing a use of that term that departs from the religious connotations traditionally attached to the phrase, "natural law," as customarily used to describe the writings of Thomas Aquinas and other scholastic philosophers.[1] In contrast, what is presented here is simply practical reasoning, analysis, and deliberation about specific circumstances absent any explicit assumptions or "theories" about the nature of right and wrong conduct, or the origins of moral obligation itself. The connection between the two meanings of the term might be connected by observing (just as Thomas's original critics feared) that once "Reason" is permitted its own realm of independent inquiry, it is no longer tied necessarily or explicitly to any requirements of faith.

---

[1] Thus, scholars familiar with contemporary natural laws debates might be tempted to mistake this proposal for that of the "new" natural law by the distinguished scholar of jurisprudence, John M. Finnis (Notre Dame University and Oxford University). His revisionist proposals, however, still fall squarely within the Catholic legal tradition defined by Aquinas, whereas the proposals here are utterly independent of any context other than basic assumptions about the nature of reason.

In the cases diagrammed below, variations that might strike novice readers as merely repetitive (the "same case" merely with different agents or circumstances) are in fact always subtly different, and it is not terribly important how "realistic" the conditions are. Anyone who has ridden the "Green Line" (The BMTA's "Commonwealth Avenue" line) in Boston probably has some bemused sense of the underlying realism of the "trolley problem" diagrammed below as Figure One. But, of course, an individual corpulent enough to stop a runaway trolley car as it collided with him on the tracks could not possibly be thrown off a bridge and onto the track by another bystander in order to accomplish that objective, as in Figure Two! Do not be deterred or confused by such odd or "counter-factual" features of contemporary moral thought experiments: in part, they are meant purely for fun and amusement. They are aids to thought, and prompts to enjoyable, and also (hopefully) disciplined and fruitful, discussions about what makes specific acts (such as deliberately taking a human life in order to achieve some other good result), right or wrong.

A particular point to be borne in mind with the cases diagrammed below, incidentally, is that each is constructed so that any attempt to use purely utilitarian calculations of benefit and harm, of lives saved versus lives lost (or taken), is by itself insufficient to explain any intuitions of difference we might have, or to explain why we might permit, approve of, or even require some of the choices diagrammed below, while condemning some of the others that look quite similar. That issue, in fact, accounts for the origins of these cases: their original proponents (Williams, Thompson, et alia) were testing the limitations of conventional utilitarian reasoning in specific instances, and wondering whether convincing evidence could be provided for an additional and independent *constraint* upon such reasoning.

The constraint discerned as operative in each of these cases is something called *the "Doctrine of Double Effect."* That doctrine (which is also central to the historical tradition of natural law reasoning) imposes constraints that are not purely utilitarian in nature upon the *means* by which otherwise morally desirable ends might be obtained: in conventional understanding, as long suspected, not all means are justifiable, even if they lead to good outcomes or ends, as all these cases illustrate. In particular, it appears that one may not propose an explicitly immoral or prohibited "means" or method (such as actively killing a bystander or patient without his prior consent) as the manner in which to achieve an otherwise-acceptable or praiseworthy goal (saving the lives of many other potential victims, for example).

II. The Cases[2]

Figure One. A runaway trolley is hurtling down a set of tracks toward a construction site on the tracks at which five railroad workers are laboring, unaware of the approaching danger. Assume that, on impact, all five railroad workers will be killed unless a bystander intervenes. His only options in this case (don't cheat by inventing others!!) are either to stand by helplessly and witness the death of five, or to intervene by throwing the switch next to him and diverting the trolley onto a siding. If he does this, however, a single pedestrian who just happens to be walking along that siding will be struck and killed. What should the bystander do? (Hint: It might be useful to ask first, what is s/he *permitted* to do? Then try examining whether s/he is *morally obligated* to either course of action. Alternatively, is s/he *morally prohibited* from either of the two options? Why?)

Figure Two. The runaway trolley is, as in Figure 1, likewise careening toward the five unsuspecting workers on the track ahead. But in this instance, as the trolley passes under a pedestrian overpass, a bystander walking overhead realizes the imminent danger of death to the five passengers further down the track. Acting swiftly, he manages to shove another, overweight bystander from the bridge, onto the tracks, directly in front of the car, resulting in the death of the second bystander, but also

---

[2]The following diagrams are pictorial representations of a sequence of six hypothetical cases or "thought experiments" portrayed in the book, *Ethics: Problems and Principles*, by John Martin Fischer and Mark Revizza (New York: Harcourt Brace & Co., 1992). The variations of what philosophers widely recognize as the "Trolley Problem" owe their origins to informal discussions by the late Bernard Williams, and first came formally to widespread public attention in an essay by Judith Jarvis Thompson entitled, "Killing, Letting Die, and the Trolley Problem" (*The Monist*, 1976). The diagrams provided here are illustrations prepared by Captain Rick Rubel.

stopping the runaway trolley. Is the first bystander *permitted* to do this? Is s/he *obligated* to do so? Is s/he *prohibited* from doing so? Why/why not?

(Hint: What similarities do you see in the two cases? What differences do you discern? How important are these differences?)

Figure Three. Five patients are threatened with death from a bacterial infection that can easily be cured by using 1/5th of the total amount of an antibiotic on hand in the physician's office. Another patient, however, has suffered a massive systemic infection (as a result of injuries sustained in a car accident) and will die quickly unless she is given all of the available supply of the antibiotic. Is the physician permitted to treat the five and save their lives, thus allowing the sixth patient to die as a result? Is the physician obligated to treat the five, rather than the one? Is the doctor strictly prohibited from either of these two courses of action (and if so, which, and why?) Again, how does this case resemble, or differ from, the preceding?

[**Addendum**: a true-life variation of this case from military medicine during the Second World War provides an interesting and troubling variation, in a dilemma confronted by many military doctors during the early days of penicillin. Penicillin was perpetually in short supply during that war. Thus: the doctor has only one vial in his possession. His five patients are soldiers returning from weekend liberty, each having contracted a socially-communicable disease, easily cured with a small dose of penicillin, but potentially fatal if left untreated. Treating all of them would exhaust his available supply of the antibiotic, but would allow all five quickly to return to their posts at the front. The sixth patient is a soldier dying of massive infections from shrapnel wounds sustained at the front. Only by using the entire vial of penicillin in his treatment can he be saved and returned home. He will not be able to fight again. Whom does the military doctor treat, and why?]

Figure Four. Five seriously ill patients are dying from serious diseases that can only be cured by organ transplants: two have diseases of the right and left lungs, respectively, while each of the remaining three suffer, respectively from liver, kidney, and heart disease. (Remember, once again, the plausibility of this scenario is not in question!) The attending physician also has under his care a patient who is recovering from a stomach ailment, and is otherwise perfectly healthy. May the physician remove or "harvest" the needed organs from the one healthy and recovering patient (and thereby causing his death) in order to save the lives of the five who will surely otherwise die? Why/why not? How does this case resemble, and/or differ from, the preceding ones?

In the final two examples, we have several swimmers in the water, about to be attacked by a shark.

Figure Five. One swimmer is directly in the path of the shark, while five others are a little further away. The shark presumably intends to eat them all. A small rowboat is nearby, in which the rower observes the danger and seeks to help. Given the limitations of time, he can row (or swim) either to save the one, or to save the five, but cannot do both. Should he save the one, or the five? Why?

Figure Six. In this scenario, there are now only five swimmers, and the shark is nearly upon them. The sixth person is a large and presumably tasty passenger in the rowboat. Our rower could save the five swimmers, but only by pushing his passenger overboard into the path of the oncoming shark in order to distract it with a snack while he quickly rows over to pick up the five swimmers. Is he permitted do this? Is he either obligated, or prohibited, from doing this?

Additional Exercises. If time permits, you may wish to examine the arrangement of the cases in the diagram. They come in pairs, arranged in horizontal rows by subject (trolleys, medical practice, and attacking sharks). But try now arranging them in vertical columns instead, according to whether the five victims can be saved, or cannot be saved, by the means or methods proposed in each case. Can you label the resulting columns? What labels would you use? Do the labels point to morally relevant differences between the two columns of cases? If so, what are these? How does the doctrine of double effect discern among these cases? Other than matters of miscellaneous detail, are there matters of substance that differ among the cases in each column, or do they merely duplicate the essential moral scenario in varying ways?

[Hint: pay particular attention to the original "trolley" case, Figure 1. Does it differ in any important ways from the others with which you have now grouped it in a column? If so, what is this difference? Can the Doctrine of Double Effect account for that distinction? If so, how does this case differ from those in the other (prohibited?) column of examples? Pay particular attention to the DDE's distinction between "intended, as the principal means to the end," and "merely foreseen, as the unintended or secondary consequence of the (otherwise permissible) means."]

Finally, equipped with these insights, you are now ready to attack some real-life cases involving choices like these. A majority of such dilemmas arise in the practices of emergency medicine, especially during wartime (as in the addendum to Figure Four above), as well as in decisions made about what targets you may be permitted or required to attack, or what means or methods you are permitted to employ in combat, versus those targets, and those methods, that you will find to be *strictly prohibited*, even situations of grave emergency. Such considerations are essential for maintaining the distinction between legitimate and permissible military combat operations, on the one hand, and acts of terrorism or war crimes, on the other. Try, for example, the case by Captain Rick Rubel: "Terror and Retaliation," found in our companion volume, *Case Studies in Military Ethics*. Even if you have already read it, you might now also wish to re-visit the case: "First Use of Nuclear Weapons" by Velasquez and Rostankowski, also in that volume.

Natural Law and the Principle of Double Effect: Six Hypothetical Cases

Natural Law Cases

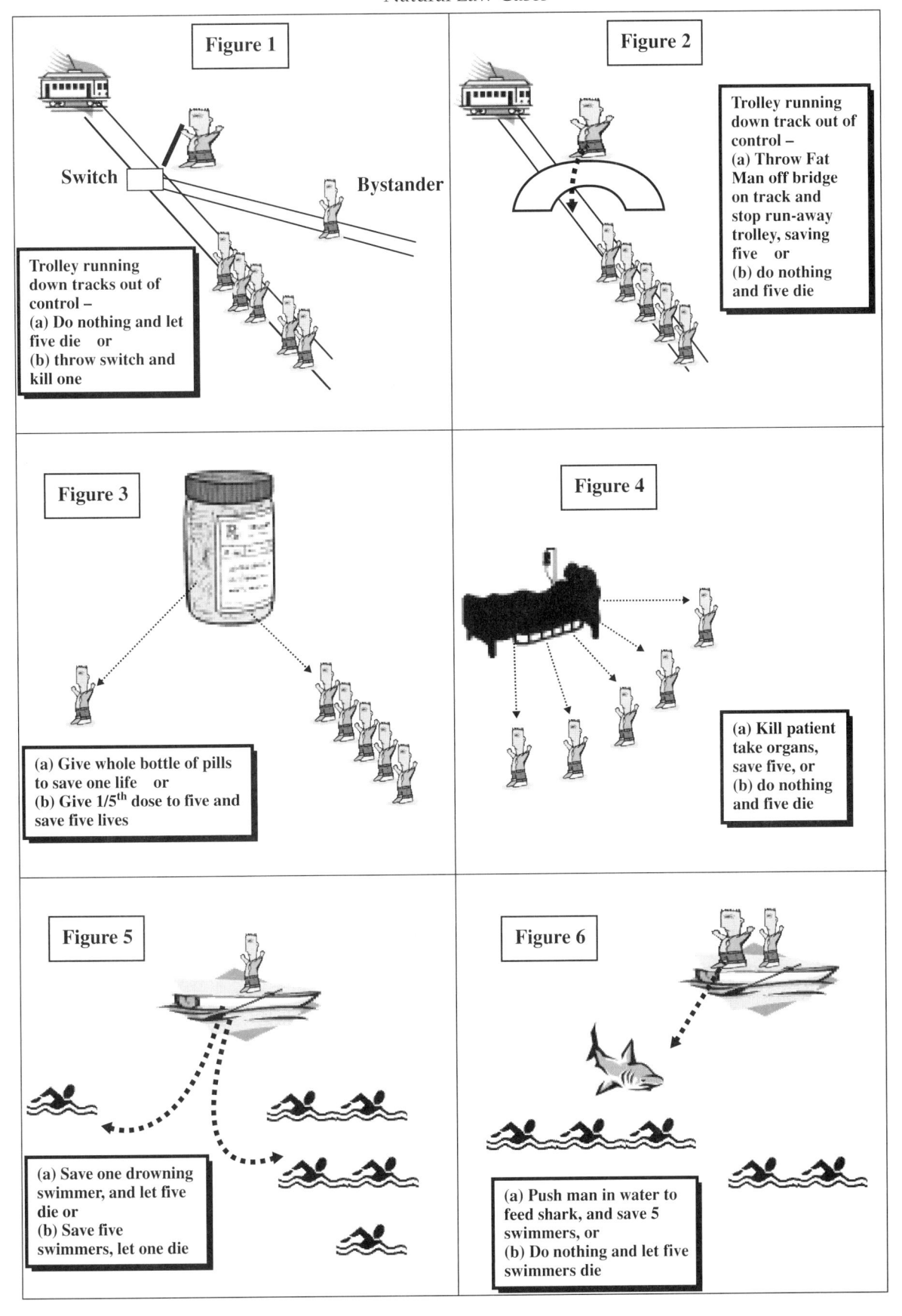

Figure 1
Switch — Bystander

Trolley running down tracks out of control –
(a) Do nothing and let five die   or
(b) throw switch and kill one

Figure 2
Trolley running down track out of control –
(a) Throw Fat Man off bridge on track and stop run-away trolley, saving five   or
(b) do nothing and five die

Figure 3
(a) Give whole bottle of pills to save one life   or
(b) Give 1/5th dose to five and save five lives

Figure 4
(a) Kill patient take organs, save five, or
(b) do nothing and five die

Figure 5
(a) Save one drowning swimmer, and let five die or
(b) Save five swimmers, let one die

Figure 6
(a) Push man in water to feed shark, and save 5 swimmers, or
(b) Do nothing and let five swimmers die

Part IV
The Moral Role of the Military Professional in International Relations

In the course of our review of the moral foundations of the military profession, and of the major traditions of moral discourse arising broadly in Western culture, we have had occasion to refer briefly in passing to a number of familiar, troubling, and perplexing moral dilemmas: murder and capital punishment, abortion and euthanasia, and racial and sexual discrimination, among others. Often students in colleges and universities are asked to take a survey course in ethics, in which the traditions of moral reflection we have covered here are also presented, and in which many of these fascinating and difficult dilemmas are read about and discussed. As Kant would be quick to point out, were he still among us, however, the most urgent, pressing, and perhaps longstanding of these great moral dilemmas is the problem of war.

In this country, the President of the United States periodically issues a National Security Strategy, designed to outline our nation's needs, priorities, strategic goals, and the security impediments to attaining them. In response, the Chairman of the Joint Chiefs of Staff, representing the military services of this country, issues a National Military Strategy (NMS), intending to set forth how this nation's military services will respond to the security threats that the President and his National Security Council have identified. Invariably, in the Foreword to that document, the Chairman of the Joint Chiefs reiterates and reaffirms the underlying purpose and intent of the military services of this country "to stand prepared for, and to fight the Nation's wars." In a recent version this document, a sitting Chairman of the Joint Chiefs, General Richard B. Meyers (U.S. Air Force) explains that

> "The National Military Strategy articulates the ways and means to protect the United States, prevent conflict and prevail against adversaries who threaten our homeland, deployed forces, allies, and friends."

The attacks by Islamic terrorists in New York City and Washington, DC on September 11, 2001 undeniably represented a "threat to our homeland." So, with equal clarity, did the attack by Japanese naval air forces on military and civilian installations in Pearl Harbor, Hawaii, on December 7, 1941. President Franklin D. Roosevelt's angry response to those unprovoked attacks in 1941 echoes with equal validity in 2001: these are days "that will live in infamy." Both events precipitated our use of military force in response, ostensibly for the purposes of self-defense.

With the exception of these two events, and possibly the War of 1812, however, the United States has not suffered an attack upon our own soil, or suffered from physical harm directed by an enemy or adversary against "the homeland." We have, however, been involved in a number of wars down through the ensuing two centuries following our founding—through an act of revolutionary war—in 1776. We have fought against French, British, and Mexicans. We have fought against one another during the Civil War. We deployed our armies to subdue indigenous peoples and to take control of much of their land during the remainder of that century. We came to the aid of "allies and friends" in Europe during the final year of the First World War, and again (once we had been ourselves attacked by the Japanese) during World War II against German Nazis and Italian fascists. While a generation now passing into eternity justifiably celebrates its tradition of service and sacrifice during that Second World War, many individuals before and after have wondered, about these other wars, "were we right to fight?" Was it absolutely necessary, and were we justified in so doing, when we as a nation chose to use our military forces to fight in these various ways, and in these numerous wars?

During the last half of the preceding century, we endured an extraordinary and sustained conflict, a "Cold War" against competing, nuclear superpowers that lasted nearly forty years, during which, in all but the last few of those years, many nations of the world (including our own) labored under the constant threat of wholesale nuclear destruction. Successful pursuit of our strategy of nuclear deterrence required that men and women in our military services be prepared, if commanded, not simply to fight wars to protect our homeland, but also to unleash unimaginable destructive force against enemy targets (in the full expectation that the enemy would, if possible, retaliate against us). The vast majority of the casualties that were foreseen in any such exchange of strategic nuclear weapons would have been non-combatants, numbering in the tens or even hundreds of millions.

This "cold war" placed a tremendous moral burden upon individual members of our armed forces. A great many individuals, from lowly enlisted corporals and airmen to Navy commanders and captains aboard sophisticated and heavily-armed nuclear submarines had to ask themselves the question: "would I, could I, if so ordered, launch the nuclear missile in this silo, drop the payload of nuclear bombs in this Strategic Air Command B-52, or fire the missiles in my Trident submarine? Could I do this, knowing what it will mean, knowing what likely will happen, both to the enemy targets (and surrounding non-combatants), and then knowing what will likely transpire as a result in my own homeland, and to my own loved ones?" Military officers certainly had to reflect on this, or else try vainly to avoid the nightmare of thinking about it. For most, it was not a pleasant experience, even when convinced beyond a doubt of the morality of our cause.

For the cold war to be successfully fought and, as it eventually was, to be won, each individual soldier, sailor, or airman had to have decided, in advance, that he or she could do as they were ordered, if and when they were so ordered. At the same time, many inside and outside the military services wondered just what kind of threat, or even actual harm, if any, could justify having to undertake such actions in response?

This is the moral dilemma that war poses for us all, and, in particular, this is the tremendous moral burden it imposes upon those we ask to undertake this activity in our behalf. Happily, the worst strains of the nuclear era of superpower rivalry seem to be behind us, and this nation has emerged, in fact, as the world's sole, remaining "superpower." We now live in an age that many characterize as "Pax Americana." Our military might, and the quality of our military personnel, vastly outstrip anything or anyone that could be placed in the field against us. Surely such invulnerability, such superiority of force and armament, should mean that our nation, along with its "allies and friends" are at last, each and every one of us, secure from the threat of all enemies, foreign or domestic, so that we can pursue our individual paths to happiness and dwell in relative peace.

Unfortunately, as we know, this is not the case. Throughout the past few years, the world has witnessed in horror as racial and ethnic groups, freed of the constraints imposed by Cold War politics, have turned upon one another with a genocidal hatred and ruthless violence not seen since the Holocaust. The time is now at hand when the greatest threat to the lives and welfare of individual citizens in many regions of the world comes, not from hostile foreign governments, but more often from their own government and fellow citizens.

The very asymmetry of American military power in this "Pax Americana" has served to provoke unpredictable forms of response. Against all reason, it would appear, we have discovered that individuals and organizations, blinded by furious hatred, are willing to sacrifice themselves individually—literally to bring about their own deaths—simply for the purpose of embarrassing this country abroad, or causing it pain and sorrow at home. Such individuals are willing to go to their own deaths in order to wound and kill enlisted men and women standing in a lunch line aboard a Navy vessel otherwise peacefully transiting the Red Sea, simply stopping to re-fuel in a foreign locale. Such individuals will commandeer a commercial aircraft and kill themselves, along with its passengers and crew, simply to take the lives of a Muslim cleaning attendant, a visiting Egyptian businessman, an Irish cop, and a large number of courageous and good-natured New York firefighters who came to the aid of these and thousands of victims like them in the New York Trade Towers, not one of whom could possibly have represented a threat to, or committed any acts of harm or disrespect to those who fomented this attack against them.

This, regrettably, is the world as we now find it. It is every bit as dangerous, unpredictable, and morally ambiguous a world as at any time in history. What is the place of the military profession, and what are the roles and moral responsibilities of military officers, within this world? When, and under what conditions, and for what purposes, are we justified in using military force in this "new world order?" How would we know? And by what criteria ought we, our fellow citizens, and our elected and appointed leaders, to decide such grave questions?

The appeal to force to settle conflict is as old as human history, and seems a regrettable part of the human condition, as philosophers from Augustine to Kant have sadly observed. *While the uses of military force are and will remain commonplace, not all such uses are morally justifiable.*

The search for, and understanding of the criteria under which, or according to which, appeal to military force is justified is one of the most enduring moral conversations in our cultural history. While a consideration of this question and the main responses to it once constituted a regular and significant part of any intelligent student's education, the topic of the morality of war and military ethics had, prior to "9/11", all but disappeared from the college and university curriculum during the past several decades. In the wake of humanitarian crises abroad, and terrorist attacks at home, the topic is resurfacing as an issue worth studying and discussing. It is, quite obviously, an important component in the proper education of any military officer.

There are several parts to this important discussion. The first is known traditionally by its Latin label, a legal idiom coined during the medieval era: *jus ad bellum*, literally, the "right" to enter into war, or the justification of war. The second important component follows from the first, and concerns how, and in what ways, military personnel are permitted or entitled to act during combat. This is likewise known by its Latin legal idiom, *jus in bello*, or the "law of war," or "law of armed conflict." We will consider both of these central concerns in this unit. Despite the use of the term "law" in these discussions, they are not primarily legal, but moral discussions. Such laws as there are suffer from the absence of a genuine overarching authority to legislate, and to enforce any resulting legislation. It is part of our consideration of these topics to discuss just what is expected and required of warriors in wartime, and when these moral obligations arise, when all normal moral and legal conventions seem to have been set aside.

One of the most significant elements of the discussion of justification, in turn, is what is termed "just cause" for pursuing armed conflict. We have already cited one such cause, which appears almost self-evident, and is enshrined in the natural law tradition with which we concluded the preceding unit: namely, self-defense. We have also observed that few if any wars are fought strictly for such purposes in any meaningful sense of the term. Thus, in the four chapters of Unit V which follows, we consider some of the other causes besides self-defense for which the deployment and use of military force is deemed acceptable, at least in principle. These include the defense of liberty, the protection of natural or "human" rights, and the enforcement of justice and the rule of law in the international arena. The latter almost exclusively (rather than self-defense, *per se*) are the causes for which military force is currently invoked: as in bringing law and order to a nation embroiled in civil strife (Haiti, or Liberia) or defending the rights or liberties of vulnerable peoples (in Rwanda, or Bosnia, or Kosovo), or in restraining "rogue states" and apprehending international criminals, regardless of their cause or grievance (ranging from the Barbary Coast pirates "on the shores of Tripoli" in the late 18th century, to Osama bin Laden and his al Qaeda terrorist network today).

It is important to remember, just as in the example of capital punishment, that *establishing the "right" of a people or a State to use force does not automatically mean that it would be wise or prudent to do so*. In no other area, perhaps, are considerations of prudence and feasibility so intertwined with the normative concerns of morality and law as in the case of declaring and fighting a war. As we will now see, some of these prudential considerations are "built into" the normative moral and criteria, particularly in the requirements for "reasonable hope of success," and the demands for "proportionality" between the injustice to be remedied by war and the destructive nature of war itself.

A. The Justification for Going to War (*Jus ad Bellum*)

Under our system of government, with civilian control of the military as a central tenet, it might seem superfluous for military officers to examine the justifications for war. The military, it might seem, is obligated in a democracy to fight when and where instructed, and *only* when and where instructed, by the elected leadership of the nation.

It would be a foolish and headstrong leadership, however, that failed to consult with its experienced and trained experts about the feasibility, desirability, and especially the morality and legality of what they propose to ask those experts to undertake. And so it is, often to their great surprise, that military officers in this country, sometimes of quite junior rank, will find themselves in a position to be consulted, or invited in other respects to participate, in policy debates, sometimes at quite high levels, as this nation's leadership evaluates the merits and prospects of going to war. Inasmuch as a good deal of such discussion centers, as it has from the outset in the second Gulf war, on whether it would be "moral" or "legally permissible" to engage in these actions, we would do well to prepare ourselves to participate responsibly in these discussions when asked to do so.

In part, this is simple prudence and good leadership: individuals and organizations invariably perform much more proficiently if they know what they are doing, why they are doing it, and most importantly, believe in the mission to which they are assigned. The converse is invariably even more true in wartime. As we learned to our enduring regret in Vietnam, warriors sent on what seems an aimless, possibly immoral, or even impossible mission, and lacking public support at home, quickly lose heart, do not perform well, and can hardly be expected to acquit themselves with a high degree of professional demeanor.

This principle translates all the way down the chain of command, and constitutes a further reason for military officers to understand the basis for the moral justification of war. Each officer will be asked, one way or another, to interpret and defend their military mission to colleagues and subordinates. They will have a much easier task of doing this if they understand the important role that morality plays in the building of morale, and in attaining something like what Locke termed "tacit consent," even if the formal consent of subordinates is not officially needed or required.

Sometimes, in an authoritarian and hierarchical organization like the military, this elementary principle of sound leadership is forgotten, and we would do well to recall it now. Consider once again the text of Lt. Col. Tim Collins' speech to his troops on the eve of the second Gulf war (Unit I, above). The speech itself is inspiring, principally because of the moral justification he develops for the necessity of fighting in Iraq, and the moral goals he proposes for British and American troops to accomplish there. That speech is all the more inspiring if the men and women under his command *believe* it to be true. That, in turn, is a much easier case for him to make to them and to sustain, if the moral justification set forth in the speech *is in fact* true!

But *how would they know,* how would *he* know, and indeed, how would any of us know, whether the case he states in that speech is valid, if we are all ignorant of the conditions under which the moral claims Lt. Col. Collins sets forth could themselves be authenticated? Authenticating such claims, and providing each military officer with the basis for his or her own evaluation of the larger public debate on such questions, is the task of what is sometimes called "just war theory," or the "just war tradition."

War and Moral Philosophy

Historically, as our reading selections indicate, this discussion takes two forms. The first is a sustained philosophical reflection dating back to Plato, Cicero, and Augustine, and given its most polished and familiar form in the writings of St. Thomas Aquinas.

War is particularly problematic in the tradition of Christian philosophy, simply because the core of Christian teaching about the moral responsibilities of the community of the faithful within the larger society is explicitly non-violent. Christianity itself is a religion, much like traditional orthodox (or *Therevada*) Buddhism, that teaches pacifism, and condemns all violence, especially the waging of war. Members of the military ranks who come from within this tradition, and who are devout and sincere in their faith, are thus confronted at the outset with a severe problem: how can they reconcile the teachings of their faith tradition with the presumed necessity of taking up arms to defend the State and their fellow citizens, let alone to go abroad to fight to enforce justice and the rule of law, or to defend vulnerable human rights against evildoers? Many Christians simply will not do this, and insist that to do otherwise would be to betray their faith, and to sin against God's commandments. Thus, in the name of separation of church and state, citizens in the U.S. have always, in principle, been provided with the right of individually declaring themselves "conscientious objectors," on religious grounds, to serving in the military, even in their own defense.

We will not pursue the interesting topics of pacifism and conscientious objection any further here, though it is well now to pause for a moment and recognize that this is why we find so many church writers, down through the ages, from Augustine and Aquinas to the American Conference of Catholic Bishops, various leaders of Protestant denominations, the Pope, the orthodox Patriarchs, the Archbishop of Canterbury, and numerous other scholars, clerics, and legions of faithful believers agonizing over this question. St. Augustine perhaps framed the dilemma for all time in his great work, *The City of God* (c. 410 C.E.). It is, he finds, regrettable but true, that living in this corrupt and "fallen, sinful" political order, the "city of Man," we are sometimes confronted with the unpleasant but unavoidable circumstance of having to choose between the lesser of two evils: either suffering the effects of chaos and monstrous injustice, or *taking up arms* to oppose it. It would not necessarily be wrong, he concludes, and under some severe circumstances it might even be required of us, as Christians, to do the latter.

The Readings (I)

It is this dilemma that leads St. Thomas (as he studies and comments on the writings of St. Augustine and others on this topic) to frame his scholastic question in our first reading selection for this unit, once again from his *Summa Theologica*: "Is it Always Sinful to Wage War?"

Note that the manner in which the question is put suggests that Thomas is prepared to answer: "Well, no, perhaps not always, *given that it usually is* sinful to wage war." The intention, that is to say, is not suddenly to give spiritual aid and comfort to war-mongers (as Kant charged the natural law tradition with doing), or to turn from the core tradition itself and blandly assert that "war is okay." Rather, the problem for Thomas, as for Augustine before him, seems to be when, if ever, to lift the presumption against the morality of war and violence. Specifically, when, and under what conditions, might we justifiably conclude that war, while regrettable, is nonetheless unavoidable. Warriors must always, and only, take up their arms reluctantly, with the "rightful intention" of rectifying injustice and restoring peace.

This leads Thomas to formulate what have come down to us since as the main criteria, the *necessary conditions*, that would have to be satisfied in order for it *not* to be "sinful," or morally wrong, for us to wage war. He proposes three important criteria, to which subsequent interpreters have added as many as four more. Thomas's original criteria, in this reading selection, are:

- War must be declared by a *legitimate authority*
- There must be a *just cause* for which the war is to be fought
- Our *intentions* in fighting must themselves be morally worthy

These are open to broad interpretation, and usually serve as a springboard for fruitful discussion. The first criterion, for example, condemns a terrorist attack in part because the terrorist, whatever the legitimacy of his grievance, is not unilaterally entitled to assume the mantle of a legitimate authority.

That leaves to debate, especially in our current situation, however, just who or what a "legitimate authority" might be. The criterion of "legitimacy" was precisely and most prominently at issue in the second Gulf war, as well as in the Kosovo campaign. Does legitimacy now, by virtue of treaty and international law, reside solely with the United Nations, or are there other regional collective entities (like NATO) that might, on occasion, assume this burden rightly? In any case, is it ever permissible for a single nation, unilaterally, to ignore the opinions of its friends and allies and wage war on its own, regardless of the justice of the cause for which it may fight?

Because this list is neither unambiguous nor exhaustive, other scholastic commentators attempted to find in Thomas's writings on this topic, or add to his discussions, some additional clarifications, such as:

- The resort to war must always be a *last resort*, after all other options have been exhausted
- There must be a *reasonable hope of success* in attaining the objectives for which the war is to be fought
- The moral value of the ends to be served, or the harms or injustices to be redressed by the war must be *proportional* to the sacrifices required and the damage that will invariably be inflicted
- Finally, an otherwise just cause must only be pursued by *just means*

For perplexed students, new to this conversation, it is important to realize that, notwithstanding the importance of this list of necessary criteria, not all writers on warfare label the conditions consistently, or agree on the items on the list. In the main, most of the first six of these criteria will appear in some form or another. Not every list of these criteria, however, will include the final one on "just means," for it is thought to entail an entirely separate consideration, what we have labeled *jus in bello*, or conduct during wartime.

This is a complex matter, and the two concepts are not so easily disentangled, since fighting in a morally justifiable war does not grant license to combatants to behave in morally unjustifiable ways. Vice-versa, the unjustifiable (illegal or immoral) conduct of combatants during wartime might well undermine the justification of an otherwise justifiable war. All of this, as we will see, is especially pertinent to wars that are not strictly wars of self defense. It would be odd, for example, if NATO troops sent to provide aid in Kosovo to potential victims of Serbian ethnic violence ended up treating the Kosovars just as badly, or even worse, than the Serbians did. Surely, at minimum, that would render what was at best a tenuous justification for war under existing criteria all but negligible. We will take this question up in more detail in the next chapter.

War and the Rise of International Law

The second mode of discourse concerning the justification of war arises in the context of what is known as international law. Compared to the classical or philosophical discussion, this treatment of the problem of war is comparatively recent, arising principally from discussions originating from the work of the seventeenth-century Dutch legal scholar, Hugo Grotius, in his work *On the Laws of War and Peace* (1625), and further prompted by the political ramifications of the Peace of Westphalia in 1648, which ushered in the nation-state system under which we live today.

The Readings (II)

The first selection by Michael Walzer, while it does not rehearse the history of this discussion, offers a summary account of its main provisions as established over the ensuing four hundred years, in what Walzer describes as "the legalist paradigm." This understanding rests on what Walzer terms the "domestic analogy," according to which the community of autonomous nation-states is considered to constitute something akin to a community of individuals living under the rule of law. Just as each individual in the domestic case is granted a degree of liberty and self-determination, and the right of self defense against aggression, so nation-states (in the Westphalian model or "legalist paradigm") may be thought to constitute entities possessed of "liberty" in the sense of sovereignty and territorial integrity.

That is, the inhabitants of these states are thought to be collectively free to pursue their common way of life inside their own borders, without undue interference from other nations or peoples. It was this constraint that brought to an end the great religious wars of the Reformation and counter-Reformation, about which Martin Cook wrote in Unit II above.

In the resulting paradigm, significantly, any act of aggression by one state against another, no matter what the reason given, constitutes (as it would in the domestic case) *a criminal act*. Note this radical departure from the "just cause" provision of the classical doctrine. There each nation is at least invited to make a moral case for resorting to violence. In the legalist paradigm, no such latitude exists, and *all such recourse constitutes criminal behavior by definition*. This is the major advantage of the Westphalian or "legalist paradigm:" it replaces the vague ambiguity of moral argument with the clear stipulation of "black-letter" law.

The criminal act of aggression, in turn, justifies two responses (as also in the domestic case): the state attacked may defend itself (just as an individual may do, in principle); and other, neighboring states may come to its aid (just as neighbors or bystanders may come to the aid of an individual victim of violence). This latter concept in the international arena is known as "collective security" and this, together with self defense, constitute the only legitimate reasons to use military force. In contrast to the possible range of defensive and even offensive instances of "just cause" in the classical just war tradition, this "baseline model" of the legalist paradigm thus *severely limits* the conceivable justifications for the use of military force.

Many readers mistakenly conclude from his summary of the main tenets of the "legalist paradigm" that Walzer's book, *Just and Unjust Wars* (1977) from which this selection is drawn, is himself an advocate of that paradigm. Quite the contrary, Walzer's own agenda is to invite us to consider how this legalist paradigm might need to be updated and amended in the late 20th century to incorporate other justifiable uses of force in extreme cases. Although written over 30 years ago, Walzer's treatment of this topic remains timely and authoritative today, and his book is owned and widely read by leading military and political figures in this country.

The Westphalian or "legalist" paradigm would have to be amended considerably to incorporate proposals in the text of our next reading. "Wars of Anticipation" sets forth another category of use of military force, which Walzer terms "preemption" or "preemptive self defense." This is controversial, less because it proposes a new policy (the U.S. and other nations have, on numerous occasions, gone to war preemptively, without direct provocation) than because it makes that tacit policy explicit, and *attempts to provide a moral justification for it*.

The U.S. invocation of this doctrine of preemption sparked a strong public outcry in opposition, even in the aftermath of terrorist attacks on this nation, because, according to the legalist paradigm (not to mention prevailing statutes of actual international law) such a proposal on its face is illegal and morally unjustifiable. This perhaps accounts for the attempt to link pre-emption with self defense: if we know we are about to be attacked, and would then be justified in defending ourselves, why wait until we are attacked? Why not pre-empt the attack, and perhaps lessen the harm? This, as Walzer notes, was Israel's dilemma in 1967, as the armies of several Arab states amassed on its borders. Under certain stringent conditions, Walzer thus proposed that the "baseline" model of international law can and should be amended to include the use of force to *prevent nations from preparing to do us harm*, especially when the danger is imminent and waiting would make things even worse.

Yet no nations were, or are currently, amassing on U.S. borders. Instead, President George W. Bush argued that shadowy, non-state actors (terrorist cells) work, sometimes with the secret cooperation of the governments of "rogue states" (such as Syria, Iran or North Korea), up until the last minute, out of sight, and then strike almost without warning, possibly using weapons of mass destruction that could kill thousands, and perhaps hundreds of thousands, of innocent people. The President accordingly proposed that we can no longer rely on concepts like nuclear deterrence, because there is no responsible enemy nation to be deterred. Instead, we must now learn to hunt these individuals, organizations, and state leaders down proactively, in order to prevent their doing mischief later. If so, however, this is less a case of pre-emption, let alone of self-defense, than of taking protective or preventative measures for our long-term security. In the past, preventative wars for the sake of such security have seemed little more than a license for perpetual conflict, and so beyond the pale of moral or legal justification.

It is interesting to note, however, that the older, classical or philosophical tradition of just war deliberation has less of a difficulty encompassing proposals such as the Bush doctrine of preemptive self defense than does prevailing international law. St. Thomas, for example, pointed out that just causes for the use of military force could include efforts "to thwart and to punish organized evildoing, or to protect innocents from harm" (*STh* II-II, Q:40). The former might well constitute counter-terrorist measures, while the latter suggests humanitarian uses of military force today.

Thomas's merely saying this does not make it so, of course, for any but the most dogmatic of scholastic theologians. Rather, his comments indicate that there is, in the scholastic tradition of just war debate, a category of "offensive war" to which counter-terrorists might appeal in making their case for the legitimacy of pre-emptive war. Once again, we might caution, however, that making such a case for moral legitimacy would not immediately translate into a case for the wisdom, political prudence, or even the tactical feasibility of such a policy.

This last point is central to the remaining articles on humanitarian intervention and "irregular warfare," yet another species of military intervention that does not entail self-defense, or even the defense of clearly-defined national interests. Many of these kinds of wars have been, are being, and likely will be fought in the lifetimes of readers of this anthology. Apart from counter-terrorism, these kinds of wars in defense of human rights are likely to be the predominant form of conflict in the post Cold-War era. Here the question, unlike pre-emptive self defense, is not whether we would be morally or legally permitted to fight such a war, but whether (as under the United Nations Genocide convention, for example) we are *obligated* to render such military assistance, even if we do not wish to do so. The lack of clear national interest in such uses of military force is only one of several dilemmas highlighted in this article. The need for such assistance, which might be likened to police or "constabulary" duties in the international arena, is increasing exponentially. As Martin Cook also noted in his article on the moral foundations of the military profession (Unit II), we have not as yet marshaled the military resources, nor provided the kinds of training and preparation, necessary for nations to respond collectively to such disasters.

What, if anything, are we prepared to do to ensure the rule of law, enforce justice, and most importantly, to protect vulnerable peoples from harm at the hands of their own governments and fellow citizens? These, after all, do seem to be reasonable questions to put before present and future military officers for thoughtful consideration, as they themselves will likely have to bear the brunt of the consequences stemming from our collective deliberations and decisions on such matters.

Is It Always Sinful to Wage War?
St. Thomas Aquinas *

In order for a war to be just, three things are necessary. First, the authority of the ruler, by whose command the war is to be waged; it is not the business of a private individual to declare war, because he can seek redress of his rights from the tribunal of his superior. Similarly, it is not the business of a private individual to summon together the people, something which has to be done in wars. But since the care of the common weal is committed to those who are in authority, it is their business to watch over the common weal of the city, kingdom, or province subject to them. And just as it is lawful for them to have recourse to the sword in defending that common weal against internal disturbances, when they punish evildoers, according to the words of the Apostle: "He bears not the sword without cause, for he is God's minister, an avenger to execute wrath upon him that does evil," so too it is their business to have recourse to the sword of war in defending the common weal against external enemies. Hence it is said to those who are in authority: "Rescue the poor and deliver the needy out of the hand of the sinner," and for this reason Augustine says, "The natural order conducive to peace among mortals demands that the power to declare and counsel war should be in the hands of those who hold the supreme authority."

Secondly, a just cause is required, namely, that those who are attacked, should be attacked because they deserve it on account of some fault. Wherefore, Augustine says, "A just war is wont to be described as one that avenges wrongs, when a nation or state has to be punished for refusing to make amends for the wrongs inflicted by its subjects or to restore what it has seized unjustly."

Thirdly, it is necessary that the belligerents should have a rightful intention, so that they intend the advancement of good or the avoidance of evil. Hence Augustine says, "True religion looks upon as peaceful those wars that are waged not for motives of aggrandizement or cruelty but with the object of securing peace, of punishing evil-doers, and of uplifting the good." For it may happen that, even if war be declared by legitimate authority and for a just cause, it is nonetheless rendered unlawful through a wicked intention. Hence Augustine says, "The passion for inflicting harm, the cruel thirst to vengeance, an unpacific and relentless spirit, the fever of revolt, the lust of power, and such like things, all these are rightly condemned in war."

As Augustine says. "To take up the sword is to arm oneself in order to take the life of someone without the command or permission of superior or lawful authority." On the other hand, to have recourse to the sword (as a private person) by the authority of the ruler or judge or (as a public person) through zeal for justice and by the authority, so to speak, of God, is not to "take up the sword" but to use it as commissioned by another; wherefore it does not deserve punishment. And yet even those who make sinful use of the sword are not always slain by the sword, yet they always perish by their own sword, because, unless they repent, they are punished eternally for their sinful use of the sword.

---

*"Is It Always Sinful to Wage War?" excerpt from *Summa Theologica* by St. Thomas Aquinas. *The Summa Theologica of St. Thomas Aquinas, Fathers of the Dominican Providence,* ed. 1911: Burns and Oates. Ltd.

Law and Order in International Society
*Michael Walzer**

Aggression

Aggression is the name we give to the crime of war. We know the crime because of our knowledge of the peace it interrupts—not the mere absence of fighting, but peace-with-rights, a condition of liberty and security that can exist only in the absence of aggression itself. The wrong the aggressor commits is to force men and women to risk their lives for the sake of their rights. It is to confront them with the choice: your rights or (some of) your lives! Groups of citizens respond in different ways to that choice, sometimes surrendering, sometimes fighting, depending on the moral and material condition of their state and army. But they are always justified in fighting; and in most cases, given that harsh choice, fighting is the morally preferred response. The justification and the preference are very important: they account for the most remarkable features of the concept of aggression and for the special place it has in the theory of war.

Aggression is remarkable because it is the only crime that states can commit against other states: everything else is, as it were, a misdemeanor. There is a strange poverty in the language of international law. The equivalents of domestic assault, armed robbery, extortion, assault with intent to kill, murder in all its degrees, have but one name. Every violation of the territorial integrity or political sovereignty of an independent state is called aggression. It is as if we were to brand as murder all attacks on a man's person, all attempts to coerce him, all invasions of his home. This refusal of differentiation makes it difficult to mark off the relative seriousness of aggressive acts—to distinguish, for example, the seizure of a piece of land or the imposition of a satellite regime from conquest itself, the destruction of a state's independence (a crime for which Abba Eban, Israel's foreign minister in 1967, suggested the name "policide"). But there is a reason for the refusal. All aggressive acts have one thing in common: they justify forceful resistance, and force cannot be used between nations, as it often can between persons, without putting life itself at risk. Whatever limits we place on the means and range of warfare, fighting a limited war is not like hitting somebody. Aggression opens the gates of hell. Shakespeare's *Henry V* makes the point exactly:[1]

> For never two such kingdoms did contend
> Without much fall of blood, whose guiltless drops
> Are every one a woe, a sore complaint
> 'Gainst him whose wrongs gives edge unto the swords
> That makes such waste in brief mortality.

---

*"Law and Order in International Society," by Michael Walzer, reprinted from *Just and Unjust Wars: A Moral Argument with Historical Illustrations,* published by Basic Books, a subsidiary of Perseus Books Group, LLC.

[1] Henry V, 112 11 24–28.

At the same time, aggression unresisted is aggression still, though there is no "fall of blood" at all. In domestic society, a robber who gets what he wants without killing anyone is obviously less guilty, that is, guilty of a lesser crime, than if he commits murder. Assuming that the robber is prepared to kill, we allow the behavior of his victim to determine his guilt. We don't do this in the case of aggression. Consider, for example, the German seizures of Czechoslovakia and Poland in 1939. The Czechs did not resist; they lost their independence through extortion rather than war; no Czech citizens died fighting the German invaders. The Poles chose to fight, and many were killed in the war that followed. But if the conquest of Czechoslovakia was a lesser crime, we have no name for it. At Nuremberg, the Nazi leadership was charged with aggression in both cases and found guilty in both.[2] Once again, there is a reason for this identity of treatment. We judge the Germans guilty of aggression in Czechoslovakia, I think, because of our profound conviction that they ought to have been resisted—though not necessarily by their abandoned victim, standing alone.

The state that does resist, whose soldiers risk their lives and die, does so because its leaders and people think that they should or that they have to fight back. Aggression is morally as well as physically coercive, and that is one of the most important things about it. "A conqueror," writes Clausewitz, "is always a lover of peace (as Bonaparte always asserted of himself); he would like to make his entry into our state unopposed; in order to prevent this, we must choose war. . . . "[3] If ordinary men and women did not ordinarily accept that imperative, aggression would not seem to us so serious a crime. If they accepted it in certain sorts of cases, but not in others, the single concept would begin to break down, and we would eventually have a list of crimes more or less like the domestic list. The challenge of the streets, "Your money or your life!" is easy to answer: I surrender my money and so I save myself from being murdered and the thief from being a murderer. But we apparently don't want the challenge of aggression answered in the same way; even when it is, we don't diminish the guilt of the aggressor. He has violated rights to which we attach enormous importance. Indeed, we are inclined to think that the failure to defend those rights is never due to a sense of their unimportance, nor even to a belief (as in the street-challenge case) that they are, after all, worth less than life itself, but only to a stark conviction that the defense is hopeless. Aggression is a singular and undifferentiated crime because, in all its forms, it challenges rights that are worth dying for.

The Rights of Political Communities

The rights in question are summed up in the lawbooks as territorial integrity and political sovereignty. The two belong to states, but they derive ultimately from the rights of individuals, and from them they take their force. "The duties and rights of states are nothing more than the duties and rights of the men who compose them."[4] That is the view of a conventional British lawyer, for whom states are neither organic wholes nor mystical unions. And it is the correct view. When states are attacked, it is their members who are challenged, not only in their lives, but also in the sum of things they value most, including the political association they have made. We recognize and explain this challenge by referring to their rights. If they were not morally entitled to choose their form of government and shape the policies that shape their lives, external coercion would not be a crime; nor could it so easily be said that they had been forced to resist in self-defense. Individual rights (to life and liberty) underlie the most important judgments that we make about war. How these rights are themselves founded I cannot try to explain here. It is enough to say that they are somehow entailed by our sense of what it means to be a human being. If they are not natural, then we have invented them, but natural or invented, they are a palpable feature of our moral world. States' rights are simply their collective form. The process of collectivization is a complex one. No doubt, some of the immediate force of individuality is lost in its course; it is best understood, nevertheless, as it has commonly been understood since the seventeenth

---

[2]The judges distinguished "aggressive acts" from "aggressive wars," but then used the first of these as the generic term: see *Nazi Conspiracy and Aggression: Opinion and Judgment* (Washington, D.C., 1947), p. 16.

[3]Quoted in Michael Howard, "War as an Instrument of Policy," in Herbert Butterfield and Martin Wight, ed., *Diplomatic Investigations* (Cambridge, Mass., 1966), p. 199, Cf. *On War,* trans Howard and Paret, p. 370

[4]John Westlake, *Collected Papers,* ed. L. Oppenheim (Cambridge, England, 1914), p. 78.

century, in terms of social contract theory. Hence it is a moral process, which justifies some claims to territory and sovereignty and invalidates others.

The rights of states rest on the consent of their members. But this is consent of a special sort. State rights are not constituted through a series of transfers from individual men and women to the sovereign or through a series of exchanges among individuals. What actually happens is harder to describe. Over a long period of time, shared experiences and cooperative activity of many different kinds shape a common life. "Contract" is a metaphor for a process of association and mutuality, the ongoing character of which the state claims to protect against external encroachment. The protection extends not only to the lives and liberties of individuals but also to their shared life and liberty, the independent community they have made, for which individuals are sometimes sacrificed. The moral standing of any particular state depends upon the reality of the common life it protects and the extent to which the sacrifices required by that protection are willingly accepted and thought worthwhile. If no common life exists, or if the state doesn't defend the common life that does exist, its own defense may have no moral justification. But most states do stand guard over the community of their citizens, at least to some degree: that is why we assume the justice of their defensive wars. And given a genuine "contract," it makes sense to say that territorial integrity and political sovereignty can be defended in exactly the same way as individual life and liberty.

It might also be said that a people can defend its country in the same way as men and women can defend their homes, for the country is collectively as the homes are privately owned. The right to territory might be derived, that is, from the individual right to property. But the ownership of vast reaches of land is highly problematic, I think, unless it can be tied in some plausible way to the requirements of national survival and political independence. And these two seem by themselves to generate territorial rights that have little to do with ownership in the strict sense. The case is probably the same with the smaller properties of domestic society. A man has certain rights in his home, for example, even if he does not own it, because neither his life nor his liberty is secure unless there exists some physical space within which he is safe from intrusion. Similarly again, the right of a nation or people not to be invaded derives from the common life its members have made on this piece of land—it had to be made somewhere—and not from the legal title they hold or don't hold. But these matters will become clearer if we look at an example of disputed territory.

The Case of Alsace-Lorraine

In 1870, both France and the new Germany claimed these two provinces. Both claims were, as such things go, well founded. The Germans based themselves on ancient precedents (the lands had been part of the Holy Roman Empire before their conquest by Louis XIV) and on cultural and linguistic kinship; the French on two centuries of possession and effective government.[5] How does one establish ownership in such a case? There is, I think, a prior question having to do with political allegiance, not with legal titles at all. What do the inhabitants want? The land follows the people. The decision as to whose sovereignty was legitimate (and therefore as to whose military presence constituted aggression) belonged by right to the men and women who lived on the land in dispute. Not simply to those who owned the land: the decision belonged to the landless, to town dwellers and factory workers as well, by virtue of the common life they had made. The great majority of these people were apparently loyal to France, and that should have settled the matter. Even if we imagine all the inhabitants of Alsace-Lorraine to be tenants of the Prussian king, the king's seizure of his own land would still have been a violation of their territorial integrity and, through the mediation of their loyalty, of France's too. For tenantry determines only where rents should go; the people themselves must decide where their taxes and conscripts should go.

But the issue was not settled in this way. After the Franco-Prussian war, the two provinces (actually, all of Alsace and a portion of Lorraine) were annexed by Germany, the French conceding German rights in the peace treaty of 1871. During the next several decades, the question was frequently asked, whether a French attack aimed at regaining the lost lands would be justified. One of the issues here is that of the moral standing of a peace treaty signed, as most peace treaties are signed, under duress, but

---
[5]See Ruth Putnam, *Alsace and Lorraine from Caesar to Kaiser:* 58 B.C.–1871 A.D. (New York, 1915).

I shall not focus on that. The more important issue relates to the endurance of right over time. Here the appropriate argument was put forward by the English philosopher Henry Sidgwick in 1891. Sidgwick's sympathies were with the French, and he was inclined to regard the peace as a "temporary suspension of hostilities, terminable at any time by the wronged state...." But he added a crucial qualification:[6]

> We must... recognize that by this temporary submission of the vanquished... a new political order is initiated, which, though originally without a moral basis, may in time acquire such a basis, from a change in the sentiments of the inhabitants of the territory transferred; since it is always possible that through the effects of time and habit and mild government—and perhaps through the voluntary exile of those who feel the old patriotism most keenly—the majority of the transferred population may cease to desire reunion.... When this change has taken place, the moral effect of the unjust transfer must be regarded as obliterated; so that any attempt to recover the transferred territory becomes itself an aggression....

Legal titles may endure forever, periodically revived and reasserted as in the dynastic politics of the Middle Ages. But moral rights are subject to the vicissitudes of the common life.

Territorial integrity, then, does not derive from property; it is simply something different. The two are joined, perhaps, in socialist states where the land is nationalized and the people are said to own it. Then if this country is attacked, it is not merely their homeland that is in danger but their collective property—though I suspect that the first danger is more deeply felt than the second. Nationalization is a secondary process; it assumes the prior existence of a nation. And territorial integrity is a function of national existence, not of nationalization (any more than of private ownership). It is the coming together of a people that establishes the integrity of a territory. Only then can a boundary be drawn the crossing of which is plausibly called aggression. It hardly matters if the territory belongs to someone else, unless that ownership is expressed in residence and common use.

This argument suggests a way of thinking about the great difficulties posed by forcible settlement and colonization. When barbarian tribes crossed the borders of the Roman Empire, driven by conquerors from the east or north, they asked for land to settle on and threatened war if they didn't get it. Was this aggression? Given the character of the Roman Empire, the question may sound foolish, but it has arisen many times since, and often in imperial settings. When land is in fact empty and available, the answer must be that it is not aggression. But what if the land is not actually empty but, as Thomas Hobbes says in *Leviathan*, "not sufficiently inhabited?" Hobbes goes on to argue that in such a case, the would-be settlers must "not exterminate those they find there but constrain them to inhabit closer together."[7] That constraint is not aggression, so long as the lives of the original settlers are not threatened. For the settlers are doing what they must do to preserve their own lives, and "he that shall oppose himself against [that], for things superfluous, is guilty of the war that thereupon is to follow."[8] It is not the settlers who are guilty of aggression, according to Hobbes, but those natives who won't move over and make room. There are clearly serious problems here. But I would suggest that Hobbes is right to set aside any consideration of territorial integrity-as-ownership and to focus instead on life. It must be added, however, that what is at stake is not only the lives of individuals but also the common life that they have made. It is for the sake of this common life that we assign a certain presumptive value to the boundaries that mark off a people's territory and to the state that defends it.

Now, the boundaries that exist at any moment in time are likely to be arbitrary, poorly drawn, the products of ancient wars. The mapmakers are likely to have been ignorant, drunken, or corrupt. Nevertheless, these lines establish a habitable world. Within that world, men and women (let us assume) are safe from attack; once the lines are crossed, safety is gone. I don't want to suggest that every boundary dispute is a reason for war. Sometimes adjustments should be accepted and territories

---

[6]Henry Sidgwick, *The Elements of Politics* (London, 1891), pp. 268, 287.

[7]*Leviathan*, ch. 30.

[8]*Leviathan*, ch. 15.

shaped so far as possible to the actual needs of nations. Good borders make good neighbors. But once an invasion has been threatened or has actually begun, it may be necessary to defend a bad border simply because there is no other. We shall see this reason at work in the minds of the leaders of Finland in 1939: they might have accepted Russian demands had they felt certain that there would be an end to them. But there is no certainty this side of the border, any more than there is safety this side of the threshold, once a criminal has entered the house. It is only common sense, then, to attach great importance to boundaries. Rights in the world have value only if they also have dimension.

The Legalist Paradigm

If states actually do possess rights more or less as individuals do, then it is possible to imagine a society among them more or less like the society of individuals. The comparison of international to civil order is crucial to the theory of aggression. I have already been making it regularly. Every reference to aggression as the international equivalent of armed robbery or murder, and every comparison of home and country or of personal liberty and political independence, relies upon what is called the *domestic analogy*.[9] Our primary perception and judgments of aggression are the products of analogical reasoning. When the analogy is made explicit, as it often is among the lawyers, the world of states takes on the shape of a political society the character of which is entirely accessible through such notions as crime and punishment, self-defense, law enforcement, and so on.

These notions, I should stress, are not incompatible with the fact that international society as it exists today is a radically imperfect structure. As we experience it, that society might be likened to a defective building, founded on rights; its superstructure raised, like that of the state itself, through political conflict, cooperative activity, and commercial exchange; the whole thing shaky and unstable because it lacks the rivets of authority. It is like domestic society in that men and women live at peace within it (sometimes), determining the conditions of their own existence, negotiating and bargaining with their neighbors. It is unlike domestic society in that every conflict threatens the structure as a whole with collapse. Aggression challenges it directly and is much more dangerous than domestic crime, because there are no policemen. But that only means that the "citizens" of international society must rely on themselves and on one another. Police powers are distributed among all the members. And these members have not done enough in the exercise of their powers if they merely contain the aggression or bring it to a speedy end—as if the police should stop a murderer after he has killed only one or two people and send him on his way. The rights of the member states must be vindicated, for it is only by virtue of those rights that there is a society at all. If they cannot be upheld (at least sometimes), international society collapses into a state of war or is transformed into a universal tyranny.

From this picture, two presumptions follow. The first, which I have already pointed out, is the presumption in favor of military resistance once aggression has begun. Resistance is important so that rights can be maintained and future aggressors deterred. The theory of aggression restates the old doctrine of the just war: it explains when fighting is a crime and when it is permissible, perhaps even morally desirable. The victim of aggression fights in self-defense, but he isn't only defending himself, for aggression is a crime against society as a whole. He fights in its name and not only in his own. Other states can rightfully join the victim's resistance; their war has the same character as his own, which is to say, they are entitled not only to repel the attack but also to punish it. All resistance is also law enforcement. Hence the second presumption: when fighting breaks out, there must always be some state against which the law can and should be enforced. Someone must be responsible, for someone decided to break the peace of the society of states. No war, as medieval theologians explained, can be just on both sides.[10]

There are, however, wars that are just on neither side, because the idea of justice doesn't pertain to them or because the antagonists are both aggressors, fighting for territory or power where they have

---

[9]For a critique of this analogy, see the two essays by Hedley Bull, "Society and Anarchy in International Relations," and "The Grotian Conception of International Society," in *Diplomatic Investigations*, chs. 2 and 3.

[10]See Vitoria, *On the Law of War*, p. 177.

no right. The first case I have already alluded to in discussing the voluntary combat of aristocratic warriors. It is sufficiently rare in human history that nothing more need be said about it here. The second case is illustrated by those wars that Marxists call "imperialist," which are not fought between conquerors and victims but between conquerors and conquerors, each side seeking dominion over the other or the two of them competing to dominate some third party. Thus Lenin's description of the struggles between "have" and "have-not" nations in early twentieth century Europe: " . . . picture to yourselves a slave-owner who owned 100 slaves warring against a slave-owner who owned 200 slaves for a more 'just' distribution of slaves. Clearly, the application of the term 'defensive' war in such a case . . . would be sheer deception. . . . "[11] But it is important to stress that we can penetrate the deception only insofar as we can ourselves distinguish justice and injustice: the theory of imperialist war presupposes the theory of aggression. If one insists that all wars on all sides are acts of conquest or attempted conquest, or that all states at all times would conquer if they could, then the argument for justice is defeated before it begins and the moral judgments we actually make are derided as fantasies. Consider the following passage from Edmund Wilson's book on the American Civil War:[12]

> I think that it is a serious deficiency on the part of historians . . . that they so rarely interest themselves in biological and zoological phenomena. In a recent . . . film showing life at the bottom of the sea, a primitive organism called a sea slug is seen gobbling up small organisms through a large orifice at one end of its body; confronted with another sea slug of an only slightly lesser size, it ingurgitates that, too. Now the wars fought by human beings are stimulated as a rule . . . by the same instincts as the voracity of the sea slug.

There are no doubt wars to which that image might be fit, though it is not a terribly useful image with which to approach the Civil War. Nor does it account for our ordinary experience of international society. Not all states are sea slug states, gobbling up their neighbors. There are always groups of men and women who would live if they could in peaceful enjoyment of their rights and who have chosen political leaders who represent that desire. The deepest purpose of the state is not ingestion but defense, and the least that can be said is that many actual states serve that purpose. When their territory is attacked or their sovereignty challenged, it makes sense to look for an aggressor and not merely for a natural predator. Hence we need a theory of aggression rather than a zoological account.

The theory of aggression first takes shape under the aegis of the domestic analogy. I am going to call that primary form of the theory the *legalist paradigm,* since it consistently reflects the conventions of law and order. It does not necessarily reflect the arguments of the lawyers, though legal as well as moral debate had its starting point here.[13] Later on, I will suggest that our judgments about the justice and injustice of particular wars are not entirely determined by the paradigm. The complex realities of international society drive us toward a revisionist perspective, and the revisions will be significant ones. But the paradigm must first be viewed in its unrevised form; it is our baseline, our model, the fundamental structure for the moral comprehension of war. We begin with the familiar world of individuals and rights, of crime and punishments. The theory of aggression can then be summed up in six propositions.

1. *There exists an international society of independent states.* States are the members of this society, not private men and women. In the absence of a universal state, men and women are protected and their interest represented only by their own governments. Though states are founded for the sake of life and liberty, they cannot be challenged in the name of life and liberty by any other states. Hence the principle of non-intervention, which I will analyze later on. The rights of private persons can be recognized in international society, as in the UN Charter of Human Rights, but they cannot be

---

[11]Lenin, *Socialism and War* (London, 1940), pp. 10–11.

[12]Edmund Wilson, *Patriotic Gore* (New York, 1966), p. xi.

[13]It is worth noting that the United Nations' recently adopted definition of aggression closely follows the paradigm: see the *Report of the Special Committee on the Question of Defining Aggression* (1974), general Assembly Official Records.

enforced without calling into question the dominant values of that society: the survival and independence of the separate political communities.

2. *This international society has a law that establishes the rights of its members—above all, the rights of territorial integrity and political sovereignty.* Once again, these two rest ultimately on the right of men and women to build a common life and to risk their individual lives only when they freely choose to do so. But the relevant law refers only to states, and its details are fixed by the intercourse of states, through complex processes of conflict and consent. Since these processes are continuous, international society has no natural shape; nor are rights within it ever finally or exactly determined. At any given moment, however, one can distinguish the territory of one people from that of another and say something about the scope and limits of sovereignty.

3. *Any use of force or imminent threat of force by one state against the political sovereignty or territorial integrity of another constitutes aggression and is a criminal act.* As with domestic crime, the argument here focuses narrowly on actual or imminent boundary crossings: invasions and physical assaults. Otherwise, it is feared, the notion or resistance to aggression would have no determinate meaning. A state cannot be said to be forced to fight unless the necessity is both obvious and urgent.

4. *Aggression justifies two kinds of violent response: a war of self-defense by the victim and a war of law enforcement by the victim and any other member of international society.* Anyone can come to the aid of a victim, use necessary force against an aggressor, and even make whatever is the international equivalent of a "citizen's arrest." As in domestic society, the obligations of bystanders are not easy to make out, but it is the tendency of the theory to undermine the right of neutrality and to require widespread participation in the business of law enforcement. In the Korean War, this participation was authorized by the United Nations, but even in such cases the actual decision to join the fighting remains a unilateral one, best understood by analogy to the decision of a private citizen who rushes to help a man or woman attacked on the street.

5. *Nothing but aggression can justify war.* The central purpose of the theory is to limit the occasions for war. "There is a single and only just cause for commencing a war," wrote Vitoria, "namely, a wrong received."[14] There must actually have been a wrong, and it must actually have been received (or its receipt must be, as it were, only minutes away). Nothing else warrants the use of force in international society—above all, not any difference of religion or politics. Domestic heresy and injustice are never actionable in the world of states: hence, again, the principle of nonintervention.

6. *Once the aggressor state has been militarily repulsed, it can also be punished.* The conception of just war as an act of punishment is very old, though neither the procedures nor the forms of punishment have ever been firmly established in customary or positive international law. Nor are its purposes entirely clear: to exact retribution, to deter other states, to restrain or reform this one? All three figure largely in the literature, though it is probably fair to say that deterrence and restraint are most commonly accepted. When people talk of fighting a war against war, this is usually what they have in mind. The domestic maxim is, punish crime to prevent violence; its international analogue is, punish aggression to prevent war. Whether the state as a whole or only particular persons are the proper objects of punishment is a harder questions, for reasons I will consider later on. But the implication of the paradigm is clear: if states are members of international society, the subjects of rights, they must also be (somehow) the objects of punishment.

---

[14]*On the Law of War*, p. 170.

Wars of Anticipation
Michael Walzer *

The first questions asked when states go to war are also the easiest to answer: who started the shooting? who sent troops across the border? These are questions of fact, not of judgment, and if the answers are disputed, it is only because of the lies that governments tell. The lies don't, in any case, detain us long; the truth comes out soon enough. *Governments lie so as to absolve themselves from the charge of aggression.* But it is not on the answers to questions such as these that our final judgments about aggression depend. There are further arguments to make, justifications to offer, lies to tell, before the moral issue is directly confronted. For aggression often begins without shots being fired or borders crossed.

Both individuals and states can rightfully defend themselves against violence that is imminent but not actual; they can fire the first shots if they know themselves about to be attacked. This is a right recognized in domestic law and also in the legalist paradigm for international society. In most legal accounts, however, it is severely restricted. Indeed, once one has stated the restrictions, it is no longer clear whether the right has any substance at all. Thus the argument of Secretary of State Daniel Webster in the *Caroline* case of 1842 (the details of which need not concern us here): in order to justify pre-emptive violence, Webster wrote, there must be shown "a necessity of self-defense . . . instant, overwhelming, leaving no choice of means, and no moment for deliberation." That would permit us to do little more than respond to an attack *once we had seen it coming* but before we had felt its impact. Pre-emption on this view is like a reflex action, a throwing up of one's arms at the very last minute. But it hardly requires much of a "showing" to justify a movement of that sort. Even the most presumptuous aggressor is not likely to insist, as a matter of right, that his victims stand still until he lands the first blow. Webster's formula seems to be the favored one among students of international law, but I don't believe that it addresses itself usefully to the experience of imminent war. There is often plenty of time for deliberation, agonizing hours, days, even weeks of deliberation, when one doubts that war can be avoided and wonders whether or not to strike first. The debate is couched, I suppose, in strategic more than in moral terms. But the decision is judged morally, and the expectation of that judgment, of the effects it will have in allied and neutral states and among one's own people, is itself a strategic factor. So it is important to get the terms of the judgment right, and that requires some revision of the legalist paradigm. For the paradigm is more restrictive than the judgments we actually make. We are disposed to sympathize with potential victims even before they confront an instant and overwhelming necessity.

Imagine a spectrum of anticipation: at one end is Webster's reflex, necessary and determined; at the other end is preventive war, an attack that responds to a distant danger, a matter of foresight and free choice. I want to begin at the far end of the spectrum, where danger is a matter of judgment and political decision is unconstrained, and then edge my way along to the point where we currently draw the line between justified and unjustified attacks. What is involved at that point is something very different from Webster's reflex; it is still possible to make choices, to begin the fighting or to arm oneself and wait. Hence the decision to begin at least resembles the decision to fight a preventive war, and it

---
*From *Just & Unjust Wars* (NY: Basic Books, 1977), 3rd ed (2000) pp. 74–85.

is important to distinguish the criteria by which it is defended from those that were once thought to justify prevention. Why not draw the line at the far end of the spectrum? The reasons are central to an understanding of the position we now hold.

Preventive War and the Balance of Power

Preventive war presupposes some standard against which danger is to be measured. That standard does not exist, as it were, on the ground; it has nothing to do with the immediate security of boundaries. It exists in the mind's eye, in the idea of a balance of power, probably the dominant idea in international politics from the seventeenth century to the present day. A preventive war is a war fought to maintain the balance, to stop what is thought to be an even distribution of power from shifting into a relation of dominance and inferiority. The balance is often talked about as if it were the key to peace among states. But it cannot be that, else it would not need to be defended so often by force of arms. "The balance of power, the pride of modern policy . . . invented to preserve the general peace as well as the freedom of Europe," wrote Edmund Burke in 1760, "has only preserved its liberty. It has been the original of innumerable and fruitless wars." In fact, of course, the wars to which Burke is referring are easily numbered. Whether or not they were fruitless depends upon how one views the connection between preventive war and the preservation of liberty. Eighteenth century British statesmen and their intellectual supporters obviously thought the connection very close. A radically unbalanced system, they recognized, would more likely make for peace, but they were "alarmed by the danger of universal monarchy."* When they went to war on behalf of the balance, they thought they were defending, not national interest alone, but an international order that made liberty possible throughout Europe.

That is the classic argument for prevention. It requires of the rulers of states, as Francis Bacon had argued a century earlier, that they "keep due sentinel, that none of their neighbors do overgrow so (by increase of territory, by embracing of trade, by approaches, or the like) as they become more able to annoy them, than they were." And if their neighbors do "overgrow," then they must be fought, sooner rather than later, and without waiting for the first blow. "Neither is the opinion of some of the Schoolmen to be received: that a war cannot justly be made, but upon a precedent injury or provocation. For there is no question, but a just fear of an imminent danger, though no blow be given, is a lawful cause of war." Imminence here is not a matter of hours or days. The sentinels stare into temporal as well as geographic distance as they watch the growth of their neighbor's power. They will fear that growth as soon as it tips or seems likely to tip the balance. War is justified (as in Hobbes' philosophy) by fear alone and not by anything other states actually do or any signs they give of their malign intentions. Prudent rulers assume malign intentions.

The argument is utilitarian in form; it can be summed up in two propositions: (1) that the balance of power actually does preserve the liberties of Europe (perhaps also the happiness of Europeans) and is therefore worth defending even at some cost, and (2) that to fight early, before the balance tips in any decisive way, greatly reduces the cost of the defense, while waiting doesn't mean avoiding war (unless one also gives up liberty) but only fighting on a larger scale and at worse odds. The argument is plausible enough, but it is possible to imagine a second-level utilitarian response: (3) that the acceptance of propositions (1) and (2) is dangerous (not useful) and certain to lead to "innumerable and fruitless wars" whenever shifts in power relations occur; but increments and losses of power are a con-

---

*The line is from David Hume's easy "Of the Balance of Power," where Hume describes three British wars on behalf of the balance as having been "begun with justice, and even, perhaps, from necessity." I would have considered his argument at length had I found it possible to place it within his philosophy. But in his *Enquiry Concerning the Principles of Morals* (Section III, Part I), Hume writes: "The rage and violence of public war: what is it but a suspension of justice among the warring parties, who perceive that this virtue is now no longer of any *use* or advantage to them?" Nor is it possible, according to Hume, that this suspension itself be just or unjust; it is entirely a matter of necessity, as in the (Hobbist) state of nature where individuals "consult the dictates of self-preservation alone." That standards of justice exist alongside the pressures of necessity is a discovery of the *Essays*. This is another example, perhaps, of the impossibility of carrying over certain philosophical positions into ordinary moral discourse. In any case, the three wars Hume discusses were none of them necessary to the preservation of Britain. He may have thought them just because he thought the balance generally useful.

stant feature of international politics, and perfect equilibrium, like perfect security, is a utopian dream; therefore it is best to fall back upon the legalist paradigm or some similar rule and wait until the overgrowth of power is put to some overbearing use. This is also plausible enough, but it is important to stress that the position to which we are asked to fall back is not a prepared position, that is, it does not itself rest on any utilitarian calculation. Given the radical uncertainties of power politics, there probably is no practical way of making out that position—deciding when to fight and when not—on utilitarian principles. Think of what one would have to know to perform the calculations, of the experiments one would have to conduct, the wars one would have to fight—and leave unfought! In any case, we mark off moral lines on the anticipation spectrum in an entirely different way.

It isn't really prudent to assume the malign intent of one's neighbors; it is merely cynical, an example of the worldly wisdom which no one lives by or could live by. We need to make judgments about our neighbor's intentions, and if such judgments are to be possible we must stipulate certain acts or sets of acts that will count as evidence of malignity. These stipulations are not arbitrary; they are generated, I think, when we reflect upon what it means *to be threatened*. Not merely *to be afraid*, though rational men and women may well respond fearfully to a genuine threat, and their subjective experience is not an unimportant part of the argument for anticipation. But we also need an objective standard, as Bacon's phrase "just fear" suggests. That standard must refer to the threatening acts of some neighboring state, for (leaving aside the dangers of natural disaster) I can only be threatened by someone who is threatening me, where "threaten" means what the dictionary says it means: "to hold out or offer (some injury) by way of a threat, to declare one's intention of inflicting injury." It is with some such notion as this that we must judge the wars fought for the sake of the balance of power. Consider, then, the Spanish Succession, regarded in the eighteenth century as a paradigmatic case for preventive war, and yet, I think, a negative example of threatening behavior.

The War of the Spanish Succession

Writing in the 1750s, the Swiss jurist Vattel suggested the following criteria for legitimate prevention: "Whenever a state has given signs of injustice, rapacity, pride, ambition, or of an imperious thirst of rule, it becomes a suspicious neighbor to be guarded against: and at a juncture when it is on the point of receiving a formidable augmentation of power, securities may be asked, and on its making any difficulty to give them, its designs may be prevented by force of arms." These criteria were formulated with explicit reference to the events of 1700 and 1701, when the King of Spain, last of his line, lay ill and dying. Long before those years, Louis XIV had given Europe evident signs of injustice, rapacity, pride, and so on. His foreign policy was openly expansionist and aggressive (which is not to say that justifications were not offered, ancient claims and titles uncovered, for every intended territorial acquisition). In 1700, he seemed about to receive a "formidable augmentation of power"— his grandson, the Duke of Anjou, was offered the Spanish throne. With his usual arrogance, Louis refused to provide any assurances or guarantees to his fellow monarchs. Most importantly, he refused to bar Anjou from the French succession, thus holding open the possibility of a unified and powerful Franco-Spanish state. And then, an alliance of European powers, led by Great Britain, went to war against what they assumed was Louis' "design" to dominate Europe. Having drawn his criteria so closely to his case, however, Vattel concludes on a sobering note: "it has since appeared that the policy [of the Allies] was too suspicious." That is wisdom after the fact, of course, but still wisdom, and one would expect some effort to restate the criteria in its light.

The mere augmentation of power, it seems to me, cannot be a warrant for war or even the beginning of warrant, and for much the same reason that Bacon's commercial expansion ("embracing of trade") is also and even more obviously insufficient. For both of these suggest developments that may not be politically designed at all and hence cannot be taken as evidence of intent. As Vattel says, Anjou had been invited to his throne "by the [Spanish] nation, conformably to the will of its last sovereign"— that is, though there can be no question here of democratic decision-making, he had been invited for Spanish and not for French reasons. "Have not these two Realms," asked Jonathan Swift in a pamphlet opposing the British war, "their separate maxims of Policy . . . ?" Nor is Louis' refusal to make promises relating to some future time to be taken as evidence of design—only, perhaps, of hope. If Anjou's succession made immediately for a closer alliance between Spain and France, the appropriate answer would seem to have been a closer alliance between Britain and Austria. Then one could wait and judge anew the intentions of Louis.

But there is a deeper issue here. When we stipulate threatening acts, we are looking not only for indications of intent, but also for rights of response. To characterize certain acts as threats is to characterize them in a moral way, and in a way that makes a military response morally comprehensible. The utilitarian arguments for prevention don't do that, not because the wars they generate are too frequent, but because they are too common in another sense: *too ordinary*. Like Clausewitz's description of war as the continuation of policy by other means, they radically underestimate the importance of the shift from diplomacy to force. They don't recognize the problem that killing and being killed poses. Perhaps the recognition depends upon a certain way of valuing human life, which was not the way of eighteenth-century statesmen. (How many of the British soldiers who shipped to the continent with Marlborough ever returned? Did anyone bother to count?) But the point is an important one anyway, for it suggests why people have come to feel uneasy about preventive war. We don't want to fight until we are threatened, because only then can we rightly fight. It is a question of moral security. That is why Vattel's concluding remark about the War of the Spanish Succession, and Burke's general argument about the fruitlessness of such wars, is so worrying. It is inevitable, of course, that political calculations will sometimes go wrong; so will moral choices; there is no such thing as perfect security. But there is a great difference, nonetheless, between killing and being killed by soldiers who can plausibly be described as the present instruments of an aggressive intention, and killing and being killed by soldiers who may or may not represent a distant danger to our country. In the first case, we confront an army recognizably hostile, ready for war, fixed in a posture of attack. In the second, the hostility is prospective and imaginary, and it will always be a charge against us that we have made war upon soldiers who were themselves engaged in entirely legitimate (non-threatening) activities. Hence the moral necessity of rejecting any attack that is merely preventive in character, that does not wait upon and respond to the willful acts of an adversary.

Pre-emptive Strikes

Now, what acts are to count, what acts do count as threats sufficiently serious to justify war? It is not possible to put together a list, because state action, like human action generally, takes on significance from its context. But there are some negative points worth making. The boastful ranting to which political leaders are often prone isn't in itself threatening; injury must be "offered" in some material sense as well. Nor does the kind of military preparation that is a feature of the classic arms race count as a threat, unless it violates some formally or tacitly agreed-upon limit. What the lawyers call "hostile acts short of war," even if these involve violence, are not too quickly to be taken as signs of an intent to make war; they may represent an essay in restraint, an offer to quarrel within limits. Finally, provocations are not the same as threats. "Injury and provocation" are commonly linked by Scholastic writers as the two causes of just war. But the Schoolmen were too accepting of contemporary notions about the honor of states and, more importantly, of sovereigns. The moral significance of such ideas is dubious at best. Insults are not occasions for wars, any more than they are (these days) occasions for duels.

For the rest, military alliances, mobilizations, troop movements, border incursions, naval blockades—all these, with or without verbal menace, sometimes count and sometimes do not count as sufficient indications of hostile intent. But it is, at least, these sorts of actions with which we are concerned. We move along the anticipation spectrum in search, as it were, of enemies: not possible or potential enemies, not merely present ill-wishers, but states and nations that are already, to use a phrase I shall use again with reference to the distinction of combatants and noncombatants, *engaged in harming us* (and who have already harmed us, by their threats, even if they have not yet inflicted any physical injury). And this search, though it carries us beyond preventive war, clearly brings us up short of Webster's pre-emption. The line between legitimate and illegitimate first strikes is not going to be drawn at the point of imminent attack but at the point of (sufficient threat.) That phrase is necessarily vague. I mean it to cover three things: a manifest intent to injure, a degree of active preparation that makes that intent a positive danger, and a general situation in which waiting, or doing anything other than fighting, greatly magnifies the risk. The argument may be made more clear if I compare these criteria to Vattel's. Instead of previous signs of rapacity and ambition, current and particular signs are required; instead of an "augmentation of power," actual preparation for war; instead of the refusal of future securities, the intensification of present dangers. Preventive war looks to the past and future, Webster's reflex action to the immediate moment, while the idea of being under a threat focuses on what we had best call simply *the present*. I cannot specify a time span; it is a span within which one can still make choices, and within which it is possible to feel straitened.

What such a time is like is best revealed concretely. We can study it in the three weeks that preceded the Six Day War of 1967. Here is a case as crucial for an understanding of anticipation in the twentieth century as the War of the Spanish Succession was for the eighteenth, and one suggesting that the shift from dynastic to national politics, the costs of which have so often been stressed, has also brought some moral gains. For nations, especially democratic nations, are less likely to fight preventive wars than dynasties are.

The Six Day War

Actual fighting between Israel and Egypt began on June 5, 1967, with an Israeli first strike. In the early hours of the war, the Israelis did not acknowledge that they had sought the advantages of surprise, but the deception was not maintained. In fact, they believed themselves justified in attacking first by the dramatic events of the previous weeks. So we must focus on those events and their moral significance. It would be possible, of course, to look further back still, to the whole course of the Arab-Jewish conflict in the Middle East. Wars undoubtedly have long political and moral pre-histories. But anticipation needs to be understood within a narrower frame. The Egyptians believed that the founding of Israel in 1948 had been unjust, that the state had no rightful existence, and hence that it could be attacked at any time. It follows from this that Israel had no right of anticipation since it had no right of self-defense. But self-defense seems the primary and indisputable right of any political community, merely because it is *there* and whatever the circumstances under which it achieved statehood.* Perhaps this is why the Egyptians fell back in their more formal arguments upon the claim that a state of war already existed between Egypt and Israel and that this condition justified the military moves they undertook in May 1967. But the same condition would justify Israel's first strike. It is best to assume, I think, that the existing cease-fire between the two countries was at least a near-peace and that the outbreak of the war requires a moral explanation—the burden falling on the Israelis, who began the fighting.

The crisis apparently had its origins in reports, circulated by Soviet officials in mid-May, that Israel was massing its forces on the Syrian border. The falsity of these reports was almost immediately vouched for by United Nations observers on the scene. Nevertheless, on May 14, the Egyptian government put its armed forces on "maximum alert" and began a major buildup of its troops in the Sinai. Four days later, Egypt expelled the United Nations Emergency Force from the Sinai and the Gaza Strip; its withdrawal began immediately, though I do not think that its title had been intended to suggest that it would depart so quickly in event of emergency. The Egyptian military buildup continued, and on May 22, President Nasser announced that the Straits of Tiran would henceforth be closed to Israeli shipping.

In the aftermath of the Suez War of 1956, the Straits had been recognized by the world community as an international waterway. That meant that their closing would constitute a *casus belli*, and the Israelis had stated at that time, and on many occasions since, that they would so regard it. The war might then be dated from May 22, and the Israeli attack of June 5 described simply as its first military incident: wars often begin before the fighting of them does. But the fact is that after May 22, the Israeli cabinet was still debating whether or not to go to war. And, in any case, the actual initiation of violence is a crucial moral event. If it can sometimes be justified by reference to previous events, it nevertheless has to be justified. In a major speech on May 29, Nasser made that justification much easier by announcing that if war came, the Egyptian goal would be nothing less than the destruction of Israel. On May 30, King Hussein of Jordan flew to Cairo to sign a treaty placing the Jordanian army

---

*The only limitation on this right has to do with internal, not external legitimacy: a state (or government) established against the will of its own people, ruling violently, may well forfeit its right to defend itself even against a foreign invasion. I will take up some of the issues raised by this possibility in the next chapter.

under Egyptian command in event of war, thus associating himself with the Egyptian purpose. Syria already had agreed to such an arrangement, and several days later Iraq joined the alliance. The Israelis struck on the day after the Iraqi annoucement.

For all the excitement and fear that their actions generated, it is unlikely that the Egyptians intended to begin the war themselves. After the fighting was over, Israel published documents, captured in its course, that included plans for an invasion of the Negev; but these were probably plans for a counter-attack, once an Israeli offensive had spent itself in the Sinai, or for a first strike at some later time. Nasser would almost certainly have regarded it as a great victory if he could have closed the Straits and maintained his army on Israel's borders without war. Indeed, it would have been a great victory, not only because of the economic blockade it would have established, but also because of the strain it would have placed on the Israeli defense system. "There was a basic assymetry in the structure of forces: the Egyptians could deploy . . . their large army of long-term regulars on the Israeli border and keep it there indefinitely; the Israelis could only counter their deployment by mobilizing reserve formations, and reservists could not be kept in uniform for very long . . . Egypt could therefore stay on the defensive while Israel would have to attack unless the crisis was defused diplomatically." *Would have to attack:* the necessity cannot be called instant and overwhelming; nor, however, would an Israeli decision to allow Nasser his victory have meant nothing more than a shift in the balance of power posing possible dangers at some future time. It would have opened Israel to attack at any time. It would have represented a drastic erosion of Israeli security such as only a determined enemy would hope to bring about.

The initial Israeli response was not similiarly determined but, for domestic political reasons having to do in part with the democratic character of the state, hesitant and confused. Israel's leaders sought a political resolution of the crisis—the opening of the Straits and a demobilization of forces on both sides—which they did not have the political strength or support to effect. A flurry of diplomatic activity ensued, serving only to reveal what might have been predicted in advance: the unwillingness of the Western powers to pressure or coerce the Egyptians. One always wants to see diplomacy tried before the resort to war, so that we are sure that war is the last resort. But it would be difficult in this case to make an argument for its necessity. Day by day, diplomatic efforts seemed only to intensify Israel's isolation.

Meanwhile, "an intense fear spread in the country." The extraordinary Israeli triumph, once fighting began, makes it difficult to recall the preceding weeks of anxiety. Egypt was in the grip of a war fever, familiar enough from European history, a celebration in advance of expected victories. The Israeli mood was very different, suggesting what it means to live under threat: rumors of coming disasters were endlessly repeated; frightened men and women raided food shops, buying up their entire stock, despite government announcements that there were ample reserves; thousands of graves were dug in the military cemeteries; Israel's political and military leaders lived on the edge of nervous exhaustion. I have already argued that fear by itself establishes no right of anticipation. But Israeli anxiety during those weeks seems an almost classical example of "just fear"— first, because Israel really was in danger (as foreign observers readily agreed), and second, because it was Nasser's intention to put it in danger. He said this often enough, but it is also and more importantly true that his military moves served no other, more limited goal.

The Israeli first strike is, I think, a clear case of legitimate anticipation. To say that, however, is to suggest a major revision of the legalist paradigm. For it means that aggression can be made out not only in the absence of a military attack or invasion but in the (probable) absence of any immediate intention to launch such an attack or invasion. The general formula must go something like this: states may use military force in the face of threats of war, whenever the failure to do so would seriously risk their territorial integrity or political independence. Under such circumstances it can fairly be said that they have been forced to fight and that they are the victims of aggression. Since there are no police upon whom they can call, the moment at which states are forced to fight probably comes sooner than it would for individuals in a settled domestic society. But if we imagine an unstable society, like the "wild west" of American fiction, the analogy can be restated: a state under threat is like an individual hunted by an enemy who has announced his intention of killing or injuring him. Surely such a person may surprise his hunter, if he is able to do so.

The formula is permissive, but it implies restrictions that can usefully be unpacked only with reference to particular cases. It is obvious, for example, that measures short of war are preferable to war itself whenever they hold out the hope of similar or nearly similar effectiveness. But what those measures might be, or how long they must be tried, cannot be a matter of *a priori* stipulation. In the case of the Six Day War, the "asymmetry in the structure of forces" set a time limit on diplomatic efforts that would have no relevance to conflicts involving other sorts of states and armies. A general rule containing words like "seriously" opens a broad path for human judgment—which it is, no doubt, the purpose of the legalist paradigm to narrow or block altogether. But it is a fact of our moral life that political leaders make such judgments, and that once they are made the rest of us do not uniformly condemn them. Rather, we weigh and evaluate their actions on the basis of criteria like those I have tried to describe. When we do that we are acknowledging that there are threats with which no nation can be expected to live. And that acknowledgment is an important part of our understanding of aggression.

From *jus ad bellum* to *jus ad pacem*: Re-Thinking Just War Criteria for Irregular War

George R. Lucas, Jr.

During the decade prior to September 11, 2001, many analysts in ethics and international relations had begun to envision a post-Cold War era in which the principal need for military force would come to be the rendering of international humanitarian assistance. Humanitarian tragedies in Somalia and Rwanda, and at least partially successful military interventions to prevent or halt atrocities in Bosnia and Kosovo, had prompted this significant new attention to the problem of using military force for humanitarian purposes in international relations.

The events of that day served as grim reminder that, humanitarian causes notwithstanding, nations equip and support military forces primarily for the purpose of defending their own borders and protecting their own citizens from unprovoked attacks from abroad. It is nonetheless a sign of the growing importance of this relatively new-found interest in the humanitarian use of military force that military intervention by the United States and Great Britain in Afghanistan (ostensibly to seek out and punish terrorists and destroy their paramilitary organizations) swiftly came to be represented to the world as a humanitarian intervention as well. The recent Afghan campaign was characterized in broad and quite plausible terms as an effort to liberate citizens from an oppressive and unrepresentative regime, restore human rights (primarily to women who had egregiously been denied them), and prevent some of the worst effects of poverty and starvation in a troubled and long-suffering region of the world.

Notwithstanding all this attention to the problem, the criteria governing the justifiable use of military force for humanitarian purposes remain quite vague. Of those analysts who have attempted to address this issue, some, like James Turner Johnson and Paul Christopher,[1] represent humanitarian interventions as an extension of traditional just-war theory because they still involve the use of military force for coercive purposes. Others, like Michael Walzer,[2] have long argued that various caveats and qualifications need to be added to the baseline legalist paradigm in international relations in order to cover extenuating or emergency situations, including massive violations of human rights within what we are now coming to call "failed states." Still others, like Stanley Hoffman, have argued that the humanitarian use of military force represents an emerging new paradigm in international relations that calls into question some of the basic assumptions regarding the sovereignty of nations, thus requiring a set of justifications all its own.[3]

I will suggest that the attempt simply to assimilate or subsume humanitarian uses of military force under traditional just war criteria *fails* because the use of military force in humanitarian cases is far closer to the use of force in domestic law enforcement and peacekeeping, and so subject to far more stringent restrictions in certain respects than traditional *jus in bello* normally entails. It is not, for example, sufficient that humanitarian military forces (any more than domestic police forces) simply refrain from excessive collateral damage, or merely refrain from the deliberate targeting of non-combatants.

In fact, the very nature of intervention suggests that the international military "police-like" forces (like actual police forces) must incur considerable additional risk, even from suspected guilty parties, in order to uphold and enforce the law without themselves engaging in violations of the law.

The second strategy for encompassing humanitarian use of military force is represented in Michael Walzer's longstanding attempts to revise and reform our understanding of international law in lieu of relying on the vagaries of moral reasoning alone. This strategy, however, does not address the underlying conceptual incoherence involved in making the autonomous nation-state the unit of analysis in international law. Humanitarian interventions are not undertaken to address solely the political problems of "failed" states (of which Rwanda and Somalia are examples), nor only to contain or discipline the behavior of "rogue" states (such as Yugoslavia and Iraq). Instead, such interventions are necessary even more frequently to address the substantial pressures placed upon the international community by the behaviors of what might be termed "inept" states (of which Afghanistan, Congo, the Sudan, and many others are examples). "Inept" states are those nations with recognizable but ineffective governments unable to provide for the security and welfare of citizens, secure the normal functioning of the institutions of civil society, or maintain secure borders sufficient to control the operations of criminal elements in their midst. None of Walzer's earliest qualifications of the baseline legalist paradigm in *Just and Unjust Wars* (1977), let alone his more recent elaborations of his reformed legalist paradigm,[4] addresses this dilemma successfully, or explains whether we have either the right or responsibility to do (for example) what the United States and Britain are currently doing in Afghanistan. As mentioned, this current exercise includes not only pursuing and destroying international terrorist networks and apprehending international criminals, but assisting in liberating—and in providing food, humanitarian assistance, and political support in nation-building—for the multi-ethnic citizens of a country held for years in virtual slavery by their own, internationally-recognized government, the Taliban. In what follows, I propose to address the unique and problematic features of *jus ad pacem* and *jus in pace*, by spelling out specific criteria and explanations and justifications for each.

Background: "On the Very Idea of *jus ad pacem* and *jus in pace*"

Jus ad pacem (or *jus ad interventionem*) refers to the justification of the use of force for humanitarian or peaceful ends.[5] The concept of this use of military force has been much discussed as incidents of it have proliferated since 1990. The discussions of justification, however, have focused mainly upon legitimacy (legality) and legitimate authority: that is, upon analysis of the sorts of entities that might theoretically have the right to use force across established national boundaries in order to restore peace, maintain order, respond to natural disasters, prevent humanitarian tragedies, or attempt to re-build so-called "failed" states.[6] Political legitimacy or "legitimate authority" was originally a paramount criterion of just war doctrine as explicated by Aquinas. It has once again been restored to its pre-eminent status[7] as nations cope with the havoc wreaked by semi-autonomous "non-state entities" (organizations like Hamas and Al Qaeda), as well as to the attempts by such entities to justify their alleged right to violate time-honored principles of *jus in bello* by targeting non-combatants who dwell in regions of the world far removed from their spheres of concern, and who are utterly innocent of any kind of involvement in the political affairs with which they claim to express grievances.

Given the extent of the interest in this topic, and the increasing demands made on military forces for this purpose, from Somalia and Rwanda to Bosnia, Kosovo, and arguably now even Afghanistan, these discussions and resulting analyses have been surprisingly unfocused, inchoate, and inconclusive. Many authors seem to treat the use of military force for humanitarian purposes as a novel development of the post-Cold War era, when in fact this use of the military has a long and noble history.[8] Other writers and analysts, suspicious of the use of military force for any purpose whatever, have been reluctant to re-consider their selective anti-military bias (forged in the aftermath of the Vietnam war), let alone embrace the emerging notion that national militaries do now, and will, for the foreseeable future, continue to have a positive and important role to play in enforcing justice, protecting individual liberty, and defending fundamental human rights, as well as in performing their more traditional role of defending national borders and protecting the welfare of their own citizens.

What is often overlooked is that *the prospective need for humanitarian military interventions is rapidly becoming the principal justification for raising, equipping, training and deploying a nation's military force.* What we might call the "interventionist imperative" lies at the core of the policy first officially formulated by former Secretary of State Madeline K. Albright in a speech at the US Naval Academy early in 1997. Secretary Albright's position at the time seemed to assert the following moral principle: "When a clearly recognizable injustice is in progress, and when we as international bystanders are in a position to intervene to prevent it, then it follows that we are under a *prima facie* obligation to do so . . . in Kantian terminology, [the interventionist imperative] amounts to an "imperfect duty" of beneficence: we have a duty to prevent harm and injustice when we are able and in a position to do so, but what actions we choose to perform or strategies we choose to devise to carry out this imperative, and the beneficiaries of our protection, are not specified."[9]

. . . Stanley Hoffman has proposed two versions of what he terms a "universal maxim" of *jus ad interventionem*:

1. collective intervention is justified whenever a nation-state's condition or behavior results in grave threats to other states' and other peoples' peace and security, and in grave and massive violations of human rights;

2. sovereignty may be overridden whenever the behavior of the state in question, even within its own territory, threatens the existence of elementary human rights abroad, and whenever the protection of the rights of its own members can be assured only from the outside.[10]

These proposals deserve careful consideration, not only on account of the distinguished credentials of their author, but because this two-part proposal constitutes the only substantive criterion thus far put forward to guide and clarify the justification for humanitarian military actions. The first version seems designed to define something akin to "just cause" in classical just-war theory, and applies to events ranging from the Holocaust to the genocidal acts in Bosnia, Kosovo, and Rwanda. The wording, however, appears to tie "threats to other states' and other peoples' peace and security" with "grave and massive violations of human rights" (presumably occurring within the affected nation's borders and not necessarily constituting an external threat). The first clause represents the traditional perspective on a state's behavior within the nation-state system; the second adds an additional provision, similar to Walzer's concern for behavior that "shocks the conscience of humanity." *Simply replacing the final "and" with "or" would clarify that one or the other objectionable behavior, and not both simultaneously, are sufficient to invoke justification for armed intervention.*

In the second version of his "universal maxim," Hoffman addresses the sovereignty problem explicitly. He seems to be attempting to address at least partially the notion of "legitimate authority." Both versions of the maxim seem to imply that legitimate authority in humanitarian interventions is restricted to "the international community" or to collectivities of some sort.

This raises in turn the most vexing aspect of intervention. Why should not a country (India, or Tanzania, or Vietnam, for example) be empowered to "invade" a neighbor (East Pakistan, Uganda, or Cambodia, respectively) engaged in massive violations of human rights carried out against its own citizens? And why should the "international community" be obliged to wait to prevent what Walzer also describes as "gross and massive" violations of human rights like this until some perceived threat to other states' freedom and security is detected? Hoffman's phrasing of (2) accurately reflects current agreements and UN policies on collective security, but for those who found the UN debacles in Bosnia and Rwanda less than satisfying, it is worth reminding ourselves that these collective humanitarian actions were carried out under the constraints imposed by such existing agreements and conventions.[11]

At present, the criterion of "legitimate authority" appears to be largely taken for granted: all legitimate humanitarian interventions, it would appear, should come about through multilateral debate and decision, and should reflect the collective will of the international community. Unilateral interventions should be prohibited. Does this, however, mean that the international community cannot appoint a single nation to act as its agent (and perhaps should have in the cases cited above)? Likewise, the role of regional security organizations, like NATO, needs to be more carefully explored as a possible legitimate agency. Interventions carried out by such regional security collectivities would neither be unilateral (and so not strictly proscribed) nor sufficiently multi-lateral to qualify as legitimate authorities under conventional understandings of that concept. Perhaps language should be included within any new *jus ad pacem* rubric to address problems, like Bosnia and Kosovo, that seem to fall as responsibilities primarily to a region (that is, Western Europe) rather than to the international community as a whole, permitting the affected region's security and cooperation organization to act as the legitimate authority in such a case.

These questions and problems suggest that it is high time to formulate a more complete list of *jus ad pacem* or *jus ad interventionem* (including some preliminary provisions for restrictions on battlefield conduct, or *jus in pace*), sufficient to govern involvement in humanitarian exercises. While there is no compelling need to require that such criteria perfectly match the seven conventional provisions of just-war doctrine, it will help guide our discussion and ensure a full and comprehensive treatment of the problem if we use the traditional provisions as guideposts for our proposed new formulations.

Criteria for (Non-defensive) Military Intervention

1. Justifiable Cause for Intervention

Let us begin with the humanitarian equivalent of "just cause":

> Humanitarian intervention is justified whenever a nation-state's behavior results in grave and massive violations of human rights.

From my comments above, it is apparent that this needs to be understood in two senses:

(a) intervention is justified when these behaviors result in grave threats to the peace and security of other states and other peoples, and

(b) intervention *need not be restricted to such cases,* but may be justified when the threats to human rights are wholly contained within the borders of the state in question.

2. Legitimate Authority

We must ask what, if anything, gives the interventionists the right to ignore international borders and nation-state sovereignty in order to respond to the clear humanitarian emergencies cited in the first provision pertaining to "just cause." Hoffman stipulates:

> Sovereignty may be overridden whenever the protection of the rights of that states' own citizens can be assured only from the outside.

Hoffman's formulation above contains two additional clarifications:

(a) sovereignty may be overridden whenever the behavior of the state in question, even within its own territory, threatens the existence of elementary human rights abroad;

(b) sovereignty may be overridden even when there is no threat to human rights outside the borders of the state in question, providing the threat to that state's own citizens are real and immediate.

This still leaves the question of *who* is to determine whether such threats are "real and immediate". Here I propose a second clause that seems to capture widespread concern on the part of most commentators on this problem that such judgments should be collective rather than unilateral, in order to ensure "right intentions" (see below) and exclude ulterior, self-interested motives.

> The decision to override sovereignty and intervene must be *made by an appropriate collective international body.*

This does not, however, mean that the intervention itself must constitute a collective military action, although there are ample grounds for finding that preferable. Instead, in light of our recent experience and the analysis above, this legitimacy provision seems to entail:

(a) The *decision* to intervene can never be undertaken unilaterally; however

(b) a unilateral agent of intervention may be authorized by an appropriate international tribunal; and also

(c) a regional security organization may be authorized by an appropriate international tribunal to undertake a military intervention for humanitarian purposes.

3. Right Intention

Much of the concern over multi-lateralism and collective action concerns the possibility of conflicted and self-interested motives. Paul Christopher notes that Hugo Grotius originally licensed military interventions for clear humanitarian purposes (such as the prevention of cannibalism, rape, abuse of the elderly, and piracy), and simultaneously warned against the likelihood of hidden and less noble agendas, such as greed, religious and cultural differences, and national self-interest, poisoning the presumptive humanitarian and disinterested motivations.[12] These considerations lead straightforwardly to the following restrictions on the use of force for humanitarian purposes:

> The intention in using force must be restricted without exception to purely humanitarian concerns, such as the restoration of law and order in the face of natural disaster, or to the protection of the rights and liberties of vulnerable peoples (as defined in the United Nations Charter and the Universal Declaration of Human Rights).

Furthermore, the intentions must be publicly proclaimed and clearly evident without conflict of interest to the international community. Intervening nations and their militaries should possess no financial, political or material interests in the outcome of the intervention, other than the publicly proclaimed humanitarian ends described above, nor should they stand to gain in any way from the outcome of the intervention, other than from the general welfare sustained by having justice served, innocent peoples protected from harm, and peace and order restored. A useful protection against abuse of this provision is for the intervening powers not only to state clearly and publicly their humanitarian ends, but also to set forth a set of conditions under which the need for intervention will have been satisfied, together with a reasonable timetable for achieving their humanitarian goals. Suspicion of possible ulterior motives might be further allayed not only by ensuring the trans-national character of the intervention, but by providing (in the publicly stated proposal to undertake it) for periodic rotation, where feasible, of the specific nationalities involved in carrying out the action.[13]

4. Last Resort

> Military intervention may be resorted to for humanitarian purposes only when all other options have been exhausted.

What this means is that good faith efforts by the international community must be made to avert humanitarian disasters within the borders of a sovereign state through diplomatic negotiation, economic sanction, United Nations censure, and other non-military means as appropriate. In practice, this is easier said than done, and could result (as in Rwanda and Bosnia) in delaying necessary deployment of force to prevent a humanitarian tragedy while the "international community" wrings its collective hands ineffectually, worrying whether all other available options have been satisfied. Paul Christopher's sensible proposal from the standpoint of just war theory, applied to humanitarian cases, is that "[this] condition is met when reasonable nonviolent efforts have been unsuccessful and there is no indication that future attempts will fare any better."[14]

5. Likelihood of Success

Johnson and Christopher tend to collapse or blend their concerns for the criterion of likely success into others treating everything from last resort and legitimate authority to the proportionality of war to its stated ends. I favor keeping this criterion distinct, as providing a unique and important constraint on the decision to deploy force for humanitarian reasons.

> Military force may be utilized for humanitarian purposes only when there is a reasonable likelihood that the application of force will meet with success in averting a humanitarian tragedy.

This seemingly obvious provision in fact imposes something like Weinberger Doctrine constraints on those whose moral outrage or righteous zeal might tempt them into military adventures for which the intervening powers are ill-prepared and unsuited, or which might make an already-bad situation even worse. Specifically:

    a. a resort to military force may not be invoked when there is a real probability that the use of such force will prove ineffective, or may actually worsen the prospects for a peaceful resolution of the crisis; and
    b. military force may not be employed, even for humanitarian ends, when the international community is unable to define or determine straightforward and feasible goals to be achieved by the application of force.

These Weinberger-like constraints are also important as reassurances to those political representatives of that camp who have been extremely reluctant to embrace what otherwise appears to be an international moral obligation to render humanitarian assistance or prevent avoidable tragedies when we as bystanders are in a position to do so (the "interventionist imperative").

    General Anthony Zinni, speaking of his experiences in Somalia,[15] warns us that this traditional criterion limiting military force has a special urgency and ambiguity in the humanitarian instance. Militaries, including the American military, are not primarily oriented or necessarily well-suited to carry out the varieties of tasks a true humanitarian exercise may require. It is difficult in advance to predict just what sorts of activities these may comprise, but they certainly transcend the straightforward projection of lethal force to include also civil engineering, police and law enforcement, and other functions of a stable civil society. In some instances, military experts on hand may perform, say, civil engineering functions (such as water purification, distribution of food rations, or bridge and road construction) as, or even more readily and ably, than civilian counterparts. In other instances, as occurred in Somalia, the need to resurrect a moribund legal system and to re-establish police, courts, and a working prison system may push the intervening forces into roles they are ill-prepared and ill-equipped to play, with disastrous consequences. Yet, as Zinni notes, any attempt to avoid engaging in these necessary nation-building exercises is likely to doom the humanitarian mission to failure.

6. Proportionality of Ends

    It is not sufficient, however, to demand that military intervention be successful, or that it merely refrain from worsening a bad situation. The NATO air campaign in Kosovo satisfied both constraints, but concerns abound regarding the consequences of the aforementioned doctrine of force protection, and the resulting civilian casualties sustained (for example, from height restrictions imposed on attacking aircraft).

> The lives, welfare, rights and liberties to be protected must bear some reasonable proportion to the risks of harm incurred, and the damage one might reasonably expect to inflict in pursuit of humanitarian ends.

In the end, the debates over this aspect of the Kosovo air campaign come down to this provision, although the question of whether to engage in that intervention initially focused on what amounted to discussion of the likelihood of success, as outlined above. Post-mortems and continuing analysis of the results of that intervention now routinely raise the question of whether the damage inflicted in an effort to stop the threatened genocide by Serbian troops against Kosovars (as a result of force-protection measures imposed, including the unwillingness to commit ground troops to the exercise) was, in the end, unduly large.[16]

    It is at this stage that discussions of justifiable military intervention for humanitarian ends shift from the actual discussion of the justification of such intervention, to discussions, similar in some respects to traditional law of armed combat (*jus in bello*), concerning the manner in which intervening forces may operate and conduct themselves. The Kosovo debate illustrates this ambiguity clearly. Given what was known of Serbian intentions within the province of Kosovo, based upon substantial prior experience elsewhere in the region (in Bosnia and Herzegovina, for example), there was every reason to expect that the anticipated casualties to be suffered by innocents in the absence of armed intervention of whatever sort would vastly outweigh any foreseeable "collateral damage" that the intervening forces might themselves inflict inadvertently. While any attempt to engage in such calculus is necessarily fraught with difficulty, it seems that most observers agree that this condition (taken as a constraint to be satisfied in the initial decision to deploy military force) was amply satisfied.

What is being debated after the fact, then, is no longer the initial justification of the intervention, but the manner in which it was ultimately carried out. *Jus ad pacem* demands that a reasonable evaluation of the likely overall outcomes (including necessary forms of military deployment and conduct during the intervention) be undertaken before deciding whether to undertake the mission. By contrast, what is now being debated is whether, during a justifiable humanitarian mission, reasonable resulting constraints on the conduct of military forces during the humanitarian mission were violated in selected instances.

7. Just Means, Moral Means (or, Proportionality of Means)

The concern that remains unaddressed is something equivalent to the traditional *jus ad bellum* requirement that justifiable wars must be prosecuted by just means. There are a variety of ways of capturing this essential insight, which may well be the most important and difficult provision to achieve in practice for otherwise justifiable, if not downright obligatory, military interventions. This would not be surprising, as strict compliance with *jus in bello,* particularly the principle of discrimination between combatants and non-combatants, remains the most elusive component of just-war theory generally.[17]

The morality of the means employed to carry out a humanitarian intervention, or to achieve its stated goals, must be commensurate with, or proportional to, the morality of the cause or ends for the sake of which the intervention is conducted. Transparently, a military intervention conducted for the sake of protecting human rights or averting a humanitarian tragedy cannot itself rely upon military means of intervention or modes of conduct by military personnel which themselves violate the very rights the interventionists sought to protect, or which provoke a humanitarian tragedy of dimensions similar to the original impending tragedy the interventionists sought to avert.

The last phrase in particular captures the concerns of critics of the NATO bombing strategy, and the concomitant decision against using low-flying Apache combat helicopters or ground forces, for the sake of force protection and the minimization of allied combat casualties in Kosovo. The critics, both military and civilian, are not quibbling about proportionality with the advantage of hindsight, so much as calling attention to this paradoxical feature of the use of deadly force for humanitarian purposes, the details of which I have collectively labeled *jus in pace* or *jus in interventione*.

Humanitarian intervention can never be pursued via military means that themselves are deemed illegal or immoral.

As I have suggested throughout this essay, the provisions and restrictions upon the conduct of military forces that this final provision imposes are not well understood, but are certainly more, rather than less, constraining than traditional *jus in bello* or law of armed combat, while including those traditional provisions as well. Specifically:

(a) captured belligerents must be treated as prisoners of war according to established international conventions, and may not be mistreated or subject to trial or sentence by the intervening forces;

(b) prisoners of war accused of humanitarian crimes and abuses may be bound over for trial by an appropriate international tribunal;

(c) civilian non-combatants must never be deliberately targeted during a humanitarian military operation;[18]

(d) military necessity during humanitarian operations can never excuse the use of weapons, or pursuit of battlefield tactics, already proscribed as illegal under established international treaties and conventional laws of armed combat;

(e) finally, military necessity during humanitarian operations cannot excuse tactics or policies, such as "force protection," *that knowingly, deliberately, and disproportionately reallocate risk of harm from the peace keeping forces and belligerents to non-combatants*. It is not sufficient that humanitarian military forces simply refrain from excessive collateral damage, or merely refrain from the deliberate targeting of non-combatants. The very nature of intervention suggests that the international military forces (like domestic law enforcement personnel) must incur considerable additional risk, even from suspected guilty parties, in order to uphold and enforce the law without themselves engaging in violations of the law.

Paragraphs (a) and (c), and (d) capture the conventional constraints characteristic of *jus in bello*. Paragraph (b), however, begins to suggest the character of law enforcement that humanitarian interventions may entail. Paragraph (b) implies that the intervening forces are not, in the name of protecting or minimizing casualties among their own personnel, permitted to turn a blind eye toward international criminals operating in their midst, but have the same obligations to apprehend criminals and enforce justice that their domestic peacekeeping counterparts do. Moreover, it explicitly states that if, during the course of an armed intervention or afterwards, an apparent perpetrator of criminal actions such as Slobodan Milosevic or Osama bin Laden is apprehended, then (as with conventional domestic criminals) the duty of the intervening forces is to ensure that the accused is properly treated and bound over for trial in a legal manner.

Why do I suggest this? Let me hasten to say that it is not because I believe that murderous Yugoslavian thugs or spoiled, vain, and destructive miscreants like bin Laden are somehow especially entitled to avail themselves of the protections of the law which they have otherwise scorned. Instead, there is an important practical element at work in this provision. *It properly classifies terrorism and its proponents as "criminals" carrying out "crimes against humanity," rather than dignifying their actions as quasi-legitimate acts of "war,"* or otherwise conferring upon their perpetrators and their shadowy, non-state organizations the status of statehood. The important domestic analogy is the continuing struggle to avoid "romanticizing" the activities of organized crime or dissident factions within a nation-state with a quasi-cultural status as "acts of war," lest we seem to be sanctioning or excusing the resulting violence and threats to legitimate and established order. No matter how legitimate the *grievances* of such individuals and organizations may otherwise be found to be, it is vital not to permit them or ourselves to fall into the fatal trap of somehow legitimating *criminal actions* (whether of Timothy McVeigh or Osama bin Laden) as if these were some sort of populist redress of grievance or otherwise-justifiable protest against the injustices they purport to cite.

It is precisely this recognition of the radically different moral status of these criminals, and of the international society against which they have set their faces and directed their actions, that imposes special burdens and responsibilities on the decorum and behavior of intervening forces, sent to enforce and uphold the law. Paragraph (e) thus directly enjoins the as-yet-unresolved paradox posed by the increasing tendency toward force protection in the course of carrying out humanitarian interventions. I suggested in the preceding pages that such tactics evolve as a result of thinking dictated by conventional just-war theory and international relations, according to which national sovereignty and national interests are the primary units of analysis. These serve to define the nature of the limitations placed upon an individual's self-sacrifice during wartime, as described in Walzer's "war convention." Since these provisions are almost always lacking in truly justifiable humanitarian interventions, the concerns they engender, while understandable, are seriously misplaced.

What we require of the intervening forces is not merely that their controlling interests and command structures lack any personal conflicts of interest in the enforcement of justice, protection of rights, and establishment of peace, but that they be *willing to incur risk and put themselves in harm's way for the sake of these moral ideals,* and with an end of securing (and certainly not themselves threatening or destroying) the blessings of rights and liberty to the vulnerable and endangered victims whose desperate plight initially prompts the international call for military intervention. This is not an imposition of lofty moral idealism, but a simple requirement for consistency of purpose that civilized society routinely imposes upon itself, and particularly upon those who choose to uphold and defend civilization's highest ideals and most essential governing principles. *In humanitarian interventions, as in domestic law enforcement, we cannot and we do not forsake our laws and moral principles in order to enforce and protect them.*

These *jus in pace* criteria, and especially this final provision, are not as strange, stringent, or unreasonable as they may at first seem, since we ask precisely these same commitments of any domestic law enforcement agency or authority. It is, I have argued, in the nature of humanitarian intervention that it not only restores a legitimate role to morality in foreign policy, but that it begins to import some of the more cherished securities and civilizing protections of domestic civil society into the international arena precisely to supplant the anarchy, ruthlessness, and terror that still too often flourish in the darker regions of our new global order.

NOTES

1. James Turner Johnson, *Morality and Contemporary Warfare* (New Haven, CT: Yale University Press, 1999); and "The Just-War Idea and the Ethics of Intervention," in *The Leader's Imperative: Ethics, Integrity and Responsibility*, ed. J. Carl Ficarrotta (Lafayette, IN: Purdue University Press, 2001); Paul Christopher, *The Ethics of War and Peace*, 2nd edition (Upper Saddle River, NJ: Prentice-Hall, 1999), hereafter cited as EWP.

2. *Just and Unjust Wars*, 3rd edition (New York: Basic Books, 2000), ch. 6.

3. Stanley Hoffman, *The Ethics and Politics of Humanitarian Intervention* (Notre Dame, IN: Notre Dame University Press, 1996), hereafter cited as EPHI.

4. Michael Walzer, "The Politics of Rescue," *Dissent* 42, no. 1 (1995), 35–41: "Emergency Ethics," in J. Carl Ficarrotta, *The Leader's Imperative*, pp. 126–39, and *Nation and Universe*, "The Tanner Lectures on Human Values, Volume XI" (Lake City, UT: University of Utah Press, 1990).

5. George R. Lucas, Jr., *Perspectives on Humanitarian Military Intervention* (Berkeley, CA: University of California/Public Policy Press, 2001), pp. 4ff.

6. A state "fails" when its ability to guarantee basic rights and liberties, provide fundamental essential services that constitute a civil society (such as basic medical care, education, banking, commerce, agriculture, and a dependable food supply), and enforce the rule of law *completely evaporates*. This contrasts with the behavior of viable but criminal states ("rogue" states) and what I am calling "inept" states. See Robert S. Litwak, *Rogue States and US Foreign Policy* (Baltimore, MD: The Johns Hopkins University Press, 2000) for an analysis of the former. An inept or incompetent state is one which does a poor or incompetent job in any or several of these categories of essential human needs.

7. In Thomas Aquinas' discussion of the morality of war in *Summa Theologica*, legitimate authority is listed as the first criterion (ahead of "just cause") to be fulfilled. It is Hugo Grotius who first reverses this priority and gives pride of place to "just cause" (specifically eliminating religious wars as eligible categories). Johnson explores this history in *The Holy War Idea in Western and Islamic Traditions* (University Park, PA: Penn State University Press, 1998).

8. Martha Finnemore, "Constructing Norms of Humanitarian Intervention," *The Culture of National Security*, ed. Peter J. Katzenstein (NY: Columbia University Press, 1996), pp. 153–85. See also Paul Christopher's discussion of Grotius on the rationale for humanitarian military intervention as early as the seventeenth century, EWP, p. 192.

9. See *Perspectives on Humanitarian Military Intervention*, p. 10. See also Julia Driver, "The Ethics of Intervention," *Philosophy and Phenomenological Research* 57, no. 4 (December, 1997), 851–70; more recently John W. Lango, "Is Armed Humanitarian Intervention to Stop Mass Killing Morally Obligatory," *Public Affairs Quarterly* (July 2001), 173–92, who argue in favor of such an imperative. It is important to recognize that imperfect duties are no less stringent than perfect duties. The term "imperfect" refers not to their stringency but their lack of specificity: such obligations do not precisely specify the nature of actions taken to fulfill the obligation ("what sort of good acts should I undertake?") nor do they always specify an obligee ("whom should I choose as beneficiary of my good actions?"). Assuming that interventionism is a species of "good Samaritanism," the obligees are specified, but the precise actions undertaken in their defense are not.

10. EPHI, 23.

11. Paul Christopher appears to agree that suitable collective bodies should be able to authorize or otherwise post facto legitimate unilateral interventions with clear humanitarian intent: see EWP, 193, 198.
12. EWP, 199.
13. This proposal is suggested by James Turner Johnson (see Ficarrotta, *The Leader's Imperative*, p. 122), but given the extraordinary logistical difficulties of coalition operations to begin with, this additional provision might add an insuperable burden to interventions justifiable on other grounds.
14. EWP, 201.
15. Lucas, *Perspectives*, pp. 53–63; note that this is an eloquent defense and rejoinder to the charges of "mission creep" that were made against that operation initially as the putative cause of its failure.
16. See, for example, Gen. Wesley K. Clark, *Waging Modern War: Bosnia, Kosovo, and the Future of Combat* (New York: Public Affairs Press, 2001) for a discussion of the differences between US and European military and civilian leadership on these questions.
17. Douglas P. Lackey, *The Ethics of War and Peace* (Upper Saddle River, NJ: Prentice-Hall, 1989), pp. 58–97.
18. For a discussion of how this standard convention plays out in humanitarian intervention, see James Turner Johnson, "Maintaining the Protection of Noncombatants," *Journal of Peace Research* 37, no. 4 (July 2000), 421–48.

Summary: *Jus ad Pacem* (Humanitarian and Counterterrorist Interventions)*

1. **(A) 'Humanitarian intervention is justified whenever a nation-state's behavior results in grave and massive violations of human rights, or poses an imminent threat of grave harm to other nations and peoples.'**
   (i) intervention is justified when these behaviors result in grave threats to the peace and security of other states and other peoples: and
   (ii) intervention need not be restricted to such cases, but may be justified when the threats to human rights are wholly contained within the borders of the state in question.
   (B) States may use military force in the face of threats of war, or impending terrorist actions, or preparations by states or non-state actors actively engaged in doing, or imminently threatening, grave harm to other nations and peoples, whenever:
   (i) there is a manifest intent on the part of such parties to injure:
   (ii) there is a degree of active preparation that constitutes a severe, imminent, and highly probable threat of such injury; and
   (iii) both of the foregoing occur in a situation in which waiting, or doing anything other than deploying military force preemptively, greatly magnifies the risk.

2. **(A) 'Sovereignty may be overridden whenever the protection of the rights of that state's own citizens can be assured only from the outside.'**
   (i) sovereignty may be overridden whenever the behavior of the state in question, even within its own territory, threatens the existence of elementary human rights abroad;
   (ii) sovereignty may be overridden even when there is no threat to human rights outside the borders of the state in question, providing the threat to that state's own citizens is real and immediate.
   (B) 'The decision to override sovereignty and intervene must finally be subject to review and approval by an appropriate collective international body.'
   (i) the *decision* to intervene, whether to protect human rights or enforce international law, *ought never* to be undertaken unilaterally; however
   (ii) a unilateral agent of intervention may be authorized by an appropriate international tribunal; and also
   (iii) a regional security organization may be authorized by an appropriate international tribunal to undertake a military intervention for humanitarian or counterterrorist purpose; and
   (iv) in the absence of prior approval, the burden of proof falls upon the intervening power to demonstrate that it has unilateral license to intervene, based upon *prima facie* compliance with all of the above.

3. **The intention in using force must be restricted without exception either to purely humanitarian concerns, such as the restoration of law and order in the face of natural disaster, or to the protection of the rights and liberties of vulnerable peoples (as defined in the United Nations Charter and the Universal Declaration of Human Rights), or to halt or prevent violations of international law by nations or non-state actors that pose a clear and imminent threat of grave harm to other nations or peoples.**
   (i) intervening nations and their militaries should possess no financial, political or material interests in the outcome of the intervention, other than achieving the publicly proclaimed humanitarian ends, enforcing international law, or averting the risk of grave and substantial harm to other nations and peoples;

---

*From G. R. Lucas "The Role of the Community in the Just War Tradition" *Journal of Military Ethics* 2, No. 2 (2003) 141–42.

(ii) the intervening nation or nations must establish a set of conditions under which the need for intervention will have been satisfied, together with a reasonable timetable for achieving their humanitarian ends or eliminating the perceived threat.

4. **Military intervention may be resorted to for humanitarian purposes, or to avert the risk of terrorism or enforce vital provisions of international law, only when all other options have been exhausted.**
(i) this condition is deemed to have been met when reasonable nonviolent efforts have been unsuccessful and there is no indication that future attempts will fare any better.

5. **Military force may be utilized for humanitarian purposes, or to avert the risk of terrorism or enforce vital provisions of international law, only when there is a reasonable likelihood that the application of force will meet with success in averting a humanitarian tragedy.**
(i) a resort to military force may not be invoked when there is a real probability that the use of such force will prove ineffective, or may actually worsen the prospects for a peaceful resolution of the crisis; and
(ii) military force may not be employed, either for humanitarian ends or for the purposes of counterterrorism and law enforcement, whenever collective, public debate and deliberation fail to determine straightforward and feasible goals to be achieved by the application of force.

6. **The lives, welfare, rights and liberties to be protected from humanitarian disaster or terrorist attacks must bear some reasonable proportion to the risks of harm incurred, and the damage one might reasonably expect to inflict in pursuit of humanitarian ends.**

7. **Military force used for humanitarian or counterterrorist purposes may never encompass the use of strategy, tactics, weapons systems or battlefield conduct that are themselves recognized as illegal or immoral.**
(i) captured belligerents must be treated as prisoners of war according to established international conventions, and may not be mistreated or subject to trial or sentence by the intervening forces;
(ii) prisoners of war accused of humanitarian crimes and abuses, or of engaging actively in planning for doing grave and indiscriminate harm to other nations and peoples, may be bound over for trial by an appropriate international tribunal;
(iii) civilian noncombatants must never be deliberately targeted during a humanitarian or counterterrorist military operation;
(iv) military necessity during humanitarian or counterterrorist operations can never excuse the use of weapons, or pursuit of battlefield tactics, already proscribed as illegal under established international treaties and conventional law of armed combat;
(v) finally, military necessity during humanitarian or counterterrorist operations cannot excuse tactics or policies, such as 'force protection', that knowingly, deliberately, and disproportionately reallocate risk of harm from the peacekeeping forces and belligerents to noncombatants.

The Concept and Practice of Jihad in Islam

Michael G. Knapp*

"All these crimes and sins committed by the Americans are a clear declaration of war on God, his Messenger, and Muslims. . . . [T]he jihad is an individual duty if the enemy destroys the Muslim countries. . . . As for the fighting to repulse [an enemy], it is aimed at defending sanctity and religion, and it is a duty. . . . On that basis, and in compliance with God's order, we issue the following fatwa to all Muslims: The ruling to kill the Americans and their allies—civilian and military—is an individual duty for every Muslim who can do it in any country in which it is possible to do it."

—Osama bin Laden et al., in "Declaration of the World Islamic Front for Jihad Against the Jews and Crusaders," 23 February 1998

The word "jihad" means "struggle" or "striving" (in the way of God) or to work for a noble cause with determination; it does not mean "holy war" (war in Arabic is *harb* and holy is *muqadassa*). Unlike its medieval Christian counterpart term, "crusade" ("war for the cross"), however, the term "jihad" for Muslims has retained its religious and military connotation into modern times. The word "jihad" has appeared widely in the Western news media following the 11 September 2001 terrorist attacks on the World Trade Center and the Pentagon, but the true meaning of this term in the Islamic world (it is sometimes called the "sixth pillar" of the faith) is still not well understood by non-Muslims.

In war, the first essential is to know your adversary—how he thinks and why he thinks that way, and what his strategy and objectives are—so that you can attempt to frustrate his plans and protect the lives of your fellow citizens. Understanding how radical Muslims see jihad and are employing it asymmetrically against us can provide us with that kind of perspective.

This article will trace the development of jihad through early Islamic history into the present day, and will focus on how jihad in concept and practice has been appropriated and distorted by Muslim extremists as part of their violent campaign against the West and their own governments. Jihad as a centerpiece of radical thought is illustrated by examining the doctrines of prominent extremist groups such as Hamas and Egyptian Islamic Jihad. Misuse of the term by prominent extremist leaders, such as by Osama bin Laden and others in the quote above, is also addressed.

---

*Michael Knapp is a Middle East/Africa analyst with the U.S. Army National Ground Intelligence Center (NGIC), in Charlottesville, Virginia. He has worked in U.S. government intelligence for over 24 years, both as a civilian and (now retired) military intelligence officer in the U.S. Army Reserve. Mr. Knapp's previous civilian assignments included analytical positions in the Defense Intelligence Agency and the Drug Enforcement Administration, and his military career consisted of active duty in Germany and Texas, and service in the Virginia Army National Guard.

The Classical Concept of Jihad

Qur'anic and Early Legal Perspectives

Muslims themselves have disagreed throughout their history about the meaning of the term jihad. In the Qur'an (or Koran), it is normally found in the sense of fighting in the path of God; this was used to describe warfare against the enemies of the early Muslim community *(ummah)*. In the *hadith*, the second-most authoritative source of the *shari'a* (Islamic law), jihad is used to mean armed action, and most Islamic theologians and jurists in the classical period (the first three centuries) of Muslim history understood this obligation to be in a military sense.[1]

Islamic jurists saw jihad in the context of conflict in a world divided between the *Dar al-Islam* (territory under Islamic control) and the *Dar al-harb* (territory of war, which consisted of all lands not under Muslim rule). The inhabitants of the territory of war are divided between "People of the Book" (mainly Jews and Christians) and polytheists. This requirement to continue jihad until all of the world is included in the territory of Islam does not imply that Muslims must wage nonstop warfare, however. Although there was no mechanism for recognizing a non-Muslim government as legitimate, jurists allowed for the negotiation of truces and peace treaties of limited duration. Additionally, extending the territory of Islam does not mean the annihilation of all non-Muslims, nor even their necessary conversion: jihad cannot imply conversion by force, since the Qur'an (2:256) states that "There is no compulsion in religion." More than a religious aim, jihad really had a political one: the drive to establish a single, unified Muslim realm justified Islam's supercession of other faiths and to allow for the creation of a just political and social order.[2]

Jihad was generally understood not as an obligation of each individual Muslim (known as *fard 'ayn*) but as a general requirement of the Muslim community *(fard kifaya)*. Only in emergencies, when the Dar al-Islam comes under unexpected attack, do all Muslims have to participate in jihad. Under normal circumstances, therefore, an individual Muslim need not take part so long as other Muslims carry the burden for all of defending the realm.[3]

Other Philosophical Perspectives

This consensus view of a restricted, defensive version of jihad was contested by Muslim legal philosopher Taqi al-Din Ahmad Ibn Taymiyya (1263–1328). He declared that a ruler who fails to enforce the shari'a rigorously in all aspects, including the conduct of jihad (and is therefore insufficiently Muslim), forfeits his right to rule. Ibn Taymiyya strongly advocated jihad as warfare against both the Crusaders and Mongols who then occupied parts of the Dar al-Islam, and most important, broke with the mainstream of Islam by asserting that a professing Muslim who does not live by the faith is an apostate (unbeliever). By going well beyond most jurists (who tolerated rulers who violated the shari'a for the sake of community stability), Ibn Taymiyya laid much of the groundwork for the intellectual arguments of contemporary radical Islamists.[4]

Islamic law condemns all warfare that does not qualify as jihad, specifically any warfare among Muslims. Thus, military action against Muslims is justified only by denying them the status of Muslims (e.g., classifying them as apostates or rebels).[5] Islamic juristic tradition is also very hostile toward terror as a means of political resistance. Classical Muslim jurists were remarkably tolerant toward political rebels by holding that they may not be executed nor their property confiscated. This tolerance vanished, however, for rebels who conducted attacks against unsuspecting and defenseless victims or who

---

[1] Bernard Lewis, *The Political Language of Islam* (Chicago: Univ. of Chicago Press, 1988), p. 72, as quoted in Douglas E. Streusand. "What Does Jihad Mean?" *Middle East Quarterly*, 4 (September 1997), 1.

[2] Streusand, p. 2.

[3] Ibid.

[4] Emmanuel Sivan, *Radical Islam: Medieval Theology and Modern Politics* (New Haven: Yale Univ. Press, 1990), p. 101; as quoted in Streusand, pp. 2–3.

[5] Fred M. Donner, "The Sources of Islamic Conceptions of War," in *Just War and Jihad: Historical and Theoretical Perspectives on War and Peace in Western and Islamic Traditions*, ed. John Kelsay and James Turner Johnson (New York: Greenwood Press, 1991), pp. 51–52, as quoted in Streusand, p. 3.

spread terror through abductions, rapes, the use of poisoned arrows and poisoning of wells (the chemical warfare of this period), arson, attacks against travelers, and night attacks. In these cases, jurists demanded harsh penalties (including death) and ruled that the punishment was the same whether the perpetrator or victim was Muslim or non-Muslim.[6]

Three main views of jihad thus coexisted in pre-modern times. In addition to the classical legal view of jihad as a compulsory, communal effort to defend and expand the Dar al-Islam, and Ibn Taymiyya's notion of active jihad as an indispensable feature of legitimate rule, there was also the *Sufi* movement's doctrine of *greater jihad*. The *Sufis* (a mystical sect of Islam) understood the greater jihad as an inner struggle against the base instincts of the body but also against corruption of the soul, and believed that the greater jihad is a necessary part of the process of gaining spiritual insight. To this day, most Muslims see jihad as a personal rather than a political struggle, while physical actions taken in defense of the realm are considered the *lesser jihad*. It is not surprising, then, that disagreement over the meaning of jihad has continued into the modern era.[7]

Origins of Radical Ideologies

Muslim reform movements in the Middle East first acquired a sense of urgency with the arrival of European imperialism in the latter part of the nineteenth century. The end of colonialism and acquisition of independence by most Muslim countries after World War II accelerated this drive. However, the massive social changes that accompanied these reforms and the simultaneous introduction of new ideas that were alien to classical Islamic tradition—such as nationalism, popular sovereignty, and women's rights—disrupted traditional ways of life and caused traumatic dislocations in these societies.[8]

Disillusionment with the path Muslim societies have taken in the modern period reached its height in the 1970s. Increasingly widespread rejection of Western civilization as a model for Muslims to emulate has been accompanied by a search for indigenous values that reflect traditional Muslim culture, as well as a drive to restore power and dignity to the community. The last 30 years have seen the rise of militant, religiously-based political groups whose ideology focuses on demands for jihad (and the willingness to sacrifice one's life) for the forceful creation of a society governed solely by the shari'a and a unified Islamic state, and to eliminate un-Islamic and unjust rulers. These groups are also reemphasizing individual conformity to the requirements of Islam.[9]

Militant Islam (also referred to as political or radical Islam) is rooted in a contemporary religious resurgence in private and public life.[10] The causes of Islamic radicalism have been religio-cultural, political, and socio-economic and have focused on issues of politics and social justice such as authoritarianism, lack of social services, and corruption, which all intertwine as catalysts. Many Islamic reform groups have blamed social ills on outside influences; for example, modernization (e.g., Westernization and secularization) has been perceived as a form of neocolonialism, an evil that replaces Muslim religious and cultural identity and values with alien ideas and models of development.[11]

Islamic militancy is still not well understood by Americans. This is partly due to the secrecy which radical Islamic groups practice to protect themselves from the authorities and from outsiders who do not share their views and aims, but also because Western public communications media frequently tend to marginalize such groups. They are dismissed as religious fanatics, anti-Western hooligans, or mindless terrorists, without making an attempt to comprehend the deep discontents that have produced these Islamic groups' violent actions or the logic of their radical cause which compels them to behave as they do.[12]

---

[6]Khaled Abou El Fadl, "Terrorism Is at Odds with Islamic Tradition," *Los Angeles Times*, 22 August 2001.

[7]Streusand, pp. 3–4.

[8]Johannes J. G. Jansen, *The Neglected Duty: The Creed of Sadat's Assassins and Islamic Resurgence in the Middle East* (New York: Macmillan, 1986), pp. xi–xii.

[9]Ibid., pp. xii–xiii.

[10]The term "fundamentalism" is also used incorrectly in conjunction with Islam to describe this phenomenon, but this concept is really more appropriate to American Christian thought, whence it originated.

[11]John L. Esposito, "Political Islam and the West," *Military Technology*, February 2001, pp. 89–90.

[12]Jansen, pp. xiii–xiv.

Differences in Sunni and Shi'a Interpretations of Jihad

Sunni and Shi'a (Shi'ite) Muslims agree, in terms of just cause, that jihad applies to the defense of territory, life, faith, and property; it is justified, to repel invasion or its threat; it is necessary to guarantee freedom for the spread of Islam; and that difference in religion alone is not a sufficient cause. Some Islamic scholars have differentiated disbelief from persecution and injustice, and claimed that jihad is justified only to fight those unbelievers who have initiated aggression against the Muslim community. Others, however, have stated more militant views which were inspired by Islamic resistance to the European powers during the colonial period: in this view, jihad as "aggressive war" is authorized against all non-Muslims, whether they are oppressing Muslims or not.

The question of right authority—no jihad can be waged unless it is directed by a legitimate ruler—also has been divisive among Muslims. The Sunnis saw all of the Muslim caliphs (particularly the first four "rightly guided" caliphs to rule after the Prophet Muhammad's death, who possessed combined religious and political authority) as legitimate callers of jihad, as long as they had the support of the realm's *ulama* (Islamic scholars). The Shi'a see this power as having been meant for the Imams, but it was wrongly denied to them by the majority Sunnis. The lack of proper authority after the disappearance of the 12th ("Hidden") Imam in 874 A.D. also posed problems for the Shi'a; this was resolved by the ulama increasingly taking this authority for itself to the point where all legitimate forms of jihad may be considered defensive, and there is no restriction on the kind of war which may be waged in the Hidden Imam's absence so long as it is authorized by a just ruler (this idea reached its zenith under Iran's Ayatollah Ruhollah Khomeini).

Both sects agree on the other prerequisites for jihad. Right intention *(niyyah)* is fundamentally important for engaging in jihad. Fighting for the sake of conquest, booty, or honor in the eyes of one's companions will earn no reward; the only valid purpose for jihad is to draw near to God. In terms of last resort, jihad may be waged only if the enemy has first been offered the triple alternative: accept Islam, pay the *jizyah* (the poll tax required for non-Muslim "People of the Book" living under Muslim control), or fight.[13]

Conditions also are placed on the behavior of combatants in jihad: discrimination of noncombatants from warriors is required, along with the prohibition of harm to noncombatants such as women, children, the disabled, monks and rabbis (unless they are involved in the fighting), and those who have been given the promise of immunity; and proportionality, meaning that the least amount of force is used to obtain the desired ends in combat.[14]

Ideas on Jihad in the Modern Era

Sayyid Abu al-A'la Mawdudi (1903–1979) was the first Islamist writer to approach jihad systematically. Warfare, in his view, is conducted not just to expand Islamic political dominance, but also to establish just rule (one that includes freedom of religion). For Mawdudi (an Indo-Pakistani who agitated for Pakistan's independence from India), jihad was akin to war of liberation, and is designed to establish politically independent Muslim states. Mawdudi's view significantly changed the concept of jihad in Islam and began its association with anticolonialism and "national liberation movements." His approach paved the way for Arab resistance to Zionism and the existence of the state of Israel to be referred to as jihad.[15]

Radical Egyptian Islamist thinkers (and members of the Muslim Brotherhood) Hasan al-Banna (1906–1949) and Sayyid Qutb (1906–1966) took hold of Mawdudi's activist and nationalist conception of jihad and its role in establishing a truly Islamic government, and incorporated Ibn Taymiyya's earlier conception of jihad that includes the overthrow of governments that fail to enforce the shari'a. This idea of revolution focuses first on dealing with the radicals' own un-Islamic rulers (the "near enemy") before Muslims can direct jihad against external enemies. If leaders such as Egyptian President Anwar

---

[13]Mehdi Abedi and Gary Legenhausen, eds., *Jihad and Shahadat: Struggle and Martyrdom in Islam* (Houston: Institute for Research and Islamic Studies, 1986), pp. 21–23.

[14]Ibid., pp. 23–24.

[15]Streusand, p. 5.

Sadat, for example, are not true Muslims, then they cannot lead jihad, not even against a legitimate target such as Israel. Significantly, radical Islamists consider jihad mandatory for all Muslims, making it an individual rather than a communal duty.[16]

The Use of Jihad by Islamic Militants

Regional Islamic Militant Groups' Perceptions

Classical Islamic criteria for jihad were based on the early unified Muslim empire. The imposition of the modern nation-state on Middle East societies, however, has made such ideas no longer applicable; this can be seen by examining contemporary Muslim militant groups' ideologies.

The Islamic Resistance Movement (commonly known as Hamas) sees its situation as similar to that of the Muslim ruler Saladin in his struggle against the Christian Crusaders, as can be seen by examining portions of its Charter. The goal of Hamas is to establish an Islamic Palestinian state in place of Israel, through both violent means (including terrorism) and peaceful political activity. Hamas argues that the current situation of the Palestinians, living under Israeli control or dispersed from their homeland, is part of an ongoing crusade by Christians to take the Holy Lands out of Palestinian hands. The loss of Palestine and the creation of Israel, the Charter continues, were brought about by the great powers of East and West, and taken together constitute a great tragedy not only for the Palestinians but for the entire Islamic community. This, Hamas proclaims, requires jihad not in the sense of expanding the territory of Islam, but of restoring it, and to recover land rather than conquer it. Nor is it a rebellion in the classical sense; rather, this is a struggle to regain a lost portion of the territory of Islam. The Hamas Charter thus provides a uniquely Islamic rationale for *al-intifada*, the "shaking off" of illegitimate rule.[17] This language thus seems to suggest defensive jihad, rather than an offensive struggle.

Since Hamas is not acting on behalf of an established government, it must find authorization elsewhere for its struggle against not only external enemies but also so-called "Muslim" governments that collaborate with the non-Muslim powers (by cooperating with Israel or allowing the basing of Western troops on their soil). The group considers Muslim governments that cooperate with the West as ignorant of the non-Muslim nations' true intentions, or corrupt. Hamas argues that it obtains its authority to declare jihad in another way: the Western powers' invasion of Islamic territory has created an emergency situation where Muslims cannot wait for authorization other than that given directly by God, so jihad is a required duty for all conscientious Muslims.[18] This exceptional situation suspends the usual lines between parties in a relationship so that every Muslim can participate in the struggle. Hamas' Charter thus relates the current situation of Muslims to the classical period, but also marks a break with that classical past. This extraordinary situation also means a change in the nature of Muslim obligation under jihad, from a collective responsibility to extend the Dar al-Islam to a duty for each individual Muslim to restore that territory.[19]

The same pattern of thinking is present in "The Neglected Duty," a pamphlet produced by Egyptian Islamic Jihad (or EIJ, the group that assassinated Anwar Sadat in 1981). This pamphlet, the group's announced "testament," is also a clear expression of the Sunni Islamist perspective on political violence as jihad. It argues that jihad as armed action is the heart of Islam, and that the neglect of this type of action by Muslims has caused the current depressed condition of Islam in the world. EIJ attempts to communicate a sense of urgency to Muslims, who are being victimized and whose territory is being divided and controlled by non-Muslim powers. The document also seeks to justify jihad against other Muslims who, because they are ignorant of this situation, actively cooperate with the unbelievers in the name of "modernization," and are worse than rebels—they are Muslim traitors and

---

[16]Sivan, pp. 16–21 and 114–16, as quoted in Streusand, p. 5.

[17]John Kelsay, *Islam and War: A Study in Comparative Ethics* (Louisville, Ky.: Westminster/John Knox Press, 1993), pp. 95–97.

[18]Kelsay bases his discussion on the translation by Muhammad Maqdsi, titled "Charter of the Islamic Resistance Movement (Hamas) of Palestine" (Dallas: Islamic Association for Palestine, 1990), pp. 17–18. Another translation of this document, by Raphael Israeli, is available on the Internet at www.ict.org.il/documents/documentdet.cfm?docid-14.

[19]Kelsay, *Islam and War*, p. 98.

apostates. Furthermore, fighting such unbelievers without the limits imposed if they were rebellious Muslims is justified, since they are worse than other unbelievers.[20]

"The Neglected Duty" defines the current rulers of the Muslim world (as Sadat was defined) as the primary enemies of Islam and apostates, despite their profession of Islam and obedience to some of its laws, and advocates their execution. This document is explicitly messianic, asserting that Muslims must "exert every conceivable effort" to bring about the establishment of truly Islamic government, a restoration of the caliphate, and the expansion of the Dar al-Islam, and that the success of these endeavors is inevitable.[21] "The Neglected Duty" cites a different historical analogy for this struggle than does Hamas' Charter, however: more appropriate than the threat posed by the European Crusaders was the struggle of Muslims against the Mongol invaders.

EIJ is raising an important issue connected with irregular war: the group is advocating mass resistance against an established government, and such revolution can be justified in Islam only where the ruler becomes an unbeliever through public displays of unbelief. The most significant of such acts is introduction of an innovation *(bid'ah)*, which is a policy, teaching, or action that violates precedents in the Qur'an or hadith. The leadership thus loses its divinely given authority when it commits apostasy, and Muslims not only must no longer obey such a ruler, but are required to revolt and depose him.

This reference to the obligation to God for the creation and maintenance of an Islamic state and the responsibilities of Muslims serves to answer the question of authorization for militant Islamic forces.[22] "The Neglected Duty" provides further justification for armed action by arguing that Egypt, like most of its neighbors, is not an Islamic state because its constitution and laws are a mix of traditional Islamic judgments and European law codes. Imposition of such a mixed legal system (non-Islamic laws that are an "innovation") by Egypt's leaders on their subjects thus means that the nation is not part of the territory of Islam, but part of the territory of war or unbelief.[23]

Shi'a radicals have a similar perspective to their Sunni extremist "brothers in arms." Ayatollah Ruhollah Khomeini (1902–1989) contended that Islamic jurists, "by means of jihad and enjoining the good and forbidding the evil, must expose and overthrow tyrannical rulers and rouse the people so the universal movement of all alert Muslims can establish Islamic government in the place of tyrannical regimes." The proper teaching of Islam will cause "the entire population to become *mujahids* [literally "strugglers for God]." Ayatollah Murtaza Mutahhari (1920–1979), a top ideologue of the Iranian Revolution, considered jihad a necessary consequence of Islam's content: by having political aims, Islam must sanction armed force and provide laws for its use. Mutahhari deemed jihad to be defensive, but his definition includes defense against oppression and may require what international law would consider a war of aggression. For example, he endorsed an attack on a country of polytheists (some Muslims see Christians as polytheists due to Christianity's belief in a God who can exist in three manifestations) with the goal simply to eliminate polytheism's evils, not to impose Islam.[24]

Another radical Shi'a perspective on the justification for jihad can be found in the words of Shaykh Muhammad Hussein Fadlallah, spiritual leader of Lebanese Hizballah. In a 1986 interview, he stated that although violence is justified only for defensive purposes and as a last resort, the contemporary situation of the people of the Middle East, in particular of Muslims, creates a scenario that breeds violence. The establishment of Israel, the dislocation of the Palestinians, and the interference of a great oppressive power (in other words, the United States) in Arab-Islamic political, economic, and social affairs leads some Muslims (e.g., militant groups) to consider themselves justified in using force to achieve their goals, and this can even sometimes lead to extreme behavior.[25] Fadlallah does clarify that terrorism *(hudna,* or violence in Arabic) is not legitimate or justified in Islam, to include the destruction of life, kidnapping, or the hijacking of airliners or ships, and suggests that militants have gone too far in the conduct of their struggle when they employ such means. Nevertheless, he concludes by

---

[20]Ibid., pp. 100–01.

[21]Jansen, p. 162, as quoted in Streusand, p. 5.

[22]Kelsay, *Islam and War*, pp. 101–02.

[23]Ibid., p. 102.

[24]Abedi and Legenhausen, p. 89, as quoted in Streusand, p. 6.

[25]Kelsay, *Islam and War*, p. 109.

informing the American people that it is up to them to improve the situation by pressing for reforms in the policies of their government.[26]

How should the West respond to Islamic militant groups? Shaykh Fadlallah suggests that the West should listen to the anger expressed by such groups. While stressing that the way to peace is through dialogue, Fadlallah said that the West must first recognize that Muslims who act in ways that are harmful to Western interests are responding to pain of their own. Islam, he added, should not be thought of as uncompromisingly hostile to the West, since militant groups do not speak for all of the community. Fadlallah adds that if the West does listen to these groups, however, it will understand that the concerns these groups have (for justice, human rights, and self-determination) are legitimate, even if their methods are excessive.[27]

Al Qaeda and Transnational Jihad: A New Twist on Old Complaints

Before his emergence as the prime suspect in the 9/11 attacks, Osama bin Laden had described his goals and grievances and the tactics of his transnational al Qaeda network in great detail in a series of statements and interviews. Taken together, these statements provide insight into an ideology that may seem abhorrent or crazy to Americans but has been carefully crafted to appeal to the disgruntled and dispossessed of the Islamic world.[28] Bin Laden's ideology, however, is really more political than religious.

At the heart of bin Laden's philosophy are two declarations of war—jihad—against the United States. The first, his *Bayan* (statement) issued on 26 August 1996, was directed specifically at "Americans occupying the land of the two holy places," as bin Laden refers to the cities of Mecca and Medina that are located in his native Saudi Arabia. Here he calls upon Muslims all over the world to fight to "expel the infidels . . . from the Arab Peninsula."[29] In his fatwa of 23 February 1998, titled "Declaration of the World Islamic Front for Jihad Against the Jews and Crusaders," which he issued along with the leaders of extremist groups in Egypt, Pakistan, and Bangladesh, bin Laden broadened his earlier edict. In the fatwa, he specifies that the radicals' war is a defensive struggle against Americans and their allies who have declared war "on God, his Messenger, and Muslims." The "crimes and sins" perpetrated by the United States are threefold: first, it "stormed" the Arabian peninsula during the Gulf War and has continued "occupying the lands of Islam in the holiest of places"; second, it continues a war of annihilation against Iraq; and third, the United States supports the state of Israel and its continued occupation of Jerusalem. The only appropriate Muslim response, according to the fatwa, is a defensive jihad to repulse the aggressor; therefore, borrowing from classical and modern Islamic scholars (because it is defensive), such a war is a moral obligation incumbent upon all true Muslims.[30]

Bin Laden's anger at the "American crusader forces" who are "occupying" his homeland stems from an injunction from the Prophet that there "not be two religions in Arabia"; the presence of foreign forces on holy soil is thus an intolerable affront to 1,400 years of Islamic tradition. In his 1996 statement of jihad, bin Laden blamed the serious economic crisis then gripping Saudi Arabia (due to falling oil prices and widespread corruption) on the presence of these Western "crusader forces." Two years later, in his 1998 fatwa, bin Laden charged that the United States was not only occupying and plundering Arabia, but was "using its bases in the peninsula as a spearhead to fight against the neighboring Islamic peoples." In bin Laden's war, the goal of expelling the "Judeo-Christian enemy" from Islamic holy lands should occur first on the Arabian peninsula, then in Iraq (which for 500 years was

---

[26]Ibid., pp. 109–10.

[27]Quoted in Kelsay, *Islam and War*, p. 108.

[28]Michael Dobbs, "Inside the Mind of Osama Bin Laden," *The Washington Post*, 20 September 2001.

[29]Ibid.

[30]Schail Hashmi, "The Terrorists' Zealotry Is Political Not Religious." *The Washington Post*, 30 September 2001. For a good analysis of bin Laden's fatwa, including its historical backgrounds, see Bernard Lewis, "License to Kill," *Foreign Affairs*, 77 (November/December 1998), 14–19. The translated text of the fatwa itself is available on the Federation of American Scientists' website at www.fas.org/irp/world/para/docs/980223-fatwa.htm.

the seat of the Islamic caliphate), and third in Palestine, site of the Al-Aqsa Mosque in Jerusalem (which is sacred to Muslims as the place from where Muhammad ascended to heaven).[31]

Although the initial attacks associated with bin Laden occurred in Saudi Arabia, Somalia, East Africa, and Yemen, he increasingly made clear that he would bring the war to the American homeland. Al Qaeda is believed to have aided the first attack against the World Trade Center in 1993, and bin Laden told an ABC News reporter in May 1998 that the battle will "inevitably move . . . to American soil."[32] Although he appears to be fired by the religious zeal of Saudi Arabia's puritanical Wahhabi movement, bin Laden's targets have not been offending religious and cultural institutions, but political, military, and economic targets. Additionally, though he quotes selective (but incomplete) passages from the Qur'an to establish the basis for the jihad, bin Laden's motivations are really not that different from the anti-imperialistic doctrines that sustain religious and nonreligious extremist groups all over the world.[33]

In return for joining the jihad against America, bin Laden has promised his followers an honored place in paradise, in accordance with a statement in the Qur'an that "a martyr's privileges are guaranteed by Allah." Bin Laden and many of the other Islamic militant groups in the Middle East are able to draw on large numbers of enthusiastic and waiting recruits for their war against the United States—impoverished youths who are ready to die simply for the idea of jihad.

"Jihad Factories": An Enduring Legacy of Hatred

It is estimated that more than one million young men from Pakistan, Afghanistan, Central Asia, and the Muslim parts of China are attending *madrassas,* or private Islamic religious schools, every year in Pakistan. Madrassa students spend most of their day in rote memorization of the Qur'an in Arabic (this is not their native language, so few understand what they are reading) and interpreting the hadith. Only theology is taught; there is no math, science, computer training, or secular history.[34] The young men at these schools are drawn from the dire poor of the societies they come from, kept in self-contained worlds that are isolated from outside influences, and indoctrinated with a powerful, not-so-academic radical message: their highest honor and duty is to wage jihad to defend Islam from its attackers, and the United States is the chief enemy of Islam.[35]

Madrassas, which have a tradition in Pakistan that dates from colonial days of promoting political independence along with their religious teaching, fill a significant gap in the underfunded public school system by offering free tuition, room, and board. Madrassas received state funding during the Afghan War when they were used to groom the mujahedin who were being sent to fight the Soviet invaders.[36] Many of these schools were emptied in the 1990s when the Taliban needed assistance in military campaigns against its Northern Alliance foes, and many students sent to the front did not return. The graduates of these madrassas have also turned up in place like Bosnia, Chechnya, and the Kashmir, and the survivors of those conflicts have taken their battlefield experience back to their home countries where it is being put to use in jihads against their own not-Islamic-enough governments and societies.

The readiness of millions of young men trained in these schools to sacrifice their lives for Islam—and their unquestioning acceptance of anti-American and pro-Islamic extremist propaganda—will continue to be a powerful and enduring weapon against the U.S.-led global war on terrorism, and one that bin Laden and other militants who are bent on attacking the United States and its allies can call on in the years ahead.

---

[31]Dobbs.

[32]Ibid.

[33]Hashmi.

[34]Jeffrey Goldberg, "Inside Jihad U.: The Education of a Holy Warrior," *New York Times Magazine,* 25 July 2000.

[35]Indira A. R. Lakshmanan, "In Some Schools, Jihad, Anger at US Are Lesson," *Boston Globe,* 4 October 2001.

[36]Ibid.

Acceptance of Militants' Ideas and Methods Is Limited

The thrust of the entire jihad tradition which Islamic radicals have "hijacked" makes it clear that not everything is permissible. Although the language in the Qur'an and hadith and in other classical Muslim sources is overwhelmingly militant in many places, this is a reflection of the Muslims' world in the seventh century, which consisted initially of resistance to a variety of more powerful non-Islamic tribes and then successful military campaigns to spread the faith. Besides containing exhortations to fight, however, Islamic sacred texts have also laid out the rules of engagement for war, which (as mentioned earlier) included prohibitions against the killing of noncombatants such as women, children, the aged, and disabled. These texts also require notice to the adversary before an attack, require that a Muslim army must seek peace if its opponent does, and forbid committing aggression against others and suicide.[37] Those who are unfamiliar with the Qur'an and hadith can miss these points when confronted with the propagandistic calls to jihad of militant Islamic groups.

The actions of rebels in the classical period of Islam encountered widespread resentment and condemnation, and this strong sentiment against rebellion remains in modern Islamic thought. Most Muslims agree with the presumption in Islamic teachings on war that individuals are innocent and therefore not subject to harm unless they demonstrate by their actions that they are a threat to the safety or survival of Muslims. On this basis, the overwhelming majority of Islamic scholars have for centuries rejected indiscriminate killing and the terrorizing of civilian populations as a legitimate form of jihad.[38] Also, at no point do Islamic sacred texts even consider the horrific and random slaughter of uninvolved bystanders that is represented by the 9/11 airliner attacks; most Muslims throughout the world were as shocked by those attacks as Americans were.

The radical message in works such as Hamas' Charter, "The Neglected Duty," and the writings of Khomeini and his fellow revolutionary Iranian Shi'a clerics nevertheless finds a lot of acceptance with contemporary Muslims. The reason is simply the poor socioeconomic circumstances and lack of human dignity that many Muslim peoples find themselves subject to, brought about by secular failures to attend to their problem.[39] Militant Islamic groups, exemplified by Hamas and the Palestinian branch of Islamic Jihad, have been able to use such poor conditions to their advantage. They provide social services (such as operating free or low-cost schools, medical clinics, sports clubs, and women's support groups), many of which the Palestinian Authority itself often cannot provide, to build public support and attract recruits in the occupied territories.[40]

Public statements over the last several months by some moderate Muslim religious authorities and commentators that Islamic extremists are corrupting a peaceful religious faith for their own twisted ends are encouraging. Equally positive is the growing recognition in the Muslim world both of bin Laden's lack of proper religious qualifications to issue any religious edicts that promote jihad, and his lack of success, on a strategic level, in forcing the United States to withdraw its military forces completely from Saudi Arabia or to give up its campaign against Islamic terrorism. A few prominent Muslim scholars have not only condemned the terrorist attacks upon the United States, but have declared the perpetrators of these attacks to be "suicides," not martyrs. This is significant, since Islam forbids suicide and teaches that its practitioners are sent not to paradise but to hell, where they are condemned to keep repeating their suicidal act for eternity.[41]

Conclusion

As described herein, jihad in Islamic thought and practice possesses a range of meanings, with Muslim radicals focusing on the physical, violent form of struggle to resist what they see as cultural, economic, military, and political assaults from outside the ummah and oppression and injustice within. So long as societal conditions within many Muslim states remain poor, with unrepresentative governments

---

[37] Teresa Watanabe, "Extremists Put Own Twist on Islamic Faith," *Los Angeles Times,* 24 September 2001.

[38] Hashmi.

[39] Jansen, p. 2.

[40] "Islamic Groups Going for Goodwill," *Daily Progress* (Charlottesville, Va.), 18 November 1998, p. A8.

[41] Bernard Lewis, "Jihad vs. Crusade," *The Wall Street Journal,* 27 September 2001.

(which are seen to be propped up by the United States) that are unwilling or unable to undertake meaningful but difficult reforms, then militant Islamic groups will continue to attract recruits and financial support. In spite of logical fallacies and inconsistencies in the doctrine of jihad of radical Islamic groups, and the fact that most of the broad constituency they are attempting to appeal to does not buy into their ideology or methods, such groups nevertheless remain as significant threats to U.S. interests everywhere in the world.

The challenge for the U.S. government over the next several years will be to encourage and support lasting reform by Muslim states who are our allies in the Middle East, while maintaining a more balanced and fair-minded foreign policy toward all key regional players. We must also do a better job of countering the Islamic extremists' widely disseminated version of jihad, while being more persuasive that our own government—and our society—are truly not anti-Islamic. Such actions will do much to deny a supportive environment to our radical Muslim foes. For its part, the U.S. military needs to better understand the religious and cultural aspects of our adversaries' asymmetric mindset—in this case, how Islamic militants conceive of and use jihad—to be successful in its global campaign against terrorism.

B. The Moral Code of the Warrior (*Jus in Bello*)

In 1539, during the golden era of the Spanish empire, students at Spain's pre-eminent university and center of liberal learning, the University of Salamanca, were working their way through some of the same texts we have considered in this volume: specifically, they were studying Thomas's *Summa Theologica*. By tradition, toward the end of the semester, a leading member of the faculty was invited to deliver lectures to the rest of the faculty and student body, offering a summary and reflection upon what had been accomplished during the school year. In that year, their most eminent scholar, the Dominican priest, Francisco de Vitoria, was selected to deliver these *"relectiones"* ("re-readings," or reflections) on their readings of St. Thomas.

Vitoria gave two lectures that semester, both of which likely surprised and startled his audience, and might well have gotten him into grave trouble with the Spanish authorities (not to mention the Spanish Inquisition). Their titles were, respectively, *De Indis*, and *De iuri belli*—"Concerning the Indians," and "Concerning the Laws of War." The first denounced the Spanish military or "Conquistadors" in the new world for their treatment of indigenous peoples (AmerIndians) in Spain's wars of conquest. The harsh treatment, enslavement, and even the military operations themselves conducted by Conquistadors against these people, Vitoria argued, were *not only morally unjust, but also illegal*.

This was an odd charge to make, especially since the "legitimate authorities," the Spanish monarchy, had itself licensed these activities. To explain these claims, the second lecture outlined, for the first time, a legal "code of the warrior" which Vitoria alleged was being routinely violated by the Spanish military. Vitoria did not deny that there were sometimes legitimate grievances on the part of the representatives of his government against native peoples, nor did he maintain that the use of Spanish military force in the New World was itself always inherently unjust. Indeed, he allowed that there are occasions, beyond self-defense, in which the use of military force would be justified, citing the passages of Thomas's *Summa* that we likewise considered in the last chapter, dealing with the punishment of "evil-doers," recovery of stolen property, and punishment or rectification of injustice.

What concerned Vitoria was both the debateable "justice" or legitimacy of the Spanish "cause" (namely, the pacification and Christianization of the population) and even more, *the manner by which it was carried out*. Of particular concern, he noted, warriors in an otherwise just war are not permitted deliberately to target non-combatants, nor to attempt to win a military victory by decimating the enemy's non-military population. Enemy combatants could, of course, be killed or taken prisoner during combat, but if the latter, then they must be treated with justice and respect, as they no longer represented a direct threat to the opposing military. These and similar provisions constitute the basis of what we now know as "the Law of Armed Conflict," or, in Latin, *jus in bello*.

The idea of declaring such actions "illegal" was unusual, as there was at the time no overarching legal authority to which to appeal in adjudicating these claims. Vitoria appealed to what he termed *ius gentium*, the "customs and practices of civilized peoples." It was these, as well as the theological reflections of St. Thomas, that established the principles he advocated with the force of law, and which thereby constrained the behavior of combatants, even in wartime. *No one, he maintained, not even soldiers fighting during a war, are beyond the reach of natural law.*[1]

Vitoria's way of conceptualizing the moral duties incumbent on combatants during wartime inaugurated a revolutionary way of thinking, some features of which we considered during the preceding chapter. In particular, his reflections dovetail with the insight of the previous chapter, that otherwise-just wars may only be pursued by just (or "legal") means. These insights came to fruition a century later in the work of the Dutch legal scholar, Hugo Grotius, in his work, *De iuri belli et pace*, "Concerning the Laws of War and Peace." Then and now, however, it has seemed counter-intuitive in the extreme to many persons to discuss "laws" governing what seems to be an inherently lawless activity—or rather, an *activity that ensues when the normal rule of law has broken down*. How can such a "lawless" activity as war nonetheless be said, at the same time, to be "law-governed?" And who or what makes, and (much more importantly) *enforces*, these laws?

The noble ideal of this tradition, in response, is to argue that even—indeed especially—when the normal rule of law has disintegrated, that human beings engaged in morally questionable acts must not lose sight of who they are, nor lose sight of the civilized and law-governed "place" from whence they have come, and (God willing) to whence they will return when hostilities cease.

So it is that these decidedly modern conceptions have fused with numerous threads of ancient tradition, drawn from our western culture, and from other cultures as well, in a "Code of the Warrior," reminding combatants of their essential humanity, warning them away from unspeakable and ultimately unforgivable behaviors, and so differentiating the pursuit of just causes through armed conflict from murder and mayhem. Absent the provisions and constraints of this Code, the pacifist is correct by default: there is then *no difference* between war and mere murder and mayhem, and war itself, as a result, is always morally unacceptable, no matter what the reason for it.

It is thus the willing adherence to the Law of War, and the Code of the Warrior, that alone differentiates a soldier from a common murderer, and a military officer from a terrorist, in very much the same sense as adherence to the constraints of domestic law, even in its enforcement, distinguish between a police officer and a common criminal. Vice versa: in neglecting these constraints on the legitimate exercise of force in behalf of lawful or morally justifiable causes, the soldier, police officer, or military officer risks losing sight of—indeed, losing altogether—his or her humanity, dignity, and the moral authority of their respective professions. The worst nightmare of both the domestic and international peace-keeper is that their honorable and very necessary public professions should, through the illicit behavior of some of their members (as happened at My Lai and again at Abu Ghraib), come to be indistinguishable from the injustice and moral corruption against which they are arrayed.

The position of so-called "realists" and skeptics, by contrast, is that "war is hell," and that when in hell, pretty much anything goes. As noted in the introduction to this book, however, the United States military, and the militaries of its allies, do not tolerate, and will not exemplify that attitude in their behavior. "War is hell," in fact, only when combatants behave as if they were devils. The profession of arms itself holds forth by tradition, and rightfully expects, something more from its members, something worthy of description as honor, courage, integrity, and self-sacrifice.[2]

---

[1] For excerpts from Vitoria's own writings, see Francisco de Vitoria, *Political Writings*, ed. Anthony Pagden and Jeremy Lawrence. (Cambridge: Cambridge University Press, 1991). Wars against tyrants ("regime change") are discussed at pp. 286-88 and 325f, as ways to protect human rights. A version of this story appears in Michael Walzer's essay (stemming in part from a lecture at the U.S. Naval Academy in November, 2002): "The Triumph of Just War Theory and the Dangers of Success," reprinted in USNA Staff Lecture Series. This version, however, relies on an earlier article on Vitoria by Gregory Reichberg, in *The Classics of Western Philosophy*, eds. Jorge Gracia, Greg Reichberg, and Bernard N. Schumacher. (London: Blackwell Publishing, 2003), 197–203.

[2] For a more detailed and richly textured treatment of this topic, see Shannon E. French, *The Code of the Warrior: Exploring Warrior Codes Past and Present*. Foreward by Senator John S. McCain. (Lanham, MD: Rowman & Littlefield, 2003). Harvard scholar Michael Ignatieff discusses contemporary versions and understanding of this code in *The Warrior's Honor: Ethnic War and the Modern Conscience* (New York: Henry Holt & Co., 1997).

Jus in bello—
Just Conduct in War
Brian Orend*

> "The greatest difficulty in the right of nations has to do precisely with right during war; it is difficult even to form a concept of this or to think of law in this lawless state without contradicting oneself."
>
> Kant[1]

"*Jus in bello*" is the Latin term just war theorists use to refer to justice *in* war—to right conduct in the midst of battle, after the war has started. Most just war theorists insist that *jus in bello* is an ethical category separate, in some sense, from *jus ad bellum*. Why? We have not finished our task of evaluating warfare once we have determined whether a community has resorted to war justly, using the principles developed in the last two chapters. For even if a state has resorted to war justly, it may be prosecuting that war in an unjustified manner. It may be deploying perverse means in pursuit of its otherwise justified end. Just war theory insists on a *fundamental moral consistency between means and ends* with regard to wartime behaviour.

Concern with consistency, however, is not the only, or even the main, reason behind our endorsement of separate rules regulating wartime conduct. Such rules are also required to limit warfare, to prevent it from spilling over into an ever-escalating, and increasingly destructive, experiment in total warfare. If just wars are limited wars, designed to secure their just causes with only proportionate force, the need for rules on wartime restraint is clear.

All that being said, and sincerely endorsed, I wish to reiterate my conviction that the so-called "separation" between *jus ad bellum* and *jus in bello* is mainly for focusing attention on different issues. It does not denote a complete split between the two, as if they had nothing to do with each other. Indeed, I assert that the three just war categories *must* morally be linked, with *jus ad bellum* setting the tone for all that follows. There is nothing worse, conceptually, than to adopt the "check list" approach to just war theory, as if all the rules and criteria were simply separate "boxes" to be checked off during the war, like ticking off the items on your grocery list as you buy them. The rules and criteria all presuppose shared values, such as: rejecting aggression; restraining warfare; and protecting the state rights of legitimate communities and the human rights of their individual residents. We'll see that we literally cannot make moral sense of some *jus in bello* rules—notably, proportionality—and probably the entire *jus post bellum* category without considering the just cause of the war to begin with.

The best metaphor—regarding just war theory as not segregated but, rather, as united into one long procedure with different phases—is probably something like surgery. You have got, quite literally, an opening phase, an operational phase and a closing phase. Different phases raise different questions and concerns, and there need to be well-founded and well-understood rules governing each.

---

*Ch 4, pp. 105–123 in *The Morality of War*, by Brian Orend (Peterboro, Ontario: Providian Press, 2006)

But everything is connected and substantially affected by what happened prior, and the ruling concern which started the whole process was why surgery was necessary in the first place.

As for the sewing up and the post-surgery rehabilitation, we have the future *jus post bellum* chapters. For now, we inaugurate the operational phase. We have diagnosed the need for surgery, have made our cut and gone in. We're in the thick of it: now what do we do? This is the topic of *jus in bello*.

Before examining and explaining these rules, it is worth stressing how responsibility for fulfilling *jus in bello* differs from the responsibility inhering in *jus ad bellum*. Responsibility for the justice of resorting to war, we saw, rests on those key members of the governing party most centrally involved in the decision to go to war, particularly the head of state or any body authorized to declare war. Responsibility for the conduct of war, by contrast, rests on the state's armed forces. In particular, responsibility for right conduct rests with those commanders, officers and soldiers who command and control the lethal force set in motion by the political hierarchy. In general, anyone involved in formulating and executing military strategy during wartime bears responsibility for any violation of *jus in bello* standards. In most cases, such violation will constitute a war crime.

I. Discrimination and Non-Combatant Immunity

The requirement of discrimination and non-combatant immunity is the most important *jus in bello* rule. It is also the most frequently, and stridently, codified rule within the international laws of armed conflict.[2] The substance of the rule is this: soldiers charged with the deployment of armed force may not do so indiscriminately; rather, they must exert every reasonable effort to discriminate between legitimate and illegitimate targets. How are soldiers to know which is which? *A legitimate target in wartime is anyone or anything engaged in harming*. All non-harming persons, or institutions, are thus ethically and legally immune from direct and intentional attack by soldiers and their weapons systems. Since the soldiers of the enemy nation, for instance, are clearly engaged in harming, they may be directly targeted, as may their equipment, their supply routes and even some of their civilian suppliers. Civilians not engaged in the military effort of their nation may not be targeted with lethal force. In general, as Michael Walzer asserts: "A legitimate act of war is one that does not violate the rights of the people against whom it is directed."[3] In response, we might ask: how is it that armed force directed against soldiers does not violate their rights, whereas that directed against civilians violates theirs? In the chaos of wartime, what exactly marks the difference?

One of the murkiest areas of Walzer's just war theory concerns the moral status of ordinary soldiers. His references to them exhibit, on the one hand, a humane sympathy for their "shared servitude" as "the pawns of war." On the other, his references occasionally display something like a glib callousness, as when he concurs with Napoleon's (in)famous remark that "soldiers are made to be killed."[4] How can soldiers be made to be killed when, as human beings, they enjoy human rights, to security amongst other things? The answer must be that soldiers do something which causes them to forfeit their rights, much as an outlaw country forfeits its state rights to non-interference when it commits aggression. One could be forgiven for inferring, from this principle, that *only soldiers of an aggressor nation* forfeit their rights, since they are the only ones engaged in the kind of rights-violating harm which grounds a violent, punitive response. Interestingly, and perhaps problematically, Walzer denies this. He believes that *all* soldiers forfeit their right not to be targeted with lethal force, whether they be of just or unjust nations, whether they be tools of aggression or instruments of defence.[5]

The Moral Equality of Soldiers?

Walzer's concept here, which is widely accepted, is of "the moral equality of soldiers." The first "war right" of all soldiers is to kill enemy soldiers: this is indeed part of international law. We do not, and should not, make soldiers pay the price for the injustice of the wars they may be ordered—perhaps even conscripted—to fight. That is the logically and morally separate issue of *jus ad bellum*, for which we have already elaborated a theory of justice, focusing on the responsibilities of political leaders. But lawyers like the chief British prosecutor during the Nuremberg trials, and philosophers like Thomas Pogge, ask: why shouldn't we hold soldiers responsible for the justice of the wars they fight? If we held soldiers responsible in this regard, wouldn't that constitute *an additional bar against aggressive war*? Wouldn't that account for the fact that, even though the war was set in motion by others, soldiers remain its essential executors? Wouldn't such a move impose and highlight an important responsibility for soldiers, namely, to refuse to participate in the prosecution of aggressive war?[6]

Walzer experiences difficulty answering this argument fully. As an opening gambit, he contends that soldiers "are most likely to believe that their wars are just." But this alone cannot justify their actions, since their beliefs may not be well-grounded, especially considering the incentive they have to believe such justification in the first place. Walzer also says that soldiers rarely fail to fight, owing to "(t)heir routine habits of law-abidingness, their fear, their patriotism [and] their moral investment in the state."[7] But the fact that soldiers rarely fail to fight does not demonstrate that they are always justified in fighting, especially if the cause is unjust. Walzer next suggests that knowledge about the justice of the wars soldiers fight is "hard to come by." This is a truly surprising claim from a just war theorist who has tried to make such knowledge more accessible and comprehensible. Perhaps, then, this is a reference to the soldier's historical lack of education, as well as to government tendencies towards secrecy. Fair enough, but ignorance at best constitutes an excuse, and not a justification, for willfully fighting in an unjust war: it seems a stretch to assert that such ignorance can morally ground a "war right" to deliberately kill enemy soldiers. Walzer's subsequent move appeals to the authority of Vitoria, who suggested that if soldiers were allowed to pick and choose the wars they were willing to fight, the result would be "grave peril" for their country.[8] (This argument has recently been endorsed by the Israeli government, which has found itself in Israeli courts being sued by a handful of its own soldiers, who do not want to fight in the ongoing struggle with the Palestinians. Israel has a system of compulsory military service for all citizens.) But this empirical generalization is speculative: why wouldn't the result actually be the preferred one, namely, that states would be seriously hampered *only* in their efforts to prosecute an aggressive war, which they could not justify to their soldiery?

Walzer then claims it is a plain fact of sociology that we do not blame soldiers for killing other soldiers in the midst of war. We blame soldiers *only* when they deliberately kill either civilians or surrendered enemy soldiers kept by them as disarmed prisoners of war (POWs). We extend to all soldiers, in the midst of battle, the right to deploy armed force on behalf of their own country. This *is* a true legal contention—equal belligerent rights seem established by international law[9]—and while it is not an implausible moral one, the latter is not quite so obvious as Walzer suggests. Do we really believe that those soldiers who fought for Hitler, for example, were utterly blameless for their bit part in the execution of his mad aggression? No doubt, we tend to exonerate conscripts like the Hitler Youth in the closing days of the war, presuming they were far too young, gullible and propagandized to have made a morally responsible choice. But what about those mature German soldiers—many with prior war experience—who invaded Poland, or France, at the war's outset? It is not so clear to me that we do not blame them for fighting on behalf of their country. Indeed, today, ex-soldiers, living abroad, who played very small roles in Hitler's army can, if found out, be stripped of their citizenship and deported back to Germany—even in spite of their very advanced age.

Walzer stresses more generally the pervasive socialization of soldiers of *any* nation, their relative youth, their frequent conscription, and their usual background as members of underprivileged classes as grounds for not holding soldiers responsible for the wars they fight. While soldiers "are not . . . entirely without volition," he says, "(t)heir will is independent and effective only within a limited sphere." This sphere contains only those tactics and manœuvres soldiers are engaged in, like training, loading their weapons, engaging in particular live-fire skirmishes, handling prisoners, and so on. It would thus constitute unfair "class legislation" for us to hold soldiers like these responsible for the justice of the wars they fight. We should focus on those most to blame, the elite and powerful leaders who set the war in motion.[10] But from the fact that political leaders are, I admit, *mostly* to blame for the crime of aggression, does it follow that they are *solely* to blame, as Walzer here insists? In my view, the more compelling alternative would be to suggest that, for the many reasons Walzer mentions, there should be *a presumption against* holding soldiers responsible for the crime of violating *jus ad bellum*. But this presumption does *not* preclude us from concluding, in particular cases based on public evidence: that some soldiers of a particular aggressor state either *did* know, or *really should have* known, about the injustice of the war they were fighting; that they could have refused to participate in it; and thus that they may be held responsible, albeit with much lesser penalties than the elites. Such soldiers would be like minor accomplices to a major crime. This is one way in which I do not believe in the absolute separateness of *jus ad bellum* and *jus in bello*.

Soldiers who fight for an aggressor do not have strict moral equality with those who fight for a victim or defender. The former are not mainly to blame for the aggression but they still *are* to blame, in a smaller but material sense. Soldiers are not automatons. Their merely *professional function does not dislodge the ordinary human duty not to inflict severe, unjustifiable harm on others.* Soldiers must inform their beliefs regarding the justice of the wars they are ordered to fight. Exceptional ones, upon seeing injustice, will refuse service, or else surrender to the other (just) side at the first non-life-threatening opportunity. Ordinary soldiers, when confronted with the fact that the war they fight is unjust, will probably still go along with the crowd of their buddies and fight—the pressures to do so are very strong. For them, we reserve the moral right to criticize and castigate them at war's end. Perhaps we will finally excuse them, on the basis of these pressures, but perhaps also, in some well-documented cases, we will have to prosecute them as minor accomplices to the one large "crime against peace" which is aggression.[11]

Engaging in Harm

Even if we agree with my proposal that some soldiers may be held responsible for *jus ad bellum* violations, can we still concur with the idea that all soldiers *generally* remain legitimate targets during wartime? After several false starts, Walzer offers us a compelling reason to do so: soldiers, whether just or unjust, are "engaged in harm." Soldiers bear arms effectively, are trained to kill for political reasons, and are "dangerous men": they pose serious threats to the lives and interests of those they are deployed against, whether for a just cause or not. Walzer suggests that an armed man trying to kill me "alienates himself from me . . . and from our common humanity" and in so doing he forfeits his right to life. This establishes, I believe, a strong *prima facie* (i.e., "first glance") case that soldiers are justified in targeting other soldiers with lethal force. Soldiers, whether for just or unjust reasons, remain among the most serious and standard external threats to life and vital interests. Only public, compelling and accessible knowledge about the injustice of the cause of his own country can undermine a soldier's entitlement, in the face of such an opposing threat, to respond in kind.[12]

The converse of this general principle, of course, is that those who are not "engaged in harm" cannot be legitimate targets during wartime. *This is the clearest sense of who is "innocent" in wartime: all those not engaged in creating harm*. The first application of this converse rule regarding harm has to do with soldiers themselves: when soldiers no longer pose serious external threats—notably by laying down their weapons and surrendering—they may no longer be targeted with force and should, in fact, be extended what international law calls "benevolent quarantine" for the duration of the war.[13]

Benevolent Quarantine and Torturing Terrorists

"Benevolent quarantine" means that captured enemy soldiers can be stripped of their weapons, incarcerated with their fellows, and questioned verbally for information. But they cannot, for example, be tortured during questioning. Nor can they be beaten, starved or somehow medically experimented on. They cannot be used as shields between oneself and the opposing side; in fact the understanding is that captured enemy soldiers are to be incarcerated far away from the front lines. Very basic medical and hygienic treatment is supposed to be offered—things like aspirin, soap, water and toothbrushes—and, while making captives engage in work projects is permitted, the Geneva Conventions actually require that, in that event, captives be paid a modest salary. I have never heard of that actually happening—the incredible detail of the Geneva Conventions means they do not always get realized—but it is fairly common for combatants to disarm, house and feed their captives, keeping them out of harm's way and ensuring their basic needs are met until the war ends. When it is all over, they are then usually freed in exchange for POWs on the other side.[14]

Controversies here focus around when, or if, aggressive questioning becomes a form of torture, and also around when non-state actors, like terrorists, are taken prisoner. The latter issue concerns whether non-state captives deserve the same quality of treatment as state captives. There's a sense that a soldier fighting for his community deserves better than a terrorist fighting for his pet cause. I think this distinction can be difficult to sustain and that, generally, non-state actors brought into capture by soldiers should be accorded the same rights as captured enemy soldiers. If soldiers fighting for an *unjust* cause deserve this treatment, then surely so do terrorists—whose method, if not the cause, is likewise unjust. In other words, if it is wrong to torture Nazi soldiers—whose cause was heinous and irredeemable—then it is wrong to torture radical Islamic terrorists (much less mere suspects). This topic has recently been highlighted regarding America's round-up of alleged terrorists, some of whom were held for intensive questioning in Guantanamo Bay, Cuba over a long period of time.[15]

The incidents in Abu Ghraib prison, in Iraq, also come to mind. In the late spring of 2004, the world saw some shocking photos of American troop conduct in that jail. Iraqi prisoners—captured during the war and the subsequent insurgency—were subjected to questionable treatment. Some of it—like deliberate, prolonged sleep deprivation, and using dogs to attack or threaten already prone and naked people—clearly violated the Geneva Conventions. Others might have been visually disturbing but do not obviously count as human rights violations, such as forcing the prisoners to wear dog collars, or having American women ridicule their private parts, or putting female panties on their faces temporarily. Combine it all, though, and you have a violation of both the letter and the spirit of the principle of benevolent quarantine. Some of the US soldiers involved have since been charged, tried and sentenced (which I would argue shows official American acceptance of the idea that non-state captives deserve the same treatment as state captives).

There is, at writing, an ongoing investigation as to whether any authorization of this treatment came from higher up the US chain of command. It does seem unlikely, after all, that some "mere" reservists would repeatedly conduct such flagrant prisoner abuse on their own, and apparently there was a vaguely-defined policy of "softening up" prisoners for subsequent anti-terrorist questioning by the CIA. Perhaps the very vagueness of such a policy was itself a moral error, even if direct orders did not explicitly authorize prisoner abuse.[16]

I suppose we might condone efforts at psychological pressure—mocking, aggressive cross-examination, ridiculing, criticizing, etc.— when the goal is getting information which might save innocent lives. Questioning is, after all, permitted under the Geneva Conventions. But the infliction of physical harm cannot be tolerated, even if it supposedly serves the questioning process and is glamourized by such TV shows as "24." Why? Because it is impossible to square the infliction of physical *harm* with the concept of *benevolent* quarantine. Benevolent quarantine may not mean actually being nice to your prisoners but it certainly cannot, logically, include things the Geneva Conventions define as torture: prolonged sleep deprivation; starvation; slapping, punching, biting or strangling; the breaking or severing of limbs or digits; urging or allowing an animal to attack; any kind of drowning-based, or electrocution-based, session; sexual assault or rape; poisoning or medical experimentation; shooting; and so on. These things are simply prohibited. Not even war—not even a *just* war—can justify the deliberate infliction of such things upon other human beings, even if such people are suspected, or even guilty, of terrible things themselves. *In domestic society, we do not permit prison guards to torture anyone—even those convicted of the very worst crimes. So why should we allow it in international society?* The thing to do with terrorist suspects is to question them within the rules, and then prosecute them for war crimes and, upon conviction, send them to jail—not to strip them, beat them and sick dogs after them. Dare I say, such activities are not only unjust but against the very self-perception of being "civilized" on the part of the forces and countries engaged in these activities. To speak more frankly, that's just not the American way. The forces of civilization must remain civilized even as they confront ruthless barbarism.[17]

As important as it is to prevent future terrorist attacks and to protect innocent lives, torture is universally agreed to be one of the very worst things a person can do. Torture is everywhere banned in the laws of war and human rights law.[18] The methods cause revulsion and disgust—as we saw in the worldwide reaction to the Abu Ghraib photos. Torture also hardens the heart and corrupts the character of the torturer. It inflicts absolutely devastating pain upon the tortured—who must endure not only the raw pain of the act(s) but also the experience and knowledge of being utterly enslaved to the whims of another. Torture inflicts severe pain, combines slavery with violence, disregards human autonomy and even renders the torturer worse off. It is, as a method, intrinsically corrupt and even evil. Moreover, torture is extremely questionable as an information-gathering device because evidence shows that people will say *anything*—even deny things they know are true, and assert things they know are false—just to get the torture to stop. Torture is not the answer to terrorism; it is a surrender to a world of brutality and barbarism—a world where the terrorist already resides. I'm not saying torture and terrorism are morally equivalent—presumably torture is more discriminating and localized than terrorism. What I am saying is that, as methods, they are both wrong—*always*—and ultimately indefensible regardless of the cause they supposedly serve.[19]

Civilian Immunity

The second application of the harm principle deals with civilians. Even though some civilians may inwardly approve, or even have voted in favour of, an unjust war effort, they nevertheless remain externally non-threatening. They do not bear arms effectively, nor have they been trained to kill, nor have they been deployed against the lives and vital interests of the opposing side. Civilians, whatever their *internal* attitude, are not in any *external* sense dangerous people. So they may not be made the direct and intentional objects of military attack.[20] Although I endorse this conclusion, it remains controversial with others.

Some people—including Osama bin Laden[21]—argue that the fact that civilians' taxes fund the military renders null and void any pretense of their being "innocent." Civilians are *causally involved* in financing the harm soldiers do. Others view nationality as shared destiny, or suggest that modern warfare is totalizing anyway and so wonder what the point of discrimination really is in our age. If you're at all in the enemy country, you're fair game.[22] These are not trivial arguments but they fail to persuade. It is hard to see, for example, how infants and young children could be anything other than innocent during wartime. Only the most dogmatic believer in collective responsibility could deny this, and then at the cost of his credibility. Just because a baby—through no fault of her own—happens to live in the enemy country, she's fair game? Even though she's not even old enough to be morally responsible for anything? It makes no sense, and amounts to stupidity for the sake of simplicity.

And the relationship between paying one's taxes and the execution of military acts remains, on an individual level anyway: 1) *coerced*, since taxes are everywhere mandatory; and 2) *extremely indirect*, with manifold agents, transfers and responsibilities intervening in-between. The degree to which the state is centrally involved in this coercion and transfer process—of taxes-to-warfare—renders *it* clearly the vastly larger, indeed the only, concrete permissible target. The state—its agents and appendages—is the main organizer and focal point of warfare. It always has been and always will be. It is thus *the* legitimate target and no attempt to "fuzzify" the difference between a government and its people can overcome this. Belligerents target civilians in warfare *not* because they think the people deserve it but, rather, because they think this will give them leverage against the state, which remains the armed, potent threat—the true adversary and target. It is sort of like kidnapping a rich man's family to force *him* to pay you blackmail: give in, or your favourite innocents get it! Just war theory and international law command, by contrast, that you may only attack the true adversary in warfare (i.e., that entity directly engaged in physical harm).

There is, moreover, little evidence that modern warfare is intrinsically totalizing: the 1991 Persian Gulf War, for instance, or the 2001–03 take-downs of the regimes in Afghanistan and Iraq, did not escalate into indiscriminate slaughters. Indeed, what seems more the recipe for ever-increasing slaughter is the idea that civilians may be targeted as readily as soldiers. No doubt there are some questions about the exact specification of "innocents" in wartime but I follow just war theory and international law in believing that it remains a crucial just war category, needed *not only to restrain violence* but also to express our strong, almost foundational, moral commitment to *punish only those who deserve it*. In the midst of "the fog of war," one of the most concrete and verifiable ways to cash out this concept is to define it in terms of external engagement in serious harm. In the midst of war, we cannot peer into the hearts of every civilian; we can only look sweepingly at external behaviour. When we do so, we notice a large and obvious difference between those engaged in harm (i.e., those in the mechanism of war) and those not. Since harm is an important concept for justice and morality, it makes sense to say that it establishes a dividing line between those who may, and those others who may not, be targeted with force.[23]

Difficulties arise, of course, when we consider those people who seem, simultaneously, to be *both* civilians *and* engaged in harming, such as civilian suppliers of military hardware. What is the status of such people? Walzer suggests that "the relevant distinction is . . . between those who make what the soldiers need to fight and those who make what they need to live like all the rest of us." So targeting farms, schools and hospitals is illegitimate, whereas targeting munitions factories is legitimate. Walzer stresses, however, that civilians engaged in the military supply effort are legitimate targets *only when they are engaged in that effort,* so to target them while at home in residential areas would be illegitimate: "Rights to life are forfeit only when particular men and women are actually engaged in war-making or national defence." Consider Thomas Nagel's further explanation that "hostile treatment of any person must be justified in terms of something *about that person* [his italics] which makes the treatment appropriate." We distinguish combatants from non-combatants "on the basis of their immediate threat or harmfulness." And our response to such threats and harms must be governed by relations of directness and relevance. It is only military—and military-related—targets which pose a *direct* and *relevant* threat of serious harm; thus, it is only they which may be resisted with lethal force.[24]

Walzer's overall judgment on targeting—which is very helpful, and essentially condenses international law—is this: soldiers may target other soldiers, their equipment, their barracks and training areas, their supply and communications lines and the industrial sites which produce their supply. Presumably, core political and bureaucratic institutions are also legitimate objects of attack, in particular things like the Defence Ministry. Illegitimate targets include residential areas, schools, hospitals, farms, churches, cultural institutions and non-military industrial sites. *In general, anyone or anything not demonstrably engaged in military supply or military activity is immune from direct attack.* Walzer is especially critical of targeting basic infrastructure, particularly food, water, medical and power supplies. He criticizes American conduct during the 1991 Persian Gulf War on this basis, since very heavy damage was inflicted on Iraq's water treatment system, and presumably would also frown upon NATO's targeting the Serbian electric power grid during its 1999 armed intervention on behalf of the ethnic Albanian Kosovars. The aim there had very little to do with bona fide military tactics; it was a raw demonstration of sheer power designed to "shock and awe" the enemy into submission with an overwhelming display of capability. And that it did: the Serbs could not turn on a toaster, watch TV, heat their homes, run their businesses, etc., unless NATO said so. You can see how this would give the Serbs rather strong incentive to please NATO.[25] While it is true that soldiers cannot fight well without food, water, medicine and electricity, those are things they—and everyone else in their society, including innocents—require *as human beings* and not more narrowly as externally threatening instruments of war. Thus, the moral need for a direct and relevant response only to the source of serious harm renders these things ethically immune from attack.

The Doctrine of Double Effect

Another serious perplexity about wartime targeting concerns the close real-world proximity of illegitimate civilian targets to legitimate military and political ones: munitions factories, after all, are often side-by-side with non-military factories, and at times just around the corner from schools and residential areas. This raises the complex issue of the Doctrine of Double Effect (DDE).[26] The core moral problem is this: even if soldiers intentionally aim *only* at legitimate targets, they can often *foresee* that taking out some of these targets will still involve *collateral* civilian casualties. And if civilians do nothing to lose their human rights, doesn't it follow that such acts will be unjust, since civilians will predictably suffer some harm or even death?

The DDE stipulates that an agent A may perform an action X, even though A foresees that X will result in *both* good (G) and bad (B) effects, *provided all* of the following criteria are met: 1) X is an otherwise morally permissible action; 2) A only intends G and not B; 3) B is not a means to G; and 4) the goodness of G is worth, or is proportionately greater than, the badness of B. The DDE, at first, can seem overly technical, and thus for some people "fishy." But it is an idea rendered complex by the complexity of the situation it deals with. Since it is a common wartime situation, however, we cannot avoid it. Let's break it down. It is a doctrine of "double effect" since it deals with actions which are going to have two effects, one good and one bad. It seeks to respond to the question: "Can one ever perform such an action?" It answers "yes," provided the criteria just listed are *all* satisfied; if even one criterion fails to be satisfied, the doctrine forbids the performance of the action. Assume now, using these criteria, that A is an army and X is an otherwise permissible act of war, like taking aim with a weapon at a military target. The good effect G would be destroying the target, the bad effect B any collateral civilian casualties. The DDE stipulates that A may still do X, provided that A only intends to destroy the military target and not to kill civilians; that A is not using the civilian casualties as means to the end of destroying the military target; and that the importance of hitting the target is worth the collateral dead.

The first objection commonly raised against the DDE concerns its controversial distinction between *intending* Z's death (or harm) and *merely foreseeing* that one's actions will result in Z's death (or harm). Some have argued that the DDE is so elastic as to justify anything: all an agent has to do, to employ its protective moral cloak, is to assert: "Well, I did not intend *that;* my aim, rather, was this. . ." It is clear, however, that agents are not free to claim whatever good intention they want in order to justify their actions, however heinous. Intentions must meet minimal criteria of *logical coherence* and, moreover, must be *connected to patterns of action* which are publicly accessible. The criminal justice system of most countries is based on these ideas: for such serious crimes as murder, the case must be made by the prosecution that the accused had *mens rea,* or the intent to kill. This is done by offering publicly-accessible evidence which is tied to the accused's actions, behaviour and assertions leading up to the time of the murder. Also needed is a consideration of whether the accused had both incentive and motive to commit the crime. Juries, as reasonable and experienced persons, are then invited to infer the accused's state of mind. The plausibility of this procedure undermines the popular academic claim that the DDE can be used to justify *any* heinous action, whether in war or peace. Walzer agrees, suggesting that *we know the intentions of agents through their actions:* "(T)he surest sign of good intentions in war is restraint in its conduct." In other words, when armies fight in strict adherence to *jus in bello*—taking aim only at legitimate targets, using only proportionate force, not employing intrinsically heinous means—they cannot meaningfully be said to intend the deaths of civilians killed collaterally. Their actions, focusing on military targets and taking due care that civilians not be killed, reveals their intentions.[27]

What exactly constitutes "due care" by armies that civilians not be killed during the prosecution of otherwise legitimate military campaigns? For Walzer, it involves soldiers accepting more risks *to themselves* to ensure that they hit only the proper targets: "We draw a circle of rights around civilians, and soldiers are supposed to accept (some) risks in order to save civilian lives." Walzer suggests we locate the limits of additional risk-taking—that soldiers can and should shoulder on behalf of those civilians they endanger—at that point where "any further risk-taking would almost certainly doom the military venture or make it so costly that it could not be repeated."[28] This principle might entail, for instance, that soldiers use only certain kinds of weapons and avoid others. America has recently pioneered the use of so-called "smart bombs" that use satellite technology to improve drastically the "hit rate" on desired targets. America also employs laser- and satellite-guided cruise missiles, which can be exceptionally precise (and are exceptionally expensive). Compare these advances with the older method of simply flying over a target, dumping one's bomb load, and beating a hasty retreat before getting shot down.

The due care principle might also mean moving in more closely on the military target to increase the likelihood of hitting it, and avoiding collateral civilian damage. America encountered criticism, in this regard, during the armed intervention in Serbia over Kosovo in 1999. Some, such as Michael Ignatieff, contended that the extensive bombing was not really that "smart," and that too much of it came from too high a distance. In other words, the participating American pilots were *willing to kill but not willing to risk dying*—and in opposition to the due care principle they actually went out of their way to increase the distance between themselves and their targets. This is permissible activity, in my view, only if one's bombing payload is not conventional explosives—with their disturbingly high rate of collateral damage—but, rather, reliable smart technology. If one's payload is otherwise, Ignatieff's criticism is both sound and scathing: it is a violation of the warrior ethos itself.[29]

Although this might sound silly at first, the due care principle might also, under certain conditions, require some kind of advance warning to nearby civilians. Consider, for example, that the Israeli military occasionally does this. Say it decides to destroy a building which it believes serves as a haven for pro-Palestinian terrorists. If this building is in a residential area, the Israeli military has sometimes been known to announce publicly its intent to destroy this target, declaring that civilians left in the area at the declared time will not be its responsibility. This warning might, admittedly, allow terrorists to escape with valuable equipment, but the Israelis still believe it important to destroy military-use targets while showing clear respect for civilians.[30]

What the due care principle implies, above all else, is this: *offensive tactics and manœuvres must be carefully planned, in advance, with a keen eye towards minimizing civilian casualties.* This demands competent and well-trained officers and soldiers, as well as the need to gather and analyze quality intelligence on the precise nature of suspected targets. While Kant thought spying morally decrepit since it involves deception, I argue that: a) with advances in technology, intelligence gathering need not involve deception; and b) in any event, intelligence gathering can serve the vital ethical need of being more precise in targeting, knowing what to hit and what is out of bounds.[31]

So Walzer, in the end, maintains that civilians are *not* entitled to some implausible kind of fail-safe, or absolute, immunity from attack; rather, they are owed neither more nor less than this "due care" from belligerent armies. Providing due care is, in fact, equivalent to "recognizing their rights as best we can within the context of war." These exact conclusions are also contained in the laws of armed conflict.[32]

We have seen explanations of the first two, of the four, criteria of the DDE. What about the third, regarding the prohibition on the use of the bad effects (like civilian casualties) to produce the good ones (like eliminating the military target)? An example of such a violation would be deliberately bombing the residential area around a munitions factory on grounds you're probably killing the workers in that factory, hence using civilian targeting to effectively eliminate a military target. It seems possible to discern whether a belligerent, such as country C, is employing civilian casualties as a means both to its immediate end of hitting the legitimate target and to its final end of victory over rival country D. If there are systemic patterns—as opposed to unavoidable, isolated cases—of civilian bombardment by C on the civilians of D, it is compelling to conclude that C is directly targeting the civilian population of D. Conversely, if the systemic pattern of C's war-fighting indicates its targeting of D's military capabilities, with only incidental and occasional civilian casualties resulting, then it is reasonable to infer that C is not trying to use civilian casualties as a pressure tactic to force D to retreat and admit defeat.

The truly difficult aspect of the DDE, in my view, is the final criterion: contending that the goodness of hitting the legitimate military target is "worth," or proportional to, the badness of the collateral civilian casualties. A pacifist, for example, will always deny this. Is the need to hit a source of military harm *sufficient* to justify killing people whom just war theory admits have done nothing to deserve death? Does the source of harm have to pass some threshold of threat before one can speak of the need for its destruction outweighing civilian claims? If so, how do we locate that threshold? More sharply, can one refer to the ultimate "worth" of hitting the military target to justify collateral civilian casualties *without* referring to the substantive justice of one's involvement in the war to begin with? Personally, I fail to grasp how it can be morally justified to foreseeably kill innocent civilians in order to hit a target which only serves the final end of an aggressive war. The *only* justification sufficient, in my mind, for the collateral civilian casualties would be that the target is materially connected to victory in an otherwise *just* war. This suggests, importantly, that aggressors not only violate *jus ad bellum*, but *in so doing* face grave difficulties meeting the requirements of *jus in bello* as well. To be as clear as possible: to satisfy the *jus in bello* requirement of discrimination, a country when fighting must satisfy all elements of the DDE. But it seems that only a country fighting a just war can fulfil the proportionality requirement in the DDE. Thus, an aggressor nation fighting an unjust war, *for that very reason*, also violates the rules of right conduct. Here too we see that the traditional insistence on the separateness of *jus ad bellum* and *jus in bello* is not sustainable. Kant was more correct when he remarked on the need for a consistent normative thread that runs through all conduct during the three phases of war: beginning, middle and end.[33]

II. Proportionality

The *jus in bello* version of proportionality mandates that soldiers deploy only proportionate force against legitimate targets. The rule is not about the war as a whole; it is about tactics within the war. Make sure, the rule commands, that the destruction needed to fulfil the goal is proportional to the good of achieving it. The crude version of this rule is: do not squash a squirrel with a tank, or swat a fly with a cannon. *Use force appropriate to the target*. Walzer is as uncertain about this requirement as he was about its *jus ad bellum* cousin, and there is reason to follow him in this regard. He notes that while the rule is rightly designed to prohibit "excessive harm" and "purposeless or wanton violence" during war, "there is [nevertheless] no ready way to establish an independent or stable view of the values" against which we can definitively measure the costs and benefits of a tactic. One case where he talks about, and endorses, a form of proportionality involves the Persian Gulf War. During the War's final days in early 1991, there was a headlong retreat of Iraqi troops from Kuwait along a road, subsequently dubbed "The Highway of Death." So congested did that highway become that, when American forces descended upon it, it was a bloodbath whose aftermath was much photographed and publicized. Although the Iraqi soldiers did not surrender, and thus remained legitimate targets, Walzer suggests that the killing was "too easy." The battle degenerated into a "turkey shoot," and thus the force deployed was disproportionate. Perhaps another example, from the other side of the same war, would be Saddam Hussein's very damaging use of oil spills and oil fires as putative means of defence against a feared amphibious invasion of Kuwait by the Allies.[34] As with the *jus ad bellum* case, we note that it is much easier to diagnose a *disproportionate* use of force than a merely proportionate one. The common sense of the abstract need for balance and moderation is clearly there, but it remains very difficult to define precisely, especially under battlefield conditions. Here, too, it may turn out that proportionality is more of a limiting factor, a negative condition, so to speak—setting outside constraints on force—than it is a positive condition which adds new content to the just war equation.

III. Prohibited Weapons

Walzer insists that the "chief concern" in wartime is the question of *who* may be targeted with lethal force. Thus, separating out civilian from military targets is all important. The question of *what means* may be employed in the targeting is, in his view, "circumstantial." He suggests that the elaborate legal rules—contained in the Hague and Geneva Conventions—defining what means may, and what others may not, be employed during war is beside the point. These rules—such as those prohibiting the use of chemical weapons on the battlefield—may be desirable, he says, but are not morally obligatory. After all, if solders may be killed, how can it matter by what means they are killed? While that is a persuasive way of putting the matter, Walzer should not be flippant about setting these rules aside, or assigning them second-place status in *jus in bello*, behind discrimination and the DDE. For the robust and elaborate set of legal rules banning the use of certain weapons in wartime seems to indicate a high level of international consensus. There is a vast number of relevant conventions and legal treaties on this issue, aside from the canonical Hague and Geneva Conventions, such as those banning the use of chemical (1925 and second protocol 1996), biological (1972) and "excessively injurious weapons" (1980). Also relevant are the conventions against genocide (1948) and against methods of warfare which alter the natural environment (1977). Prohibiting weapons also puts an added restriction upon belligerents and, as such, is consistent with the deepest aim of *jus in bello*, namely, to limit war's destruction. It is simply not enough, in my view, to let weapons development proliferate, as if to say: "Develop and deploy whatever weapon you want, just do not aim it at the suburbs." So I think international law corrects a weakness in Walzer by adding this further rule regarding weapons prohibition.[35]

The legal conventions regarding *jus in bello* are much more detailed, specific and thickly textured than those mentioning *jus ad bellum*. It is interesting to reflect on the disparity. I suggest it has to do with two factors. First, usually those with the war power—like the executive branch—also negotiate international treaties, and they want maximum latitude in connection with reasons for going to war. Second, *jus in bello* seems to have developed first,[36] coming out of ancient and medieval conventions regarding chivalrous ways of battling, rules for knights' fighting tournaments, and so on. Indeed, even the Old Testament talks about not laying waste to fruit trees when attacking enemy cities.[37] So *jus in bello* has had more development than *jus ad bellum*, and so we should expect more detail. Just to sample the flavour of a *jus in bello* rule in international law, consider the following snippet on "booby traps" from the convention banning excessively injurious weapons:

1. Without prejudice to the rules of international law applicable in armed conflict relating to treachery and perfidy, it is prohibited in all circumstances to use:
    (a) Any booby trap in the form of an apparently harmless portable object which is specifically designed and constructed to contain explosive material and to detonate when it is disturbed or approached, or
    (b) Booby-traps which are in any way attached to or associated with:
        (i) Internationally recognized protective emblems, signs or signals;

(ii) Sick, wounded or dead persons;
(iii) Burial or cremation sites or graves;
(iv) Medical facilities, medical equipment, medical supplies or medical transportation;
(v) Children's toys or other portable objects or products specially designed for the feeding, health, hygiene, clothing or education of children;
(vi) Food or drink;
(vii) Kitchen utensils or appliances except in military establishments, military locations or military supply depots;
(viii) Objects clearly of a religious nature;
(ix) Historic monuments, works of art or places of worship which constitute the cultural or spiritual heritage of peoples;
(x) Animals or carcasses.

2. It is prohibited in all circumstances to use any booby trap which is designed to cause superfluous injury or unnecessary suffering.[38]

In addition to these remarkably detailed legal conventions, one might suggest that there is a widely shared moral convention which stipulates that even though soldiers may be targeted with lethal force, some kinds of lethal force—such as burning them to death with flame-throwers, or asphyxiating them with nerve gas—inflict so much suffering, and involve such cruelty, that they are properly condemned. Moreover, the reasoning which distinguishes between legitimate and illegitimate weapons is very similar to the reasoning which generates the combatant/non-combatant distinction. For example, there is a legal ban on using bullets which contain glass shards. These shards are essentially impossible to detect. If the soldier survives the shot, and the bullet is removed by surgery, odds are that some glass shards will still remain in his body. These shards can produce massive internal injuries long after the soldier has ceased being "a dangerous man" to the other side. Parallel reasoning was behind the 1999 passing into law of the International Treaty Banning Land Mines: land mines, too frequently, remain weapons of destruction long after the conflict is over. Finally, restrictions on weapons can play a causal role in reducing destruction and suffering in wartime, something which *jus in bello* as a whole is designed to secure.

Walzer does not even explicitly object to particular weapons on grounds that they are more likely than not to have serious spillover effects on civilians, and thus run afoul of discrimination. Biological weapons would fall under this category, as would many land mines. Such a stance is consistent with the fact that one does not hear from him any criticism of America's extensive use of napalm and Agent Orange in Vietnam, which inflicted long-term damage to Vietnamese agriculture. The aim, at the time, was to defoliate all the jungle vegetation which was providing such effective cover for the Viet Cong. The consequences include abnormally low soil fertility for staple crops like rice and continue to the present. Walzer is curiously unreflective about these considerations, and I cannot agree in this regard: in addition to the general *jus in bello* principles, all belligerents must adhere to the applicable international laws regarding weaponry.[39]

Walzer recovers his reflectiveness about weaponry only when he considers nuclear arms, which for a number of reasons have not been declared illegal by ratified international treaty. The main reason, of course, is that the major powers are all nuclear powers, and they would never agree to such a treaty, as it would eliminate a major military advantage they all share. "Nuclear weapons explode the theory of just war," Walzer famously declares.[40] This is a graphic and gripping, but unfortunate, formulation since it seems to endorse the popular academic view that just war theory is out of date in the post-Hiroshima world. But Walzer cannot believe this, for he believes that the atomic bombing of Japan was unjust. Thus, what his dramatic declaration must really mean is that nuclear weapons can never be employed justly. Why not? First, and most crucially for Walzer, they are radically *indiscriminate* weapons. Perhaps only a handful of the most volatile biological weapons are more uncontrollable in their effects. Second, nuclear weapons are unimaginably destructive, not just in terms of short-term obliteration but also longterm radiation poisoning and climate change, so that their use will always run afoul of *proportionality*. Finally, there is the hint in Walzer that, owing to these two factors combined, deliberate use of nuclear weapons—and emphatically an all-out nuclear war—is an act evil in itself.[41]

Some people, perhaps first starting with Henry Kissinger, believe it might be possible to use nuclear weapons in a discriminate way. Their notion is to use very low-yield nuclear bombs directly on a battlefield against the enemy. I suppose in theory that might be possible, but then: 1) there's still the question of proportionality; and 2) what would be the real difference between such bombs and conventional explosives? Perhaps it would just be the expense and strategy of five low-yield nukes versus 50 very high-yield conventional bombs. But I suspect the real strategy behind having nukes in the first place is the massive intimidation factor, in which case the low-yield nukes are useless. Low-yield nukes might be more discriminating—though it would be comforting to hear nuclear scientists say this rather than foreign policy strategists—but then their deterrent and intimidation factors would be undercut. The most useful aspect of nuclear weapons is precisely the deterrent and intimidation properties brought about by their highly destructive yield. Yet it is that exact yield which is morally objectionable, owing to the incredible destruction and lack of discrimination. In other words, if nukes can be made discriminate they will not be much different from conventional missiles—they might save a few bucks on transportation. The really useful nukes, so to speak and by contrast, are precisely the ones it is never morally permissible to use—wildly powerful and indiscriminately destructive.[42]

What about building, testing and threatening to use such nukes but never actually using them (except twice on Japan)? Generally, moral thinkers agree that it is wrong to *threaten* to do something which it is *actually* wrong to do (e.g., threaten to murder someone), and so there are deep moral questions to be raised even about this. (Our legal systems reflect this: it is a separate crime merely to utter a death threat.) On the other hand, some politicians and historians credit America's nuclear deterrence with having stopped Soviet expansion and with having won the Cold War. These latter are sweeping claims, and probably untrue. Communism still expanded massively in the 1945–85 era after all, so what exactly did American nuclear deterrence stop? Thus, I think that, from a just war point of view, nuclear weapons are going to be dimly viewed no matter how they are interpreted.

Notes

1. I. Kant, *The Metaphysics of Morals*, trans. and ed. M. Gregor (Cambridge: Cambridge University Press, 1995), 117.
2. W. Reisman and C. Antoniou, eds. *The Laws of War* (New York: Vintage, 1994).
3. M. Walzer, *Just and Unjust Wars* (New York: Basic Books, 3rd ed., 2000), hereafter "Wars," 42–3, 135.
4. Walzer, *Wars*, 37, 40, 136; J. Dubik, "Human Rights, Command Responsibility and Walzer's Just War Theory," *Philosophy and Public Affairs* (1982), 354–71.
5. Walzer, *Wars*, 135.
6. Walzer notes the British prosecutor's arguments in his *Wars*, 38. For the trials, see Chapter 6.
7. Walzer, *Wars*, 127, 39.
8. Walzer, *Wars*, 39.
9. Walzer, *Wars*, 128; Reisman and Antoniou, eds. *The Laws of War*, 41–57.
10. Walzer, *Wars*, 40, 138.
11. J. MacMahan, "Preventive War," a chapter in a forthcoming collection on just war theory to be edited by David Rodin and published by Oxford University Press. Thanks to the Press for the advance reading.
12. Walzer, *Wars*, 142.
13. Walzer, *Wars*, 142, 46.
14. Reisman and Antoniou, eds. *The Laws of War*, 35–230.
15. E. Saar, *Inside the Wire* (New York: Penguin, 2005); M. Ratner and E. Ray, *Guantanamo: What The World Should Know* (New York: Chelsea Green, 2004).

16. S. Hersh, *Chain of Command: The Road from 9/11 to Abu Ghraib* (New York: Harper Collins, 2004); M. Danner, *Torture and Truth: America, Abu Ghraib and The War on Terror* (New York: New York Review of Books, 2004). In late 2005, allegations surfaced about the CIA running secret prisons in various locations, off American soil, for the sole purpose of aggressive anti-terrorist detainment and questioning.
17. A. Dershowitz, *Why Terrorism Works* (New Haven, CT: Yale University Press, 2003).
18. Reisman and Antoniou, eds., *Laws*, 153–393; I. Brownlie, ed. *Basic Documents in International Law* (Oxford: Oxford University Press, 4th ed., 1995), 255–388.
19. B. Innes, *The History of Torture* (New York: St. Martin's, 1998); J. Glover, *Humanity* (New Haven, CT: Yale University Press, 2001).
20. Walzer, *Wars*, 146–51.
21. R. Gunaratna, *Inside Al-Qaeda* (New York: Berkley Group, 2003).
22. M. Gelven, *War and Existence* (Philadelphia: Penn State University Press, 1994).
23. R.K. Fullinwinder, "War and Innocence," *Philosophy and Public Affairs* (1976), 90–7.
24. Walzer, *Wars*, 146, 219; T. Nagel, "War and Massacre," *Philosophy and Public Affairs* (1971/72), 123.
25. Walzer, *Wars*, xx.
26. P. Woodward, ed. *The Doctrine of Double Effect* (Notre Dame: University of Notre Dame Press, 2001).
27. Walzer, *Wars*, 106.
28. Walzer, *Wars*, 151, 157.
29. M. Ignatieff, *Virtual War: Kosovo and Beyond* (Toronto: Viking, 2000).
30. Journalists of Reuters, *The Israeli-Palestinian Conflict* (New York: Reuters Books, 2002).
31. T. Erskine, "Moral Agents and Intelligence," *Intelligence and National Security* (Spring 2004), 38–54.
32. Walzer, *Wars*, 152 and 156, in the note; Reisman and Antoniou, eds. *The Laws of War*, 80–4.
33. B. Orend, *War and International Justice: A Kantian Perspective* (Waterloo, ON: Wilfrid Laurier University Press, 2000).
34. Walzer, *Wars*, 129; xxi.
35. Walzer, *Wars*, 42 and 215; Reisman and Antoniou, eds. *The Laws of War*, 35–132.
36. James T. Johnson, *Ideology, Reason and The Limitation of War* (Princeton, NJ: Princeton University Press, 1975).
37. *Deuteronomy* 20:19.
38. Convention II, Article 6, of the Convention on Prohibitions or Restrictions on the Use of Certain Conventional Weapons which may be Deemed to be Excessively Injurious or to have Indiscriminate Effects. This Convention was ratified in 1980. See Reisman and Antoniou, eds. *The Laws of War*, 53.
39. R. Regan, *Just War: Principles and Cases* (Washington, DC: Catholic University of America Press, 1996), 87–99, 136–50.
40. Walzer, *Wars*, 282. While there have been two UN General Assembly resolutions, in 1961 and 1972, banning the use of nuclear weapons (see Reisman and Antoniou, eds. *The Laws of War*, 66–7), and a 2004 one calling for the ultimate dismantling of them all, these do not carry the binding force of a ratified international treaty.
41. Walzer, *Wars*, 263–83.
42. Regan, *Just War*, 100–22; H. Kissinger, *Diplomacy* (New York: Harper Collins, 1995).

Guerilla War
Michael Walzer

Resistance to Miltary Occupation

A Partisan Attack

Surprise is the essential feature of guerrilla war; thus the ambush is the classic guerrilla tactic. It is also, of course, a tactic in conventional war; the concealment and camouflage that it involves, though they were once repugnant to officers and gentlemen, have long been regarded as legitimate forms of combat. But there is one kind of ambush that is not legitimate in conventional war and that places in sharp focus the moral difficulties guerrillas and their enemies regularly encounter. This is the ambush prepared behind political or moral rather than natural cover. An example is provided by Captain Helmut Tausend, of the German Army, in Marcel Ophuls' documentary film *The Sorrow and the Pity*. Tausend tells of a platoon of soldiers on a march through the French countryside during the years of the German occupation. They passed a group of young men, French peasants, or so it seemed, digging potatoes. But these were not in fact peasants; they were members of the Resistance. As the Germans marched by, the "peasants" dropped their shovels, picked up guns hidden in the field, and opened fire. Fourteen of the soldiers were hit. Years later, their captain was still indignant. "You call that 'partisan' resistance? I don't. Partisans for me are men that can be identified, men who wear a special armband or a cap, something with which to recognize them. What happened in that potato field was murder."

The captain's argument about armbands and caps is simply a citation from the international law of war, from the Hague and Geneva conventions, and I shall have more to say about it later on. It is important to stress first that the partisans had here taken on a double disguise. They were disguised as peaceful peasants and also as Frenchmen, that is, citizens of a state that had surrendered, for whom the war was over (just as guerrillas in a revolutionary struggle disguise themselves as unarmed civilians and also as loyal citizens of a state that is not at war at all). It was because of this second disguise that the ambush was so perfect. The Germans thought they were in a rear area, not at the front, and so they were not battle-ready; they were not preceded by a scouting party; they were not suspicious of the young men in the field. The surprise achieved by the partisans was of a kind virtually impossible in actual combat. It derived from what might be called the protective coloration of national surrender, and its effect was obviously to erode the moral and legal understandings upon which surrender rests.

Surrender is an explicit agreement and exchange: the individual soldier promises to stop fighting in exchange for benevolent quarantine for the duration of the war; a government promises that its citizens will stop fighting in exchange for the restoration of ordinary public life. The precise conditions of "benevolent quarantine" and "public life" are specified in the law books; I need not go into them here. The obligations of individuals are also specified: they may try to escape from the prison camp or to flee occupied territory, and if they succeed in their escape or flight, they are free to fight again; they have regained their war rights. But they may not resist their quarantine or occupation. If a prisoner kills a guard in the course of his escape, the act is murder; if the citizens of a defeated country attack the occupation authorities, the act has, or once had, an even grimmer name: it is, or was, "war treason" (or "war rebellion"), a breaking of political faith, punishable, like the ordinary treason of rebels and spies, by death.

But "traitor" does not seem the right name for those French partisans. Indeed, it is precisely their experience, and that of other guerrilla fighters in World War II, that has led to the virtual disappearance of "war treason" from the law books and of the idea of breaking faith from our moral discussions of wartime resistance (and of peacetime rebellion also, when it is directed against alien or colonial rule). We tend to deny, today, that individuals are automatically subsumed by the decisions of their government or the fate of its armies. We have come to understand the moral commitment they may feel to defend their homeland and their political community, even after the war is officially over. A prisoner of war, after all, knows that the fighting will go on despite his own capture; his government is in place, his country is still being defended. But after national surrender the case is different, and if there are still values worth defending, no one can defend them except ordinary men and women, citizens with no political or legal standing. I suppose it is some general sense that there are such values, or often are, that leads us to grant these men and women a kind of moral authority.

But though this grant reflects new and valuable democratic sensibilities, it also raises serious questions. For if citizens of a defeated state still have a right to fight, what is the meaning of surrender? And what obligations can be imposed on conquering armies? There can be no ordinary public life in occupied territory if the occupation authorities are subject to attack at any time and at the hands of any citizen. And ordinary life is a value, too. It is what most of the citizens of a defeated country most ardently hope for. The heroes of the resistance put it in jeopardy, and we must weigh the risks they impose on others in order to understand the risks they must accept themselves. Moreover, if the authorities actually do aim at the restoration of everyday peacefulness, they seem entitled to enjoy the security they provide; and then they must also be entitled to regard armed resistance as a criminal activity. So the story with which I began might end this way (in the film it has no end): the surviving soldiers rally and fight back; some of the partisans are captured, tried as murderers, condemned, and executed. We would not, I think, add those executions to the list of Nazi war crimes. At the same time, we would not join in the condemnation.

So the situation can be summed up: resistance is legitimate, and the punishment of resistance is legitimate. That may seem like a simple standoff and an abdication of ethical judgment. It is actually a precise reflection of the moral realities of military defeat. I want to stress again that our understanding of these realities has nothing to do with our view of the two sides. We can deplore the resistance, without calling the partisans traitors; we can hate the occupation, without calling the execution of the partisans a crime. If we alter the story or add to it, of course, the case is changed. If the occupation authorities do not live up to their obligations under the surrender agreement, they lose their entitlements. And once the guerrilla struggle has reached a certain point of seriousness and intensity, we may decide that the war has effectively been renewed, notice has been given, the front has been reestablished (even if it is not a *line*), and soldiers no longer have a right to be surprised even by a surprise attack. Then guerrillas captured by the authorities must be treated as prisoners of war—provided, that is, they have themselves fought in accordance with the war convention.

But guerrillas don't fight that way. Their struggle is subversive not merely with reference to the occupation or to their own government, but with reference to the war convention itself. Wearing peasant clothes and hiding among the civilian population, they challenge the most fundamental principle of the rules of war. For it is the purpose of those rules to specify for each individual a single identity; he must be either a soldier or a civilian. The British *Manual of Military Law* makes the point with special clarity: "Both these classes have distinct privileges, duties, and disabilities . . . an individual must definitely choose to belong to one class or the other, and shall not be permitted to enjoy the privileges of both; in particular . . . an individual [shall] not be allowed to kill or wound members of the army of the opposed nation and subsequently, if captured or in danger of life, pretend to be a peaceful citizen." That is what guerrillas do, however, or sometimes do. So we can imagine another conclusion to the story of the partisan attack. The partisans successfully disengage, disperse to their homes, and go about their ordinary business. When German troops come to the village that night, they cannot distinguish the guerrilla fighters from any other of the villagers. What do they do then? If, through searches and interrogations—police, not soldier's work—they seize one of the partisans, should they treat him as a captured criminal or a prisoner of war (leaving aside now the problems of surrender and resistance)? And if they seize no one, can they punish the whole village? If the partisans don't maintain the distinction of soldiers and civilians, why should they?

The Rights of Guerrilla Fighters

As this example suggests, the guerrillas don't subvert the war convention by themselves attacking civilians; at least, it is not a necessary feature of their struggle that they do that. Instead, they invite their enemies to do it. By refusing to accept a single identity, they seek to make it impossible for their enemies to accord to combatants and noncombatants their "distinct privileges... and disabilities." The political creed of the guerrillas is essentially a defense of this refusal. The people, they say, are no longer being defended by an army; the only army in the field is the army of the oppressors; the people are defending themselves. Guerrilla war is "people's war," a special form of the *levée en masse*, authorized from below. "The war of liberation," according to a pamphlet of the Vietnamese National Liberation Front, "is fought by the people themselves; the entire people... are the driving force... Not only the peasants in the rural areas, but the workers and laborers in the city, along with intellectuals, students, and businessmen have gone to fight the enemy." And the NLF drove the point home by naming its paramilitary forces *Dan Quan*, literally, civilian soldiers. The guerrilla's self-image is not of a solitary fighter hiding among the people, but of a whole people mobilized for war, himself a loyal member, one among many. If you want to fight against us, the guerrillas say, you are going to have to fight civilians, for you are not at war with an army but with a nation. Therefore, you should not fight at all, and if you do, you are the barbarians, killing women and children.

In fact, the guerrillas mobilize only a small part of the nation—a very small part, when they first begin their attacks. They depend upon the counter-attacks of their enemies to mobilize the rest. Their strategy is framed in terms of the war convention: they seek to place the onus of indiscriminate warfare on the opposing army. The guerrillas themselves have to discriminate, if only to prove that they are really soldiers (and not enemies) of the people. It is also and perhaps more importantly true that it is relatively easy for them to make the relevant discriminations. I don't mean that guerrillas never engage in terrorist campaigns (even against their fellow countrymen) or that they never take hostages or burn villages. They do all those things, though they generally do less of them than the anti-guerrilla forces. For the guerrillas know who their enemies are, and they know where they are. They fight in small groups, with small arms, at close quarters—and the soldiers they fight against wear uniforms. Even when they kill civilians, they are able to make distinctions: they aim at well-known officials, notorious collaborators, and so on. If the "entire people" are not really the "driving force," they are also not the objects of guerrilla attack.

For this reason, guerrilla leaders and publicists are able to stress the moral quality not only of the goals they seek but also of the means they employ. Consider for a moment Mao Tse-tung's famous "Eight Points for Attention." Mao is by no means committed to the notion of noncombatant immunity (as we shall see), but he writes as if, in the China of the warlords and the Kuomintang, only the communists respect the lives and property of the people. The "Eight Points" are meant to mark off the guerrillas first of all from their predecessors, the bandits of traditional China, and then from their present enemies, who ravage the countryside. They suggest how the military virtues can be radically simplified for a democratic age.

1. Speak politely.
2. Pay fairly for what you buy.
3. Return everything you borrow.
4. Pay for anything you damage.
5. Do not hit or swear at people.
6. Do not damage crops.
7. Do not take liberties with women.
8. Do not ill-treat captives.

The last of these is particularly problematic, for in the conditions of guerrilla war it must often involve releasing prisoners, something most guerrillas are no doubt loath to do. Yet it is at least sometimes done, as an account of the Cuban revolution, originally published in the *Marine Corps Gazette*, suggests:

> That same evening, I watched the surrender of hundreds of *Batistianos* from a small-town garrison. They were gathered within a hollow square of rebel Tommy-gunners and harangued by Raul Castro:
>
> "We hope that you will stay with us and fight against the master who so ill-used you. If you decide to refuse this invitation—and I am not going to repeat it—you will be delivered to the custody of the Cuban Red Cross tomorrow. Once you are under Batista's orders again, we hope that you will not take arms against us. But, if you do, remember this:
>
> "We took you this time. We can take you again. And when we do, we will not frighten or torture or kill you . . . If you are captured a second time or even a third . . . we will again return you exactly as we are doing now."

Even when guerrillas behave this way, however, it is not clear that they are themselves entitled to prisoner of war status when captured, or that they have any war rights at all. For if they don't make war on noncombatants, it also appears that they don't make *war* on soldiers: "What happened in that potato field was murder." They attack stealthily, deviously, without warning, and in disguise. They violate the implicit trust upon which the war convention rests: soldiers must feel safe among civilians if civilians are ever to be safe from soldiers. It is not the case, as Mao once suggested, that guerrillas are to civilians as fish to the ocean. The actual relation is rather of fish to other fish, and the guerrillas are as likely to appear among the minnows as among the sharks.

That, at least, is the paradigmatic form of guerrilla war. I should add that it is not the form such war always or necessarily takes. The discipline and mobility required of guerrilla fighters often preclude a domestic retreat. Their main forces commonly operate out of base camps located in remote areas of the country. And, curiously enough, as the guerrilla units grow larger and more stable, their members are likely to put on uniforms. Tito's partisans in Yugoslavia, for example, wore distinctive dress, and this was apparently no disadvantage in the kind of war they fought. All the evidence suggests that quite apart from the rules of war guerrillas, like other soldiers, prefer to wear uniforms; it enhances their sense of membership and solidarity. In any case, soldiers attacked by a guerrilla main force know who their enemies are as soon as the attack begins; ambushed by uniformed men, they would know no sooner. When the guerrillas "melt away" after such an attack, they more often disappear into jungles or mountains than into villages, a retreat that raises no moral problems. Battles of this sort can readily be assimilated to the irregular combat of army units like Wingate's "Chindits" or "Merrill's Marauders" in World War II. But this is not what most people have in mind when they talk about guerrilla war. The paradigm worked out by guerrilla publicists (together with their enemies) focuses precisely on what is morally difficult about guerrilla war—and also, as we shall see, about anti-guerrilla war. In order to deal with these difficulties, I shall simply accept the paradigm and treat guerrillas as they ask to be treated, as fish among the ocean's fish. What then are their war rights?

The legal rules are simple and clear-cut, though not without their own problems. To be eligible for the war rights of soldiers, guerrilla fighters must wear "a fixed distinctive sign visible at a distance" and must "carry their arms openly." It is possible to worry at length about the precise meaning of distinctiveness, fixity, and openness, but I do not think we would learn a great deal by doing so. In fact, these requirements are often suspended, particularly in the interesting case of a popular rising to repel invasion or resist foreign tyranny. When the people rise *en masse*, they are not required to put on uniforms. Nor will they carry arms openly, if they fight, as they usually do, from ambush: hiding themselves, they can hardly be expected to display their weapons. Francis Lieber, in one of the earliest legal studies of guerrilla war, cites the case of the Greek rebellion against Turkey, where the Turkish government killed or enslaved all prisoners: "But I take it," he writes, "that a civilized government would not have allowed the fact that the Greeks . . . carried on mountain guerilla [war] to influence its conduct toward prisoners."

The key moral issue, which the law gets at only imperfectly, does not have to do with distinctive dress or visible weapons, but with the use of civilian clothing as a ruse and a disguise.* The French partisan attack perfectly illustrates this, and it has to be said, I think, that the killing of those German soldiers was more like assassination than war. That is not because of the surprise, simply, but because of the kind and degree of deceit involved: the same sort of deceit that is involved when a public official or party leader is shot down by some political enemy who has taken on the appearance of a friend and supporter or of a harmless passer-by. Now it may be the case—I am more than open to this suggestion—that the German army in France had attacked civilians in ways that justified the assassination of individual soldiers, just as it may be the case that the public official or party leader is a brutal tyrant who deserves to die. But assassins cannot claim the protection of the rules of war; they are engaged in a different activity. Most of the other enterprises for which guerrillas require civilian disguise are also "different." These include all the possible varieties of espionage and sabotage; they can best be understood by comparing them to acts carried out behind enemy lines by the secret agents of conventional armies. It is widely agreed that such agents possess no war rights, even if their cause is just. They know the risks their efforts entail, and I see no reason to describe the risks of guerrillas engaged in similar projects any differently. Guerrilla leaders claim war rights for all their followers, but it makes sense to distinguish, if this is possible, between those guerrillas who use civilian dress as a ruse and those who depend upon camouflage, the cover of darkness, tactical surprise, and so on.

The issues posed by the guerrilla war paradigm, however, are not resolved by this distinction. For guerrillas don't merely fight *as* civilians; they fight *among* civilians, and this in two senses. First, their day-to-day existence is much more closely connected with the day-to-day existence of the people around them than is ever the case with conventional armies. They live with the people they claim to defend, whereas conventional troops are usually billeted with civilians only after the war or the battle is over. And second, they fight where they live; their military positions are not bases, posts, camps, forts, or strongholds, but villages. Hence they are radically dependent on the villagers, even when they don't succeed in mobilizing them for "people's war." Now, every army depends upon the civilian population of its home country for supplies, recruits, and political support. But this dependence is usually indirect, mediated by the bureaucratic apparatus of the state or the exchange system of the economy. So food is passed from the farmer to the marketing co-op, to the food processing plant, to the trucking company, to the army commissary. But in guerrilla war, the dependence is immediate: the farmer hands the food to the guerrilla, and whether it is received as a tax or paid for in accordance with Mao's Second Point for Attention, the relation between the two men is face-to-face. Similarly, an ordinary citizen may vote for a political party that in turn supports the war effort and whose leaders are called in for military briefings. But in guerrilla war, the support a civilian provides is far more direct. He doesn't need to be briefed; he already knows the most important military secret; he knows who the guerrillas are. If he doesn't keep this information to himself, the guerrillas are lost.

Their enemies say that the guerrillas rely on terror to win the support or at least the silence of the villagers. But it seems more likely that when they have significant popular support (which they don't always have), they have it for other reasons. "Violence may explain the cooperation of a few individuals," writes an American student of the Vietnamese war, "but it cannot explain the cooperation of a whole social class [the peasantry]." If the killing of civilians were sufficient to win civilian support, the guerrillas would always be at a disadvantage, for their enemies possess far more fire power than they do. But killing will work against the killer "unless he has already pre-empted a large part of the population and then limits his acts of violence to a sharply defined minority." When the guerrillas succeed, then, in fighting among the people, it is best to assume that they have some serious political support among the people. The people, or some of them, are complicitous in guerrilla war, and the war would

---

* The case is the same with the wearing of civilian clothing as with the wearing of enemy uniforms. In his memoir of the Boer War, Deneys Reitz reports that Boer guerrillas sometimes wore uniforms taken from British soldiers. Lord Kitchener, the British commander, warned that anyone captured in khaki would be shot, and a considerable number of prisoners were later executed. While he insists that "none of us ever wore captured uniforms with the deliberate intention of decoying the enemy, but only out of sheer necessity," Reitz nevertheless justifies Kitchener's order by telling of an incident in which two British soldiers were killed when they hesitated to shoot at guerrillas dressed in khaki. (*Commando*, London, 1932, p. 247)

be impossible without their complicity. That doesn't mean that they seek out opportunities to help. Even when he sympathizes with the goal of the guerrillas, we can assume that the average civilian would rather vote for them than hide them in his house. But guerrilla war makes for enforced intimacies, and the people are drawn into it in a new way even though the services they provide are nothing more than functional equivalents of the services civilians have always provided for soldiers. For the intimacy is itself an additional service, which has no functional equivalent. Whereas soldiers are supposed to protect the civilians who stand behind them, guerrillas are protected by the civilians among whom they stand.

But the fact that they accept this protection, and depend upon it, doesn't seem to me to deprive the guerrillas of their war rights. Indeed, it is more plausible to make exactly the opposite argument: that the war rights the people would have were they to rise *en masse* are passed on to the irregular fighters they support and protect—assuming that the support, at least, is voluntary. For soldiers acquire war rights not as individual warriors but as political instruments, servants of a community that in turn provides services for its soldiers. Guerrillas take on a similar identity whenever they stand in a similar or equivalent relationship, that is, whenever the people are helpful and complicitous in the ways I have described. When the people do not provide this recognition and support, guerrillas acquire no war rights, and their enemies may rightly treat them when captured as "bandits" or criminals. But any significant degree of popular support entitles the guerrillas to the benevolent quarantine customarily offered prisoners of war (unless they are guilty of specific acts of assassination or sabotage, for which soldiers, too, can be punished.)[*]

This argument clearly establishes the rights of the guerrillas; it raises the most serious questions, however, about the rights of the people; and these are the crucial questions of guerrilla war. The intimacies of the struggle expose the people in a new way to the risks of combat. In practice, the nature of this exposure, and its degree, are going to be determined by the government and its allies. So the burdens of decision are shifted by the guerrillas onto their enemies. It is their enemies who must weigh (as we must) the moral significance of the popular support the guerrillas both enjoy and exploit. One can hardly fight against men and women who themselves fight among civilians without endangering civilian lives. Have these civilians forfeited their immunity? Or do they, despite their wartime complicity, still have rights vis-a-vis the anti-guerrilla forces?

The Rights of Civilian Supporters

If civilians had no rights at all, or were thought to have none, it would be a small benefit to hide among them. In a sense, then, the advantages the guerrillas seek depend upon the scruples of their enemies—though there are other advantages to be had if their enemies are unscrupulous: that is why anti-guerrilla warfare is so difficult. I shall want to argue that these scruples in fact have a moral basis, but it is worth suggesting first that they also have a strategic basis. It is always in the interest of the anti-guerrilla forces to insist upon the soldier/civilian distinction, even when the guerrillas act (as they always will if they can) so as to blur the line. All the handbooks on "counter-insurgency" make the same argument: what is necessary is to isolate the guerrillas from the civilian population, to cut them off from their protection and at the same time to shield civilians from the fighting. The last point is more important in guerrilla than in conventional war, for in conventional war one assumes the hostility of "enemy civilians," while in a guerrilla struggle one must seek their sympathy and support. Guerrilla war is a political, even an ideological conflict. "Our kingdoms lay in each man's mind," wrote T. E. Lawrence of the Arab guerrillas he led in World War I. "A province would be won when we had taught the civilians in it to die for our ideal of freedom." And it can be won back only if those same civilians are taught to live for some counter-ideal (or in the case of a military occupation, to acquiesce

---

[*]The argument I am making here parallels that made by lawyers with reference to "belligerent recognition." At what point, they have asked, should a group of rebels (or secessionists) be recognized as a belligerent power and granted those war rights which customarily belong only to established governments? The answer has usually been that the recognition follows upon the establishment of a secure territorial base by the rebels. For then they actually function like a government, taking on responsibility for the people who live on the land they control. But this assumes a conventional or near-conventional war. In the case of a guerrilla struggle, we may have to describe the appropriate relation between the rebels and the people differently: it is not when the guerrillas look after the people that they acquire war rights, but when the people "look after" the guerrillas.

in the re-establishment of order and ordinary life). That is what is meant when it is said that the battle is for the "hearts and minds" of the people. And one cannot triumph in such a battle by treating the people as so many enemies to be attacked and killed along with the guerrillas who live among them.

But what if the guerrillas cannot be isolated from the people? What if the *levée en masse* is a reality and not merely a piece of propaganda? Characteristically, the military handbooks neither pose nor answer such questions. There is, however, a moral argument to be made if this point is reached: the anti-guerrilla war can then no longer be fought—and not just because, from a strategic point of view, it can no longer be won. It cannot be fought because it is no longer an anti-guerrilla but an anti-social war, a war against an entire people, in which no distinctions would be possible in the actual fighting. But this is the limiting case of guerrilla war. In fact, the rights of the people come into play earlier on, and I must try now to give them some plausible definition.

Consider again the case of the partisan attack in occupied France. If, after the ambush, the partisans hide in a nearby peasant village, what are the rights of the peasants among whom they hide? German soldiers arrive that night, let's say, seeking the men and women directly involved or implicated in the ambush and looking also for some way of preventing future attacks. The civilians they encounter are hostile, but that doesn't make them *enemies* in the sense of the war convention, for they don't actually resist the efforts of the soldiers. They behave exactly as citizens sometimes do in the face of police interrogations: they are passive, blank, evasive. We must imagine a domestic state of emergency and ask how the police might legitimately respond to such hostility. Soldiers can do no more when what they are doing is police work; for the status of the hostile civilians is no different. Interrogations, searches, seizures of property, curfews—all these seem to be commonly accepted (I will not try to explain why); but not the torture of suspects or the taking of hostages or the internment of men and women who are or might be innocent. Civilians still have rights in such circumstances. If their liberty can be temporarily abridged in a variety of ways, it is not entirely forfeit; nor are their lives at risk. The argument would be much harder, however, had the troops been ambushed as they marched through the village itself, shot at from the cover of peasant homes and barns. To understand what happens then, we must look at another historical example.

The American "Rules of Engagement" in Vietnam

Here is a typical incident of the Vietnam war. "An American unit moving along Route 18 [in Long An province] received small arms fire from a village, and in reply the tactical commander called for artillery and air strikes on the village itself, resulting in heavy civilian casualties and extensive physical destruction." Something like this must have happened hundreds, even thousands of times. The bombing and strafing of peasant villages was a common tactic of the American forces. It is a matter of special interest to us that it was permitted by the U.S. Army's "rules of engagement," worked out, so it was said, to isolate the guerrillas and minimize civilian casualties.

The attack on the village near Route 18 looks as if it was intended to minimize only army casualties. It looks like another instance of a practice I have already examined: the indiscriminate use of modern fire power to save soldiers from trouble and risk. But in this case, the trouble and risk are of a sort very different from anything encountered on the front line of a conventional war. It is most unlikely that an army patrol moving into the village would have been able to locate and destroy an enemy position. The soldiers would have found . . . a village, its population sullen and silent, the guerrilla fighters hiding, the guerrilla "fortifications" indistinguishable from the homes and shelters of the villagers. They might have drawn hostile fire; more likely, they would have lost men to mines and booby traps, the exact location of which everyone in the village knew and no one would reveal. Under such circumstances, it was not difficult for soldiers to convince themselves that the village was a military stronghold and a legitimate target. And if it was known to be a stronghold, surely it could be attacked, like any other enemy position, even before hostile fire was encountered. In fact, this became American policy quite early in the war: villages from which hostile fire might reasonably be expected were shelled and bombed before soldiers moved in and even if no movement was planned. But then how does one minimize civilian casualties, let alone win over the civilian population? It was to answer this question that the rules of engagement were developed.

The crucial point of the rules, as they are described by the journalist Jonathan Schell, was that civilians were to be given warning in advance of the destruction of their villages, so that they could break with the guerrillas, expel them, or leave themselves. The goal was to force the separation of combatants and noncombatants, and the means was terror. Enormous risk was attached to complicity in guerrilla war, but this was a risk that could only be imposed on whole villages; no further differentiation was possible. It is not the case that civilians were held hostage for the activities of the guerrillas. Rather, they were held responsible for their own activity, even when this activity was not overtly military. The fact that the activity sometimes was overtly military, that ten-year-old children threw hand grenades at American soldiers (the incidence of such attacks was probably exaggerated by the soldiers, in part to justify their own conduct toward civilians) blurs the nature of this responsibility. But it has to be stressed that a village was regarded as hostile not because its women and children were prepared to fight, but because they were not prepared to deny material support to the guerrillas or to reveal their whereabouts or the location of their mines and booby traps.

These were the rules of engagement: (1) A village could be bombed or shelled without warning if American troops had received fire from within it. The villagers were presumed able to prevent the use of their village as a fire base, and whether or not they actually were able, they certainly knew in advance whether it would be so used. In any case, the shooting itself was a warning, since return fire was to be expected—though it is unlikely that the villagers expected the response to be as disproportionate as it usually was, until the pattern had become familiar. (2) Any village known to be hostile could be bombed or shelled if its inhabitants were warned in advance, either by the dropping of leaflets or by helicopter loudspeaker. These warnings were of two sorts: sometimes they were specific in character, delivered immediately before an attack, so that the villagers only just had time to leave (and then the guerrillas could leave with them), or they were general, describing the attack that might come if the villagers did not expel the guerrillas.

> The U.S. Marines will not hesitate to destroy immediately any village or hamlet harboring the Vietcong . . . The choice is yours. If you refuse to let the Vietcong use your villages and hamlets as their battlefield, your homes and your lives will be saved.

And if not, not. Despite the emphasis on choice, this is not quite a liberal pronouncement, for the choice in question is very much a collective one. Exodus, of course, remained an individual option: people could move out of villages where the Vietcong had established itself, taking refuge with relatives in other villages, or in the cities, or in government-run camps. Most often, however, they did this only after the bombing had begun, either because they did not understand the warnings, or did not believe them, or simply hoped desperately that their own homes would somehow be spared. Hence it was sometimes thought humane to dispense with choice altogther and forcibly to deport villagers from areas that were considered under enemy control. Then the third rule of engagement went into effect. (3) Once the civilian population had been moved out, the village and surrounding country might be declared a "free fire zone" that could be bombed and shelled at will. It was assumed that anyone still living in the area was a guerrilla or a "hardened" guerrilla supporter. Deportation had stripped away civilian cover as defoliation stripped away natural cover, and left the enemy exposed.

In considering these rules, the first thing to note is that they were radically ineffective. "My investigation disclosed," writes Schell, "that the procedures for applying these restraints were modified or twisted or ignored to such an extent that in practice the restraints evaporated entirely . . ." Often, in fact, no warning was given, or the leaflets were of little help to villagers who could not read, or the forcible evacuation left large numbers of civilians behind, or no adequate provision was made for the deported families and they drifted back to their homes and farms. None of this, of course, would reflect on the value of the rules themselves, unless the ineffectiveness were somehow intrinsic to them or to the situation in which they were applied. This was clearly the case in Vietnam. For where the guerrillas have significant popular support and have established a political apparatus in the villages, it is unrealistic to think that the villagers will or can expel them. This has nothing to do with the virtues of guerrilla rule: it would have been equally unrealistic to think that German workers, though their homes were bombed and their families killed, would overthrow the Nazis. Hence the only protection the rules provide is in advising or enforcing the departure not of guerrillas from peaceful villages but of civilians from what is likely to become a battlefield.

Now, in a conventional war, removing civilians from a battlefield is clearly a good thing to do; positive international law requires it wherever possible. Similarly in the case of a besieged city: civilians must be allowed to leave; and if they refuse (so I have argued), they can be attacked along with the

defending soldiers. But a battlefield and a city are determinate areas, and a battle and a siege are, usually, of limited duration. Civilians move out; then they move back. Guerrilla war is likely to be very different. The battlefield extends over much of the country and the struggle is, as Mao has written, "protracted." Here the proper analogy is not to the siege of a city but to the blockade or strategic devastation of a much wider area. The policy underlying the American rules of engagement actually envisaged the uprooting and resettlement of a very substantial part of the rural population of Vietnam: millions of men, women, and children. But that is an incredible task, and, leaving aside for the moment the likely criminality of the project, there was never more than a pretense that sufficient resources would be made available to accomplish it. It was inevitable then, and it was known to be inevitable, that civilians would be living in the villages that were shelled and bombed.
What happened is quickly described:

> In August 1967, during Operation Benton, the "pacification" camps became so full that Army units were ordered not to "generate" any more refugees. The Army complied. But search and destroy operations continued. Only now the peasants were not warned before an air-strike was called on their village. They were killed in their villages because there was no room for them in the swamped pacification camps.

I should add that this sort of thing doesn't always happen, even in anti-guerrilla war—though the policy of forced resettlement or "concentration," from its origins in the Cuban Insurgency and the Boer War, has rarely been carried out in a humane manner or with adequate resources. But one can find counter-examples. In Malaya, in the early 1950s, where the guerrillas had the support of only a relatively small part of the rural population, a limited resettlement (to new villages, not concentration camps) seems to have worked. At any rate, it has been said that after the fighting was over, few of the resettled villagers wanted to return to their former homes. That is not a sufficient criterion of moral success, but it is one sign of a permissible program. Since governments are generally thought to be entitled to resettle (relatively small numbers of) their own citizens for the sake of some commonly accepted social purpose, the policy cannot be ruled out altogether in time of guerrilla war. But unless the numbers are restricted, it will be difficult to make the case for common acceptance. And here, as in peacetime, there is some requirement to provide adequate economic support and comparable living space. In Vietnam, that was never possible. The scope of the war was too wide; new villages could not be built; the camps were dismal; and hundreds of thousands of displaced peasants crowded into the cities, forming there a new *lumpen* proletariat, miserable, sick, jobless, or quickly exploited in ill-paid and menial jobs or as servants, prostitutes, and so on.

Even had all this worked, in the limited sense that civilian deaths had been avoided, the rules of engagement and the policy they embodied could hardly be defended. It seems to violate even the principle of proportionality—which is by no means easy to do, as we have seen again and again, since the values against which destruction and suffering are to be measured are so readily inflated. But in this case, the argument is clear, for the defense of resettlement comes down finally to a claim something like that made by an American officer with reference to the town of Ben Tre: we had to destroy the town in order to save it. In order to save Vietnam, we had to destroy the rural culture and the village society of the Vietnamese. Surely the equation does not work and the policy cannot be approved, at least in the context of the Vietnamese struggle itself. (One can always shift, I suppose, to the higher mathematics of international statecraft.)

But the rules of engagement raise a more interesting question. Suppose that civilians, duly warned, not only refuse to expel the guerrillas but also refuse to leave themselves. Can they be attacked and killed, as the rules imply? What are their rights? They can certainly be exposed to risks, for battles are likely to be fought in their villages. And the risks they must live with will be considerably greater than those of conventional combat. The increased risk results from the intimacies I have already described; I would suggest now that it is the only result of those intimacies, at least in the moral realm. It is serious enough. Anti-guerrilla war is a terrible strain on conventional troops, and even if they are both disciplined and careful, as they should be, civilians are certain to die at their hands. A soldier who, once he is engaged, simply fires at every male villager between the ages of fifteen and fifty (say) is probably justified in doing so, as he would not be in an ordinary firefight. The innocent deaths that result from this kind of fighting are the responsibility of the guerrillas and their civilian supporters;

the soldiers are cleared by the doctrine of double effect. It has to be stressed, however, that the supporters themselves, so long as they give only political support, are not legitimate targets, either as a group or as distinguishable individuals. Conceivably, some of them can be charged with complicity (not in guerrilla war generally but) in particular acts of assassination and sabotage. But charges of that sort must be proved before some sort of judicial tribunal. So far as combat goes, these people cannot be shot on sight, when no fire-fight is in progress; nor can their villages be attacked merely because they might be used as firebases or because it is expected that they will be used; nor can they be randomly bombed and shelled, even after warning has been given.

The American rules have only the appearance of recognizing and attending to the combatant/noncombatant distinction. In fact, they set up a new distinction: between loyal and disloyal, or friendly and hostile noncombatants. The same dichotomy can be seen at work in the claims American soldiers made about the villages they attacked: "This place is almost entirely V.C. controlled, or pro-V.C." "We consider just about everyone here to be a hard-core V.C., or at least some kind of supporter." It is not the military activities of the villagers that are being stressed in statements of this sort, but their political allegiance. Even with reference to that, the statements are palpably false, since at least some of the villagers are children who cannot be said to have any allegiance at all. In any case, as I have already argued in the example of the villagers of occupied France, political hostility does not make people enemies in the sense of the war convention. (If it did, there would be no civilian immunity at all, except when wars were fought in neutral countries.) They have done nothing to forfeit their right to life, and that right must be respected as best it can be in the course of attacks against the irregular fighters the villagers both resemble and harbor.

It is important to say something now about the possible shape of those attacks, though I cannot talk about them like a military strategist; I can only report on some of the things that strategists say. Bombing and shelling from a distance have undoubtedly been defended in terms of military necessity. But that is as bad an argument strategically as it is morally. For there are other and more effective ways of fighting. Thus a British expert on counter-insurgency writes that the use of "heavily armed helicopters" against peasant villages "can only be justified if the campaign has deteriorated to the extent where it is virtually indistinguishable from conventional war." I doubt that it can be justified even then, but I want to stress again what this expert has grasped: that counter-insurgency requires a strategy and tactics of discrimination. Guerrillas can be defeated (and, similarly, they can win) only at close quarters. With regard to peasant villages, this suggests two different sorts of campaigns, both of which have been extensively discussed in the literature. In areas of "low intensity operations," the villages must be occupied by small units specially trained for the political and police work necessary to seek out guerrilla supporters and informants. In areas where the guerrillas are effectively in control and the fighting intense, the villages must be encircled and entered in force. Bernard Fall has reported in some detail on a French attack of this sort in Vietnam in the 1950's. What is involved here is an effort to bring numbers, expertise, and technology directly to bear, forcing the guerrillas to give battle in a situation where fire can be relatively precise, or driving them into a surrounding net of soldiers. If the soldiers are properly prepared and equipped, they need not accept unbearable risks in fighting of this sort, and they need not inflict indiscriminate destruction. As Fall points out, a very considerable number of men are required for this strategy: "No sealing off of an enemy force could be successful unless the proportion of attackers to defenders was 15 to 1 or even 20 to 1, for the enemy had in its favor an intimate knowledge of the terrain, the advantages of defensive organization, and the sympathy of the population." But these proportions are frequently achieved in guerrilla, war, and the "surround and storm" strategy would be eminently feasible were it not for a second and more serious difficulty.

Since the villages are not (or should not be) destroyed when they are stormed, and since the villagers are not resettled, it is always possible for the guerrillas to return once the specially assembled task force has moved on. Success requires that the military operation be followed by a political campaign—and this neither the French in Vietnam nor the Americans who followed them were able to mount in any serious fashion. The decision to destroy villages from a distance was a consequence of this failure, which is not at all the same thing as the "deterioration" of guerrilla into conventional war.

At some point in the military progress of the rebellion, or in the decline of the political capacity of the government that opposes it, it may well become impossible to fight the guerrillas at close quarters. There aren't enough men or, more likely, the government, though it can win particular battles, has no staying power. As soon as the fighting is over, the villagers welcome back the insurgent forces. Now the government (and its foreign allies) face what is in effect, or rather what has become, a people's war. This honorific name can be applied, however, only after the guerrilla movement has won very substantial popular support. It is by no means true all the time. One need only study Che Guevara's abortive campaign in the jungles of Bolivia to realize how easy it is to destroy a guerrilla band that has no popular support at all. From there, one might trace a continuum of increasing difficulty: at some point along that continuum, guerrilla fighters acquire war rights, and at some further point, the right of the government to continue the struggle must be called into question.

This last is not a point which soldiers are likely to recognize or acknowledge. For it is an axiom of the war convention (and a qualification on the rules of war) that if attack is morally possible, counterattack cannot be ruled out. It cannot be the case that guerrillas can hug the civilian population and make themselves invulnerable. But if it is always morally possible to fight, it is not always possible to do whatever is required to win. In any struggle, conventional or unconventional, the rules of war may at some point become a hindrance to the victory of one side or another. If they could then be set aside, however, they would have no value at all. It is precisely then that the restraints they impose are most important. We can see this clearly in the Vietnam case. The alternative strategies I have briefly outlined were conceivably a way of winning (as the British won in Malaya) until the guerrillas consolidated their political base in the villages. That victory effectively ended the war. It is not, I suppose, a victory that can be distinguished in any definitive fashion from the political and military struggle that preceded it. But one can say with some assurance that it has occurred whenever ordinary soldiers (who are not moral monsters and would fight by the rules if they could) become convinced that old men and women and children are their enemies. For after that, it is unlikely that the war can be fought except by setting out systematically to kill civilians or to destroy their society and culture.

I am inclined to say more than this. In the theory of war, as we have seen, considerations of *jus ad bellum* and *jus in bello* are logically independent, and the judgments we make in terms of one and the other are not necessarily the same. But here they come together. The war cannot be won, and it should not be won. It cannot be won, because the only available strategy involves a war against civilians; and it should not be won, because the degree of civilian support that rules out alternative strategies also makes the guerrillas the legitimate rulers of the country. The struggle against them is an unjust struggle as well as one that can only be carried on unjustly. Fought by foreigners, it is a war of aggression; if by a local regime alone, it is an act of tyranny. The position of the anti-guerrilla forces has become doubly untenable.

Just War Criteria and the New Face of War: Human Shields, Manufactured Martyrs, and Little Boys with Stones

Michael Skerker *

Abstract: This paper applies *jus in bello* criteria to a relatively novel tactic in asymmetrical warfare: the attempt by a conventionally weaker force to shape the conditions of combat so that the (morally scrupulous) stronger force cannot advance without violating or appearing to violate the rules of war. The weaker side accomplishes this by placing its own civilian population in front of the attacking force: by encouraging or forcing civilians to be human shields; by launching attacks from civilian areas; by provoking reprisal massacres; by creating humanitarian disasters; and by secreting military targets in civilian neighborhoods. This set of tactics is introduced with historical examples taken from recent conflicts in the Balkans and the Middle East. The paper argues that the doctrine of double effect is largely inapplicable to these tactics due to their publicity-seeking nature and that enemy war crimes do not reciprocally release the attacker from his moral obligations. Specific tactical recommendations are generated for situations where the deployment of this tactic can be anticipated; for situations where the attacker is and is not immediately imperiled by its use; and in situations where attempts at discrimination are futile.

The last two decades have seen great armies deploy against technologically inferior opponents. At a time when America in particular has a peerless technological advantage in conventional war making ability, military tacticians and defense procurers in the West are ironically forced to grapple with dilemmas created by this asymmetry in war fighting ability. So long as this asymmetry persists, weaker forces will seek to neutralize their enemies' superior technology through the use of guerilla or terrorist tactics, the decentralization of command structures, the exploitation of civilian populations and refugee migrations, and other unconventional means (Joint Forces Urban Operations Manual, III-27). A tactic adopted by some conventionally overmatched groups is the deliberate manipulation of the moral scruples of the stronger side or of the wider world (now able to witness war-time atrocities nearly in real time) by attempting to force violations or apparent violations in the rules of war.[1] In what follows, I will first seek to draw attention to the use of this tactic in recent conflicts and then analyze its moral ramifications in light of the Western just war tradition.

1.

In basketball, it is a common and accepted practice to "draw the foul," to position oneself so an opposing player cannot avoid initiating illegal contact. The contact results in a penalty, and possession of the ball, or one or two uncontested shots at the basket for the fouled player. The tactic of "drawing the foul" is a good analogue for the unconventional and deliberate tactic described in the examples to

---

*Journal of Military Ethics, 3, no. 1 (March 2004), pp. 27–39.

[1] This fact is recognized in the recent Joint Forces Urban Operations Manual. "Recent operations have shown potential adversaries may try to take advantage of the fact that US military forces will comply with the requirements of the law of armed conflict." III-8.

follow. This tactic involves a relatively weak military force's attempt to neutralize the technological and/or numerical advantage of its enemy by shaping the conditions of combat so that the enemy cannot act without violating or appearing to violate the rules of *jus in bello*. The weaker force's tactics may succeed at protecting vital assets by forestalling enemy aggression outright. If the enemy instead proceeds and causes significant civilian casualties, the accompanying global outrage and disgust of the enemy's own people may preclude comparable future assaults, or require retreats as effectively as would have the weaker force's conventional victory.[2] "Drawing the foul" enables the weaker force to change the dynamic from a game the stronger force could not lose, to one it cannot afford to play, or it can only lose, by winning.

It is counter-intuitive to think that inviting massacres on one's own people could serve as a military tactic. It is not an application of conventional military thinking to protect bunkers and command posts with handcuffed tourists; to snipe on enemy patrols with the intention of provoking reprisal massacres in nearby villages; nor to send little boys armed with slingshots to fight APCs and tanks. From the standpoint of traditional military tactics, these avenues are absurd, and from a just war perspective, they are immoral. One must rather try to isolate civilians (especially one's own) from the battlefield. Yet these counter-intuitive acts have been deliberately undertaken by the conventionally weaker side in Iraq, Kosovo, and the Palestinian territories, and are arguably afoot in all of al-Qaeda's operations.

In the months prior to the start of Operation Desert Storm, Saddam Hussein revealed he was "hosting" foreign "guests"—tourists, businesspeople, and other foreign nationals—at various strategic sites throughout the country (Bulloch & Morris 1991: 111). There were reports of Iraqis "volunteering" for such roles before the most recent conflict in Iraq. Adopting a similar strategy, Serbian forces chained unarmed UN monitors to military targets in Bosnia to forestall allied airstrikes during the siege of Sarajevo (Dunlop 1999: 29). In 1996, amidst suspicions that American military planners were considering a strike against a suspected underground chemical weapons facilities, Libyan leader Muammar Quaddafi threatened to surround the plant with "millions of Muslims."[3]

A variation of the same tendency involves the placement of command posts or other military targets in residential neighborhoods. This tactic has been undertaken by both the Iraqis, Serbs, and Palestinians (Ignatieff 2000: 200). Unable to defend targets in a conventional manner against superior air power, the use of human shields is clearly meant by the defenders to deter attack. Media coverage is a crucial factor, as unseen and unsuspected human shields obviously form no deterrent to high altitude bombing. Tellingly, foreign journalists were invited to film or report on all of the above-mentioned events, the media coverage itself being used as a weapon to cow and shame the more powerful side.

Civilians killed in assaults can be deemed martyrs and exploited to deter further attacks. Iraq quickly let foreign journalists view the devastated remains of the Amiriyek bunker in Baghdad after a US air strike killed over 300 civilians on February 13, 1991. The Pentagon asserted that it had not known that civilians had sought shelter there. The result of the attack was noted by Iraqis and others: after February 13, bombing in downtown Baghdad effectively ceased (Ignatieff 2000: 192). In present day Afghanistan, the Taliban are zealous videographers, posting video of civilian casualties to the internet within hours of airstrikes. The outcry over these casualties has led NATO forces to severely restrict airstrikes in Afghanistan since 2009. This point has to be underlined: a ragtag militia with no air force and no advanced anti-aircraft assets has all but grounded the most advanced air forces in the world through the reporting of civilian casualties. Yet military forces may do more than take advantage of enemies' errors, and actually "manufacture" martyrs for propaganda purposes.

There is suspicion among surviving Serbian journalists that during the Kosovo War, Slobodan Milosevic insisted that the TV station in downtown Belgrade be manned round the clock despite his belief that it would be bombed. His hope was to use civilian deaths to splinter NATO's resolve (Igantieff 2000: 195). Saddam Hussein refused to buy medicines allowed his regime under the U.N.'s

---

[2] It should be noted that the tactic I'm describing fails if the stronger army and/or its civilian population lacks such compunctions. Chechen rebels apparently hid amongst the civilian population of Grozny and conducted ambushes from this cover, but this tactic did not forestall the leveling of Grozny by Russian forces, nor did subsequent accounts of the bombing and continued scorched earth policies by Russian forces create much outrage amongst the Russian people.

[3] "Libyans to Form Shield at Suspected Arms Plant," Baltimore Sun 5/17/96, p.14.

"oil for food" program, presumably in order to fault the allies with killing Iraqi children.[4] It has been reported that his regime stored children's corpses in morgues for months—in contravention of Muslim funereal law—in order to hold anti-sanction rallies with sufficiently impressive numbers of tiny coffins.

Another tactic involves the deliberate provocation of a more powerful military force in the hope that disproportionate vengeance will be wrought on the weaker force's civilian population. Michael Ignatieff (2000: 28) reports of talk that the secret strategy behind the Kosovar Liberation Army's ambushes of Serbian police outposts in 1998 was to provoke Serbian massacres in Kosovo awful enough to compel NATO to intervene. I would argue that during the second Intifada, elements in the Palestinian Authority have similarly used suicide bombings and shootings partly to provoke heavily-armed Israeli raids into the West Bank and Gaza Strip. There have been numerous reports from Operation Iraqi Freedom of Sadaam Fedayeen, dressed as civilians, firing on Coalition soldiers and marines, and then hiding in crowds of civilians when Coalition troops return fire.[5] Since he founded al-Qaeda, Osama bin Laden has argued that Western involvement in the Muslim world justifies a defensive jihad against the West; the invasions of Afghanistan and Iraq, ostensibly launched in response to the attacks of 9/11, have provided compelling fodder for his propaganda operation. Michael Walzer (180) argues this sort of provocation is characteristic of guerilla movements: a small revolutionary force attempts to "place the onus of indiscriminate warfare on the opposing army" in order to mobilize the entire population against the opposing force. In this vein, Mao wrote, "With the common people of the whole country mobilized, we shall create a vast sea of humanity and drown the enemy in it" (Katzenbach 1962: 14). In contemporary times, media coverage enables mobilization on a global scale. Bin Laden makes this purpose plain in his electronic communiqués calling on *all* Muslim men to rise up in defense of their religion.

2.

The *jus in bello* demand for discrimination between combatants and non-combatants limits the stronger force's options in responding to the tactics I described above (I will refer to the stronger force as "the agent," and the weaker force as "the enemy" from hereon). Given a desire to strike at a legitimate military target, or to advance troops or armor, the agent appears now to be faced with two choices: withdraw, or forgo discrimination to achieve those goals. The nature of the situation may even render sincere attempts at discrimination futile so that the results of discriminate action would be similar or identical to an action in which no attempt at discrimination was made. Forgoing discrimination, or being unable to effectively discriminate means knowingly killing or injuring noncombatants, either in front of cameras, or with the certainty that camera crews will shortly be invited to the morgues.

Many just war theorists would invoke the doctrine of double effect in relation to the possibility of harming noncombatants in the course of legitimate military activities. The medieval scholastics concluded that it is morally permissible for an agent to do something he foresees will have a simultaneous and inseparable good and bad effect, provided that the bad effect is an unintended side effect of the good done, and that the good done outweighs the bad done (Schüller 1978: 175).

The doctrine may not be appropriate for many of the situations described above. The doctrine of double effect only applies to situations where an action will foreseeably have a good and bad effect, and the effects are inseparable. The publicity-oriented nature of a tactic meant to "draw the foul" will often make it appear that good and bad effects *are* separable; via enemy-sanctioned media reports, we can *see* that we only need to untie the hostages, or chase demonstrators away with fire hoses or tear gas, etc. The agent's imaging technology—for example, as deployed on spy planes and UAVs—will reveal the same. In many of these situations, it will be clear to warfighters and the civilian public that alternatives to lethal action remain and so double effect does not apply. (By contrast, the situation would seem more ambiguous without advanced imaging technology.) Further, advanced militaries

---

[4] Barbara Crossette, "Children's Death Rates Rising in Iraqi Lands, Unicef Reports," New York Times, 8/13/99, p.6.

[5] Peter Maas, "Good Kills," The New York Times Magazine, Apr. 20, 2003; Oliver Poole, "Time to Stop Being Mr. Nice Guy," Telegraph, Mar. 25, 2003; Alex Perry, "Lamenting a Civilian Casualty," Time, Apr. 14, 2003, p.64; Tony Karon, "Painting Iraq as the West Bank," Time, Apr. 1, 2003; "As ITN reporter Terry Lloyd is Laid to Rest…" Dominic Kennedy, The Times (London), Apr. 26, 2003.

now in large measure have the ability to *act* on their discriminatory impulses due to the precision afforded by laser- and satellite-guided munitions.[6] Therefore, the situations in which double effect might apply—in which good and bad effects are truly inseparable—will be rarer than before the creation of these systems.

Bolstering our reservations about the applicability of the doctrine of double effect is a tactical appraisal of the enemy's action. The enemy is cynically playing a game involving its own civilian population, the agent's conscience, and the opinion of the outside world. The nature of this action is two-fold, meant first to deter, and then failing deterrence, punish, so that even if the agent attacks, it may still ultimately lose the tactical advantage when public outcry eventually forces it to withdraw,[7] stand-down,[8] or a third, more powerful force takes its victory from him.[9] (With regard to deterrence, it is unimportant whether the outside world actually does blame the agent for noncombatant casualties so long as the agent suspects that blame is likely.) This is to reassert that "drawing the foul" is a *tactic*, a deliberate enemy action, potentially as effective as any conventional maneuver, and not (or not only) a complication that occurs because of the agent's advance. So even if the principle of double effect might justify noncombatant deaths in a necessary action, this new tactic might render even morally excusable actions counter-productive.

The agent's predicament involves the following elements. The agent does not want to kill innocents. It also does not want the enemy's tactic to succeed:[10] It does not wish to lose outright, by withdrawing (to avoid *jus in bello* violations), and it does not want to lose the battle after winning it, by being forced by an intervening third power to retreat. The agent may further wish to deter the enemy from using this tactic again (Ramsey [1968] 2002: 428; Johnson 1984: 58). If, for example, it can afford to retreat or re-task an assault in a particular instance to avoid injuring human shields, it does not want this action to encourage the enemy to repeat the tactic in a situation of greater importance.[11] In what follows, I will attempt to apply *jus in bello* criteria to this novel tactic by first addressing the question of whether the agent is under any special obligations since it is the *enemy's* action which directly endangers civilians. Arguing that the agent is still under *jus in bello* obligations, I will then consider how the agent should meet those obligations first, when he is not immediately endangered because of the enemy action, and second, when he *is* so endangered. Finally, I will consider what to do when fighting well—meeting *jus in bello* criteria—appears to be impossible.

3.

Is the agent absolved of responsibility for noncombatants' welfare in situations where, barring the enemy's unconventional tactic, the agent's actions *would* be considered discriminate? We might ask why the agent should be put at risk when its enemy is callous or fanatical. After all, the enemy government may have a long history of brutalizing its people. Why, for example, do invading American troops have to treat Iraqi civilians with more care than Saddam treats them?

---

[6] See (Dunlop 1999) for the moral questions raised by the availability of PGMs. The sphere of the possible will be further increased over the next decade given the Pentagon's new interest in the development of non-lethal and photon-based weapons, the latter of which could disable targets (scuds, for instance) without fires (including secondary detonation) or damage to surrounding structures.

[7] Arguably, the sympathy created by images of heavily armed Israeli troops firing on stone-throwing teenagers created the conditions for the Madrid peace conference and subsequent withdrawals from occupied lands.

[8] For example, the halt to Allied air strikes in downtown Baghdad after the Amiriyek bunker bombing.

[9] For example, NATO's forcing of a Serbian retreat from Kosovo.

[10] IDF commandos for instance, displayed an acute sensitivity to how their spring '02 offensive in the West Bank was being portrayed by and to the Palestinians—how their own action was in effect being manipulated by the Palestinian militants for propaganda purposes. Scott Anderson, "An Impossible Occupation," New York Times Magazine, 5/12/02.

[11] A New York Times reporter documented a pattern during the early stage of the second intifada, where several hours of stone throwing would be ignored by Israeli troops in fortified positions or armored vehicles, until, apparently without any change in the tactical situation, troops would shoot one boy, in violation of IDF rules of engagement. This may have been done purely out of frustration and/or according to an informal policy of making the Palestinians pay a price for their harassment of the outposts. Michael Finkel, "Playing War," New York Times Magazine, 12/24/00.

To meet this question in its most general form first: there is consensus in the just war tradition that actions *malum in se* (evil in themselves) are illicit regardless of circumstance. For example, intentionally killing civilians (apart from any military target) is never allowable. These actions are therefore not allowed in retaliation for the enemy's in-kind just war violations, or anything else he does. Even if foregoing discrimination is not *malum in se*, the point to retrieve here is that the enemy's violation of the rules of war does not give the agent blanket license to do the same. Two wrongs don't make a right.

According to Roger Williamson, a situation in which an enemy threatens to kill or cause to be killed civilians if he is attacked (rather than threatening the agent directly) is comparable to situations in which police confront kidnappers with hostages, or nurses try to calm psychotic patients. In both scenarios, it could be argued that the police and nurses are responsible for the actions of the ruthless or psychotic parties, because it is reasonable to assume that the actions of the police or nurses will lead to the kidnappers killing the hostages or the patients harming themselves or others. It is fairly clear in these examples, Williamson writes, that force ought to be used as a last resort (Williamson 1991: 65).

According to Williamson, the agent is partially responsible for the things his irrational or amoral opponent does, because they are in reaction to the agent's action. If I understand Williamson's point, he is saying that the nature of some situations allows the agent to conceive of another's likely (albeit irrational, callous) reaction with a degree of probability similar to that regarding the agent's own action. Given that a person is responsible for the foreseeable effects of his own actions, he is partially responsible even for secondary effects foreseeably tied to those actions.[12] Applying this mode of reasoning to a military context where a dictator threatens his own people, Williamson concludes that the agent is morally culpable for the actions of its enemy if there is an alternative to force (sanctions, for instance), even if such a route does not "provide such a clear result as military victory"(1991: 66).

A just war theorist can go so far as to argue that responsibility for the relatively powerless devolves to anyone who has the power to offer protection (Johnson 1984: 22). One might object to this argument that the military and domestic hostage situations are different, because the protection the policemen owe the hostages is an extension of their normal duty to protect law-abiding citizens. The police ought properly to make every effort to save the hostages, mitigated only by the greater threat the kidnappers might pose to the wider community (e.g. if they are members of a terrorist group). By contrast, the military agent is normally obligated to avoid, or at least, not directly threaten foreign noncombatants only insofar as they are potentially endangered by his advance. The military agent should not target civilians; should attempt to route attacks around them; and should give civilians the opportunity and means to flee. However, the military agent does not have to take the sort of positive steps the civilians' own civil defense forces should take such as helping to build sandbag berms in front of their homes or building air raid shelters for them prior to the military assault. The military agent's duty to "protect" noncombatants is qualified because it is not consequentially motivated in the same sense as the comparable duty of the police: it is rather a mitigation of the agent's right to attack legitimate military targets in service of defending its own nation in a just war. Further (the argument continues), the mitigated responsibility the agent owes to foreign noncombatants is ultimately secondary to the full-fledged consequential responsibility it owes to its citizenry.

The noncombatants the agent encounters certainly are owed full-time protection by their *own* nation's armed forces. In "drawing the foul" situations, the noncombatants' usual protectors are abdicating their duty, and so the question is what, if any, extra protection the noncombatants are owed by the agent. We need not enter into a larger conversation here about a state agent's responsibilities toward foreigners as opposed to his fellow citizens. The discriminatory imperative of *jus in bello* calls for the agent to pick tactics it thinks will *actually* discriminate between combatants and noncombatants in the anticipated battlespace. It won't do to design rules of engagement for an *imaginary* battlespace—for

---

[12] For the sake of clarity, there is a distinction between foreseeing a consequence to one's action and intending that consequence. Following G.E.M. Anscombe's definition of intention, the agent does not *intend* the hostages' death, even though he foresees it as a possible consequence to his action, because the agent would elect to subdue the kidnappers without the hostages' death if such a course was available.

Rommel's North Africa, for instance, when the battle is in Baghdad. These civilians living amidst the snipers, or chained to the munitions factory are "weaponized" by the enemy—turned into unconventional obstacles before the agent's conventional forces—*but they remain noncombatants*. They have not chosen this path; their rights are no more expendable than if they were far from the action.[13] To be sure, the agent must address a more challenging context with its rules of engagement if the enemy has placed military assets in residential neighborhoods or deployed human shields. Regardless, the agent needs to strive to protect noncombatants from harm to the extent that they are endangered by the agent's advance in the anticipated environment; whether this environment is formed by geographic features or by the enemy's likely defenses and counter-tactics makes no difference.

We might ask if noncombatant immunity is compromised if noncombatants consent to remain in harm's way. Walzer writes, "[w]hen we judge the unintentional killing of civilians, we need to know how those civilians came to be in a battle zone in the first place. This is, perhaps, only another way of asking who put them at risk…" ([1977] 1992:159) If the noncombatants chose to be present, Walzer argues, their consent absolves the agent of guilt if they die as an unintended consequence of hostilities (they still cannot be deliberately targeted) (169). This principle leads him to recommend that civilians be given the option of leaving a besieged city (168).

Should we treat someone's decision to be a human shield or other sort of propaganda-ready martyr differently than a decision to remain in a besieged city to help in the city's defenses, or to show solidarity with the city's defenders? Is the choice for martyrdom prima facie irrational, and as such, on par with the "choices" made by Williamson's psychotic patient? These are questions that cannot be answered here, though I think an ethicist might investigate the criteria for martyrdom in various religious traditions for insights into this question. If we beg for now an evaluation of the content of the choice, perhaps something can be said about the actor or the conditions in which he makes his decision. Of course, children cannot consent to sacrifice their own lives (this rules out treating stone-throwing youths as combatants, particularly if they are harassing troops in fortified positions or armored vehicles). If adults (if anyone) could be said to have the right to make the decision to sacrifice themselves, how then do we assess the decision for an adult to martyr himself as a passive noncombatant for the sake of his state, if the decision is made in an environment of pervasive state-sponsored propaganda calling on him to do the same? Before answering too quickly, we might ask if there is a difference between propaganda that urges this sort of action and that which urges citizens to join the conventional armed forces, or which glorifies those who die in service of their country.

These difficult questions can also be begged given the other interests at play when an advancing force confronts noncombatants who, willingly or not, have been deployed as obstacles. The conscientious agent does not want to kill the noncombatants; his conscience may be soothed—the justification of double effect actually felt—when the collateral damage is only to be gauged in probabilities. For example, there is probably an operationally-significant psychological difference for warfighters in acknowledging that there will likely be noncombatant casualties in an airstrike (even if only the secretarial or custodial staff in a command center), on the one hand, and seeing these certain-to-be casualties through the lens of a gun camera, on the other.[14] Even if these noncombatants have been "weaponized," in a sense, by the enemy's tactical manipulation of them, and even if this status change is compounded by their consent, civilians still look like civilians on the television screen, from the eye of a Predator drone, or through a sniper's scope. Double effect, of course, provides for the certainty

---

[13] Here I am in disagreement with Paul Ramsey, who takes a hard line with respect to civilian rights in guerilla situations. The insurgents have violated the rule of discrimination by not only hiding amongst civilians, but in enlisting them in the struggle. As such, civilians in these contexts may be considered combatants and exposed to force (subject to due proportion). (Ramsey [1968] 2002: 435-436). Daniel Statman argues along similar lines with respect to the first intifada (1997: 146).

[14] Psychological stress has been shown to have an adverse effect on operations. Joint Force Manual for Urban Combat, III-25.

that harm will be done, and military discipline may well compel even the morally troubled warfighter to move forward in his assault. Yet if jus in bello rules are adhered to in part because of the agent's needs—in order for warfighters to retain some solidarity with their peace-time character, to distinguish (in their own minds) their actions from those of sociopaths—the obvious and announced deployment of noncombatants as human shields may erode the practical efficacy of even the most robust version of double effect.

If the moral equation is potentially complicated on the one hand, by the noncombatants' consent to their use as shields and propaganda tools—arguably changing their status to combatants of some sort—and on the other, by warfighters' actual confrontation with the collateral deaths they are about to incur, the tactical interest of the agent supports the voice of the warfighters' consciences. Civilians who embraced their deaths as martyrdom look no different on TV than those unwillingly corralled, nor, for that matter, do those dead whom the agent anguished over killing. The enemy can still utilize their deaths in a tactical manner.

I conclude that even if noncombatants are in harm's way due to the direct actions of the enemy, and/or due to the adult noncombatants' own choices, the agent must alter its tactics from those otherwise recommended, if those tactics will foreseeably result in noncombatant casualties. In a siege situation, provisions ought to be taken to allow noncombatants to leave. Targets shielded by noncombatants may need to be re-prioritized, or if judged to be vital, special operations troops may be needed to separate human shields from targets. Non-lethal weapons may be necessary in cases where the human shields have consented to their role (as in the vigils held on the bridges of Belgrade during the Kosovo War), in cases of civilian rioting, or again, in cases where combatants and noncombatants are mixed together.[15] (Following guerilla attacks by the Sadaam Fedayeen during Operation Iraqi Freedom, Coalition troops had no means other than lethal ones to respond to potentially hostile situations involving crowds and approaching civilian vehicles.)[16] Scenarios of these sorts should be incorporated into troops' training regimens. Anti-guerilla operations may require police-style tactics: systematic, troop-intensive house to house searches (with care taken not to antagonize civilians and compensate them for damage, etc.), the use of informants, the imposition of curfews, etc (Paget 1967: 168). If it can be done in a humane manner, it may be prudent to relocate civilians to a safe-zone, with safeguards in place (adequate funding, for one) to prevent Vietnam-era abuses of this practice.

4.

The previous section dealt with situations in which the agent was not in immediate danger due to the enemy's tactics. I will now consider situations in which the agent is so endangered. Robert Nozick raises the issue of human shields and what he calls "innocent threats" as possible exceptions to noncombatant immunity. He asks whether violence against either could be justified through self-defense. He does not answer the question of shields, but suggests with his characteristic wit that if the agent were trapped at the bottom of a well and his enemy seized an innocent person and hurled him at the agent, the agent would be justified in using his "ray gun" to disintegrate the innocent, but weaponized, person (1974: 35). Consider a more plausible example: a civilian is intentionally infected with Ebola or smallpox and forced to run towards the enemy's encampment, a bio suicide-bomber. To illustrate an "innocent shield of a threat," we can conceive of a human shield protecting, albeit unwillingly, a weapon that is directly threatening the agent (e.g. a civilian strapped to a tank or protecting a sniper nest). The shield is then incorporated into the threat—he in fact makes the weapon more menacing to

---

[15] Statman defends the IDF's widespread use of rubber-coated bullets, tear-gas, and beatings during the first intifada as preferable to the use of lethal force. (1997: 148). For statistics on Palestinian casualties, see (Graff 1997).

[16] See Peter Maas, "Good Kills," The New York Times Magazine, Apr. 20, 2003. There is a moral argument to be made for the development of non-lethal weapons. Moral obligation is limited by what is possible for the agent. So warfighters act in a morally excusable way when, for instance, out of fear of car bombs, they use lethal force to stop a civilian vehicle, following warnings, when lethal force is their only means of so doing. However, if developing non-lethal means of doing the same thing are within the realm of technological possibility, it is immoral to leave warfighters with no recourse but the use of lethal force.

the advancing soldier because of the soldier's presumed reluctance to kill the human shield. These contrivances are meant to illustrate a point by making the parameters as plain as possible: the agent must choose between his own death and using lethal force against someone normally immune to attack. The intuitively appealing argument Nozick makes is that one has the right, in war or peace, to defend himself with force sufficient to stop an attack, regardless of the motive of the attacker.

Yet this argument is not wholly persuasive, because in Nozick's examples, the threat has not chosen his status, and so has not ceded his right to life (by attempting to violate the reciprocal right to life of others). Our dilemma then produces two apparently incommensurable claims, both of which I imagine meet with intuitive approval: *one has a right to protect and preserve one's own life*, and *an innocent person cannot be intentionally killed*.

There are two aspects to be noted for otherwise parallel situations of "innocent threats" and "innocent shields of threats" arising in combat. First, the speed and distance at which an attack or ambush occur might render the moral dilemma moot. A soldier returning fire may never see, nor have the time to register that his attacker is protected by a human shield. Secondly, if a soldier has a responsibility to discriminate between combatant and noncombatant, he also has a (sometimes competing) responsibility to protect and support his comrades. One's fellow soldiers are *vocationally* owed protection, but then again, so are noncombatants: a soldier is vocationally bound to use force only against other combatants and to separate or protect noncombatants from the fighting to the extent that they are endangered by the soldier's advance.

Walzer seeks to balance these competing interests in his emendation of the doctrine of double effect. It is not enough, he writes, for a soldier to not intend to harm noncombatants; he must take active steps to protect noncombatants (or so order his troops), even at risk to himself ([1977] 1992: 157). The acceptable level of risk ought to be set at that point where mission success would not likely be compromised by the tactical modification. With respect to tactical situations that can be anticipated, this directive would practically call for weapons and tactics affording more precision, and in some cases, less lethality than what is afforded by more conventional tactics. Such a directive might call for low rather than high altitude bombing,[17] concrete- rather than high explosive-filled ordnance,[18] precision-guided rather than unguided munitions,[19] ground attack rather than aerial bombardment, special operation troops rather than general ground troops, the use of non-lethal weapons, etc. With respect to unanticipated tactical situations involving human shields, etc., I would suggest that situations where a unit commander has the time, cover, vantage point, and material options to formulate alternate tactics are *prima facie* also those where mission success does not hinge on the killing of noncombatants. Under these circumstances, a concern for fighting well indicates that these alternate tactics should be employed.

It remains that the most discriminating tactics may still leave the agent unable to avoid civilian casualties. If there is no way to fight but to conduct a scorched earth, or general counter-population campaign, for instance, then the criteria for fighting well have exhausted themselves, and it is no longer appropriate to restrict the moral question to *jus in bello* criteria. Under these circumstances, warfighters and warmakers have to ask, in ways appropriate to military discipline and democratic government: is this war worth winning? This is to say that the question of justice once again becomes a political, and robustly moral one, a question of *jus ad bellum*, rather than one linked strictly to the morality of certain tactics. Most likely, if there is no way to fight a war justly, the war itself is unjust. This may be

---

[17] NATO has rightly been criticized for depending exclusively on high altitude bombing in the Kosovo campaign, for fear that low-flying aircraft, while capable of more accurate and more efficacious assaults, would be vulnerable to anti-aircraft fire. See Ignatieff, *Virtual War*.

[18] Allied warplanes were equipped with concrete-filled bombs in 2001 to limit civilian casualties in the Iraqi "No-Fly Zones."

[19] With the added cost of the PGMs factored into the calculus of probable mission or campaign success, to answer a question asked by Dunlop (1999).

because the insurgents truly represent the interests of the people and the agent does not (Walzer [1977] 1992: 196), (or conversely, the state and its military represent the people's interest and the insurgents do not). The war's injustice may also follow for a more complicated reason, which meets the obvious rejoinder to the last point: that a people's support for a guerilla movement or the official government may merely be the result of brainwashing and fear. It is a point true to the Augustinian roots of the Western just war tradition, and partly accounts for Augustine's extreme reserve with regard to military action. When a society is as corrupted as many we could name, where desperate straits and pervasive propaganda have led whole nations to nihilistic despair, a just claim against it may lead to greater injustice when it is prosecuted through force.

REFERENCES

Bulloch, John and Morris, Harvey, 1991. *Saddam's War.* London: Faber & Faber.

Dunlop Jr., Charles J., 1999. 'Technology: Recomplicating Moral Life for the Nation's Defenders,' *Parameters*, 39, 3: 24-53.

Graff, James A., 1997. "Targeting Children," in Tomis Kapitan, ed., *Philosophical Perspectives on the Israeli-Palestinian Conflict.* Armonk, NY: M.E. Sharpe (157-184).

Ignatieff, Michael, 2000. *Virtual War: Kosovo and Beyond.* New York: Metropolitan Books.

Johnson, James T., 1984. *Can Modern War Be Just?* New Haven: Yale University Press.

Katzenbach Jr., E.L., 1962. 'Time, Space, and Will: The Politico-Military Views of Mao Tse-Tung,' in T. N. Greene, ed., *The Guerilla—and How to Fight Him.* New York: Praeger.

Nozick, Robert, 1974. *Anarchy, State, and Utopia.* New York: Basic Books.

Paget, Julian, 1967. *Counter-Insurgency Campaigning.* London: Faber & Faber.

Ramsey, Paul, [1968] 2002. *The Just War.* Lanham, MD: Rowman & Littlefield.

Schüller, Bruno, 1978. 'The Double Effect in Catholic Thought: A Reevaluation,' in Richard McCormick & Paul Ramsey, eds, *Doing Evil to Achieve Good.* Chicago: Loyola Univ. Press (165-192).

Statman, Daniel, 1997. "*Jus In Bello* and the Intifada," in Tomis Kapitan, ed., *Philosophical Perspectives on the Israeli-Palestinian Conflict.* Armonk, NY: M.E. Sharpe (133-156).

Walzer, Michael, [1977] 1992. *Just and Unjust Wars.* New York: Basic Books.

Williamson, Roger, 1991. "Engulfed in War: On the Ambivalence of the Just War Tradition during the Gulf War," in B. Hallet, ed., *Engulfed in War.* Honolulu: Spark Matsunage Institute for Peace (45-78).

Ethical Issues in the Use of Military Force in Irregular Warfare
George R. Lucas, Jr. *

In contrast to the conventional deployment of force in collective security operations, "irregular" military interventions may entail radically different rules of engagement and expectations regarding the lawful conduct of military personnel than anything they have previously trained for or encountered.

In a widely publicized incident on the night of May 20, 1997, a young Hispanic-American goatherd, 18 year-old Esequiel Hernandez, Jr., was killed in an altercation with a four-man Marine unit assigned, at the request of the U.S. Border Patrol, to guard a well-known drug smuggling route across the Rio Grande. Hernandez carried a .22 caliber rifle which he allegedly fired several times into the darkness in the general direction of the camouflaged Marines. Apparently believing the unit under fire from a suspected drug smuggler, 22 year-old Marine Cpl. Clemente Banuelos returned fire, killing the lad.[1] I certainly do not wish either to revisit or to pass judgment on this unfortunate incident which, in any case, is not itself an example of humanitarian intervention *per se*. I stress the age, ethnicity, and circumstances of the incident, however, to emphasize that the young Marine in question *behaved exactly as he was trained to do* under the standard rules of engagement during combat. Unlike civilian-trained domestic peacekeeping and law enforcement personnel, a soldier in combat is not required to identify himself or give warning to an armed enemy, certainly not when under fire. Choosing to deploy these young Marines for a reasonable and important mission for which they had not been trained inadvertently created a situation in which such a tragedy was quite likely to occur.

Similarly, in decrying the allegedly "Widening Gap between the Military and Society," *Washington Post* columnist (and author of *Making the Corps*) Thomas E. Ricks finds evidence of this gap in the example of Marine battalion asked to participate in restoring order during the Los Angeles riots of 1992. Domestic police officers asked some of the Marines to "cover them" while they confronted an armed suspect barricaded in his residence. The Marines (Ricks reports) shot approximately 30 rounds of covering fire into the building before the police stopped them.[2] Such events are indicative of what we

---

*Prepared for the Carnegie Conference on "Ethics and Warfare in the 21st Century" (April, 1998); delivered to the NROTC brigade of midshipmen, University of California (Berkeley), April 1999.

[1] "U.S. Probes Marine's Killing of Texas Teen," *Baltimore Sun*, Saturday, 16 August 1997, p. 3A.

[2] *Atlantic Monthly, July* 1997, p. 77; see his book upon which this article is based, *Making the Corps* (NY: Scribners, 1997), which is replete with examples of these sorts, and discusses the growing problem with using this socially distinctive military organization increasingly for humanitarian political purposes abroad.

might label the "Bertolt Brecht" syndrome,[3] in which specific sorts of conduct—licensed, encouraged, and even rewarded during wartime under combat conditions—become, in sudden or radically altered peacetime situations, criminally negligent.

These incidents merely point to a conflict between quite distinctive organizational cultures: in both examples, the Marines involved were responding exactly as they had been trained since boot camp to respond. The problem was not their actions, but the radically altered context in which they transpired—a context with which they were understandably unfamiliar, and for which they had not been primarily trained. Incredibly thus far the only lesson to be derived from such incidents is that one must exercise caution in deploying American military personnel in domestic operations!

The implications for humanitarian and peacekeeping activities abroad should, however, be just as plain and as potentially problematic. Using military forces alongside, or in place of, domestic civil authorities in any situation likewise confronts military personnel with a radically altered context of operation, in which the familiar and long habituated rules of engagement or laws of armed conflict may no longer apply. Hence, individuals trained as warriors under one set of ROEs will be expected to make a seamless transition to activities much more like those of domestic peacekeeping and law enforcement authorities, who (as the foregoing examples dramatically illustrate) operate under a very different set of constraints governing acceptable conduct under fire. If sophisticated political pundits and journalists, with the benefit of both leisure and hindsight from which to reflect on such issues, persistently misdiagnose the nature of the dilemma this presents (as the foregoing cases also unintentionally illustrate) it seems rather much to expect of a young Lance Corporal, or even a more thoroughly trained junior officer, that these should, in the midst of armed conflict, immediately intuit the relevant differences of context, call to mind the appropriate ROEs, and adhere without significant probability of error to the modes of professional conduct appropriate to wholly unfamiliar situations absent the training, and the supporting institutions of civil society. Police protect citizens of their own country from aberrations in the lawful order of that society, and do so with the consent and support of the laws and civil institutions of that society. By contrast, humanitarian interventions aimed not just at peacekeeping but at peace-*making* between hostile belligerents (as in Bosnia) inject military forces into a largely unordered society lacking wholly in the guidance and support that normal peacekeeping forces rely upon for guidance. In many instances, we who issue these orders are at a loss even to define clearly the problem that we want the military to fix.

This is not meant to excuse gross negligence, unprofessional behavior, or absence of reasonable common sense on the part of military personnel. Rather, I mean to suggest that in carrying out the "Albright Doctrine" we have recently committed, and will likely continue in the future committing, military personnel to the adjudication and resolution of circumstances of conflict that would challenge the common sense of the most seasoned and experienced professional. It would be well to reflect on how we ought to revise enlisted and officer training, and mission briefing procedures, to anticipate such problems.

Thus far, beyond discussing the principles of justifiable intervention (*jus ad interventionem*) that should govern the commitment of military force, our discussion of the practical problems of deployment and ROEs has been scanty, and consisted almost exclusively in case-based or case-driven analysis of past mistakes. From the Somalian intervention, the principle case-study lesson drawn has been as follows: when, if ever, should a humanitarian peacekeeping force, interposed to insure delivery of food and medical aid to victims of famine, permit itself to become involved in the competition between rival warlords and gangs whose own military incursions endanger both the rescue workers and the humanitarian coalition military force installed to protect them? The focus on so-called "mission creep" as the alleged "mistake" made by humanitarian forces there ignores a number of salient features of that case. Given the initial humanitarian mission, and reasonably competent assessments of the difficulties imposed by internal civil strife in realizing those initial mission objectives, it was certainly within the

---

[3]In reference to Brecht's classic, anti-war play, "Mother Courage and Her Children," a persistent theme is the shifting context-dependence of certain act-descriptions: one of the heroine's sons is decorated as a hero for committing actions in support of the army during wartime that when (unbeknownst to him) peace is suddenly declared, become classifiable as theft and murder, leading to his arrest and execution.

bounds of common sense and professional experience to extend the military aspects of that mission to include forcible cessation of military hostilities and even attempted interdiction of the principals responsible for the disruptive behavior. Rival warlords were not only disrupting the production and distribution of local food supplies, they were intimidating relief workers and stealing the donated food and medical supplies from abroad. It was not unreasonable to seek to put a stop to this behavior. The "mistake" was less in the "creep"—which in fact represented a logical extension of the activities necessary to support the mission, to be carried out proactively by personnel better equipped and trained for that sort of activity than for the mission to which they had been originally assigned—than in the subsequent failure to recognize and obtain prior international approval and adequate support for the increased needs and costs in personnel and *materiele* that the additional activities would exact from mission participants.

On the other hand, as illustrated in the journalist Mark Bowden's dramatic account, *Blackhawk Down*, the overall assessment of just what the nature of the conflict was in that country was far from competent or complete. Hence, military professionals representing a wide range of preparedness and battle-hardened combat expertise were drawn together into an altercation in behalf of a goal that may have been dramatically distorted. In any case, the nature of the conflict they would face, as that dramatic account indicates, was far more complex and confusing than anything for which they had prepared.

In any case, what lesson are we to draw from the Somalian experience to guide collective military intervention in the former Yugoslavia? Is there any logical reason to suppose that *ad hoc* and *post facto* principles drawn from the former experience will apply *a priori* to the subsequent crisis? Yet this is precisely the assumption that governed our response to the Bosnian crisis, beginning with our reluctance to become involved militarily at all (despite the obvious and desperate need for military intervention to halt the bloodshed), and continuing with the decision of UNPROFOR and NATO forces to maintain strict neutrality with regard to the warring factions. In light of Somalia, both reactions are reasonable. The net result, however, was disastrous: unnecessary prolongation of genocidal activities, followed by the spectacle of NATO forces turning a blind eye while individuals indicted and charged under international law with direct involvement in some of the most monstrous crimes since the Holocaust roamed the countryside with impunity, held jobs in the newly established civil authority, and in some cases even continued to threaten innocent civilians without fear of reprisal.[4] Are there any obvious lessons to draw from these experiences to govern our conduct in the next humanitarian altercation? How would we even begin to know?

The problem with the case-study approach to discerning the types of moral dilemmas attendent upon the use of military force for humanitarian purposes is that the methodology itself is (in these current instances, at least) hopelessly nominalistic and *ad hoc,* with no general lessons or principles yet emergent from the welter of particular and unique details. Analysis and adequate explanation, of the sort that leads to preparation for, and recognition in advance of, possible pitfalls inherent in these new military activities depends upon the emergence through induction of some general or "universal" principles from the particular cases. Thus far, I am suggesting, no reliable "covering laws" have emerged or been discerned amid the myriad details. Even worse, in the absence of such general principles or covering laws, we are reduced to a purely reactive (rather than proactive) stance: desiring to avoid the moral pitfalls encountered in the previous engagement, we constantly risk backing into new and wholly unforeseen, unrecognized pitfalls of a different variety in the latest altercation.

That is not to say that discussion and analysis of case studies have no value; only that, at present, we must use them as best we can and guard against false generalization. That very caution—that we cannot, in principle, know in advance precisely what we are doing, or how effectively we can conduct ourselves in humanitarian missions of these sorts—would be a tremendous advance upon the minimal training, reflection, and preparation that military personnel are given at present. For example, what ought to be the reaction of humanitarian peacekeeping forces installed in an unstable democracy such as Haiti if they discovered representatives of the local government whom they were installed to protect committing acts of unspeakable atrocity against the civilian population?[5] Should

---

[4]"Wanted for Murder," *Baltimore Sun,* Friday 11 July 1997, p. 2A.

[5]Supporters of Aristide, for example, were prone to continue the unfortunate practice of throwing burning tires around the necks of their helpless opponents, actions which undermine the fundamental humanitarian intent of the initial U.S. intervention in his behalf.

the peacekeepers intervene to protect the victims as Capt. Lawrence Rockwood attempted—even in violation of rules of engagement? How, in what way, and to what degree (especially when, as in this instance, the related structures of civil society—e.g., legal and judicial systems—are either absent or suspect)?

Some (but by no means all) of the kinds of problems encountered in this fashion may turn out to be highly similar to the kinds of moral dilemmas routinely encountered in the use of force for domestic peacekeeping purposes. In short, in our search for relevant universal principles and applicable professional standards of conduct, the thoughtful use of case study methodology should lead us to examination of the professional codes, experiences, and well documented problems of our colleagues in domestic peacekeeping professions. These are general principles and guiding professional codes of conduct devised by persons who likewise volunteer to surrender their own individual autonomy and subject themselves to rigorous discipline in a chain of command, incurring risk and placing themselves in harm's way for the sake of public protection (police, firefighters, the, DEA, FBI, and perhaps the Coast Guard). We would do well to consult such professional colleagues for insights, and even for advice and assistance, as we attempt to ascertain with some degree of foresight the problems our military personnel can anticipate routinely encountering in the new international context that Secretary Albright envisions.

Questions of Professional Military Ethics

I conclude by considering some nagging questions of professional ethics that the foregoing developments collectively portend. Unlike what has gone before, the following considerations, while significant, have received almost no formal analysis.

The Albright Doctrine formalizes a foreign policy gradually put into place by Presidents Reagan, Bush, and Clinton, in response to increasing demands from the international community and the United Nations. In effect, Ms. Albright suggests that, in the future, the military will serve as a kind of global police force, rather than as a political instrument for protecting U.S. interests or furthering U.S. foreign policy exclusively. This constitutes a direct contradiction and refutation of the earlier Weinberger doctrine, which specifically, proscribes (as does the international "legalist paradigm") most such uses of military force. The so-called "Weinberger doctrine," however, was developed through a lengthy process of careful and considered reflection and discussion by senior military and civilian defense analysts in this nation's service academies and war colleges, all of whom were tasked to establish more reliable guidelines for the use of American military forces in the aftermath of Vietnam.

While it is the privilege in principle of the civilian authorities at the pinnacle of the military chain of command to establish, revoke, and reformulate foreign policy at their own discretion, this new change of direction could be seen as an act of bad faith, a violation of contractual trust and agreement established painstakingly through dialogue and consensus. Whatever its other moral attributes, the Albright doctrine decidedly lacks these features—it represents a repudiation of earlier policy without a similar process of consultation, consensus, or "buy in" on the part of those most intimately affected by it. Under the Constitution, this is, of course, perfectly legal and proper. The Albright Doctrine, despite difficulties of interpretation and appropriate enforcement, may be not only morally correct, but even morally superior to the policies it supercedes. It may represent the best and most politically astute and realistic response to international conflict in the "new global order." Yet, I am prepared to argue, it still may not be prudent, and is perhaps not fair, inasmuch as those being asked to effect the policy have had little say in its formulation.

This concern may seem strange or even incongruous when applied to the military services whose communities of professionals pride themselves on their being accustomed to shouldering their mission and doing their duty, no matter what. To civilian authorities inclined to distrust the motives of the military, moreover, my intimation of some sort of thinly-disguised reluctance to embrace Secretary Albright's policy may seem to license insubordination, or to pander to what Thomas Ricks worries are increasing tendencies toward political polarization and resistance within military culture itself.[6] I take a sharply divergent view of these tendencies, based in part on considerations arising from the following kinds of examples:

---

[6]See "The Widening Gap," pp. 72–75; see also Gregory D. Foster, "Confronting the Crisis in Civil-Military Relations," *The Washington Quarterly* (Autumn, 1997), pp. 20ff, where he complains of the reluctance of the military to acquiesce to civilian authority in the matter of humanitarian intervention in Bosnia.

(1) A bright and able young individual with deep community spirit enlists in a local police academy, and spends years training and working her way up the career ladder of her local police force. Her desire, her abilities, and her training all make her an ideal police officer: fearless, physically able, fair, reliable, and entirely competent. Along the way she performs many different kinds of duties: cop on the beat, detective, desk watch—she likes some duties more than others, does not excel equally at all, and harbors no illusions about the morality or omnicompetence of the local government, or the purity of its citizens. But she clearly perceives the various duties she performs and the obstacles and frustrations she encounters as part of the work of a dedicated police officer. That is, until one day she and her colleagues are told that owing to changing community circumstances, their local unit is being reassigned for the foreseeable future to handle refuse collection. While crime has abated in their community, the problem of uncollected trash has mounted to crisis proportions, threatening the health and the beauty of the neighborhood. Our police officer, both in uniform and as a private citizen, has participated in neighborhood cleanups before; besides, she is loyal and dedicated, and willing in principle to do whatever is necessary for the health and welfare of the community she swore to defend and protect. This new professional transformation, however, forces her to consider whether, after all, she entered the right profession. Some of her police colleagues view it as outrageous that, without their advice or consent, the city government has chosen to solve its problem in this fashion at the cost of considerable negative impact on their professional careers.

(2) An exceptionally courageous and physically strong young man dreams all his life of becoming a professional firefighter. He eagerly goes through the years of requisite, arduous training, and is fortunate to be selected for a coveted position in his community's professional firefighter's unit. He amasses years of experience on the job, facing many harrowing encounters, saving lives, preventing damage to property, all at considerable risk to himself. Part of his job is fire prevention and community education; he and his fellow firefighters are so successful, that the radically reduced rate of fires occasions a decision by the local government to reallocate and redeploy the firefighters themselves. As it happens, the postal service has, for years, been unable to recruit sufficient numbers of postal workers, and delays in mail delivery have reached crisis proportions. Without so much as a "by your leave," the government authorities order the firefighters to deliver the mail, not merely temporarily, but for the foreseeable future. When some of the firefighters complain that they did not volunteer, nor were they trained for or interested in such work, the authorities reply coldly that each had pledged himself or herself to following orders, doing their duty, and serving the community, subject to the authority of the elected governing officials. This is now the community need, and for the foreseeable future, notwithstanding their prior training and professional expectations, delivering the mail will become their job.

I certainly do not mean to equate trash collection or mail delivery with the protection of the lives and liberties of defenseless and vulnerable populations in other nations. These analogies are obviously far from exact. But I have deliberately chosen illustrations of professional communities whose members have volunteered for rigorous mental and physical training and agreed to place themselves at risk for the sake of protecting life, property, and preserving the security of their communities, and who pride themselves on character, integrity, and devotion to mission and duty over personal welfare. The questions these cases raise are, first, do the individual police or fire fighters have reason to complain of their circumstances, to feel puzzled, angered, cheated perhaps by circumstances, or even misused by their government supervisors? I think it self evident that an array of such feelings would hardly be unreasonable under these circumstances, and might even be justified. In fact, I suspect the reactions of the professionals themselves would be quite mixed and at variance with one another: some would see these events as an interesting evolution with unfortunate side effects, and think of retiring or choosing another career. Others would be angered that their oaths of office were so liberally interpreted, or that they themselves would not have been invited to provide inputs into the problems for which they had become the unwilling solution. A few might wonder whether they would, in fact, "be any good" at collecting trash or delivering the mail, both of which turn out to require considerable skill. Still others would berate the first group as "quitters," and the second and third as "whiners," citing their sworn duty to do whatever necessary to protect and defend the community. "Your job," this fourth group of police or firefighters might argue, "is basically to do what you're told! If collecting trash or delivering the mail are what needs doing urgently, and we've been ordered to do it, then our duty is to comply without comment or complaint, and learn the requisite new skills on the job!"

It is hardly a secret that just such a range of conversations has been provoked by the gradual evolution of the nature of military service since 1989. I think it entirely reasonable to expect such mixed reactions, and thoroughly unhealthy to deny or surpress them. Many members of different branches of the all-volunteer military service have argued with some justification that they were not asked, nor did they agree, to volunteer for the missions now being proposed as central to the profession of arms, and also that they, and military forces in general, are not in principle well equipped or properly trained to carry out the kind of global police and peacekeeping work that the Albright Doctrine envisions as the military's principal new task. These, including the implicit charge of violation of informed consent, are legitimate concerns that deserve thoughtful consideration rather than rebuke.

It bears mention that U.S. Army General (and former Green Beret) Henry H. Shelton was invited, during his confirmation hearings to serve as head of the Joint Chiefs of Staff, to comment on Bosnia in particular and on the appropriateness of humanitarian and peacekeeping activities on the part of the military in general. While he did not find such use of military forces inappropriate, Gen. Shelton did suggest that "specially trained police" or "elite commandos" should be tasked with particular, unusual, and delicate actions such as interdicting war criminals in Bosnia - implying that such activities exceed the range of duties that the average soldier or sailor might normally be trained or expected to perform.[7]

Gen. Shelton's comments run parallel to concerns raised by Professor Martin L. Cook, at the conclusion of his 1994 Reich lecture at the USAir Force Academy,[8] addressing the "'social contract" aspect of the commitment made by individuals in the military. Cook argues in effect that a Principle of Informed Consent (familiar in other professional contexts such as law, business, and medicine) also comes into play in the context of an all volunteer military force, especially since individuals volunteer to join these organizations with certain understandings. Military service, Cook suggests, of necessity involves the "unlimited liability clause" that one may be in a position where the loss of one's life is required. One would therefore assume that thoughtful individuals are willing to acquire that risk because they believe that such sacrifice will *not* be asked of them except in circumstances where the cause truly warrants it. The customary expectations of the volunteers are that such sacrifices are reasonable to incur in the defense of one's nation, or of its vital interests. Only rarely, Cook observes, would the defense of the lives and property of foreign nationals be connected clearly to these causes. Like Shelton, Cook suggests that a solution might be the recruitment of individuals into such a peacekeeping force—either as a specific branch of the existing military (such as the Rapid Deployment Forces) or as an additional command structure parallel to the existing three branches. Put forthrightly and straightforwardly, it is hard to imagine that such an opportunity for national and international service would not appeal on its own merits to a great many idealistic, dedicated, and patriotic young people, much as traditional military service or service in the Peace Corps did in times past. The difference in this case is that the matter is out in the open, and the knowledgeable consent of the would-be volunteer has been obtained in advance of assignment, deployment, and the incurring of risk.

The relevant point I would wish to stress is that some such process of deliberation is crucial for compliance with the principle of informed consent. This does not supplant the tradition of following orders, doing one's duty, and respecting the chain of command—nor is it a call for a "softening" or democratization of that tradition. Rather, my point is that this fundamental element of trust is the foundation for that tradition, and it is this necessary foundation which is jeopardized by unilateral, dramatic changes in fundamental professional policy such as the Albright doctrine proposes.

---

[7] "Shelton Stresses Importance of Soldiers," *Baltimore Sun*, Wednesday 10 September 1997, p. 5A.
[8] Martin L. Cook, "Why Serve the State?" 7th Annual Reich Lecture (Colorado Spring, CO: USAFA, 1994); 2.

Perspectives on Irregular Warfare: The Case of Somalia
—A Response to Professor Lucas

Gen Anthony C. Zinni, USMC (Retired)*

It is possible to go through a list of things that I feel are lessons learned by the military in dealing with problems such as those discussed in the foregoing essays by Dr. Lucas. I propose, however, to take a different approach. I propose to outline the story of one intervention, because, in that case study of a single American intervention, we will see all the problems and issues about which Professor Lucas spoke very clearly illustrated.

The case I will use is that of Somalia. I have been involved in very probably close to a dozen of these kinds of operations in one way or another. I was deeply involved in Somalia from 1992 to 1995. Recall that President Bush (the elder) had been defeated in the election in November, 1992. In December of '92, we had this tremendous disaster in the Horn of Africa, partly natural (drought, starvation) and partly—in places like Somalia—man-made. The government had collapsed. There was a series of factions or warlords that were running what used to be Somalia. These militias—some of them were rogue militias—were brutalizing the people.

So we had the combination of a natural and a humanitarian disaster, overlaid with fighting, overlaid with a collapsed infrastructure of nonexistent government. These kinds of operations have become known as "Complex Emergency Operations." They are termed "complex" in that there isn't just one dimension. A purely humanitarian problem might be defined as say, simple earthquake relief, famine, or hurricane relief. In Somalia, by contrast, we faced the multiple problems of peace enforcement, collapsed infrastructure, and humanitarian disaster.

I think it was clear that President Bush made the decision to intervene not based on any political motivation. Obviously, you couldn't find anything that you could even closely or remotely tie to a vital national interest in our intervention in Somalia. You certainly had a President who was leaving office, probably forever, so there was nothing to be gained personally for him. I think truly what has become known as the "CNN effect," seeing the horrors, and terror, and the half-million people to 700,000 people that were on the verge of death, motivated then-President Bush to attempt to do something good.

We were told when we intervened—when the decision was made to commit military force—that we would go in and lead a coalition of willing nations. We were also told that this operation would last a month to a month and a half, and that, basically, what we would do would be very limited. We would just go in and provide humanitarian relief. We would get on the ground, and we would try to relieve the problems that were causing the deaths, and the starvation, and the problems of their region.

---

*from: *Perspectives on Humanitarian Military Intervention*, by George R. Lucas, Jr. (Berkeley, CA: Berkeley Public Policy Press, 2001), pp. 53–63.

I volunteered to go to Somalia. I wasn't part of the force that was chosen to lead the operation, the First Marine Expeditionary Force down in Southern California. But because I had experience in these operations, I was ordered to go as the Director of Operations. And I knew that none of these things are simple and easy. It isn't a matter of just cramming MRE's down the throat of some sick people somewhere, and expecting them to come around. As a matter of fact, we usually do more damage in our approach because we don't understand, obviously, some of the complexities even for dealing with some of the simple humanitarian conditions and requirements.

We understood when we went in that we would have the cooperation of the factions and the warlords. However, this was very reluctant cooperation. There had been a small UN operation ongoing, but it had been a failure, basically paralyzed because it was too small and unable to deal with the gangs, and the factions, and the road militias that were on the streets. The military component of that U.N. "force" was one battalion of Pakistani soldiers, and some observers who were trying to maintain a line between some of the warring factions, especially in the city of Mogadishu.

When the U.S. forces came into the environment, we wanted to get out to where the need was greatest. And the need was greater out in the hinterlands. Somalia is like a desert. There is nothing there. When you leave the coast and go further inland, it is really terrible. We arrived there, fortunately, at a time of year when it wasn't at the peak of the heat and the most terrible conditions. But, obviously, it was bad, and it was going to get worse as the summer months came in.

When we first got there we realized in order to help the people it was going to require a tremendous amount of logistics to support ourselves. We had about 38,000 American troops, and we had countries already committed to be involved with us. They ranged from the Canadians and the French, to the Italians, to three African nations, and three Arab nations. Before this operation ended in that first phase, we had the forces of 26 nations on the ground.

That sounds good, but I've got to tell you, most of those troops arrived ill-equipped and without the right kind of support. They became a large drain on the resources of the U.S. force that has to provide for their support. Some come politically unable, or incapable, not willing to take on harder missions. So these harder missions tend to fall to forces like the U.S., especially if they involve fighting, or if they involve the tougher end of the business of peace enforcement.

When we arrived we realized we were under tremendous pressure. The media was all over the beach. When the marines and the SEALS first came in, the lights and the media, that had already been present, were shining on our people landing on the beach! We had to get out to the hinterlands and bring food. We had to get out there and provide for our own security—and we provided it in a way that insured that if we met any resistance, we could handle it.

The political decision was made that we would not take any sides. We were not going to judge right or wrong. We were not going to determine whether one faction was good or bad. We were simply there for humanitarian purposes. In about three weeks we were able to get food and medicines to every place that needed them, providing at least the beginnings of humanitarian assistance, although the media was pressuring us to work even faster. It was tremendously difficult.

We were operating in an area the size of the State of Texas and, of course, with a population that was traumatized, and an infrastructure that was nonexistent—meaning no roads, or anything like that you could use. We literally had to build our way out to these people. We stabilized the population fairly quickly. There were a lot of NGOs on the ground trying to do good work but, obviously, under tremendous pressure. Many of them had hired some of these militias and gangs for their own protection. These militias and gangs were actually ripping them off. We had NGOs that were killed off by their own security guards, were robbed of their money—their pay. The "guards" were stealing two thirds or three quarters of the food to sell themselves. The NGOs, in order to get some food through, were accepting this.

We now found ourselves working with the "security forces" for these NGOs, consisting of the kinds of elements whom we didn't trust, and with whom we would have preferred not to do business. We ran into the problem of how to deal with this. Were we there to provide security for every NGO organization—over 120 of them—with 500 or so of their people inside Mogadishu's warehouses, residences, those sorts of things, and living a kind of lifestyle, if you will, that was very different from a military lifestyle. They came and went as they pleased. They expected to have their own squad to protect them, which was not the way we did business. Obviously, we were much more orderly, and much more precise, scheduled, and careful. So we had an automatic clash of cultures between the military and the NGOs. Here I am, a tough-looking general in a jacket with a cigar hanging out of his mouth, and I find myself looking at a young 20-year-old girl with a "Save the Whales" T-shirt—that's the NGO. I look at her and say, "You remind me of my daughter." And she looks at me and says, "You remind me of my old man." And we probably aren't going any further in that discussion. Cooperation becomes extremely difficult. Both groups are well-meaning, but both come from different cultures and have a different approach to what needs to be done.

It became evident to us after a while that we were trying to put a band-aid on a sucking chest wound. We had thought we were going to fix the problem in the short term, but we weren't going to solve the problem in Somalia. As soon as we left this problem was just going to start up again and require humanitarian assistance. We military types tend to be not the best force for that purpose. The military is great in an emergency. It brings a lot of logistics. It can freeze a situation in a security sense. But it isn't the right kind of organization to allow for a society to be rehabilitated or reconditioned. It is not what we do. We can give plenty of food and water. We can give victims plenty of medicine, but it is not always done in the best way.

Nevertheless we were stuck with this situation. We had been told by Inauguration Day, the 20th of January, 1993, we would be gone. We were not hearing anything about anybody coming in after us. We thought maybe the U.N. would come in with a bigger mission, or somebody else would follow up behind us after we had set the conditions for success. President Bush came out to see us right before he left office. I remember sitting there thinking, "He is going to tell us now that we are getting out and what the plan is for who follows us." And I remember briefing him. Our headquarters was in the bombed-out, destroyed building that was the former U.S. Embassy. The walls were knocked out, walls all blackened from fire and everything. And that is what we were using as headquarters. And he was sitting there. And we asked him the question, "Well, Mr. President, we have come here. We have done the mission. We have stopped the dying and the suffering. And we are here in big numbers now. What is next?" And he turned to Brent Scowcroft, the retired Air Force general who served as national security adviser, and he said, "Brent, what is the plan to follow-up on this?" And Gen. Scowcroft gave the President a look of confusion. There was no plan.

When the President left we were almost in tears because we had known that we had picked this patient up. In Dr. Lucas's imagery, we had picked the dying skateboarder up. The trouble was the skateboarder did not have a home; did not have anyone to care for him. And there wasn't anybody else that was going to tend to the very serious patient for whom we had provided this initial first aid. Suddenly it dawns on you: *you* have the obligation. Once you begin to treat the patient, you own him. And we owned this patient called Somalia.

We were going nowhere. We had no political objectives. We had a new administration coming in. We had a change of Chairman of the Joint Chiefs of Staff. And we were sitting out there now on top of a powder keg. The political situation was not very good. Extreme violence was going on. All this time we are trying to save and protect people, and feed them. But we had sort of stagnated. We had reached the phase where we weren't going forward. We were just trying to keep our head above water. We had become the government of Somalia in many ways, and the security forces of Somalia.

As a result, we proceeded to undertake a classic thing, that now has become known as "mission creep." You can't put a military force in there to see these problems with a very limited mission, and ask them not to step up beyond that mission. So, we decided in our own clumsy way to start fixing things. The first thing we decided to do was to bring a little law and order to Somalia.

So we re-established the police force. I became the Chief of Police of Mogadishu. I actually still have my hat and my police force uniform. And we actually put Somali police on the street and armed them. The old Somali policemen came back. They were revered. They were not part of the oppressive regime that had been overthrown. Every place we saw a Somali policeman, and he had dusted off his police hat, and his old uniform, and had come back to his street corner, it seemed like that corner seemed to thrive. And the little market stalls would come into being. So, we re-established this police force. And we tried to back it up with our own military and our own force.

Now, this is touchy business because the U.S. military has very strict laws we have to follow that don't allow us to get involved in police training. So, we had to be very careful what we did. And we asked some of our coalition allies, like the Italians and others, to assist us. We set up this police force.

And we were very proud of this police force. And our policemen went out dutifully and arrested people. And when they arrested them they asked us, "What do we do with these people that we arrest?"

And so we decided to set up a prison system in Somalia. I became a warden for the prison system in Mogadishu. And we actually re-opened the jails and the prisons. We provided food, got the old wardens, and the old guards. And now we had a prison system. So we had policemen that arrested people, and we had the prisons to keep them in. And then the old wardens came to us and said, "You know, they are arrested. They are in prison, but we can't keep them there forever. I mean, what do we do with them now? Some people are here for minor offenses, not crimes that we should throw them away in a cell forever for." And, of course, we monitored these very carefully.

So, we decided to set up a judicial system with judges and courts. And I was on the judiciary committee. And we rehired some of the old judges and lawyers. We brought them out of hiding, put on their robes, and we began to try cases. And the judges came to us in our judicial committee and said, "What law would you like us to enforce here now?" Since it was complete anarchy, there was no more Somali law.

So, we dusted off an old Italian code book, and went through it. There were a couple Latin phrases we recognized in there: *Habeas corpus*, and a couple of others. And we thought, "This looks pretty good. Let's use this law."

Now we were quite proud of ourselves. We had a police force and a prison system. We had judges, a law, a code to live by, one they had known from their past, and seemed fairly just. And then one day one of our judges passed down a death sentence, and my police executed somebody. Suddenly we had our lawyers coming to me saying, "You are way over the line now. We have gone way beyond our mission. We have gone way beyond our charter, maybe even way beyond what we legally are supposed to be able to do in trying to fix the situation."

We were attempting to bring law and order. We were attempting, at the same time, to bring political reform. We were holding sessions with the 15 faction leaders to try to get some sort of national government, some sort of agreement on a government, some sort of agreement on a cooperative multi-representational government. We were actually hauling them off to meetings to get them out of Somalia. Some of my officers had written home and had their spouses send them their civilian clothes, suits and all. When we went to these meetings, we weren't in uniform.

Here we were in the military, involved in processes like these. And there were many, many others that were going on that were way beyond our initial mission because we were stuck with this task, and utterly without a guiding policy. By the time the U.N. was convinced to take this over, they looked at what we had done and did not accept a lot of what we had done. They didn't accept the police force, and many of the other things we had put into effect, things I personally thought were beginning to work. But they decided to take another approach.

Our first phase was fairly successful. We brought a degree of order to Somalia. We brought a degree of humanitarianism and survivability to the countryside. The NGOs were beginning to piece together a few long-term structures, such as markets and other things. And the U.N. decided it was time to rebuild the nation. And they actually came in with a plan that in two or three years they could install democracy into Somalia, and that the warlords, like General Aidid, would be excluded from this process.

Let me mention a bit about General Mohamed Farrah Aidid. Those who may remember reading about Somalia very likely formed the impression, as did most Americans, that Gen. Aidid was some sort of evil warlord out there who was utterly ruthless and purely out to dominate the country at the expense of his people. Well, he was ruthless, but he was a general in his own right. He had been a general in the Somali army. He had been educated in Rome. He had been off to war college in India. He was the ambassador to India from Somalia. He was a member of the Cabinet. He had been imprisoned for six years and was held in a dungeon and forced to read nothing but the Koran. So he was a hero to his people for having suffered because he was subject to the dictatorship.

He rose up against the dictator. When he was free he led a revolution. His own people thought of him much as we would think of George Washington. He became the military leader of the revolutionaries who overthrew the dictator. He was a prominent head of his own clan, a respected clan elder. He was highly educated, published a number of books, spoke a number of languages including English very fluently, had a son in the United States—in fact, Aidid's son was part of our United States Marine

Corps contingent there. He was a corporal in the United States Marine Corps. He has since taken his father's place, and is running the faction there. But he was with us when we were there.

Aidid was tough, and he felt that he had earned the right to lead his country. His political party, the Somali National Congress, had elected him as their President. And he was the one they felt, as the most powerful party, should be President of that country. But the U.N., in all its wisdom, led by U.S. policy, decided these warlords will not be part of the process. Before that we didn't make that judgment. We felt the Somalis ought to decide who should lead them.

We were being pressured to disarm the Somalis. You couldn't possibly take the weapons away in Somalia. And bright ideas came out of Washington to pay the Somalis for their weapons. However, we simply couldn't afford to buy the weapons because there were so many them. We were being told to physically take them away. Well, if we physically took them away, we were going to be at war. And we were going to kill a lot of Somalis, and they were going to kill a lot of us.

We had come to an agreement with the militias on weapons, agreeing that they would have no weapons visible in the street. This was an agreement that allowed us not to get into violent confrontations, and to minimize them except with the rogue elements out there. And it allowed them to save face, and to maintain their weapons and their militias until they could politically sort out their future. The U.N. went down a path of attempting to install a democracy over two or three years. And they attempted to do it by excluding certain factions, especially those that had the weapons and the guns. They ended up in a running gun battle with the faction leaders, especially General Aidid. And I think most of you know the story of what happened there.

I went back right after the battle of Mogadishu, during which nineteen of America's special operations forces people were killed. I went there to see General Aidid, whom I knew well, accompanied by Ambassador Robert Oakley, who had been with us in the first phase of Somalia—he had been the previous ambassador to Somalia—to get the prisoners out. We hoped to get a cease fire and get the situation turned around, which we were able to do.

I returned two years later as the Commander of the Task Force that protected the U.N. withdrawal when the U.N. intervention forces finally left Somalia. We withdrew under fire off the beaches of Somalia. And I turned around and saw a devastated Mogadishu in worse condition than when we had come two-and-a-half years before in our first involvement.

What went wrong? Let me offer a short list of some of the most important things. The first is—and I think this was touched on by Dr. Lucas earlier—a lack of policy on our part about what we will do, when we will make a commitment, and how we will make that commitment. We have no policy or strategy for our involvement in these sorts of operations. There is no sense as to which ones we pick to get involved in and which ones we decide not to get involved in. Our decisions often make no larger strategic sense as to how much we put into a certain operation, or how little.

We do a lousy job, first of all, identifying the political objectives we have there, and then a terrible job, consequently, of defining the military tasks. Military people have different verbs than politicians. Politicians have very broad, fine-sounding verbs that are usually deliberately obscure so that they can sell it and receive consensus. Military people can't deal with that kind of vagueness. We need specifics.

I remember when Cyrus Vance and Richard Owen were working on a peace plan between the factions in Bosnia. And one of the elements of that plan was that the military contingent that would go in—U.S./NATO/U.N.—would "monitor heavy weapons owned by each side." Now, as a military man, I don't know what "monitor heavy weapons" means. Do I blow them up? Do I watch them shoot and tell somebody? Do I take them away and lock them up? Do I trust the other guy to lock them up and inspect them once in a while? What does that mean? If that is a task someone is going to give to a military organization, then it needs further refinement and further definition.

That sort of thing happens all the time. Our military gets thrown into these situations without that specific translation of the political objectives into the exact military tasks on the ground. In Somalia there were expectations the NGOs had of us that were not in keeping with the mission we were given by our political leaders. There were expectations that the Somalis had of us. I had been on the ground a week, and I was asked to meet with a woman's group in Mogadishu. They complained to me that we had been there for one week, and we had not started a jobs program yet for the Somalis. I said, "We don't do job programs in the United States Marine Corps."

Here was an expectation they had that we weren't living up to. We do a lousy job defining what success is. Is success rebuilding a nation and molding it in our model where maybe the concept of democracy has never existed? Or perhaps we are to seek a more limited kind of objective, a more realistic, basic, law and order that would be sufficient, and would be the best we might hope for. Can we stand to have a form of government that runs contrary to what we believe in? What if General Aidid—as much as we may have disliked him personally and disliked the idea of a dictator or authoritarian ruler—what if General Aidid, at least in the short term, could have brought law and order to Mogadishu, some structure, some infrastructure? But this would stick in our craw because it wasn't a democracy, and this was an authoritarian/totalitarian leader who might prove to be oppressive and ruthless.

If we can't settle for half measures, what does that mean about the commitment we have to make? Are we willing to spend the lives, and the treasury resources, and the time, and the duration of the commitment of our forces to see our mission through to some grand altruistic goal that may not be within the realm of reality? Rarely does the military itself represent a complete or long-term solution to such problems. We need the other elements there in full force. We need political and diplomatic support. We need those that understand humanitarian intervention better than we do. We need those that understand economic development and rehabilitation better than we do. Oftentimes all of that is left to us. I have spent a lot of my career, lately, doing negotiations, diplomatic work. I find myself at times more worried about economic conditions, and not just the pure military dimension of what we do.

Is that the appropriate role for our people in uniform? Have we figured out how to blend all the dimensions to make it work on the ground? Our military, as Dr. Lucas suggested, has been resistant to these kinds of humanitarian missions. This isn't what we train, organize, and equip for. We can do some of this as a secondary, additional duty. We have never spent the time to develop the doctrine, the tactics, the techniques, the procedures, the equipment to do this really well, and concentrate on it. Some nations have. The Canadians are a good example. They really view their military as, primarily, a peacekeeping force. But we never told our military in the U.S., "That is what you will do. And it will be a primary mission," because as soon as we do that, we will find that there is a lot that follows from that change of policy. Every one of these humanitarian missions comes out of our military "lunch money."

Every one of these missions has to be paid for. The military pays for these missions out of the money that is supposed to go into the training, organizing, and equipping our forces to meet the threats against our nation. Every time we bear the brunt of the expense of one of these operations, which are very expensive, that is less fuel; that is fewer rounds that went down range; that is a tank that is not purchased, or an airplane that doesn't get on the flight line. Sometimes Congress will come back later and give us a supplemental reimbursement, but by then we may have lost the training opportunity. We may not get fully reimbursed. And sometimes if the Congress objects to the President's commitment, as a punishment, the military may not get reimbursed at all. Once again, the military bears the burden or the brunt of these decisions.

There are a few countries out there that have militaries like ours. There is a desire to have international credibility and legitimacy. Our political leaders want to see as many flags as possible on the ground in this effort to show international resolve. But that costs us. If you have a battalion from a third-world nation that decides to join, who gets them there? Who maintains them while they are there? Who protects them if they get in over their head? Who provides the equipment, the food, the training they need? That will always fall to the U.S. contingent who invited them in.

Then we take on another responsibility and another burden, rather than an asset that can help us towards a solution to the problem. We haven't always developed the right skills within the military to deal with this. I mentioned negotiating. There are many kinds of things like this that we simply don't train our people to do. We are not that aware, that sensitive on how to operate in strange cultures. We don't do the kind of work necessary to become extremely proficient in these kinds of missions.

To be sure, some of the things we do and capabilities we have are directly applicable to humanitarian relief. If I have a young marine who operates a reverse osmosis water purification unit, whether he does it in a combat environment, or whether he does it in a humanitarian operation, it doesn't matter. That is a directly translatable skill. Other skills have to be twisted and bent around. If we go into cities, we aren't going to go in like it is Stalingrad. We may have to go in like it is Northern Ireland and Belfast. And so it is a hostile environment, but it is a different kind of hostility. It is not war. It is a semi, quasi war. There is an existing, functioning civilian population, but there is also a threat on the street.

It is a different kind of urban patrolling. It is a different kind of urban environment than we would find in a combat situation.

So, skills have to be turned around. And then there are those skills that we just don't train our people in, like the ability to bring people to a table and to negotiate, and to get them to do what we would like them to do, or what they should do—the right thing. We never make these interventions early, when the problems are just starting, when they're easy. We invariably wait until they are catastrophes. We wait until that CNN effect drives us into full involvement, when our public recognition of the horror of what went on forces us into it. By then, however, it is already too late, and it is tough to do much lasting good.

I think if we are going to take on this humanitarian project as a mission, and it is going to be given to our military, we need to stand up to the mission, and recognize it. We need someone to say, "We will do these, under these conditions, with this sort of policy. And we will provide the resources and direction to our military to do it. And we will force the other components of government that should be involved with us to provide the economic, political, and humanitarian dimensions that we need to cooperate and form the right kind of organization we need. And we will structure ourselves to make the right decisions; to be honest and up-front about what level of commitment we are willing to make, or not make; to know when we are willing to get out and say, 'we have had enough;' to put realistic objectives on these missions, and not try to have utopian dreams about what we can do, when we can't succeed in these situations."

I think we are going to face this situation many, many more times in the near future. And I would tell you, we have only begun to make marginal improvements on how we can do this, not significant improvements. I think most of us in the military would just like to see the humanitarian intervention task go away. But every indication is, it will not just "go away." Therefore we have got to come to grips, as a nation, as to what our responsibility is.

I would close by suggesting that we have to take a deep look inside. We are the most powerful nation in this world; economically, the greatest nation the world has ever seen. We are certainly the land of the haves. We demand a lot of ourselves. Would we accept it if our own community didn't have a 911 response for us? If something went wrong, would we accept it that we couldn't simply pick up the phone and get something done? We demand energy. We demand medical care. We demand response if there is a threat to us. Most of the world, however, has no 911 number. Most of the world looks to us for some sort of help and guidance. And we don't provide it. We have got to make some hard decisions about the moral obligation we have for the rest of the world, and how we are going to take on that significant obligation.

Why Warriors Need a Code
Shannon E. French [*]

At the United States Naval Academy I teach a military ethics course called "The Code of the Warrior." My students are all midshipmen preparing for careers as officers in the U.S. Navy or Marine Corps. Each semester, on the first day the class meets, I ask the midshipmen to reflect on the meaning of the word "warrior" by giving them an exercise that requires them to identify whether any of a list of five words are perfect synonyms for "warrior." They are then asked to write a brief explanation of why each word succeeds or fails as a synonym. The time constraints keep their responses relatively raw, yet they are often surprisingly earnest and at times even impassioned.

The words I offer my students for their consideration are "murderer," "killer," "fighter," "victor," and "conqueror." I have found that most of them reject all five. The reasons that they give to account for why they wish to dismiss each of these as synonyms for "warrior" consistently stress the idea that a true "warrior" has to be morally superior in some way to those who might qualify for the other suggested epithets. Consider these representative comments from a variety of midshipmen:

Murderer
"This word has connotations of unjust acts, namely killing for no reason. A warrior fights an enemy who fights to kill him."

Killer
"A warrior may be required to kill, but it should be for a purpose or cause greater than his own welfare, for an ideal."

Fighter
"Simply fighting doesn't make a warrior. There are rules a warrior follows."

Victor
"Warriors will lose, too—and the people who win aren't always what a warrior should be."

Conqueror
"A conqueror may simply command enough power to overcome opposition. He can be very lacking in the ethical beliefs that should be part of a warrior's life."

Almost without exception, my students insist that a "warrior" is *not* a "murderer." They can even become emotional in the course of repudiating this (intentionally provocative) potential synonym. It

---
[*] Excerpted from *The Code of the Warriors: Exploring Warrior Values, Past and Present* by Shannon E. French, New York: Rowman and Littlefield Publishers, 2003.

is very important to them that I understand that while most warriors do *kill* people, they never *murder* anyone. Their remarks are filled with contempt for mere murderers:

- "Murder is committed in cold-blood, without a reason. A warrior should only kill in battle, when it is unavoidable."
- "Murderers have no noble reason for their crimes."
- "While a murderer often kills innocent or defenseless people, a warrior restricts his killing to willing combatants. He may stray, but that is an error, not the norm."
- "A murderer is someone who kills and enjoys it. That is not a warrior."
- "This term has very negative connotations associated with it because a murderer is one who usually kills innocent, unarmed people—while a warrior has honor in battle and does not take advantage of the weak."
- "A murderer murders out of hate. A warrior does not. He knows how to control his anger."
- "A warrior is not a murderer because a warrior has a code that he lives by which is influenced by morals which must be justified."
- "Warriors fight other warriors. Therefore they kill, not murder."
- "'Murderer' lacks any implication of honor or ethics, but rather calls to mind ruthlessness and disregard for human life."
- "A murderer kills for gain, or out of anger. He does not allow victims a fair fight."
- "The term 'murder' represents an act done with malice. Warriors kill people in an honorable way."
- "A murderer has no honor."

Clearly, my students do not regard the distinction between a warrior and a murderer as a trivial one. Nor should they. In fact, the distinction is essential, and understanding it is one of the primary aims of this text.

Murder is a good example of an act that is cross-culturally condemned. Whatever their other points of discord, the major religions of the world agree in the determination that murder (variously defined) is wrong. According to the somewhat cynical 17th Century philosopher Thomas Hobbes, the fear of our own murderous appetites is what drove us to form societies in the first place. We eagerly entered into a social contract in which certain rules of civilized behavior could be enforced by a sovereign power in order to escape the miserable, anarchic State of Nature where existence is a "war of every man against every man," and individual lives are "solitary, poor, nasty, brutish, and short."[1] In other words, people want to live under some sort of system that at least attempts to make good the guarantee that when they go to sleep at night, they will not be murdered in their beds.

Unfortunately, the fact that we abhor murder produces a disturbing tension for those who are asked to fight wars for their tribes, clans, communities, cultures or nations. When they are trained for war, warriors are given a mandate by their society to take lives. But they must learn to take only certain lives in certain ways, at certain times, and for certain reasons. Otherwise, they become indistinguishable from murderers and will find themselves condemned by the very societies they were created to serve.

Warrior cultures throughout history and from diverse regions around the globe have constructed codes of behavior, based on that culture's image of the ideal warrior. These codes have not always been written down or literally codified into a set of explicit rules. A code can be hidden in the lines of epic poems or implied by the descriptions of mythic heroes. One way or another, it is carefully conveyed to each succeeding generation of warriors. These codes tend to be quite demanding. They are often closely linked to a culture's religious beliefs and can be connected to elaborate (and frequently death defying or excruciatingly painful) rituals and rites of passage.

---

[1] Thomas Hobbes, *The Leviathan*.

In many cases this code of honor seems to hold the warrior to a higher ethical standard than that required for an ordinary citizen within the general population of the society the warrior serves. The code is not imposed from the outside. The warriors themselves police strict adherence to these standards; with violators being shamed, ostracized, or even killed by their peers. One historical example comes from the Roman legions, where if a man fell asleep while he was supposed to be on watch in time of war he could expect to be stoned to death by the members of his own cohort.

The code of the warrior not only defines how he should interact with his own warrior comrades, but also how he should treat other members of his society, his enemies, and the people he conquers. The code restrains the warrior. It sets boundaries on his behavior. It distinguishes honorable acts from shameful acts. Achilles must seek vengeance for the death of his friend Patroclus, yet when his rage drives him to desecrate the corpse of his arch nemesis, he angers the gods. Under the codes of chivalry, a medieval knight has to offer mercy to any knight who yields to him in battle. In feudal Japan, samurai are not permitted to approach their opponents using stealth, but rather are required to declare themselves openly before engaging combat. Muslim warriors engaged in offensive *jihad* cannot employ certain weapons, such as fire, unless and until their enemies use them first.

But why do warriors need a code that ties their hands and limits their options? Why should a warrior culture want to restrict the actions of its members and require them to commit to lofty ideals? Might not such restraints cripple their effectiveness as warriors? What's wrong with, "All's fair in love and war?" Isn't winning all that matters? Why could any warrior be burdened with concerns about honor and shame?

One reason for such warriors' codes may be to protect the warrior himself (or herself) from serious psychological damage. To say the least, the things that warriors are asked to do to guarantee their cultures' survivals are far from pleasant. There is truth in the inescapable slogan, "War is hell." Even those few who, for whatever reason, seem to feel no revulsion at spilling another human being's guts on the ground, severing a limb, slicing off a head, or burning away a face are likely to be affected by the sight of their friends or kinsmen suffering the same fate. The combination of the warriors' own natural disgust at what they must witness in battle and the fact that what they must do to endure and conquer can seem so uncivilized, so against what they have been taught by their society, creates the conditions for even the most accomplished warriors to feel tremendous self-loathing.

In the introduction to his valuable analysis of Vietnam veterans suffering from post-traumatic stress disorder (PTSD), *Achilles in Vietnam: Combat Trauma and the Undoing of Character*, psychiatrist and author Jonathan Shay stresses the importance of "understanding . . . the specific nature of catastrophic war experiences that not only cause lifelong disabling psychiatric symptoms but can *ruin* good character."[2] Shay has conducted countless personal interviews and therapy sessions with American combat veterans who are part of the Veterans Improvement Program (VIP). His work has led him to the conclusion that the most severe cases of PTSD are the result of wartime experiences that are not simply violent, but which involve what Shay terms the "betrayal of 'what's right.'"[3] Veterans who believe that they were directly or indirectly party to immoral or dishonorable behavior (perpetrated by themselves, their comrades, or their commanders) have the hardest time reclaiming their lives after the war is over. Such men may be tortured by persistent nightmares, may have trouble discerning a safe environment from a threatening one, may not be able to trust their friends, neighbors, family members, or government, and many have problems with alcohol, drugs, child or spousal abuse, depression, and suicidal tendencies. As Shay sorrowfully concludes, "The painful paradox is that fighting for one's country can render one unfit to be its citizen."[4]

Warriors need a way to distinguish what they must do out of a sense of duty from what a serial killer does for the sheer sadistic pleasure of it. Their actions, like those of the serial killer, set them apart from the rest of society. Warriors, however, are not sociopaths. They respect the values of the society in which they were raised and which they are prepared to die to protect. Therefore it is important for

---

[2]Jonathan Shay, M.D., Ph.D., *Achilles in Vietnam: Combat Trauma and the Undoing of Character*, New York: Simon and Schuster, 1994, p. xiii.

[3]Op. cit.

[4]Ibid p. xx.

them to conduct themselves in such a way that they will be honored and esteemed by their communities, not reviled and rejected by them. They want to be seen as proud defenders and representatives of what is best about their culture: as heroes, not "baby-killers."

In a sense, the nature of the warrior's profession puts him or her at a higher risk for moral corruption than most other occupations because it involves exerting power in matters of life and death. Warriors exercise the power to take or save lives, order others to take or save lives, and lead or send others to their deaths. If they take this awesome responsibility too lightly—if they lose sight of the moral significance of their actions—they risk losing their humanity and their ability to flourish in human society.

In his powerful work, *On Killing: The Psychological Cost of Learning to Kill in War and Society*, Lt. Col. Dave Grossman illuminates the process by which those in war and those training for war attempt to achieve emotional distance from their enemies. The practice of dehumanizing the enemy through the use of abusive or euphemistic language is a common and effective tool for increasing aggression and breaking down inhibitions against killing:

> It is so much easier to kill someone if they look distinctly different than you. If your propaganda machine can convince your soldiers that their opponents are not really human but are "inferior forms of life," then their natural resistance to killing their own species will be reduced. Often the enemy's humanity is denied by referring to him as a "gook," "Kraut," or "Nip."[5]

Like Shay, Grossman has interviewed many U.S. veterans of the Vietnam War. Not all of his subjects, however, were those with lingering psychological trauma. Grossman found that some of the men he interviewed had never truly achieved emotional distance from their former foes, and seemed to be the better for it. These men expressed admiration for Vietnamese culture. Some had even married Vietnamese women. They appeared to be leading happy and productive post-war lives. In contrast, those who persisted in viewing the Vietnamese as "less than animals" were unable to leave the war behind them.

Grossman writes about the dangers of dehumanizing the enemy in terms of potential damage to the war effort, long-term political fallout, and regional or global instability:

> Because of [our] ability to accept other cultures, Americans probably committed fewer atrocities than most other nations would have under the circumstances associated with guerrilla warfare in Vietnam. Certainly fewer than was the track record of most colonial powers. Yet still we had our My Lai, and our efforts in that war were profoundly, perhaps fatally, undermined by that single incident.
>
> It can be easy to unleash this genie of racial and ethnic hatred in order to facilitate killing in time of war. It can be more difficult to keep the cork in the bottle and completely restrain it. Once it is out, and the war is over, the genie is not easily put back in the bottle. Such hatred lingers over the decades, even centuries, as can be seen today in Lebanon and what was once Yugoslavia.[6]

The insidious harm brought to the individual warriors who find themselves swept up by such devastating propaganda matters a great deal to those concerned with the warriors' own welfare. In a segment on the "Clinical Importance of Honoring or Dishonoring the Enemy," Jonathan Shay describes an intimate connection between the psychological health of the veteran and the respect he feels for those he fought. He stresses how important it is to the warrior to have the conviction that he participated in an *honorable* endeavor:

> Restoring honor to the enemy is an essential step in recovery from combat PTSD. While other things are obviously needed as well, the veteran's self-respect never fully recovers so long as he is unable to see the enemy as worthy. In the words of one of our

---

[5] Lt. Col. Dave Grossman, *On Killing: The Psychological Cost of Learning to Kill in War and Society*, Boston: Little, Brown and Company, 1996, p. 161.

[6] Ibid. p. 163.

patients, a war against subhuman vermin "has no honor." This is true even in victory; in defeat, the dishonoring absence of human *themis* [shared values, a common sense of "what's right"] linking enemy to enemy makes life unendurable.[7]

Shay finds echoes of these sentiments in the words of J. Glenn Gray from Gray's modern classic on the experience of war, *The Warriors: Reflections on Men in Battle*.[8] With the struggle of the Allies against the Japanese in the Pacific Theater of World War II as his backdrop, Gray brings home the agony of the warrior who has become incapable of honoring his enemies and thus is unable to find redemption himself:

> The ugliness of a war against an enemy conceived to be subhuman can hardly be exaggerated. There is an unredeemed quality to battle experienced under these conditions, which blunts all senses and perceptions. Traditional appeals of war are corroded by the demands of a war of extermination, where conventional rules no longer apply. For all its inhumanity, war is a profoundly human institution. . . . This image of the enemy as beast lessens even the satisfaction in destruction, for there is no proper regard for the worth of the object destroyed. . . . The joys of comradeship, keenness of perception, and sensual delights [are] lessened. . . . No aesthetic reconciliation with one's fate as a warrior [is] likely because no moral purgation [is] possible.[9]

By setting standards of behavior for themselves, accepting certain restraints, and even "honoring their enemies," warriors can create a lifeline that will allow them to pull themselves out of the hell of war and reintegrate themselves into their society, should they survive to see peace restored. A warrior's code may cover everything from the treatment of prisoners of war to oath keeping to table etiquette, but its primary purpose is to grant nobility to the warriors' profession. This allows warriors to retain both their self-respect and the respect of those they guard.

Some may prefer to establish the importance of a warrior's code without reference to the interests of the warriors themselves. It is in fact more conventional to defend the value of a warrior's code by focusing on the needs of society, rather than the needs of warriors as individuals. These are well-intentioned attempts to provide warriors with an external motivation to commit to a code. One such approach has been presented in military ethics circles as "the function argument."

Lt.Col. (Retired) J. Carl Ficarrotta, USAF, provides an extremely lucid exposition of the function argument in his essay, "A Higher Moral Standard for the Military."[10] The central thesis of the function argument, the proponents of which include Sir John Winthrop Hackett and Malham M. Wakin (see *War, Morality, and the Military Profession*, Boulder: Westview Press, 1986), is that men and women of bad character cannot function well as soldiers, sailors, airmen, or marines. This claim is based on the unique demands of military service. Those who support the function argument point out that comrades-in-arms must be able to trust one another in order to be effective; they must be willing to behave selflessly and sacrifice themselves for the good of the mission; and they must embody "the virtues of courage, obedience, loyalty, and conscientiousness"[11] when the stakes are at their highest. In other words, "...if one thinks (for whatever reason) that it is important to have a military that functions as well as it can, one also is committed for these same reasons to thinking military professionals are more strictly bound to exhibiting these functionally necessary virtues and behaviors."[12]

The function argument is useful, as far as it goes. It highlights the unique demands of military service that seem to require special virtues or moral commitments. Ficarrotta argues persuasively, however, that the range of valid conclusions that can be derived from the function argument is limited.

---

[7]Shay p. 115.

[8]Op. cit.

[9]J. Glenn Gray, *The Warriors: Reflections on Men in Battle*, New York: Harper and Row, 1970, pps. 152-153.

[10]Lt. Col. (ret.) J. Carl Ficarrotta, "A Higher Moral Standard for the Military," in *Ethics for Military Leaders*, George R. Lucas, Col. (ret.) Paul E. Roush, USMC, Lawrence Beyer, Shannon E. French, and Douglas MacLean, editors, Boston: Pearson Custom Publishing, 2001, pps. 61-71.

[11]Ibid p. 64.

[12]op. cit.

Because it links the motive for ethical behavior to military effectiveness, the function argument cannot, by itself, provide reasons for the warrior to behave well in situations where bad behavior does not seem to have a negative impact on the function of the military:

> ... All this argument leads us to are higher standards *in the military context*. Military people must be scrupulously honest with each other when there is some military issue at hand. They must be selfless when it comes to the demands of military work. They must be courageous when there is some military task to be performed.
>
> What the function line does *not* establish is that the military professional has special reasons to be "good" through and through. The argument gives a soldier who would never even think about lying in his unit no *special* reason not to lie to his spouse or cheat on his income tax. The military function will be no worse off if a sailor always puts the needs of the service above her own, but still gives nothing to charity. As long as a pilot is courageous in combat or in dealing with his fellow professionals, he might just as well be a coward with a burglar for his father or his wife. We might well be (and should be) disappointed about these non-military moral failures, but the function line doesn't give us *special* reasons to be strict outside the military context.[13]

To stretch this objection even further, the function argument (again, considered *by itself*) gives no guarantee against the conclusion that it makes no difference how warriors behave *even in the military context*, so long as their behavior does not in fact cause them to fail to function effectively in their specific martial roles. That moral failings such as selfishness or a tendency to manipulate the truth could lead to functional failure is irrelevant. Only the actual consequences matter. The argument does not hinge on the acceptance of specific concepts of good character or moral absolutes. It is contingent upon the validity of certain empirical claims about the real world. If a particular warrior were to prove that he can function effectively and get his job despite having despicable character flaws, the function argument alone would not present him with any reason to improve himself.

The ancient Greek philosopher Aristotle, writing about the virtue of courage in the 4th Century BC, speculated that military effectiveness might indeed be distinct from personal virtue. "It is quite possible," he wrote, "that the best soldiers may be not men [who possess virtues such as courage] but those who are less brave but have no other good; for these are ready to face danger, and they sell their lives for trifling gains."[14] Regrettably, the answer to the question "Can warriors still get the job done if they do not have a code?" might be yes.

Suppose we allow that the function argument is enough to prove that warriors need to maintain at least some moral standards and exhibit certain virtues, in site-specific ways (e.g. to be courageous on the battlefield, not to lie to peers when doing so could jeopardize their safety or cause them to lose faith in you, etc.). Even so, it cannot ensure against the very moral failings that most worry those who want warriors to have a code for society's sake. It is possible, for instance, that the U.S. military might function more effectively if those in uniform routinely misled Congress (perhaps to manipulate them into increasing the defense budget or to maintain the secrecy of covert operations, as in the Iran-Contra scandal). Nor would such behavior necessarily erode trust within the ranks, if it were seen as a case of "us" versus "them."

The counter-claim could be that since the overriding function of the U.S. military is to uphold and defend the Constitution, strictly speaking, any unconstitutional behavior—and intentionally subverting legitimate Congressional authority over the military certainly qualifies as that—would be prohibited, if one were reasoning along the lines of the function argument. This suggests that a code derived from the function argument could be construed to require the warrior to commit to a wider range of values, all stemming from the Constitution. These would include respecting the rights of individuals, treating persons equally regardless of race, gender, ethnicity, or religious beliefs, supporting a system of checks and balances to prevent abuses of power, and promoting democratic ideals such as liberty and justice for all.

One potential pitfall to this otherwise appealing argument for American warriors upholding Constitutional values is that it might lose force for the warriors themselves when they believe the actual

---

[13] Ibid p. 65.

[14] Aristotle, *Nicomachean Ethics*, 1117b. 15-20, in Richard McKeon, editor, *Introduction to Aristotle*, Chicago: University of Chicago Press, 1973, p. 406.

survival of their nation is threatened. When asked whether they would hesitate to violate the Constitution if their country's very continued existence as an independent power were on the line, I have heard several of my students respond, "We've all taken an oath to uphold and defend the Constitution of the United States, but what is the point of upholding and defending the Constitution if the United States itself no longer exists?" In other words, if the choice were between staying true to the values of the Constitution and seeing the nation fall or violating the Constitution in order to save the country from annihilation, they would choose to commit the violation, save the nation, and try to restore Constitutional order when peace returned. It would be vanity to assume no crisis so extreme could ever face a superpower like the United States. What is evident from this is that if we are going to link a warrior's moral obligations to his or her function, we need to be very clear about what that function is. Warriors who think their job is to defend the nation, understood in terms of preserving territorial integrity and protecting the population, will have a different code than those who think their duty is to preserve Constitutional values at all costs (including the cost of the nation itself), even if both subscribe to the function argument.

A further concern I have regarding the function argument is that it only considers warriors as means to an end, namely the end of protecting the nation. I realize that this is due to the argument's structure, and not the result of any lack of compassion on the part of its authors or proponents. Yet it is a fault nonetheless. The highly influential 18th Century German philosopher Immanuel Kant charged that every rational being is bound by a categorical imperative "to treat humanity, whether in your own person or in the person of another, always as an end in itself and never merely as a means."[15] The word "merely" in this formulation must not be overlooked. Of course warriors are the means by which the nation is defended. To treat them as *mere* means, however, would be to fail to recognize that they are also citizens of the nation and human beings whose value is not limited to their utility as warriors. Although they may enjoy fewer liberties than their civilian counterparts, warriors retain their inalienable rights and deserve to be granted a full measure of dignity and respect.

Both the leaders and members of the general population of a nation that shows no concern for what price its warriors must pay for its defense stand in violation of the categorical imperative. It is possible to show proper respect for individuals whom you intend to send into harm's way or even to certain death. This requires first of all acknowledging the profundity of their sacrifice. The warrior is not a mere tool; he or she is a complex, sentient being with fears, loves, hopes, dreams, talents and ambitions—all of which may soon be snuffed out by a bomb, bullet, or bayonet. Secondly, those who send them off to war must make an effort to ensure that the warriors themselves fully understand the purpose of and need for their sacrifice. Those heading into harm's way must be given sincere assurances that their lives will not be squandered, and their leaders must not betray their trust.

Finally, the state must show concern for what will happen to its warriors after the battles are won (or lost). The dead should be given decent burials (if it is possible) and appropriate memorials. Those wounded in body should be given the best medical care, and treatment should be made available for those with psychological wounds. Former warriors must be welcomed back into the communities that spawned them and sent them away to do what needed to be done. If these conditions are met, even those warriors who lose their lives for the cause were not *mere* means to an end.

This brings us back to my earlier line of reasoning. It is not enough to ask, "Can our warriors still get the job done if they do not have a code?" We must also consider the related question: "What will getting the job done do to our warriors if they do not have a code?" Accepting certain constraints as a moral duty, even when it is inconvenient or inefficient to do so, allows warriors to hold onto their humanity while experiencing the horror of war—and, when the war is over, to return home and reintegrate into the society they so ably defended. Fighter who cannot say, "this far but no farther," who have no lines they will not cross and no atrocities from which they will shrink, may be effective. They may complete their missions, but they will do so at the loss of their humanity.

Those who are concerned for the welfare of our warriors would never want to see them sent off to face the chaotic hell of combat without something to ground them and keep them from crossing over into an inescapable heart of darkness. A mother and father may be willing to give their beloved son or daughter's *life* for their country or cause, but I doubt they would be as willing to sacrifice their child's

---

[15] Immanuel Kant, *Groundwork for the Metaphysics of Morals*.

soul. The code is a kind of moral and psychological armor that protects the warrior from becoming a monster in his or her own eyes.

Nor is it just "see-the-whites-of-their-eyes" front-line ground and Special Forces troops who need this protection. Men and women who fight from a distance—who drop bombs from planes and shoot missiles from ships or submarines—are also at risk of losing their humanity. What threatens them is the very ease by which they can take lives. As technology separates individuals from the results of their actions, it cheats them of the chance to absorb and reckon with the enormity of what they have done. Killing fellow human beings, even for the noblest cause, should never feel like nothing more than a game played using the latest advances in virtual reality.

In his book *Virtual War: Kosovo and Beyond*, the highly regarded international journalist and author Michael Ignatieff airs his concerns about the morality of asymmetric conflicts in which one side is able to inflict large numbers of casualties from afar without putting its own forces at much risk (e.g. by relying primarily on long-range precision weapons and high-altitude air assaults). In such a mismatched fight, it may be easy for those fighting on the superior side to fail to appreciate the true costs of the war, since they are not forced to witness the death and destruction first-hand. Ignatieff warns modern warriors against the "moral danger" they face if they allow themselves to become too detached from the reality of war:

> Virtual reality is seductive.... We see war as a surgical scalpel and not a bloodstained sword. In so doing we mis-describe ourselves as we mis-describe the instruments of death. We need to stay away from such fables of self-righteous invulnerability. Only then can we get our hands dirty. Only then can we do what is right.[16]

I have argued that it can be damaging for warriors to view their enemies as sub-human by imagining them like beasts in a jungle. In the same way, modern warriors who dehumanize their enemies by equating them with blips on a computer screen may find the sense that they are part of an honorable undertaking far too fragile to sustain. Just as societies have an obligation to treat their warriors as ends in themselves, it is important for warriors to show a similar kind of respect for the inherent worth and dignity of their opponents. Even long-distance warriors can achieve this by acknowledging that some of the "targets" they destroy are in fact human beings, not demons or vermin or empty statistics.

More parallels can be drawn between the way that societies should behave towards their warriors and how warriors should behave towards one another. Societies should honor their fallen defenders. Warriors should not desecrate the corpses of their enemies, but should, whenever possible, allow them to be buried by their own people and according to their own cultural traditions. Among his therapy patients, Jonathan Shay found several veterans suffering from "the toxic residue left behind by disrespectful treatment of enemy dead."[17] And while societies must certainly show concern for the after-effects of war on their own troops, victorious warriors can also maintain the moral high ground by helping to rebuild (or in some cases create) a solid infrastructure, a healthy economy, an educational system, and political stability for their former foes.

These imperatives I have put forward apply to relations among warriors and nations defended by warriors. The moral requirements become much murkier when warriors must battle murderers. This is a topic I will tackle in the concluding chapter of this book (Chapter Nine, "The Warrior's Code Today: Are Terrorists Warriors?").

There seems little need to advise warriors from within the same culture to treat each other as ends—at least, not after they have fought together against a common enemy. The motif that echoes in nearly every war memoir or veteran's interview is that comrades-in-arms come to feel an intense love for one another. In his outstanding war novel, *Gates of Fire*, chronicling the courageous stand by the Spartans at Thermopolyae, Steven Pressfield makes a compelling case that "the opposite of fear is love."[18] The love a warrior has for his comrades is what gives him the strength to stand against the dreadful tide of a heavily armed, charging host of enemies. Pressfield presents this point eloquently through the words of a character named "Suicide," a slave fighting beside the Spartan peers:

---

[16]Michael Ignatieff, *Virtual War: Kosovo and Beyond*, New York: Picador USA (Metropolitan Books, Henry Holt and Company), 2000, pps. 214-215.

[17]Shay p.117.

[18]Steven Pressfield, *Gates of Fire: An Epic Novel of the Battle of Thermopylae*, New York: Bantam Books, 1998, p. 380.

"When a warrior fights not for himself, but for his brothers, when his most passionately sought goal is neither glory nor his own life's preservation, but to spend his substance for them, his comrades, not to abandon them, not to prove unworthy of them, then his heart truly has achieved contempt for death, and with that he transcends himself and his actions touch the sublime. This is why the true warrior cannot speak of battle save to his brothers who have been there with him. This truth is too holy, too sacred, for words."[19]

Writing about his own experiences in World War II, renowned biographer William Manchester conveys matching sentiments in his memoir *Goodbye, Darkness*:

I understand, at last, why I jumped hospital that Sunday thirty-five years ago and, in violation of orders, returned to the front and almost certain death.

It was an act of love. Those men on the line were my family, my home. They were closer to me than I can say, closer than any friends had been or ever would be. They had never let me down, and I couldn't do it to them. I had to be with them, rather than let them die and me live with the knowledge that I might have saved them. Men, I now knew, do not fight for flag or country, for the Marine Corps or glory or any other abstraction. They fight for one another. Any man in combat who lacks comrades who will die for him, or for whom he is willing to die, is not a man at all. He is truly damned.[20]

The warriors' willingness to fight alongside their comrades and perhaps die for them is the most undeniable evidence there can be for their mutual love and respect. In the famous St. Crispin's Day speech from *Henry V*, the ever-perceptive William Shakespeare has his King Henry acknowledge his shared humanity with even the most "vile" commoners among those warriors with whom he intends to make a stand at Agincourt:

> . . .he which hath no stomach to this fight,
> Let him depart; his passport shall be made,
> And crowns for convoy put into his purse:
> We would not die in that man's company,
> That fears his fellowship to die with us.
> This day is call'd—the feast of Crispian:
> He that outlives this day, and comes safe home,
> Will stand a tip-toe when this day is nam'd,
> And rouse him at the name of Crispian.
> He that outlives this day, and sees old age,
> Will yearly on the vigil feast his friends,
> And say, To-morrow is saint Crispian:
> Then will he strip his sleeve, and show his scars,
> And say, These wounds I had on Crispin's day.
> Old men forget; yet all shall be forgot,
> But he'll remember, with advantages,
> What feats he did that day. Then shall our names,
> Familiar in their mouths as household words,—
> Harry the king, Bedford and Exeter,
> Warwick and Talbot, Salisbury and Gloster,—
> Be in their flowing cups freshly remember'd.
> This story shall the good man teach his son;
> And Crispin Crispian shall ne'er go by
> From this day to the ending of the world,
> But we in it shall be remembered,—
> We few, we happy few, we band of brothers;

---

[19]Ibid p. 379.

[20]William Manchester, *Goodbye, Darkness: A Memoir of the Pacific War*, New York: Dell Publishing, 1979, p. 451.

> For he to-day that sheds his blood with me,
> Shall be my brother; be he ne'er so vile,
> This day shall gentle his condition.
> And gentlemen in England, now a-bed,
> Shall think themselves accurs'd, they were not here;
> And hold their manhoods cheap, whiles any speaks,
> That fought with us upon saint Crispin's day.[21]

Sharing a code, along with the sheer force of shared experience, is what binds warriors together in the crucible of combat. That is why most warriors' codes come from within the warrior culture itself, they are not imposed upon it by some external source (such as a fearful civilian population). Of course, there are many rules that govern the lives of modern warriors that were put in place by the societies that they serve. Some of these exist to protect against abuses of military power. Others are to make sure a given nation's warriors do not violate international standards of conduct. In the United States, specific Rules of Engagement (ROE) and the Uniform Code of Military Justice (UCMJ) spell out much of what is expected of our warriors. But a code is much more than just rules of this type.

In his critical analysis of the problem of motivating ethical behavior among combat troops, *Obeying Orders: Atrocity, Military Discipline and the Law of War*, Mark Osiel, a law professor at the University of Iowa, wrestles with the complex subject of how to control warriors' conduct in the "fog of war." His research goes beyond traditional academic and legal scholarship to include first-hand interviews with war criminals and their victims.

The central thesis of *Obeying Orders* is that "...the best prospects for minimizing war crimes (not just obvious atrocity) derive from creating a personal identity based on the virtues of chivalry and martial honor, virtues seen by officers as constitutive of good soldiering."[22] In other words, Osiel asserts that the best way to ensure that, for instance, a young U.S. Marine will not commit a war crime even if given (illegal) orders to do so by a superior officer is not to drill the said Marine on the provisions of international law and the UCMJ, but rather to help him internalize an appropriate warrior's code that will inspire him to recognize and reject a criminal direction from his officer.

He tells the story of a young enlisted Marine in the Vietnam war whose judgment concerning the distinction between combatants and non-combatants was compromised after he had seen one too many of his buddies blown away. An officer found the youth "with his rifle at the head of a Vietnamese woman."[23] The officer could have tried barking out the relevant provisions of military law. Instead, he just said, "Marines don't do that." Jarred out of his berserk state and recalled to his place in a long-standing warrior tradition, the Marine stepped back, and lowered his weapon. As Osiel notes, the statement, "Marines don't do that," is "surely a simpler, more effective way of communicating the law of war than threatening prosecution for war crimes, by the enemy, an international tribunal, or an American court-martial."[24]

Osiel makes a strong case for this psychologically powerful code- and character-based approach to the prevention of war crimes. He connects it to Aristotle's virtue ethics, which stresses the importance of positive habituation and the development of certain critical virtues, such as courage, justice, benevolence, and honor, over the rote memorization of specific rules of conduct. Simply staying within the bounds of a rulebook, as Osiel observes, can often be less demanding than consistently upholding high standards of character and nobility:

> The manifest illegality rule merely sets a floor, and a relatively low one at that: avoid the most obvious war crimes, atrocities. It does not say, as does the internal ideal of martial honor: always cause the least degree of lawful, collateral damage to civilians, consistent with your military objectives. By taking seriously such internal conceptions of martial honor, we may be able to impose higher standards on professional soldiers

---

[21] William Shakespeare, *Henry V*, Act IV, Scene III, in *The Globe Illustrated Shakespeare: The Complete Works*, edited by Howard Staunton, New York: Greenwich House, 1986, p. 850.

[22] Mark Osiel, *Obeying Orders: Atrocity, Military Discipline, and the Law of War*, New Brunswick and London: Transaction Publishers, 1999, p. 23.

[23] Op. cit.

[24] Op. cit.

than the law has traditionally done, in the knowledge that good soldiers already impose these standards upon themselves.[25]

Osiel comments on the importance of shaming tactics (especially so-called "reintegrative shaming" which aims to reform, not permanently ostracize, the offender) to motivate modern warriors' dedication to the ideals of martial honor. He also defends the value of presenting persons entering the military culture with role models who remained true to their codes of honor even in the face of overwhelming challenges or temptations. As further support for his position, he points out that this approach of reinforcing desirable character traits among military professionals in no way undermines a rule-following approach, but rather provides additional motivation to obey rules when they are clear ("bright-line rules") while giving much-needed guidance when the rules are not enough:

> In cases that are legally easy (but otherwise stressful, dangerous, or physically demanding), then, martial honor contributes to having the proper inclinations and emotions, those conducive to skillful performance of one's duties. In legally hard cases, however, professional character reveals itself more in virtuosity of perception, deliberation, and choice.[26]

A warriors' code of the type advocated by Osiel cannot be reduced to a list of rules. "Marines don't do that" is not merely shorthand for "Marines don't shoot unarmed civilians; Marines don't rape women; Marines don't leave Marines behind; Marines don't despoil corpses, etc," even though those firm injunctions and many others are part of what we might call the Marines' Code. What Marines internalize when they are indoctrinated into the culture of the Corps is an amalgam of specific regulations, general concepts (e.g. of honor, courage, commitment, discipline, loyalty, teamwork, etc.), history and tradition that adds up to a coherent sense of *what it is to be a Marine*. To remain "Semper Fidelis," or forever faithful to the code of the Marine Corps is never to behave in a way that cannot be reconciled with that image of "what it is to be a Marine."

What it is to be a Marine is not the same as what it is to be an Air Force pilot or what it is to be an Army Ranger. For this reason, specific sub-cultures within the U.S. military, though equal under the UCMJ, have different warriors' codes. Their codes necessarily have some common ground, given that they are all American warriors, sworn to uphold and defend the values of the Constitution. This shared foundation notwithstanding, even within the U.S. Navy, for example, the code that governs the Navy SEAL community is quite distinct from the code of a Naval aviator, and neither could be confused with the code of a Surface Warfare Officer or that of a submariner.

A code that encompasses all of what it is to be a particular kind of warrior may help the warrior who has internalized it to determine the proper course of action in a situation the rule-writers could never have foreseen. The motivation for individual warriors to remain true to their code often comes from their desire not to betray the memory of the warriors who came before them as well as from their determination not to let down the warriors alongside whom they are now fighting. When future U.S. Army officers attend the United States Military Academy at West Point, they are taught to remember the "Long Gray Line" of former cadets that stretches back through the centuries and includes such giants as Generals Robert E. Lee, Ulysses S. Grant, Dwight D. Eisenhower, Omar Bradley, George S. Patton, and Roscoe Robinson Jr. (the first African-American four-star general). At times, the reverence modern warriors feel for their illustrious predecessors almost resembles the ancestor worship that is found in so many of the world's older religions. Warriors are proud to receive the legacies of the past and wish to remain worthy of them.

While unique aspects of the codes of each warrior's own culture or sub-culture provide the final flourishes to his or her self-definition, most warriors also feel themselves a part of an even longer line; a line of men and women from diverse cultures throughout history who are deserving of the label "warrior." This is a legacy that spans not just centuries, but millennia.

When the warriors of today want to improve their understanding of what it is to be a warrior, they naturally seek out the most accurate information about the harsh realities of war. But in addition they are often drawn to war-themed poetry, literature, and myth. These sources can convey the emotional

---

[25] Ibid p. 32.
[26] Ibid p. 37.

dimension of the warrior's existence more profoundly than purely factual accounts of battles, tactics, casualties and military careers. And the inspiration to strive for high ideals can come even from a culture's unrealized visions of the perfect warrior. When a television recruiting advertisement for the U.S. Marine Corps shows a knight in shining armor morphing into a Marine in Dress Blues, the specter of medieval chivalric perfection it evokes may never have existed in the flesh—yet this in no way reduces the impact.

In the following chapters, I will offer a sampling of some of the most influential interpretations of the code of the warrior and constructions of warrior archetypes from around the globe that have helped shape our broader conceptions of what it is to be a warrior. Though my selections do have some breadth—ranging from the Romans to Native Americans to Chinese warrior monks—this is by no means an exhaustive survey. Even restricting myself to the historical rather than the current, I found the number of significant warrior cultures available for study absolutely staggering. In the end, my selections were dictated by several considerations.

First of all, I wanted to give examples that were likely to be familiar but which might not have considered in this context or in relation to one another. A number of my students when asked to conjure up an archetypal warrior picture a Knight of the Round Table. Others see a samurai, still others a Viking in a horned helmet (although a few already know it was the Celts, not the Vikings, who wore horned helmets). These are at least a small selection of the cultures with which a modern Western warrior might identify. Secondly, I tried to avoid too much overlap. The culture of the Spartans is deeply fascinating, but it contains key elements that I have explored here by looking at Homeric Greeks warriors and the soldiers of the Roman legions. Each of the warrior cultures I touch upon introduces features that are not seen in the others I cover, although those distinctive features can be found in many additional cultures that I was not able to consider in this volume. Finally, I confess to simply having a personal interest in the particular warrior values and ideals featured in this book.

A gentleman I have been honored to call my friend, Mr. Andrew H. Hines, Jr., was once a B-17 navigator for the U.S. Army Air Corps in World War II. Andy flew several successful missions out of Foggia airbase in Italy, including raids over Belgrade, but was eventually shot down and spent the remainder of the war in Germany prison camps Stalag Luft III (site of the notorious "Great Escape") and Moosburg before being liberated by Patton's Third Army. Many years later, as a retired businessman and poet, Andy tried to capture in verse the experience of what it was, for him, to be a warrior. One of his poems, entitled "The Somme, 1914-1918," contains his reflections on the meaning of a warrior's life, and death:

> We drove across the Somme today
> Where war's flail once struck hard,
> And all was peaceful and serene
> But graves stood as on guard.
>
> The endless rows of sugar beets
> Bespoke a fertile earth
> But richer dust is buried here
> Beneath the dark green turf.
>
> And battles sometimes futile seem
> Not worth the heavy cost,
> But each must stand in his own time
> Or see his freedom lost.[27]

Why do warriors fight? What is worth dying for? How should a warrior define words like "nobility," "honor," "courage," or "sacrifice?" What are the duties and obligations of a warrior and to whom are they owed? How can you measure a warrior's commitment? What should bring a warrior honor, and what should bring a warrior shame? [. . .]My hope is that . . . you [will take] the opportunity to critically evaluate and compare different warriors' codes, so that you can judge what aspects of them should be rejected and which should be preserved or revived for the future in order to create an ideal code of the warrior for this new millennium.

---

[27] Andrew H. Hines, Jr., *Time and The Kite*, St. Petersburg, Florida, 1989, p. 23.

Part V
Upholding Truth, Enforcing Justice, Defending Liberty and Rights

Throughout the preceding sections, we have alleged that justifications for the use of military force routinely rely upon appeals to concepts like "justice," or the defense of "liberty and rights." What is the moral force of such concepts, than renders them of sufficient gravity to warrant violence, destruction, and loss of life in their pursuit or protection? What are these ideals, that human beings come to prize them so highly, or regard them as so precious that we are prepared to kill, or to die, in their defense?

Amazingly, it is all too often the case that those prepared to order the killing, and those likewise prepared to undertake the dying, do so with little knowledge or direct acquaintance with these moral cornerstones of western culture. To be sure, a police officer need not also be a lawyer (God forbid!) or even a scholar of jurisprudence, to willingly subject himself or herself to threat of harm in order to ensure the rule of law domestically. Likewise, we needn't require military officers to be authorities in matters of justice and human rights in order to provide for their defense.

Still it would seem incumbent upon the society that these officers protect to offer them something in the way of an explanation for the importance our society attaches to these moral ideals. Likewise, the police officer or the military officer would have not only a right to understand the causes for which they are asked to wield force or place themselves in harm's way, but also the obligation for coming to terms with these issues as part of their orientation to their respective professions. Somehow—all too often in fact—these particular "rights" of prospective combatants are overlooked, and these obligations to them remain unfulfilled.

It may seem odd or even mistaken to include a consideration of "truth," or the importance of truthfulness, alongside these other, seemingly more political, moral ideals. To be sure: telling the truth, or being persons who are truthful and therefore trustworthy, are among the moral characteristics of a good military officer, and accordingly constitute a requirement or expectation well established within the profession of arms. But specifying an essential characteristic or "virtue" required of members of the profession is different in kind from setting forth, and examining, the sorts of political or social ideals for which that professional might be expected to struggle and sacrifice.

Upholding the truth is of greater political consequence than this comparison might suggest. Following decades of internal, often violent struggle over the legal and political system known as "apartheid," the nation of South Africa finally, in the latter part of the last century, achieved a relatively peaceful transition to majority rule. Those decades of racial violence, however, left a bitter, simmering, and dangerous legacy of injustice that threatened to spawn a thirst for vengeance. Left unacknowledged, such legacies in other nations have inspired the kinds of ethnic tensions and violence that have spilled over into genocide. Knowing this, South Africa's spiritual leader, Archbishop Desmond Tutu, formed a "Truth Commission" designed to elicit—voluntarily, without punishment or recrimination—explanations and acknowledgement of guilt for the secret atrocities that had been perpetrated in the past in the name of racial supremacy.

The effort was ridiculed at first as pointless and ineffective. Yet gradually, as members of that society came forward in shame and sorrow to describe the often terrible things they had done, to acknowledge complicity and guilt, and to ask forgiveness for past wrongs, a spirit of healing and reconciliation took hold in that troubled land. Bishop Tutu's effort was, and is, far from complete, the effects far from perfect, and no doubt the confessions were not always fully accurate or (as critics complained) were they entirely sincere. Yet all have come to acknowledge, in that instance, the restorative importance of "knowing the truth." While it remains to be seen, it is quite possible that these acts of voluntary acknowledgement and contrition, in a spirit of mercy and forgiveness, have saved South Africa from the fate of Bosnia, Rwanda, and Kosovo.

Many nations and peoples, including our own, by contrast, have been less than forthcoming regarding our own violent pasts. Perhaps we have much to learn from the gentle Bishop from South Africa. If he is correct, then understanding, acknowledging, and respecting Truth may turn out to be every bit as important, politically, as understanding and establishing Justice, or protecting basic human Rights. Perhaps we will discover what the Bishop's own mentor, a simple carpenter from Nazareth, once proclaimed, that: "you will know the Truth, and the Truth will set you Free."

A. Liberty and Rights

The rhetorical appeal to "liberty and rights" is often made, as we have observed, as a justification or "just cause" for using military force and for committing the nation to war. At the same time, almost paradoxically, military life itself is structured by duties, responsibilities, regulations, and, most especially *restrictions* on the customary "rights and liberties" that civilian citizens routinely enjoy. Accordingly, military officers need to understand why those rights and liberties are worth defending, and at the same time, why some of these rights and liberties must be somewhat restricted in the military world. It is helpful, also, to know why some rights must be respected absolutely, even in military life.

Thus it would be helpful to have answers to questions, such as: "what is this thing called 'Liberty,' and what are, and from whence come, these things called 'Rights?'"

We have seen that the notion of "natural rights," central to our nation's founding documents and prevailing political philosophy, has its origins in the blending of Stoic philosophy and natural law. The same is true of liberty, understood perhaps as the most singular and precious of these rights, that of political and social self-determination, free (insofar as reasonable) from external political constraint.

Despite scattered examples of "popular uprisings" in the ancient world (as in the city-state of Athens, or as led by the Roman slave, Spartacus) in behalf of political prerogatives that might justifiably be characterized as "essential liberties" or "basic human rights," however, it is somewhat anachronistic to read our contemporary historical understanding of these concepts back into such events. Such concepts are barely in evidence (if operative at all) in the "biblical" cultures of the ancient near east, or in ancient Greece or Rome. No condemnation of slavery, for example, is to be found in the writings of Plato or Aristotle, nor explicitly (apart from the Exodus narrative, perhaps) in the literature of Judaism, Christianity, or Islam. Nor is there, within these otherwise-rich and diverse cultural resources, any condemnation, for example, of "paternalism" toward adults about which Mill and Gerald Dworkin write in our reading selections for this chapter. Nor, finally, do these ancient authorities insist on establishing "rights" or basic civil liberties for all citizens, such as freedom of opinion, speech, press, assembly, or religious observance.

Rather, it is modern ethical thought that is marked by the emergence to prominence of the concept of individual *human rights*. And likewise, it is modern social and political thought that is dominated by a concern with the particular rights, or "negative claims," of *liberty*: that is, political guarantees respecting the freedom of individuals, and further protecting them from unwarranted, coercive interference by the state, society, or other individuals. These concepts are intertwined and closely related, and enjoy a kind of evolutionary history in western culture, from scattered and nascent recognition in the ancient world, to a position of dominance in the political and social thought of the Reformation, and especially the European Enlightenment. This assessment reveals a bit about how, and from where these concepts emerged, but still does not shed much light on what they are, or what they mean, let alone why such ideas ever "caught on," or why they are (and whether they properly should be) of such great importance to individuals today.

In part this is due to the fact that an understanding of "rights" or of "natural" or "human rights" is not unlike the problem that we have already encountered of understanding the role of "virtues" in moral theory. Both are broad generalizations of concepts that usually have clarity and purchase only in specific social, historical, or legal contexts, or in very specific relationships established (as in work, or marriage) between individual human beings.

For example: if I agree to undertake labor for hire for another individual, perhaps my neighbor, I then take upon myself a duty to perform certain chores in certain, mutually agreeable ways for that individual, and he has a "right" (a justifiable claim upon me) to perform these activities as agreed. Upon satisfactory completion, I then have a "right" to be paid for my work, and my neighbor has a duty to pay me. We may both assume, in this context, that neither of us is under a special compulsion to enter into these arrangements. Both of us have the freedom—we are said to be "at liberty"—to act (or refrain from acting) in this manner as we individually choose. Our liberty is bound only, in this instance, by the claims our mutual agreement subsequently places upon each of us. In this simple example, the notion of liberty, certain "rights" as justifiable claims (and reciprocally, the corresponding duties or obligations laid upon the other agent to perform in ways that fulfill these rights) seems reasonably well-defined in principle.

We can likewise understand the notions of liberty (acting freely, without compulsion) and "rights" (justifiable claims laid by one individual upon another, or upon society in general) within a distinct legal or political system. The State may establish laws governing the behavior of its citizens that grant "rights" (certain privileges or prerogatives) to each of them. In a famous example from the ancient world, the Christian apostle, Paul of Tarsus, was placed under arrest by local Roman authorities in Jerusalem, and convicted of inciting civil unrest. Paul, however, had earlier been conferred the status of "citizen of Rome," which status entitled him to certain "legal rights," among which was the "right" to appeal his conviction directly to Caesar (a vague antecedent of our modern right of legal appeal). He produced evidence of his citizenship, demanded this "right," and the local authorities accordingly sent him off to Rome (where, according to what little detail is known, things did not go all that well at trial).

We enjoy similar, but a far wider variety of, specific prerogatives or special status under the law within political society today. These "political rights," of course, will be found to vary from nation to nation. In the U.S., and in many other democracies, for example, the reigning legal system guarantees the "right" to a speedy trial, and usually also the "right" to trial by jury. The right is a prerogative, granted by law, pertaining to citizens of that nation living under that law. These rights may, but need not necessarily, be granted to non-citizens: to visitors or immigrants from other countries, for example. The corresponding duty (and these two rights engender considerable duties and financial obligations to maintain a legal system and to organize a qualified pool of peers as jurors) fall upon the State, or "society" as a whole. So, for example: we are all obliged to pay taxes to support the functioning of the legal system, and to serve as jurors if summoned.

These latter obligations, flowing as a consequence of the "rights," prerogatives, or special status granted to all citizens under some specific rule of law, represent corresponding limitations upon our freedom to act. They establish restraints upon our political "liberty." In this instance, presumably, we willingly take on or accept these restrictions of liberty as a kind of price tag associated with the granting and the enjoyment of protection, under the law, of these political rights. Again, the definition and symmetry in this instance seem reasonably straightforward, and the understanding of the meaning of "liberty," and of "rights" in this specific context, correspondingly clear.

Now we encounter the difficulty, similar to the difficulty encountered in generalizing from the specific virtues or excellences associated with well-defined practices, professions, or specific cultures, to those characteristic of "human beings" as such. In the case of liberty and rights, the problem is: can we understand or generalize these concepts, as well, to apply to human beings as such, quite apart from any specific circumstances or legal or political situations in which these concepts are ordinarily defined? Is there some property of human existence that we might define as individual "liberty" *per se*, or are there a list of "justifiable claims" that each individual may make upon others, and upon society as a whole, that we might define as "basic human rights?"

Historically, for example, very few states or governments granted the rights we have examined, or afforded citizens the liberties we have described. Apart from Athenians, and some fortunate Roman citizens, few individuals were permitted to move about and pursue their lives and interests freely. Few enjoyed the full range of "rights" or protections afforded by the prevailing legal system. The vast majority of ancient populations, as slaves, foreigners, or females, were excluded from consideration, or from the protection, of "legal rights," and their liberty was almost non-existent.

A cultural relativist might be tempted to observe, of course, "that was then, this is now." Certainly it is trivially true that we understand our political and social world differently today than did ancient Babylonians, Romans, or Spartans and Callatians. But is this all there is to say on this matter? Might we arguably claim to have made some "moral progress" since then, or to have learned some "moral lessons" precisely about concepts like liberty and rights, and the essential roles they do, and always have, in principle, played in making for an authentically human life? Might we, after all, discern a narrative historical thread of humanity, linking the citizens of Athens, Hebrew slaves in Egypt, Spartacus, William Wallace, and other dramatic figures and periods lifted up from time to time in books and cinema to illustrate a "story" about how and why we no longer think it right, or just, to deny individuals their liberty? And would this common narrative validate our belief that it is not now right, nor was it ever, to enslave them, or to disenfranchise or debase them on the basis of their race or gender, or otherwise deny to these others what we (along with those ancient tyrants, kings, and nobles) desire so fervently for ourselves: namely, basic human rights?

Let us set aside that difficult question for a moment. It is still the case, and until recently was the case in this nation, that not all inhabitants, or even full-fledged and native-born "citizens," enjoy the full range of protections ("rights") or enjoy a degree of political freedom and liberty to act as they choose. Is it possible, following the procedures and traditions of natural law, to generalize (as did Jefferson, and later King) and argue that there are a list of "basic rights" or "natural rights" whose justification is "self-evident," and which are inalienable (in Jefferson's phrase), presumably just because we are human beings or rational, moral agents?

If so, what are these rights? Are they restricted to the realm of what might be called "negative liberties," claims by the individual for non-interference by the State, or do they include some "positive" rights, such as the right to a basic education, health care, trial by jury, employment, equal opportunity, or other requisite political and biological needs? If so, *who or what serves as guarantor of these rights*, especially if they are not recognized by the particular nation or State within which some of the biological individuals or moral agents may find themselves living? Pre-eminently, would the right of political self-determination—Liberty—be foremost among these? Or, would our relativist still stubbornly argue that "liberty and rights" are merely historically-conditioned, and politically context-dependent: that is, that these rights and liberties only "exist" if granted within a historical culture or a specific political or legal system, and are meaningless apart from such systems.

This difficult question is one which has long been addressed by the natural law tradition, as we have seen. However, our other, more recent moral perspectives have a great deal to add to the difficult deliberation over whether political liberty is a self-evident and inalienable, absolute right, as well as to the accompanying question of what other "human rights" (if any) should be added to the list. Our reading selections in this chapter address different aspects of this complex problem.

The Readings

Given what was said earlier about Jeremy Bentham's utter contempt for the concept of human rights ("the concept of 'rights' is nonsense," Bentham fumed, "and that of 'natural rights,' nonsense on stilts!") it may be surprising to find one of the most brilliant defenses of the fundamental right of liberty in the entire history of western culture was written by fellow-utilitarian philosopher, John Stuart Mill. Had we read further in the final chapter of Mill's pamphlet on utilitarianism, however, we would have found that he offered a spirited interpretation and defense of justice and basic human rights, quite apart from natural law. These ideals are not, he thought, conceptions that stand apart from a consideration of the common good. Rather, they represent "rules of thumb:" that is, particularly important and specific formulae, discerned in human history and in the totality of human experience, precisely for the purpose of guaranteeing, and bringing about, the common good. First and foremost among these *prima facie* "rules of thumb," he argued, is respect for individual liberty.

On Liberty (1859) is arguably Mill's most outstanding single work. In our selection, Mill confronts the problem of liberty, and distinguishes between what he terms "self-regarding" (private) and "other-regarding" (public) behavior, including behavior that might constitute harm to others. He then formulates what he terms the "Liberty Principle," or the "Harm Principle," namely: that the only justification for interference by society or the state in the affairs of individuals is *to protect others from harm*.

Otherwise, with respect to one's purely private behavior, Mill maintains, "over himself, over his body and mind, the individual is sovereign."

Unlike the declarations of Thomas Jefferson emanating from the natural law tradition, however, this famous utterance of Mill's is not simply a "self-evident" truth or unsubstantiated political rallying cry. That is the importance of this seminal essay. Instead, Mill adduces reasons, grounded in utility, for acknowledging this claim regarding individual freedom. Liberties of speech and press, for example, aid in discerning truth and exposing error, both of which are essential to responsible self-government and the pursuit of happiness in a democratic society. The protections and limitations afforded by the "harm principle" are sufficient to prevent any genuine abuses of this freedom, and unless there is harm to others from my speech or writing, I ought (Mill maintains) to be free to pursue it as in the overall best interests of society, even if others are on occasion annoyed or offended by the positions I advocate.

Among the other principle argument advanced for extending the maximum amount of liberty possible to each individual, concurrent with a like amount of liberty to all, is that individuals are usually far better at discerning their interests, and pursuing their individual goods, than is the State or society, however well meaning and well intentioned it may be. "In general," Mill writes, "government should avoid interfering with the private lives of citizens, since they invariably do a poor job of regulation, and cause more harm than good, even when well intentioned."

This is an argument against what is termed "paternalism," the concept that society and the State have a vested interest in controlling the behavior of individuals. Mill rather clearly does not believe there are a great many occasions when such a belief is justified, and that, in general, the less the State interferes in our private affairs, the better off everyone will be. This is not unambiguously the case, however, with respect to matters like health and safety, for example. And we are left, in any case, to ponder the question raised by David Rodin, whether and when the *defense* of our liberties constitutes a "right" to go to war. Thus, following a somewhat elliptical path, we are back where we began, but are now better equipped to consider, with greater awareness and understanding, just how difficult the specification of a list of valid "human rights" might be. Perhaps we can now also appreciate just how important a concept this is for our political and individual well being, and why these concepts might be worthy of our respect, our loyalty, and, if necessary, our defense.

On Liberty*
John Stuart Mill

© 2008 Jonathan Bennett

Chapter 1: Introduction

The object of this Essay is to assert one very simple principle and to argue that it should absolutely govern how society deals with its individual members in matters involving compulsion and control, whether through physical force in the form of legal penalties or through the moral coercion of public opinion. The principle is this:

> The only end for which people are entitled, individually or collectively, to interfere with the liberty of action of any of their number is self-protection. The only purpose for which power can be rightfully exercised over any member of a civilized community, against his will, is to prevent harm to others.

The person's own good, whether physical or moral, isn't a sufficient ground ·for interference with his conduct·. He cannot rightfully be compelled to do (or not do) something because doing it (or not doing it) •would be better for him, •would make him happier, •would be wise (in the opinions of others), or •would be right. These are good reasons for protesting to him, reasoning with him, persuading him, or begging him, but not for compelling him or giving him a hard time if he acts otherwise. To justify that—·i.e. to justify compulsion or punishment·—the conduct from which it is desired to deter him must be likely to bring harm to someone else. The only part of anyone's conduct for which he is answerable to society is the part that concerns others. In the part that concerns himself alone he is entitled to absolute independence. Over himself, over his own body and mind, the individual is sovereign.

I hardly need say that this doctrine is meant to apply only to human beings when they have reached the age of maturity. We aren't speaking of children, or of young persons below the age that the law fixes as that of manhood or womanhood. Those who still need to be taken care of by others must be protected against their own actions as well as against external injury. For the same reason, Liberty, as a principle, doesn't apply to any state of affairs before mankind have become capable of being improved by free and equal discussion. Until then, there is nothing for them but implicit obedience to an Akbar or a Charlemagne, if they are so fortunate as to find one—·i.e. to find a despot so wise·. But in all the nations with which we need to concern ourselves here, the people long ago became able to be guided to self-improvement by conviction or persuasion; and once that stage has been reached, compulsion—whether direct physical compulsion or compulsion through penalties for non-compliance—is no longer admissible as a means to their own good, and is justifiable only for the security of others.

---

* Copyright © 2010–2015 All rights reserved Jonathan Bennett. Last amended: April 2008 [Brackets] enclose editorial explanations. Small ·dots· enclose material that has been added, but can be read as though it were part of the original text. Occasional •bullets, and also indenting of passages that are not quotations, are meant as aids to grasping the structure of a sentence or a thought. Every four-point ellipsis . . . . indicates the omission of a brief passage that seems to present more difficulty than it is worth. Longer omissions are reported between square brackets in normal-sized type. Further editing by Lawrence Lengbeyer.

It might seem easier for me to defend my position if I took this stance:

> 'It is just objectively abstractly *right* that people should be free; never mind what the consequences of their freedom are.'

But I don't argue in that way, because I hold that the ultimate appeal on all ethical questions is to *utility*—i.e. to 'what the consequences are'. However, it must be utility in the broadest sense, based on the permanent interests of man as a progressive being. Those interests, I contend, make it all right to subject individual spontaneity to external control *only* in respect to those actions of each individual that concern the interests of *other* people. If anyone does something harmful to others, there is a prima facie case for punishing him—either by law or, where legal penalties are not safely applicable, by general disapproval. There are also many positive acts for the benefit of others that an individual may rightfully be compelled to perform:

> to give evidence in a court of justice,
> to do his fair share in the defence of his country, or
> any other joint work necessary to the interests of the society whose protection he enjoys;

and to perform certain acts of individual beneficence. For example, a man may rightfully be held to account by society for not saving a fellow-creature's life, or not protecting a defenceless person against ill-treatment, in situations where it was obviously his duty to do this. A person may cause harm to others not only by his •actions but by his •inaction, and either way he is justly accountable to them for the harm. The •latter case, it is true, requires a much more cautious exercise of compulsion than the •former. To make someone answerable for *doing harm* to others is the rule; to make him answerable for *not preventing harm* is, comparatively speaking, the *exception*. Yet there are many cases clear enough and serious enough to justify that exception. In everything concerning the external relations of the individual, he is legally answerable to those whose interests are concerned, and if necessary to society as their protector. There are often good reasons for not holding him to that responsibility; but these reasons must arise from special features of the case: either

- it is a kind of case where he is likely to act better when left to himself than when controlled in any way that society could control him; or
- the attempt to exercise control would have bad effects greater than those that it would prevent.

When such reasons as these rule out the enforcement of responsibility, the person's own conscience should move into the vacant judgment-seat and protect those interests of others that have no external protection; judging himself all the more severely because the case doesn't admit of his being made accountable to the judgment of his fellow-creatures.

But there is a sphere of action in which the interests of society, as distinct from those of the individual, are involved only indirectly if they are involved at all: it is the sphere containing all the part of the individual's life and conduct that affects only himself, or affects others but only with their free, voluntary, and undeceived consent and participation. When I say 'affects only himself' I am talking about the direct and immediate effects of his conduct. ·This has to be stipulated·, for whatever affects himself may affect others *through* himself. (Conduct may be objected to on that ground; I'll consider this later.)

So this is the appropriate region of human liberty. ·I map it as containing three provinces·. **(1)** The inward domain of consciousness, demanding •liberty of conscience in the broadest sense, •liberty of thought and feeling, absolute •freedom of opinion and sentiment on all subjects, practical or theoretical, scientific, moral, or theological, and •liberty of expressing and publishing opinions. This last may seem to belong under a different principle, since it involves conduct of an individual that affects other people; but it can't in practice be separated from the liberty of thought—it is almost as important as the latter and rests in great part on the same reasons. **(2)** Liberty of •tastes and pursuits, of •shaping our life to suit our own character, of •doing what we like. . . .—all this without hindrance from our fellow-creatures, so long as what we do doesn't harm them even though they may think our conduct foolish, perverse, or wrong. **(3)** Following from the first two domains of liberty, there is the liberty, within the same limits, •of individuals to come together, their freedom to unite for any purpose not involving harm to others—always supposing that the people in question are of full age and aren't being forced or deceived.

No society in which these liberties are not •mainly respected is free, whatever form of government it has; and none is *completely* free in which they don't exist •absolute and unqualified. The only freedom that deserves the name is the freedom to pursue our own good in our own way, so long as we don't try to deprive others of their good or hinder their efforts to obtain it. Each is the proper guardian of his own health of body, mind, and spirit. Mankind gain more from allowing each other to live in the way that seems good to themselves than they would from compelling each to live in the way that seems good to the rest.

Chapter 2: Liberty of Thought and Discussion

Let us suppose that the government is entirely in harmony with the people, and never thinks of coercing anyone except in ways that it thinks the people want. But I deny the right of the people to exercise such coercion, whether directly or through their government. The power ·of coercion· itself is illegitimate. The best government has no more right to it than the worst. It is *at least* as noxious when exerted •in accordance with public opinion as when it is exerted •in opposition to it. If all mankind minus one were of one opinion, and that one had the contrary opinion, mankind would be no more justified in silencing that one person than he would be in silencing them if he could. ·You might think that silencing *only one* couldn't be so *very* wrong, but that is mistaken, and here is why·. If an opinion were a personal possession of no value except to the person who has it, so that being obstructed in the enjoyment of it was simply a private injury, it *would* make some difference whether the harm was inflicted on only a few persons or on many. But the special wrongness of silencing the expression of an opinion is that it is robbing

- not one individual, but· the human race,
- posterity as well as the present generation,
- those who dissent from the opinion as well as those who hold it.

Indeed, those who dissent are wronged more than those who agree. •If the opinion in question is right, they are robbed of the opportunity of exchanging error for truth; and •if it is wrong, they lose a benefit that is almost as great, namely the clearer perception and livelier impression of truth that would come from its collision with error.

We need to consider these two cases separately; each has a distinct branch of the argument corresponding to it. We can never be sure that the opinion we are trying to suppress is false; and even if we were sure of its falsity it would still be wrong to suppress it.

First: the opinion the authorities are trying to suppress may be true. Those who want to suppress it will deny its truth, of course; but they aren't infallible. They have no authority to decide the question for all mankind, and exclude every other person from the means of judging. To refuse a hearing to an opinion because they are sure that it is false is to assume that their certainty is the same thing as absolute certainty—i.e. that *their being sure* that P is the same as *its being certainly true* that P·. All silencing of discussion is an assumption of infallibility.

An objection that is likely to be made to this argument runs somewhat as follows:

> There is no greater assumption of infallibility in forbidding the propagation of error than there is in anything else done by public authority on its own judgment and responsibility. *Judgment* is given to men to be used. Because it may be used erroneously, are men to be told that they oughtn't to use it at all? To prohibit something they think to be pernicious is not •to claim exemption from error, but •to perform their duty to act, fallible though they are, on their conscientious convictions. If we were never to act on our opinions because they may be wrong, we would leave all our interests uncared for and all our duties unperformed. An objection that applies to *all* conduct can't be a valid objection to any conduct in particular.

> Governments and individuals have a duty to form the truest opinions they can; to form them carefully, and never impose them on others unless they are quite sure of

being right. But when they are sure, it is not *conscientiousness* but *cowardice* to shrink from acting on their opinions, and to allow doctrines that they honestly think dangerous to the welfare of mankind—either in this life or in another—to be scattered abroad without restraint, just because other people in less enlightened times have persecuted opinions that are now believed to be true! Let us take care not to make the same mistake; but governments and nations have made mistakes in other things that are not denied to be fit subjects for the exercise of authority, such as imposing bad taxes and making unjust wars. Ought we therefore to impose no taxes, and whatever the provocation to make no wars? Men and governments must act to the best of their ability. There's no such thing as absolute certainty, but there is assurance sufficient for the purposes of human life. We may—we *must*—assume our opinion to be true for the guidance of our own conduct; and that's all we are assuming when we forbid bad men to pervert society by spreading opinions that we regard as false and pernicious.

I answer: No, it is assuming very much more than that. There is the greatest difference between •presuming an opinion to be true because, with every opportunity for contesting it, it hasn't been refuted, and •assuming its truth as a basis for not permitting its refutation. Complete liberty of contradicting and disproving our opinion is the very condition that justifies us in assuming its truth for purposes of action; and on no other terms can a human being have any rational assurance of being right.

Look at the history of •what people have believed; or look at the ordinary •conduct of human life. Why are each of these no worse than they are? It is certainly not because of the inherent force of the human understanding! Take any proposition that isn't self-evident: for every person who is capable of judging it, there are ninety-nine others who aren't; and the 'capability' of that one person is only comparative; for the majority of the eminent men of every past generation held many opinions now known to be erroneous, and did or approved many things that no-one would now defend. Well, then, why is it that there is on the whole a preponderance among mankind of •rational opinions and •rational conduct? If there really is this preponderance—and there must be, unless human affairs are and always were in an almost desperate state—it is owing to the fact that *the errors of the human mind can be corrected*. This quality of the human mind is the source of everything worthy of respect in man, whether as a thinking or as a moral being. He is capable of correcting his mistakes by discussion and experience. Not by experience alone: there must be discussion, to show how experience is to be interpreted. Wrong opinions and practices gradually give way to fact and argument; but facts and arguments can't have any effect on the mind unless they are brought before it. Very few facts are able to tell their own story, without comments to bring out their meaning. So: because the whole strength and value of human judgment depends on a single property, namely that it can be set right when it is wrong, it can be relied on only when the means of setting it right are kept constantly at hand. Consider someone whose judgment really is deserving of confidence— how has it become so? Through his conducting himself as follows:

- He has kept his mind open to criticism of his opinions and conduct.
- He has made it his practice to listen to all that could be said against him, to profit by as much of it as was sound, and expound to himself—and sometimes to others—the fallacy of what was fallacious.
- He has felt that the only way for a human to approach knowing the whole of a subject is by hearing what can be said about it by persons of every variety of opinion, and studying all the ways in which it can be looked at by every kind of mind.

No wise man ever acquired his wisdom in any way but this; and the human intellect isn't built to become wise in any other manner. The steady habit of correcting and completing his own opinion by comparing it with those of others, so far from causing doubt and hesitation in acting on the opinion, is the only stable foundation for a sound reliance on it. Knowing everything that can, at least obviously, be said against him, and having taken up his position against those who disagree, knowing that he has looked for objections and difficulties instead of avoiding them, and has shut out no light that can be thrown on the subject from any direction—he has a right to think his judgment better than that of any person or crowd of them that hasn't gone through a similar process.

Let us now pass to the second branch of the argument. Dismissing the thought of falsehood in the publicly accepted opinions, let us assume them to be true; and ·on that basis· let us look into the value of *how* they are likely to be held, given that their truth is not freely and openly discussed, pro and con. However unwilling a person who has a strong opinion may be to admit that his opinion *might* be false, he ought to be moved by this thought: however true it may be, if it isn't fully, frequently and fearlessly discussed, it will be held as a dead dogma rather than as a living truth.

There are people (fortunately not quite as many as there used to be) who will be satisfied if you assent undoubtingly to something that they think is true, even if you have no knowledge whatever of the grounds for the belief in question and couldn't defend it decently against the most superficial objections. When such people get their creed to be taught as authoritative, they naturally think that no good and some harm will come from allowing it to be questioned. Where their influence is dominant, they make it nearly impossible for the publicly accepted opinion to be rejected •wisely and considerately. It may still be rejected •rashly and ignorantly; for it is seldom possible to shut off discussion *entirely*, and once discussion gets started, beliefs that are held as creeds rather than being based on reasons are apt to give way before the slightest semblance of an argument. Set aside that possibility, and take the case where the true opinion remains in the person's mind, but sits there as a prejudice, a belief that owes nothing to argument and isn't vulnerable to argument—this isn't the way truth ought to be held by a rational being! This is not *knowing the truth*. Truth when accepted in that way is merely one more superstition, accidentally clinging to words that enunciate a ·genuine· truth.

Here is something that might be thought:

> When the publicly accepted opinions are true, if the harm done by the absence of free discussion of them were merely that men are left ignorant of the grounds of those opinions, this may be an intellectual evil but it isn't a moral one; it doesn't affect the value of the opinions so far as their influence on character is concerned.

In fact, however, the absence of discussion leads men to forget not only the •grounds for an opinion but too often also its •meaning. The words in which it is expressed cease to suggest ideas, or suggest only a small portion of the ideas they were originally used to communicate. Instead of a vivid conception and a living belief, there remain only a few phrases learned by heart; or if any part of the meaning is retained it is only the shell and husk of it, the finer essence being lost. This fact fills a great chapter in human history—one that cannot be too earnestly studied and meditated on.

It is illustrated in the experience of almost all ethical doctrines and religious creeds. They are all full of meaning and vitality to those who originate them, and to their immediate disciples. As long as a doctrine or creed is struggling for ascendancy over other creeds, its meaning continues to be felt as strongly as—and perhaps even more strongly than—it was at the outset. Eventually either •it prevails and becomes the general opinion or •its progress stops, in which case it keeps possession of the ground it has gained but doesn't spread any further. When either of these results has become apparent, controversy about the doctrine slackens and gradually dies away. The doctrine has taken its place as a publicly accepted opinion or as one of the recognized sects or divisions of opinion; most of its present adherents have inherited it rather than being convinced of it by reasons, and they don't give much thought to the idea of anyone's being converted from their doctrine to some other, because such conversions happen so rarely. At first they were constantly on the alert, either to defend themselves against the world or to bring the world over to their side; but now they have subsided into a passive state in which they •don't (if they can help it) listen to arguments against their creed, and •don't trouble dissentients (if there are any) with arguments in its favour. It is usually at about this stage in its history that the doctrine starts to lose its living power. We often hear the teachers of all creeds lamenting the difficulty of keeping up in the minds of believers a lively awareness of the truth to which they pay lip-service . . . . [W]hen the creed has come to be hereditary, and to be accepted passively rather than actively—when the mind is no longer as compelled as it once was to exercise its vital powers on the questions that its belief presents to it—there's a progressive tendency to forget all of the belief except the words expressing it, or to give it a dull and lethargic assent, as though by taking it on trust one freed oneself from any need to make it real in one's mind, or to test it by personal experience; until it comes to have almost no connection with the person's inner life.

No doubt there are •many reasons why doctrines that are the badge of one particular sect retain more of their vitality than do ones that are common to all recognized sects, and why teachers take more trouble to keep the meaning of the former alive; but •one reason certainly is that a doctrine that is special to one particular sect is more questioned, and has to be oftener defended against open opponents. Both teachers and learners go to sleep at their post when there is no enemy in the field.

There is yet another powerful reason why diversity of opinion can be advantageous—a reason that will hold good until mankind advances to an intellectual level that at present seems incalculably far off. We have so far considered only two possibilities: that •the publicly accepted opinion is false, and therefore some other opinion is true; and that •the publicly accepted opinion is true, but a conflict with the opposite error is essential to a clear grasp and deep feeling of its truth. However, there is a commoner case than either of these, namely: •the conflicting doctrines. . . .share the truth between them, and the minority opinion is needed to provide the remainder of the truth, of which the publicly accepted doctrine captures only a part. On matters other than plain empirical fact, popular opinions are often true but are seldom or never the whole truth. They are a part of the truth—sometimes a large part, sometimes a small—but exaggerated, distorted, and torn apart from other truths that ought to accompany them and set limits to them. Minority opinions, on the other hand, are generally some of these suppressed and neglected truths, bursting the bonds that kept them down and either •seeking reconciliation with the truth contained in the common opinion or •confronting it as enemies and setting themselves up as the whole truth. The latter case has always been the most frequent, because in the human mind one-sidedness has always been the rule and many-sidedness the exception. Hence, even in •revolutions of opinion one part of the truth usually sets while another rises [the comparison is with the setting and rising of the sun in the daily •revolutions of the earth]. Even *progress*, which ought to add truth to truth, usually only substitutes one partial and incomplete truth for another; and any improvement this brings comes mainly from the fact that the new fragment of truth is more wanted, better suited to the needs of the time, than was the one it displaces. Even when a prevailing opinion is basically true, it will be so partial that every ·rival· opinion that embodies some part of the truth that the other omits ought to be considered precious, no matter how much error and confusion it blends in with its portion of truth. No reasonable person will be indignant because those who force on our notice truths that we would otherwise have overlooked do themselves overlook some of the truths that we see. Rather, a reasonable person will think that so long as popular truth is one-sided, it is more desirable than regrettable that unpopular truth should have one-sided defenders too, because they—the one-sided ones—are usually the most energetic and the most likely to compel reluctant attention to the fragment of wisdom that they proclaim as if it were the whole.

[I]n the existing state of the human intellect the only chance of fair play for all sides of the truth is through diversity of opinion. Even when the world is almost unanimous on some subject and is *right* about it, people who disagree—if there are any to be found—probably have something to say for themselves that is worth hearing, and truth would lose something by their silence.

I have argued that freedom of opinion, and freedom of the expression of opinion, are needed for the mental well-being of mankind (on which all other kinds of well-being depend). I now briefly repeat my four distinct reasons for this view. **1** An opinion that is compelled to silence may, for all we can certainly know, be true. To deny this is to assume our own infallibility. **2** Even when the silenced opinion is an error, it can and very commonly does contain a portion of the truth; and since the general or prevailing opinion on any topic is rarely if ever the whole truth ·about it·, it is only through the collision of conflicting opinions that the remainder of the truth has any chance of being supplied. **3** Thirdly, even if the publicly accepted opinion is not only true but is the whole truth ·on the subject in question·, unless it is vigorously and earnestly disputed most of those who accept it will have it in the manner ·merely· of a prejudice, with little grasp or sense of what its rational grounds are. **4** And also (this being my fourth argument), the meaning of the doctrine itself will be in danger of being lost or weakened, and deprived of its vital effect on character and conduct. It will become a mere formal pronouncement, effective not in doing any good but only in cluttering up the ground and preventing the growth of any real and heartfelt conviction from reason or personal experience.

Chapter 3: Individuality—One of the Elements of Well-Being

No-one claims that actions should be as free as opinions. On the contrary, even opinions lose their immunity when the circumstances in which they are expressed are such that merely expressing them is a positive incitement to some harmful act. The opinion that *corn-dealers are starvers of the poor* ought to be allowed to pass freely when it is simply presented to the world in print; but someone can justly be punished for announcing it orally or passing it out on a placard to an excited mob that has gathered in front of a corn-dealer's house. (Another example might be the opinion that *private property is robbery*.) Acts of any kind that harm others without justifiable cause may be—and in the more important cases absolutely must be—brought under control, either by adverse opinion or (when necessary) by active interference. The liberty of the individual must be limited by this:

He Must Not Adversely Affect Other People.

[Mill's actual words: 'He must not make himself a nuisance to other people.'] But if he refrains from interfering with others in things that are their own concern, and merely acts according to his own inclination and judgment in things that concern himself, he should be freely allowed to put his opinions into practice at his own cost; and the reasons for this are the very ones that show that opinion should be free:

- mankind are not infallible;
- their truths are mostly only half-truths;
- uniformity of opinion is not desirable unless it results from the fullest and freest comparison of opposite opinions;
- diversity is a good thing, not a bad one, until mankind become much more able than at present to recognize all sides of the truth.

These principles apply as much to men's conduct as to their opinions. Just as it is useful that while mankind are imperfect there should be different opinions, so is it that •there should be different experiments in living; that •different kinds of personal character should be given free scope as long as they don't injure others; and that •the value of different ways of life should tried out in practice when anyone wants to try them. It is desirable, in short, that in matters that don't primarily concern others individuality should assert itself. When a person's conduct is ruled not by *his* character but by the traditions or customs of *others*, one of the principal ingredients of human happiness—and *the* chief ingredient of individual and social progress—is lacking.

The trouble is that in the thinking of most people individual spontaneity is hardly recognized as having any intrinsic value, or as deserving any respect on its own account. The majority are satisfied with the ways of mankind as they now are (·of course·, for it is they who *make* them what they are!), and they can't understand why those ways shouldn't be good enough for everybody. As for moral and social reformers—·who by definition are *not* satisfied with the ways of mankind as they now are·—the majority of *them* don't have spontaneity as any part of their ideal; rather, they look on it with resentment, as a troublesome and perhaps rebellious obstruction to the general acceptance of what these reformers themselves think would be best for mankind. Wilhelm von Humboldt, so eminent both for his learning and as a politician, based one of his works [*The Sphere and Duties of Government*] on the thesis that

- 'the end of man, or that which is prescribed by the eternal or immutable dictates of reason and not suggested by vague and transient desires, is the highest and most harmonious development of his powers to a complete and consistent whole';

that therefore,

- 'that towards which every human being must ceaselessly direct his efforts, and on which especially those who design to influence their fellow-men must ever keep their eyes, is the individuality of power and development';

that •for this two things are needed, 'freedom, and variety of situations'; and that •from the combination of these arise 'individual vigour and manifold diversity', which combine themselves in 'originality'. Few people outside Germany even understood what he meant!

Nobody denies that people should be taught and trained in youth so that they can know what has been learned from human experience and can benefit from it. But when a human being has arrived at the maturity of his faculties, it is his privilege—and indeed his proper role—to use and interpret experience *in his own way*. It is for him to find out what part of recorded experience is properly applicable to his own circumstances and character. The traditions and customs of other people provide some evidence of what their experience has taught *them*—evidence that has some weight, and thus has a claim to his deference. But ·there are three reasons for not giving it the final decision about how he should live his life·. In the first place, •those people's experience may be too narrow, or they may not have interpreted it rightly. Secondly, •their interpretation of their experience may be correct but unsuitable to him. Customs are made for customary circumstances and customary character; and *his* circumstances or his character may be uncustomary. Thirdly, •even when the customs are good in themselves and are suitable to him, still ·he ought not· to conform to custom as such, ·because that· doesn't educate or develop in him any of the qualities that are the distinctive endowment of a human being. The human faculties of perception, judgment, discriminative feeling, mental activity, and even moral preference are exercised only in making choices. He who does something *because it is the custom* doesn't make a choice. He gains no practice either in seeing what is best or in wanting it. Like our muscular powers, our mental and moral powers are improved only by being used. You don't bring your faculties into play by •doing something merely because others do it, any more than by •believing something only because others believe it. If the reasons for an opinion are not conclusive in *your* way of thinking, your reason can't be strengthened by your adopting the opinion, and is likely instead to be weakened; and if the reasons for acting in a certain way are not in harmony with *your* feelings and character, acting in that way is contributing towards making your feelings and character inert and slack rather than active and energetic.

He who lets the world (or his own portion of it) choose his plan of life for him doesn't need any faculty other than the ape-like ability to *imitate*. He who chooses his plan for himself employs all his faculties. He must use

>observation to see,
>reasoning and judgment to foresee,
>activity to gather materials for decision,
>discrimination to decide,

and, when he has decided,

>firmness and self-control to keep to his deliberate decision.

And *how much* he requires and uses these abilities depends directly on *how much* of his conduct is determined according to his own judgment and feelings. He *might* be guided in some good path, and kept out of harm's way, without any of these things. But ·in that case· what will be his comparative worth as a human being? It really does *matter* not only what men do but also what sort of men they are that do it. Among the works of man that human life is rightly employed in perfecting and beautifying, surely the most important is man himself. Human nature is not a machine to be built on the basis of a blueprint, and set to do exactly the work prescribed for it; rather, it is a tree that needs to grow and develop itself on all sides, according to the tendency of the inward forces that make it a living thing.

A person whose desires and impulses are his own—expressing his own nature as it has been developed and modified by his own culture—is said to have a *character*. (One whose desires and impulses are not his own doesn't have a character, any more than a steam-engine does.) If the impulses are not only his but are strong, and are under the control of a strong will, then he has an energetic character. If you think that individuality of desires and impulses shouldn't be encouraged to unfold itself, you must maintain that society doesn't need strong natures—that it isn't the better for containing many people who have much character—and that a high average level of energy is not desirable.

In some early states of society, these forces ·of high-level individual energy· were too far ahead of society's power at that time to discipline and control them. There was a time when the element of spontaneity and individuality was excessive, and social forces had a hard struggle with it. But society *now* has the upper hand over individuality. In our times, from the highest class of society down to the lowest, everyone lives as though under the eye of a hostile and dreaded censorship. Not only in what concerns others but in what concerns *only themselves*, the individual or the family don't ask themselves:

>what do I prefer? or
>what would suit my character and disposition? or
>what would allow the best and highest in me to have fair play, and enable it to grow and thrive?

They ask themselves:

>what is suitable to my position?
>what is usually done by people in my position and economic level?

or (worse still)

>what is usually done by people whose position and circumstances are superior to mine?

I don't mean that they choose what is customary in preference to what suits their own inclination. It doesn't occur to them to *have* any inclination except to do what is customary. Thus the mind itself is bent under the yoke. Even in what people do *for pleasure*, conformity is the first thing they think of; they like in crowds; they exercise choice only among things that are commonly done; they shun peculiarity of taste and eccentricity of conduct as much as they shun crimes. Eventually, by not following their own nature they come to have no nature to follow: their human capacities are withered and starved; they become incapable of any strong wishes or natural pleasures, and are generally without either •opinions or •feelings that are home-grown and properly *theirs*. Now is *this* the desirable condition of human nature?

The way to make •human beings become something noble and beautiful to see and think about is not by wearing down into uniformity all that is individual in them but rather by cultivating it and enabling it to grow, within the limits imposed by the rights and interests of others. . . . And that is also the way to make •human life become rich, diversified, and animating, furnishing more abundant nourishment for high thoughts and elevating feelings, and strengthening the tie that binds every individual to the ·human· race by making the race infinitely better worth belonging to. The more each person develops his individuality, the more valuable he becomes to himself, and thus the more capable of being valuable to others.

It's true that persons of genius are and probably always will be a small minority; but in order to have them we must preserve the soil in which they grow. [In Mill's time, 'genius' meant something like 'high intelligence combined with creative imagination'—something like what it means today, but not quite as strong.] Genius can breathe freely only in an *atmosphere* of freedom. Persons of genius are by definition more individual than other people—and therefore less able to squeeze themselves, without being harmed, into any of the small number of moulds that society provides in order to save its members the trouble of forming their own character. If out of timidity they consent to be forced into one of these moulds, and to let all that part of themselves that can't expand under the pressure remain unexpanded, society won't gain much from their genius. If they are of a strong character and break their fetters, they become a target for the society that hasn't succeeded in reducing them to something commonplace, to point at with solemn warning as 'wild', 'erratic', and so on; like complaining against the Niagara river because it doesn't flow smoothly between its banks like a Dutch canal.

I have said that it is important to give the freest possible scope to •uncustomary things, so that in due course some of these may turn out to be fit to be converted into •customs. But independence of action and disregard of custom don't deserve encouragement only because they may lead to better ways of action and customs more worthy of general adoption; and people of decided mental superiority are not the only ones with a just claim to carry on their lives in their own way. There is no reason why all human lives should be constructed on some one pattern or some small number of patterns. If a person has even a moderate amount of common sense and experience, his own way of planning his way of life is the best, not because it is the best in itself but because it is *his own way*. Human beings are not like sheep; and even sheep aren't indistinguishably alike. A man can't get a coat or a pair of boots to fit him unless they are either made to his measure or he has a whole warehouse full or coats or boots to choose from. Well, is it easier to fit him with a *life* than with a *coat*? Are human beings more like one another in their whole physical and spiritual make-up than in the shape of their feet? If it were only that people differ in their *tastes*, that would be reason enough for not trying to fit them all into one mould. But different people also need different conditions for their spiritual development; they can't all exist healthily in the same •moral atmosphere and climate any more than all plants can flourish in the same •physical climate. The very things that help one person to develop his higher nature hinder another from doing so. A way of life that is a healthy excitement to one person, keeping all his faculties of action and enjoyment in their best order, is to another a distracting burden that suspends or crushes all his inner life. Such are the differences among human beings in their sources of pleasure, their susceptibilities to pain, and the operation on them of different physical and moral forces, that unless there is a corresponding variety in their ways of life they won't get their fair share of happiness and won't rise to the mental, moral, and aesthetic level that they are naturally capable of. Why then should tolerance, as far as the public attitude is concerned, extend only to tastes and ways of life that *have* to be accepted because so many people have them? Nowhere (except in some monastic institutions) is diversity of taste *entirely* unrecognized: a person may without blame either like or dislike rowing, or smoking, or music, or athletic exercises, or chess, or cards, or study, because those who like each of these things are too numerous to suppress, and so are those who dislike them. But the man—and still more the woman—who can be accused either of doing 'what nobody does' or of not doing 'what everybody does' is criticized as much as if he or she had committed some serious moral offence.

If the claims of individuality are ever to be asserted, the time is *now*, when the enforced assimilation is still far from complete. It is only in the earlier stages that any defence can be successfully mounted against the attack. When mankind spend some time without seeing diversity, they quickly become unable even to conceive it.

Chapter 4: The Limits to The Authority of Society Over The Individual

What, then, is the rightful limit to the sovereignty of the individual over himself? Where does the authority of society begin? How much of human life should be assigned to individuality, and how much to society?

Each will receive its proper share if each has that which more particularly concerns it. To individuality should belong the part of life in which it is chiefly the individual that is interested; to society, the part which chiefly interests society. [Here and throughout, your being 'interested in' something means that your interests are involved in it; this is not 'interested in' as the opposite of 'bored by'.]

Though society is not founded on a contract, and though no good purpose is served by inventing a contract in order to infer social obligations from it, everyone who receives the protection of society owes society something in return for this benefit, and the sheer fact that they *have* a society makes it indispensable that each should be bound to conform to a certain line of conduct towards the rest. This conduct consists ·in two things·.

> **(1)** Not harming the interests of one another; or, rather, not harming certain ·particular· interests which ought to be classified (by explicit law or tacit understanding) as *rights*.
>
> **(2)** Doing one's share (to be fixed by some fair principle) of the labours and sacrifices incurred for defending the society or its members from injury and harassment.

Society is justified in enforcing these conditions at the expense of those who try to avoid fulfilling them. Nor is this all that society may do. The acts of an individual may be hurtful to others, or lacking in proper consideration for their welfare, without going so far as violating any of their constituted rights. The offender may then be justly punished by opinion, though not by law. As soon as any part of a person's conduct has a negative effect on the interests of others, society has jurisdiction over it, and the question whether or not the general welfare will be promoted by interfering with it becomes open to discussion. But there is no room for raising any such question when a person's conduct affects the interests of no-one but himself, or needn't affect others unless they want it to (all the persons concerned being adults with the ordinary amount of understanding). In all such cases there should be perfect freedom, legal and social, to perform the action and accept the consequences.

Many people will deny that we can distinguish the part of a person's life that concerns only himself from the part that concerns others. They may say:

> How can any part of the conduct of a member of society be a matter of indifference to the other members? No person is an entirely isolated being; it is impossible for a person to do anything seriously or permanently hurtful to himself without harm coming at least to those closely connected with him and often far beyond them. If he injures his property, he does harm to those who directly or indirectly derived support from it, and usually lessens somewhat the general resources of the community. If he worsens his physical or mental abilities, he not only brings evil on all who depended on him for any portion of their happiness but makes himself unable to render the services that he owes to his fellow-creatures generally; perhaps becomes a burden on their affection or benevolence....
>
> Finally, if by his vices or follies a person does no direct harm to others, he nevertheless does do harm by the example he sets, and he ought to be compelled to control himself, for the sake of those whom the sight or knowledge of his conduct might corrupt or mislead.
>
> And even if the consequences of misconduct could be confined to the vicious or thoughtless individual himself, ought society to abandon to their own guidance those who are manifestly unfit for it? We all agree that children and young people should be protected against themselves; so isn't society equally bound to protect against themselves adults who are equally incapable of self-government? If gambling, or drunkenness, or sexual licence, or idleness, or uncleanliness, are as injurious to happiness and as great a hindrance to improvement as many or most of the acts prohibited by law, why shouldn't the law try to put *them* down also (as far as practicability and social convenience allow)? And as a supplement to the unavoidable imperfections of law, oughtn't *public opinion* at least to organize a powerful guard against these vices, and rigorously apply social penalties on those who are known to practise them? There is no question here of restricting individuality or blocking trials of new and original experiments in living. Nothing is being prevented except things that have been tried and condemned from the beginning of the world until now—things that experience has shown not to be useful or suitable to any person's individuality. There must be *some* length of time and amount of experience after which a moral or prudential truth can be regarded as established! All that is proposed here is to prevent generation after generation from falling over the same precipice that has been fatal to their predecessors.

I fully admit that the harm a person does to himself may seriously affect (both through their sympathies and their interests) those closely connected with him, and may in a lesser degree affect society in general. When by conduct of this sort a person is led to violate a distinct and assignable obligation to one or more others, the case is no longer in the self-regarding category and becomes amenable to *moral* condemnation in the proper sense of the term. [By an 'assignable' obligation, Mill means an obligation to someone in particular, as distinct from (say) an obligation to keep yourself fit in case *someone or other* comes to need your help.] For example: if through intemperance or extravagance a man becomes unable to pay his debts, or unable to support and educate his family, he is deservedly condemned and might be justly punished; but it is for the breach of duty to his family or creditors, not for the extravagance. If the resources that ought to have been devoted to them had been diverted from them for the most prudent investment, the moral culpability would have been the same.... Again, if (as often happens) a man causes grief to his family •by his addiction to bad habits, he deserves reproach for his unkindness or ingratitude; but he may deserve it just as much if he causes grief to his family •by cultivating habits that are not in themselves vicious. Someone who

>fails in the consideration generally due to the interests and feelings of others, without •being compelled by some more imperative duty or •justified by allowable self-preference,

is a subject of moral disapproval for that failure, but not for •the cause of it and not for •any errors that are merely personal to himself and may have indirectly led to it. Similarly, when a person disables himself through purely self-regarding conduct from the performance of some definite duty he has towards the public, he is guilty of a social offence. No-one ought to be punished simply for being drunk, but a soldier or a policeman should be punished for being drunk on duty. In short, whenever there is definite damage, or a definite risk of damage, either to an individual or to the public, the case is taken out of the domain of liberty and placed in that of morality or law.

But with regard to the merely contingent. . . .harm that a person causes to society by conduct that doesn't violate any specific duty to the public or bring harm to any assignable individual except himself: this inconvenience is one that society can afford to bear for the sake of the greater good of human freedom.

But the strongest of all the arguments against the public's interfering with purely personal conduct is that when it does interfere the odds are that it interferes •wrongly and •in the wrong place. On questions of *social* morality—of duty to others—the opinion of the overruling majority is likely to be right oftener than it is wrong, because on such questions they are only required to judge how a given mode of conduct, if allowed to be practised, would affect *their* interests. But the opinion of a similar majority imposed as a law on the minority on questions of *self-regarding* conduct is quite as likely to be wrong as right; for in these cases public opinion means at best *some* people's opinion of what is good or bad for *other* people; while very often it doesn't even mean that, because the public consider only their own preference and don't pay the slightest regard to the pleasure or convenience of those whose conduct they censure. There are many who regard any conduct that they have a distaste for as an insult to themselves, and resent it as an outrage to their feelings; as a religious bigot, when accused of disregarding the religious feelings of others, has been known to reply that *they* disregard *his* feelings by persisting in their abominable worship or creed! But •a person's feeling for his own opinion is not on a par with •the feeling of someone else who is offended at his holding it; any more than •a person's desire to keep his purse is on a par with •a thief's desire to take it. Someone's *taste* is as much his own particular concern as is his opinion or his purse. It is easy for anyone to imagine an ideal public which leaves individuals free to choose in all matters where there are two sides to the question, and only requires them to abstain from kinds of conduct that •universal experience has condemned. But whoever saw a public that *did* set any such limit to its censorship? And when does the public trouble itself about •universal experience? In its interferences with personal conduct the public is seldom thinking of anything but the dreadfulness of anyone's acting or feeling *differently from itself* ; and this standard of judgment is what ninety percent of all moralists and moral theorists hold up to mankind as the dictate of religion and philosophy. The standard in question is thinly disguised in the hands of these people. What they openly teach is that things are right because they are right—because we feel them to be so. They tell us to search in our own minds and hearts for laws of conduct binding on ourselves and on all others. What can the poor public do but apply these instructions and make their own personal feelings of good and evil, if they are reasonably unanimous in them, obligatory on all the world?

Chapter 5: Applications

·Free Trade·

Someone who undertakes to sell goods of any kind to the public is doing something that affects the interests of other people and of society in general; and so his conduct does *in principle* come within the jurisdiction of society. Restrictions on trade, or on production for purposes of trade, are indeed restraints; but the restraints ·on trade that are· in question here affect only that part of conduct that society is in principle entitled to restrain. So the principle of individual liberty is not involved in most of the •questions that arise concerning for example, •how much public control is admissible to prevent fraud by adulteration; •how far sanitary precautions, or arrangements to protect people working in dangerous occupations, should be enforced on employers. Questions like these involve the liberty issue only ·in a marginal way·, through the general thesis that

> leaving people to themselves is always better, other things being equal, than controlling them.

It can't be denied that people may be legitimately controlled for ends such as the ones I have just mentioned. On the other hand, some questions relating to interference with trade are centrally questions of liberty; such as the prohibition on importing opium into China, the restriction of the sale of poisons; all cases—in short, where the object of the interference is to make it hard or impossible to obtain a particular commodity. These interferences are objectionable as infringements on the liberty not of the producer or seller but of the buyer.

·Selling poisons·

One of these examples, that of the sale of poisons, raises a new question—How far can liberty legitimately be invaded for the prevention of crime or of accidents? It is one of the undisputed functions of government to take •precautions against crime before it has been committed, as well as to •detect and punish it afterwards. The preventive function of government, however, is far more liable to be abused at the expense of liberty than is its punitive function; for almost every part of the legitimate freedom of action of a human being could be represented, and fairly too, as increasing the facilities for *some* kind of misconduct. (·Someone earns his living making hammers; now think about the crimes that can be committed using a hammer!·) Still, if a public authority or even a private person sees someone evidently preparing to commit a crime, they aren't bound to stay out of it until the crime is committed, but may interfere to prevent it. If poisons were never bought or used for any purpose except •to commit murder, it would be right to prohibit their manufacture and sale. In fact, however, they may be wanted for •purposes that are not only innocent but useful, and restrictions can't be imposed in •one case without operating also in •the other. Again, it is a proper part of the duty of public authority to guard against accidents. If either a public officer or anyone else saw a person starting to cross a bridge that was known to be unsafe, and there was no time to warn him of his danger, they might seize him and pull him back without any real infringement of his liberty; for liberty consists in doing what one desires, and he doesn't desire to fall into the river. Nevertheless, when there is not a certainty of trouble but only a risk of it, no-one but the person himself can judge whether in this case he has a strong enough motive to make it worthwhile to run the risk; and so I think he ought only to be warned of the danger, not forcibly prevented from exposing himself to it. (This doesn't apply if he is a child, or delirious, or in some state of excitement or preoccupation that won't let him think carefully.) Similar considerations, applied to such a question as the sale of poisons, may enable us to decide which possible kinds of regulation are contrary to principle and which are not. For example, a precaution such as labelling the drug with some word warning of its dangerous character can be enforced without violation of liberty: the buyer can't want *not to know* that the stuff he has bought has poisonous qualities.

·Selling Alcohol·

Society's inherent right to ward off crimes against itself by antecedent precautions suggests the obvious limitations to the maxim that purely self-regarding misconduct cannot properly be meddled with in the way of prevention or punishment. For example, drunkenness isn't a fit subject for legislative interference; but if someone had once been convicted of an act of violence to others under the influence of drink, I think it legitimate that *he* should be placed under a special legal restriction, personal to himself; that if he were ever again found drunk he would be liable to a penalty, and that if when in that state he committed another offence, the punishment he would be liable to for that other offence should be increased in severity. In a person whom drunkenness excites to do harm to others, making *himself* drunk is a crime against *others*. Another example: if an *idle* person isn't receiving support from the public or breaking a contract, it would be tyranny for him to be legally punished for his idleness; but if he is failing to perform his legal duties to others, as for instance to support his children, it is not tyrannical to force him to fulfil that obligation—by forced labour if no other means are available. This applies whether the source of the trouble is his idleness or some other avoidable cause.

Again, there are many acts which, being directly harmful only to the agents themselves, ought not to be legally prohibited, but which when done publicly are a violation of good manners. That brings them within the category of *offences against others*, and so they may rightfully be prohibited. Offences against decency come into this category, but I shan't spend time on them, especially since they are connected only indirectly with our subject. ·Indecent actions are thought of as wrong in themselves, whether or not done publicly; but· the objection to publicness—·which is our subject·—is equally strong in the case of many actions that aren't in themselves condemnable and aren't thought to be so by anyone.

·Dissuasion·

A further question: when the state regards certain conduct as contrary to the best interests of the agent, should it without *forbidding* that conduct nevertheless *discourage* it? For example, should the state take measures to make the means of drunkenness more costly, or make them harder to get by limiting the number of the places where they are sold? On this as on most other practical questions, many distinctions need to be made. To tax stimulants solely so as to make them more difficult to obtain is a measure differing only in degree from prohibiting them entirely; and it would be justifiable only if prohibition were justifiable. Every increase of cost is a prohibition to those who can't afford the newly raised price; and to those who can afford it, the increase is a penalty inflicted on them for gratifying a particular taste. Their choice of pleasures, and their way of spending their income (after satisfying their legal and moral obligations to the state and to individuals), are their own concern and must be left to their own judgment. These considerations may seem at first sight to condemn the selection of alcohol as a special subject of taxation for purposes of revenue. But it must be remembered •that taxation for fiscal purposes is absolutely inevitable; •that in most countries a considerable part of that taxation *has to* be indirect; and therefore •that the state can't help imposing penalties on the use of some articles of consumption—penalties that may prevent some people from buying such articles. So the state has a duty to consider, in the imposition of taxes, what commodities the consumers can best spare; and that points very clearly to commodities that it thinks are positively harmful when used in more than very moderate quantities. (Something that will *harm* people is certainly something they can *spare*!) So taxation of stimulants—up to the point that produces the largest amount of revenue (supposing that the state needs so much)—is not only admissible but to be approved of.

·Contracts—Slavery·

I pointed out in an early part of this work that the liberty of •the individual, in matters that concern him alone, implies a corresponding liberty in •any number of individuals to regulate by mutual agreement such matters as involve them jointly and don't involve anyone else. [Yet i]n this and most other civilized countries, for example, an engagement by which a person sells himself (or allows himself to be sold) as a slave would be null and void—not enforced by law or by public opinion. The ground for thus limiting his power of voluntarily disposing of his own course of life is obvious, and is very clearly seen in this extreme case. Here it is.

> The reason for *not* interfering with a person's voluntary acts except for the sake of others is consideration for his liberty. His voluntary choice is evidence that what he so chooses is desirable to him, or at least endurable by him, and his good is on the whole best provided for by allowing him to pursue it in his own way. But by selling himself as a slave he *abdicates* his liberty; he forgoes any future use of it after that single act. He therefore defeats the very purpose that is the justification for allowing him to dispose of himself. From now on he won't be free. . . . The principle of freedom can't require that he should be free not to be free! Being allowed to give up his freedom is not *freedom*.

These reasons, the force of which is so conspicuous in this special case, are evidently of far wider application; yet a limit is everywhere set to them by the necessities of life, which continually require not that we should give up our freedom but that we should consent to this or that limitation of it [by contract, for example]. But the principle that demands uncontrolled freedom of action in all that concerns only the agents themselves requires that those who have become obliged to one another in things that don't concern any third party should be able to release one another from their engagement. . . .

·Contracts—Marriage·

Baron Wilhelm von Humboldt, in the excellent Essay from which I have already quoted, asserts his view that engagements involving personal relations or services should never be legally binding beyond a limited duration of time; and that the most important of these engagements, marriage, having the special feature that its objectives are defeated unless the feelings of *both* the parties are in harmony with it, should require nothing more than the declared wish of *either* party to dissolve it. This subject is too important and too complicated to be discussed in an aside, and I touch on it only so far as I need it for purposes of illustration. When a person, either by explicit promise or by conduct, has encouraged someone else to rely on his continuing to act in a certain way—to build expectations and plans, and to stake any part of his plan of life on that supposition—a new series of moral obligations arises on his part towards that person. They may possibly be overruled, but they can't be ignored. Again, if the relation between two contracting parties •has had consequences for others, if it •has placed third parties in any special position or even (as in the case of marriage) •has brought third parties *into existence*, then both the contracting parties come to have •obligations towards those third persons; and the choice of whether to maintain the original contract must have a great effect on whether—or at least on *how*—those •obligations are fulfilled. It doesn't follow—and I don't believe—that these obligations extend to requiring the fulfilment of the contract at all costs to the happiness of the reluctant party, but they are a necessary element in the question. And even if, as von Humboldt maintains, they ought to make no difference to the •legal freedom of the parties to release themselves from the engagement (and I also hold that they oughtn't to make *much* legal difference), they necessarily make a great difference to the parties' •moral freedom. A person is ·morally· bound to take all these circumstances into account before deciding on a step that may affect such important interests of others; and if he doesn't allow proper weight to those interests he is morally responsible for the wrong ·he does to the third parties·.

War and Self-Defense*
David Rodin

Rights

Self-defense is referred to as a right. Why? What kind of right is it? What is its normative origin? And how is it related to other rights?

Building Blocks

The most intuitive notion with which to commence is that of a **duty**. A duty is simply an obligation owed by some party to perform or abstain from performing a certain act. A duty has three distinct elements: a subject (the party whose duty it is), a content (what it is his or her duty to do), and an object (the party to whom the duty is owed).

The correlate of a duty is a **claim**. Thus, if you are indebted to me, then you have a *duty* to me to pay the specified sum of money. The correlate of this is my *claim* against you that you pay me the money. The claim and the duty are logically equivalent descriptions of the same normative relation. Each describes the relation from a different perspective—that of the subject and object of the duty respectively.

If you have **no claim** against me that I do the dishes, then I have the **liberty** with respect to you, not to do them. My liberty consists simply in the absence of an obligation to you to act otherwise.

The Logical Structure of Rights

[It might plausibly seem that] **rights** are complex packages which must necessarily include some claims against others. To be effective, even rights principally directed towards protecting liberties require duties on the part of others to assist with and not to obstruct the exercise of that liberty. For example, my right to vote is in the first instance a liberty, but it could not be an effective right if it did not contain claims against others, for example, not to obstruct my voting and against certain officials to assist me by providing voting papers.

This analysis is certainly persuasive for many moral and political rights. However, I shall argue below that not all rights contain claims against others. For, as we shall see, there are certain simple liberties which are properly described as rights. I shall argue that the right of self-defense and defensive rights more generally do not contain claims, and yet are genuine rights.

Having a Right and Being in the Right

Rights are not the whole of morality, and ascribing a right is importantly different to asserting an all-things-considered moral judgment.

---

* Excerpts from *War and Self-Defense* (Oxford University Press, 2003), edited by Lawrence Lengbeyer (incl. text & footnotes omitted, bold & underline emphases added).

Consider the following examples. If someone breaks in the door of your house to rescue a child from poisonous fumes, then he violates your claim-right that he not damage your property. This is the case even though it is obviously right to act as he did in the circumstances. Thus rights may be violated and a right-bearer wronged by an action which, all-things-considered, is the right thing to do. Conversely, a person may act wrongly even though he has an undeniable right to act as he does. For example a rich moneylender who needlessly causes the ruination of a poor family by refusing to extend a loan has the right to act as he does, though his action is, all-things-considered, wrong because it is heartless and uncharitable. It follows that having a right to do X and being in the right in doing X are not logical equivalents.

When we use the vocabulary of rights, we are appealing to a form of moral consideration distinct from, and potentially opposed to, other aspects of morality such as consequentialist considerations (as in the case of breaking the door to rescue the child) and virtues (as in the case of the miserly moneylender). **How are we to distinguish a right from other forms of moral consideration?**

Although this is a complex and difficult question, there are **three characteristics** which can be readily identified. **First**, rights have a distinctive stringency, such that they generally override competing moral considerations. Secondly, they have a particular role in standing against, and placing limits upon, what may be called 'goal-based' moralities such as consequentialism. Thirdly, rights are moral considerations which have a unique relation to individual moral subjects.

Ronald Dworkin highlights two of these features in his well-known characterization of rights: 'Rights are best understood as trumps over some background justification for political decisions that states a goal for the community as a whole.' The peculiar stringency of rights is powerfully illustrated by Dworkin's metaphor of a trump card. This metaphor suggests that rights are absolute ethical considerations which always override competing considerations. However, as our two examples above suggest (and Dworkin himself acknowledges), rights do not always override competing ethical considerations. Sometimes it is wrong, all things considered, to do what one has a right to do, and there may be circumstances in which it is morally right to violate a person's genuine rights. For this reason I prefer to think of rights as analogous, not to a trump card, but to a strong **'breakwater.'** On the breakwater view, the function of a right is to erect a normative barrier against the infringement of individual interests and liberties. It is in the nature of this barrier to be sufficient to defeat the great majority of competing claims. However, it is always conceivable that circumstances will arise in which an individual's right is simply overwhelmed by the gravity of competing moral considerations.

The **second** idea implicit in Dworkin's description of rights is that they function specifically to set limits on what Dworkin has called 'goal-based' moralities which are based on some interest of society taken as a whole rather than simply those of individual agents. The most prominent form of goal-based morality is utilitarianism. Dworkin is right to identify this as an important feature of the morality of rights, and indeed the tension between rights and consequentialist considerations is one of the most important in moral theory. However, as the example of the miserly moneylender demonstrates, rights do not only place limits on goal-based moralities—they may also conflict with other aspects of morality such as the virtues.

The **third** way in which rights are distinguished from other aspects of morality is the special way in which they attach to particular moral subjects. As Jeremy Waldron has said: 'When a person's rights are violated, we say not only that something wrong has been done, but that the right-bearer himself has been wronged.'

Rights are necessarily directed towards the protection of individual subjects.

Justification and Excuse

Our preliminary investigation into the logical structure of rights has yielded sufficient conceptual tools to proceed with the examination of defensive rights.

What is self-defense? In legal terms the plea of self-defense is a defense against a charge of murder (or lesser offenses such as assault) leading, if successful, to a full acquittal. Self-defense is therefore a form of exculpation that serves to remove or mitigate the blame attributable to an agent for the performance of a proscribed action.

Exculpations may be of two different kinds: they may constitute an **excuse** for a wrongful action, or they may constitute a full **justification**. Excuse and justification are quite distinct normative conceptions and each serves to make an agent *ex culpa* in different ways. According to the philosopher J. L. Austin: 'In the one defence [justification] we accept responsibility but deny that it was bad: in the other [excuse], we admit that it was bad but don't accept full, or even any, responsibility.' George Fletcher's legal definition asserts that claims of justification 'concede that the definition of the offence is satisfied, but challenge whether the act is wrongful,' whereas claims of excuse, 'concede that the act is wrong but seek to avoid the attribution of the act to the actor.'

A successful **excuse** leads to the withholding of punishment or blame without conferring approval. Excuses themselves may be of various different kinds each of which operates in a slightly different way. Some excuses such as physical compulsion, automatism, and mistake invoke the claim that the agent did not intend to perform the proscribed act, or did not have the intention to perform the action under the proscribed definition. Other excuses concede that the action was intentional but deny that it was voluntary. Duress, necessity, and provocation are excuses of this form. A third group of excuses involve cases in which the agent is either incapable of forming a criminal intent or lacks the capacity for full deliberation of the moral or legal issues involved; this group includes the excuses of infancy, insanity and involuntary intoxication. Excuses may be of varying degrees of strength: they may partially mitigate fault for an action, or they may serve to remove it completely.

With **justification**, however, the situation is very different. Justification is a much stronger form of exculpation than even a fully mitigating excuse, for in contrast to excuses it concerns the rightness of the action itself. A justified action is one that would normally be wrong, but which, given the circumstances, is either fully permissible or a positive good. It is not a forgivable lapse or a regrettable action that one is none the less excused for performing; it is rather something that one is either fully entitled to do, or not prohibited from doing.

We may draw a distinction between justifications grounded in consequentialist considerations and those deriving from the rights of the persons concerned. A classic example of a consequentialist justification is the case of the farmer who burns his neighbour's field to prevent a wild fire from engulfing a town. The neighbor has a right not to have his field destroyed, but the farmer's action is justified in the circumstances because it is overwhelmingly the lesser evil. What is distinctive about this case, however, is that the farmer does not have a simple liberty to burn the field, since the neighbour's claim-right against having his property destroyed does not disappear in the face of the justification. This manifests itself in the feeling that despite the justified nature of the farmer's act, the neighbour is still owed some compensation, redress, or apology. What this implies is that justifications arising from consequentialist considerations (in particular 'lesser evil' justifications) [do not eliminate the duties and claims that they override.]

In contrast to this, consider the case of a person whose justification for taking my car is that I have given him permission to drive it. There is no sense in which I have a residual right that my car not be taken, nor is there any implication that I should be owed redress or apology. The justification consists simply in his liberty to take my car, in the context of the background presumption of a duty to abstain from acts of that kind. More generally, justifications of this form may be defined as the liberty to perform a specified act when the act would, in normal circumstances, have been forbidden. Justifications resulting from claims of right frequently take this form.

Now is self-defense a justification or an excuse, and if it is a justification is it of the kind which may be reduced to a simple liberty?

I take the central case of self-defense to have the following features: (i) an aggressor makes an intentional attack on the victim's life which will succeed unless the victim uses lethal force against the aggressor; (ii) the attack is objectively unjust in the sense that the aggressor has no legitimate right to make the attack, for example, he is not a law enforcement officer acting within his duty; (iii) the aggressor is fully culpable in making the attack; (iv) the victim is wholly innocent with respect to the attack (for example, he has not provoked the aggressor in any faultworthy way). I do not mean to suggest that these features are necessary conditions for self-defense, but they are clearly sufficient, in the sense that anyone who believes that there is a right of self-defense must certainly believe that the right is effective in these circumstances.

If we restrict ourselves to the central case, it seems clear that self-defense is a justification not an excuse. This is clearly recognized in most legal jurisdictions. Someone who kills in self-defense does not commit a murder for which we exempt him of liability; rather, the defensive nature of the act makes it fail to be an instance of murder at all. We do not consider self-defense to be a wrongful form of action for which the defender ought not to be held fully responsible (as we might think about someone who flies into a violent rage after being cruelly provoked). It is rather that the victim may strike back—it is right and proper for him to do so.

Moreover, when someone justifiably kills an aggressor in self-defense we say that the aggressor has been harmed but not that he has been wronged. The aggressor (or his estate) is not in a position to demand compensation or apology for the act. This strongly suggests that self-defense is a form of justification which consists in a simple liberty to use lethal defensive force without the presence of any residual duty to refrain from so acting.

Self-defense is best analysed as a simple liberty to commit homicide [or a lesser assault] in the defense of life.

[But] why is there a liberty to kill in self-defense? In what context is the liberty operative? What is the scope of its limitations and its permissions?

A Model of Defensive Rights

The right to commit homicide in self-defense is not *sui generis*, a case alone unto itself. It is rather one case within a range of morally and legally justified defensive actions. It is a range which might properly include defending one's position in a queue by delivering some sharp words to an interloper, defending a valuable art work by striking a thief who is about to steal it, through to defending one's life by shooting and killing an assailant who is about to kill you. A model which explains the full range of justified defensive acts in terms of an underlying moral structure will have great value, especially when we come to discuss war.

A Three-Legged Stool

We have seen in the last chapter that self-defense is a specific liberty to commit homicide in the context of a background presumption against such acts. [A liberty is a] relation between three elements, namely: **subject** (the party who possesses the liberty), **content** (what the subject has a liberty to do), and **object** (the party with respect to whom the liberty is held). In order to construct a working explanatory model of defensive rights I propose to add to this schema a further element which we may the call the *end* of the right, the good or value which a defensive action is intended to preserve or protect.

The normative structure of defensive rights may be represented diagrammatically.

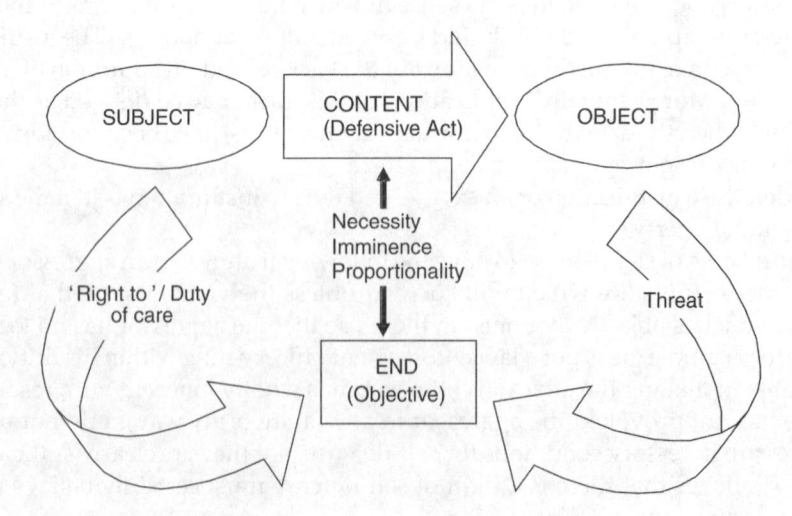

Figure 2.1. The Structure of defensive rights

Defense as a Derivative Right

Let us begin with the relationship between the subject and end of defensive rights. The content of a defensive right is the performance of an act which is frequently not an intrinsic good and which is sometimes a grave harm, such as assault or in extreme cases homicide. There are **three different relationships** [between the subject and end of the defensive action] which may serve as a **grounding** for **a defensive right.**

The first case is that in which the subject has **a right *to* the good whose protection is the end of the defensive action**. Defensive rights seem to be entailed in a very basic way by rights *to* things. Thus if I have a right to X, then it seems to follow as a simple corollary that I have the right to take measures to prevent my right to X from being violated. There are different ways in which one may have a right *to* something. Someone may have the right to an object if he owns it and therefore has property rights over it. Alternatively it may be a good to which he enjoys a liberty right or right of access, for example, the right to free speech, freedom of movement, or economic rights. We say that a person has a right *to* life and this right is in some ways analogous to a property right: it has as a central element the claim against each and every other person that they not take the agent's life, or interfere with his bodily integrity.

Not all defensive rights arise, however, from property rights and rights *to* things. Consider a parent whose child is in danger from an assailant. The parent certainly has the right to defend the child, and yet there is no question of the parent having property rights over the child or of the parent's right of defense deriving from any kind of right *to* the child. Rather it would seem that the parent's right of defense arises from the duty of care that they owe towards it; the parents are obligated to protect the child and for this reason, they are at liberty to undertake the defense. Thus a second normative basis for defensive rights would seem to be **an established duty of care towards a certain person or object**. In an analogous way, leaders are obligated and permitted to safeguard the welfare of their subordinates, museum curators to protect the artifacts entrusted to their custody, and friends to care for each other's interests. In each of these cases a right of defense may arise out of a duty of care embodied in an established moral relationship. Such relationships are a powerful source of defensive rights.

Sometimes, however, one may acquire a duty to protect or preserve a person or thing even though one has no established obligation of trust or care with regard to it. We might call these 'duties of rescue' and they arise simply from the fact that **the end (a particular good or value) is in danger, combined with the fact that one is situated so as to be able to assist**. One can find oneself the subject of such duties quite unexpectedly and through no explicit action or choice of one's own, as when one discovers a drowning man or a victim of an attack. This form of duty is capable of providing a ground for defensive rights. It is from this source that the duty and right to defend the lives of third parties who are strangers arise.

The right to defend one's own life derives from one's right *to* that life, whereas the right to defend a third party derives from more general considerations concerning the duty to protect the good and valuable. There is at least one material difference between the two forms of the defensive right.

If a right of defense is derived from a right *to* the end of defensive action, then generally one will be free to defend the end or not as one sees fit. The reason is that rights *to* things are generally discretionary [, so that one may] **waive** one's rights if one so wishes. Thus defense of one's own life is generally thought to be discretionary rather than obligatory. What is more there may be cases in which the decision not to defend one's life would be not only permissible, but a laudable act of supererogation. Indeed Hugo Grotius held that in general: 'while it is permissible to kill him who is making ready to kill, yet the man is more worthy of praise who prefers to be killed rather than to kill.' It has been an element of certain Christian teachings that we are under a duty to preserve our own life ([making] suicide a sin) & that self-defense is therefore obligatory. However, the most natural way to view this is to see the duty to protect our own lives as a duty owed to God. These cases are therefore best considered as falling under the second source of defensive rights; obligations deriving from explicit relationships of care, that is, the duty of care, owed to God, for the life he has entrusted to us.

Rights of defense which arise from specific or general duties of care towards the end of the defensive action, on the other hand, [include] a duty to act in defense. The duty in question will be of varying degrees of strength depending on the circumstance. Duties deriving from relationships of care such as those of a parent are generally the strongest, but even these are not absolute. For example, if a parent faced a high risk of death in an uncertain attempt to defend his child's life, then it is not clear that he would be under an obligation to act. Equally, the general duty to protect goods and values can result in extremely strong duties to act if the end in question is extremely valuable (such as human life) and the costs of action to the subject very low. However, when the risks to the subject are high, the chances of success doubtful, and the relationship between the subject and the end of the defensive action tenuous, the duty to act may become diminished.

Limits on the Right: Necessity, Imminence, Proportionality

Defensive rights are not absolute, but exist within determinate moral and legal bounds. The basic problematic of defensive action can be defined as the question of how far the permission to defend a particular good properly extends where the defensive measures are themselves harmful, or would otherwise be impermissible. The legal and philosophical literature on self-defense has identified **three intrinsic limitations to the right**. These limitations are necessity, imminence, and proportionality.

What do these limitations mean? **Necessity** in this context refers to indispensability and unavoidability rather than inevitability. It expresses the requirement that one may only take a harmful measure to protect one's legitimate right if there is no less costly course of action available that would achieve the same result. For example, <u>if one can protect oneself from harm by shooting an aggressor in the arm</u>, then one is not permitted to shoot him through the heart, for such use of lethal force would not be necessary.

A corollary of the requirement of necessity is that there is a general duty to <u>retreat</u> from an aggressor, if it is possible (that is to say, readily feasible) to avoid harm in this way. It should be noted that necessity does not require that the defensive measure taken be the only possible means to avoid the violation of one's rights, but merely that there be no less harmful alternative measure that would achieve the same result. The rationale behind the limitation of necessity is easy to discern. The liberty to perform defensive action is grounded in the fact that it is intended to protect a good or value. But general considerations of value will require us to choose the least costly course to that defense; in other words, the one which is least destructive of the good. This in turn requires us to inflict a harm only if it is necessary.

The second requirement on defensive action is that the defense may only be undertaken when the threatened act of harm is **imminent**. As George Fletcher puts it: 'A pre-emptive strike against a feared aggressor is illegal force used too soon; and retaliation against a successful aggressor is illegal force used too late. Legitimate self-defence must be neither too soon or too late.' In legal discussions imminence is often treated as an independent requirement for self-defense; however, it would appear on reflection that imminence is conceptually derivative from necessity. The point is that we cannot know with the required degree of certainty that a defensive act is necessary until the infliction of harm is imminent. If, *per impossible,* we could know that a certain defensive act was necessary to prevent some harm long before the harm was to be inflicted, would we still have to wait until the harm became imminent before acting? It does not seem to me that we would: necessity is enough. Imminence, like the duty to retreat, is simply a component and corollary of the requirement of necessity.

If necessity and imminence require a particular relationship between the defensive act (the content of the right) and the aggressive act, **proportionality** requires us to balance the harmful effects of the defensive action against the good to be achieved. A classic legal formulation requires that: 'the mischief done by, or which might reasonably be anticipated from, the force used be not disproportioned to the injury or mischief which it is inflicted to prevent.' Commentators sometimes explicate the idea of proportionality by speaking of 'proportionate force' implying that what must be balanced in self-defense is the means of force employed. But this is misleading, for the proportionality that is required is between the harm inflicted and the good preserved, not between the type[s] of force employed. For instance, <u>if you are about to kill me with a knife and all I have to defend myself with is a gun or a bazooka</u>, then my use of it is proportionate even though a bazooka is a far more forceful a weapon than a knife.

Although it may seem that the requirements of proportionality and necessity converge, they are in fact logically independent of one another. Thus, if I see from a hilltop that you are about to wrongfully read through my personal papers, and I shoot you with my high-powered rifle, this action may be necessary to protect my right to privacy, in the sense that I possess no less harmful means to prevent the violation of my right, but it would clearly be a disproportionate use of force. On the other hand, if I am a champion runner and can easily escape from an assailant who is trying to kill me, then shooting him would be a proportionate use of force, but it would not be necessary. Necessity, imminence, and proportionality are not sufficient conditions for an action to be justified defense, but they are necessary conditions. In both of the above examples the use of defensive force would be impermissible.

The requirements of necessity, imminence, and proportionality are the limits imposed on defensive action by justice. In practice, however, judgements about necessity, imminence, and proportionality must be such as to be capable of being made by frightened victims facing situations of extreme stress and danger. As Justice Oliver Wendell Holmes famously said, 'detached reflections cannot be demanded in the presence of an uplifted knife.' The standards of necessity and proportionality must thus be interpreted with a degree of latitude and allowance for reasonable error. It is important to understand, however, that the standard of action remains an objective and not a subjective one. In other words, it is not sufficient for a victim to honestly believe that his action is necessary and proportionate; his belief must also be objectively reasonable. What constitutes a reasonable belief, however, must be sensitive to the strains and stresses implicit in the circumstances of a conflict situation.

B. Justice

Nothing is of greater importance to individuals in society, or within an organization, than that opportunities for admission or employment, promotion or advancement in rank, and for rewards as well as punishments associated with their individual performance, be undertaken "fairly." To do otherwise, to show unjustifiable preference for (or prejudice against) individuals on the basis of characteristics or criteria that are wholly irrelevant to the benefit (or burden) in question, or that differentiate in a wholly arbitrary fashion between the degree of punishment administered to two different individuals for a given, specific offense, is said to constitute "injustice," to be "unfair." Societies, or organizations, that behave arbitrarily in these ways quickly anger and alienate their members or citizens, foster deep resentment among their constituents, and risk losing the loyalty, support, and trust necessary to function or to govern effectively. But what does this concept of "justice" or "injustice" mean, in fact?

Justice, Aristotle observed, is largely a matter of treating equals with equality, and giving to each what he is "due." Aristotle incidentally used, and meant to use, the masculine pronoun here, inasmuch as, unlike his teacher, Plato, he did not regard women (or for that matter, slaves, or "barbarians") as "equals," and thought them to be "due" (or "owed by right") very little. In fact, Aristotle himself thought society was naturally stratified, divided into distinct groups or classes on the basis of different individuals' "natures," with some kinds of human beings (nobles, aristocrats, the well-educated, or wealthy, or politically powerful) owed or due more "by their nature" in the way of benefits than other, "lower" strata of society. Women and slaves are "deficient" in important respects "by nature," and so enjoy less status than free men in Athens, for example.

That presumption of class structure, privilege, and social stratification was also a widespread feature of ancient political societies. The problem of justice, for Aristotle, is then to determine what each member of a distinctive class *deserves*, or is entitled to, on the basis of his own merits within his distinctive social class. It would be wrong to accord an individual less than the share of wealth, power, or social respect than he might properly be due on the basis of these factors—or, for that matter, wrong to accord others (like women) more than they are due "by nature."

Figuring out this system of proportionalities within a society characterized by such a "natural hierarchy" was understood as the problem of "justice." Justice, on Aristotle's analysis, includes at least two distinct concepts:

- equality and appropriate proportionality in the *distribution* of society's opportunities, benefits, and also its burdens (taxes, military service, and so forth).
- equality and appropriate proportionality in the administration of the law, and, in particular, in *punishing* those who disobey it.

Aristotle labeled this first category *"distributive justice,"* and labeled the second conception *"retributive justice."*

There are some interesting and even slightly amusing, implications of this stratified, and highly patronizing view of human beings. For example, we might conclude that it is "just" in such a society to punish women and children less harshly for a given infraction of the law than we would ordinarily punish free, adult males for the very same infraction, on the grounds that the former "couldn't help it," or couldn't be expected to know any better (unlike the men). In many societies, the distinction with regard to children might be maintained, but women, and others, like slaves, were usually in fact punished more severely than men for a given infraction, on the grounds that they were less worthy, and would only respond to the harsh control that the threat of punishment inflicts.

In such cases, distribution and punishment function less to support a conception of justice than simply to exercise power, establish control, and to impose a given *status quo* on individuals who would otherwise be unlikely to acquiesce in, or support it. Our old friend, Glaucon, from Plato's *Republic* (in our opening chapter) in fact thought "justice" to be about little else than this raw exercise of power by the elite, simply for the purpose of maintaining control of society. Many skeptics and "moral realists" continue to hold this position today.

Whatever the case in such instances, the distinctions that Aristotle first introduced—the problems of a "fair and equitable distribution" of benefits and burdens, and that of appropriate punishment for infractions of the law—both continue to serve well in defining the problem of justice.

We have seen in the preceding chapter, however, what a great divide in other key respects has developed between this ancient way of understanding the social and political world, and our own conception of politics and society in the modern era. Looking back upon Aristotle's assumptions about "natural endowments" and social stratification with the benefit of hindsight, we might now (in direct opposition to Glaucon) be tempted to characterize many if not all of those ancient societies as inherently *unjust*, if it could be shown that the benefits enjoyed by living in such societies (and exemptions from strict punishment for legal infractions) fell disproportionately to the high-born, powerful, and privileged among them (without regard for merit or worthiness in other respects), while the burdens (and the punishment meted out for infractions of the laws) fell disproportionately on the "lower" castes or strata (regardless of individual effort, merit, or worth in other respects).

A cultural or historical relativist, as in the last chapter, might once again object that we are making inappropriate judgments about such cultures and ancient societies, which were simply organized differently than our own. That observation about underlying differences in cultural (or historical) presuppositions that contribute, in turn, to the noticeable differences in organization is, of course, quite true. We might try to explain in response, however, that our complaint against these earlier societies and cultures is not simply that they are *organized differently*, or that their partisans made what are now *unfamiliar assumptions* about human nature and human worth. Rather, our objection to their practices is that the resulting *criteria* by which the division of benefits and burdens was calculated, and punishments administered, in such societies was inherently unjust *simply because those criteria (one's race, gender, wealth, birthright, or class or caste membership alone) are wholly irrelevant to determining what each of them is "due"* or "deserves" in the way of both benefits and burdens, and those criteria are likewise irrelevant to determining how harshly each should be (or should have been) punished for a given infraction of the law.

Yet if we complain that these past criteria, and the schemes for distributing benefits and administering punishment that resulted from them were seriously flawed, we are left with the problem of determining what would be, instead, a fair and equitable scheme for distributing benefits and administering punishment, and what would be the correct or relevant criteria by which these tasks of society should properly be carried out. America's "founding fathers" had, of course, much to say on these very topics, central to democratic rule under the Constitution. Likewise, Bentham, Mill, and their followers in England took their own social order to task precisely on the grounds of insufficiency in these respects. Kant and his Enlightenment followers also had much to contribute to the discussion of the sorts of social and political arrangements that would treat individual citizens with the proper degree of dignity and respect. This problem (like that of liberty, rights, or the scourge of war) is a perennial problem in all cultures, in all times and places in history.

Equality, fairness, impartiality, and even-handedness in distributing benefits and meting out punishment: these, then, are the twin conceptions of justice. Both are important to everyone. Handing out punishment and discipline, however, is a specific responsibility of military officers in a chain of command. Nothing, it seems, fosters resentment and ill-will among those under command than the perception of inconsistency and favoritism in punishing infractions of a legal system or disciplinary code. Thus, striving for fairness and equity here, and for consistency, is a goal of great importance in the military profession.

The Readings

The famous novella of Herman Melville, *Billy Budd* (1891), dramatically portrays the issues of innocence and guilt, and of infraction and punishment, in a distinctly military setting, aboard a 19th-century British Man o'War. The story, however, is loosely based upon an actual historical circumstance in the U.S. Navy earlier in that same century. The court martial and execution at high sea of a midshipman, allegedly for conspiring to commit mutiny, led to a public outcry: the midshipmen in question happened to be the son of the sitting Secretary of War, who was not altogether pleased with the charges, the proceedings, or the conduct of the ship's captain and officers. Public debate over this scandal led directly to the founding of the United States Naval Academy in 1845, in large part to improve the moral and intellectual formation of young officers for the new Republic's navy. It is preceded by a brief outline of the basic conceptions behind the practice of punishment by R. A. Duff.

Our main reading, however, is a classic essay, "Justice as Fairness," drawn from the work of the late Harvard philosopher John Rawls, *A Theory of Justice* (1971). His essay deals with the other problem of justice: liberty, equality of opportunity, and the social differences that invariably arise in the distribution of benefits and burdens in society. The distribution of punishment, we might argue, is merely a special case or example of that larger problem.

Understanding what it actually means for an organization or for a political order to be "fair" is a very vexing notion. It is far easier to spot and denounce injustice than it is to say with precision what a valid formula for distributive justice might be. A number of deceptively simple formulae have been suggested through the centuries, such as:

- "to each and every one an *equal* share" (egalitarianism),
- "to each according to his or her *effort*" or
- "to each according to his or her *merit*" (meritocracy)
- "from each according to his *ability*, and to each according to his *need*" (communism), and
- "for each an *equal opportunity*, and to each according to his or her resulting *success*" (*laissez-faire* capitalism).

Every one of these formulae embodies one or more criteria of great moral significance. Individuals are thought to be equal under the law, regardless of their other differences, for example, but the egalitarian formula of justice as a more general formula for distribution beyond legal consideration paradoxically results often in great resentment, jealousy, inequality, and can lead to perceptions of great injustice.

The same is true with respect to all the other formula, including the last, which resembles, but is not identical with the principles on which this nation was founded. Both of the last formulae, in particular, are open to the corruption of special interests, which then inhibits the distribution of benefits each is designed to encourage. The widespread corruption and cynicism fostered in Soviet society under that formula for justice is now well documented, and led to the internal decay, and ultimately the collapse of that economic and political order.

Meanwhile, the last formula, a "pure" and unregulated form of capitalism, often rewards good and bad luck as much or more than it does honest effort or genuine merit. Think, for example, of the agricultural victims of the "dust bowl" during the Great Depression, compared to a 24-year-old computer "geek" who happens to stumbles upon a meaningless computer game that nets him millions. Would we honestly wish to say that a system that offered unimaginable financial rewards to the "geek," while punishing the hard-working farmers of that earlier era with poverty, starvation, and despair was, on its face, just?

In a different sense, consider the case of "Cadet Smith," handsome, gifted as an athlete, perhaps a nationally-acclaimed quarterback on the Air Force or Army football team, for example. In what sense do these disproportionate natural endowments of birth, which he did nothing to "deserve," nonetheless entitle him to receive, as he invariably does, most favored status or additional rewards? Or consider "Midshipmen Jones," likewise a gifted athlete, stunningly beautiful, and exceptionally gifted at mathematics and chemistry, as well as at gymnastics. Why is she entitled to be blessed, as she is, by a showering of public attention, affection, and reward, simply on the basis of exercising talents or capacities endowed at birth? This is especially frustrating if others, with less natural endowment, can be shown to practice harder or to study more diligently than these gifted individuals, but still have their labors unrecognized and unrewarded. Is the resulting system, within which rewards, punishments, duties, rank, and privileges are distributed, "fair?"

These are the kinds of intractable dilemmas, particularly those arising from the inevitable inequalities that arise in the otherwise normal operation of most social and political systems, that John Rawls set himself to analyze and evaluate. That analysis draws directly from Rawls' own understanding of the implications of Kant's "third form" of the Categorical Imperative, the "Kingdom of Ends." There we noted that a system of morality, in general, might be conceived as a system of laws laid down for the organization of a society by all of its members, each of whom would then be (voluntarily) constrained to live under the laws they made. Rawls recognizes that there are some difficulties with this conception of a democratic social order, and so offers a more limited theory of justice designed to evaluate the inherent fairness of *the basic structure of a society* (for example the structure established by the U.S. Constitution).

One problem of Kant's conception is that actual, individual legislators often are motivated by the interests and needs attached to their position in society. White males might tend to favor laws that would grant them special privileges, or result in a more favored position in the resulting society, if they could. The inherent problem of the ancient societies that we considered earlier, and indeed, with any actual society, is that the laws and social institutions are largely formulated by those who hold positions of power and privilege, and are designed to protect those powers and privileges rather than to promote a fair distribution of benefits and burdens.

We could make a transition from any actual, historically-conditioned situation (open to these kinds of corruption and abuse) to the ideal "moral kingdom" that Kant envisioned, only if we could imagine each "legislator" in the Kingdom temporarily devoid of any particular knowledge about his or her own situation or place in the community. That is, no one could know, while making the laws or organizing the social and economic institutions, whether they themselves would be rich or poor, black or white, male or female, Protestant, Jew, Catholic, Muslim, atheist . . . in short, each "rational legislator" would be deprived of the special knowledge of his or her circumstances that would lead to conflicts of interest in passing laws and organizing social institutions that were inherently "fair." Only when shielded behind what Rawls terms a "veil or ignorance," deprived of any but the most general rational knowledge about the workings of human psychology, economics, and basic political principles, could the rational deliberators or "legislators" design a system of laws and social and economic institutions that would judged "fair" by all living under them.

Rawls terms this hypothetical situation, "the original position." This conception draws heavily on the political tradition of social contract theory, but with an interesting twist. The "original position" can be considered a powerful thought experiment, a hypothetical situation from which vantage point we can test our intuitions about fairness, and compare the hypothetical principles derived from such a vantage point with the actual, historically conditioned laws, institutions, and principles under which we and others live. Rawls forged this conception at the height of political upheavals in our country over civil rights and racial inequality, and simultaneously during a period of great international debate about the comparative justice of "capitalist" and "communist" economic systems. Critics of America's institutions at this time pointed to both debates, and denounced this country's fundamental laws and institutions as "unjust," owing to the *radical inequalities* resulting from both.

In response, Rawls is able to demonstrate (against communism or egalitarianism, for example) that *not all inequalities are inherently unjust*. On the other hand, one can legitimately denounce as unjust the legal and political systems of "apartheid" that existed inside a number of state and local governments, or which resulted from the settled practices of many of those same social and economic institutions, such as schools, colleges, and businesses. Why are these (racial) inequalities unjust, if not all inequalities (such as economic ones) are unjust?

The answer is found in the "reflective equilibrium" afforded by evaluation of these practices from the vantage point of Rawls' "original position." Rawls argued that rational individuals in the "original position" would always tend to promote two basic principles (as much an act of considered, and risk-aversive self interest, as from altruistic motives): namely, *liberty, and equality*. That is, rational beings under these constraints would be willing to accord a degree of liberty to each individual in the resulting society commensurate with a like amount of liberty to every other individual in the society.

The concept of "equality" has two parts. As to the economic and social (as well as political) institutions that would constitute the society (that is, schools, jobs, the right to vote and participate in the political process, and so forth), these "offices" would be equally open to the participation of any and all the members of the society. The resulting participation, and success in so doing, however, would depend upon effort, merit, natural endowments (like those of our outstanding service academy students), and often (as in computer games, or the "dust bowl") just plain luck. Differences, therefore, in social and economic status would invariably arise. These resulting inequalities would not be unjust, Rawls concluded, provided that they arose (as specified) from the operation of offices equally open to the participation of all, *and provided, in addition, that the resulting inequalities could be shown to work to the benefit of even the least advantaged under the system.*

This last constraint is termed "the Difference Principle," and is the most controversial aspect of Rawls' theory. Recall, however, that Jeremy Bentham proposed just such a constraint on utilitarian benefits: "every individual affected by a policy must be shown to have derived at least some measure of benefit from it." That was the constraint that prevented slavery from gaining approval as an acceptable practice under utilitarianism. Rawls now incorporates this utilitarian insight to identify the injustice of racial inequalities, while defending the economic inequalities that may otherwise arise under the normal operation of democratic capitalism. "Jim Crow" laws, discriminating on the basis of race, are unjust because they do not satisfy the constraints of the "original position:" their offices are not equally open to all, and the inequalities that arise as a result do *not* work to any perceptible benefit to the least advantaged under the system (the victims of racism). A similar argument would apply to laws and practices (such as college admissions or employment) that discriminated solely on the basis of gender.

Likewise, our intuition that the Oklahoma "dust bowl" farmers were unfairly disadvantaged in the normal workings of the economic system are supported: victims of tragic bad luck, even in a system that provides open access and equal opportunity in principle, do not derive any perceptible benefit from the normal workings of such a system. Hence, we have moved in this nation and in many others to provide minimal forms of insurance and support for such victims, as a rational policy of insuring against undue hardship that might come to be imposed inadvertently on any one of us during the otherwise-normal operation of the system.

By contrast, there is nothing inherently unfair in the normal inequalities of wealth and position that arise in this system, in that everyone has an opportunity to participate in it, and even the least advantaged derive considerable benefits from these inequalities (in the form of general economic prosperity and range of free consumer choice, education and public works paid for by taxation, cultural institutions arising from philanthropy, and so forth).

Thus, while Rawls' adaptation of Kant's original political insight is complex, far from perfect, and surely open to criticism on specific provisions, it masterfully exploits that original insight of the great Enlightenment philosopher to address our intuitions concerning economic and political justice. In particular, Rawls helps identify what it is about racial and sexual prejudice that is unfair, while defending some of this nation's most important basic institutions from inappropriate and misleading criticism. Perhaps most significantly, the theory provides a basis from which each of us might evaluate our own and our society's practices to discern when and where these might, on occasion, go awry, and so guide our common actions toward that goal that the Preamble of our Constitution so elegantly sets forth: "to form a more perfect Union, *establish justice* . . . promote the general Welfare, and secure the Blessing of Liberty to ourselves and our Posterity."

Excerpts from
A Theory of Justice (1971)
*John Rawls**

JUSTICE AS FAIRNESS

The Role of Justice

Justice is the first virtue of social institutions, as truth is of systems of thought. A theory however elegant and economical must be rejected or revised if it is untrue; likewise laws and institutions no matter how efficient and well-arranged must be reformed or abolished if they are unjust. Each person possesses an inviolability founded on justice that even the welfare of society as a whole cannot override. For this reason justice denies that the loss of freedom for some is made right by a greater good shared by others. It does not allow that the sacrifices imposed on a few are outweighed by the larger sum of advantages enjoyed by many. Therefore in a just society the liberties of equal citizenship are taken as settled; the rights secured by justice are not subject to political bargaining or to the calculus of social interests. The only thing that permits us to acquiesce in an erroneous theory is the lack of a better one; analogously, an injustice is tolerable only when it is necessary to avoid an even greater injustice. Being first virtues of human activities, truth and justice are uncompromising.

These propositions seem to express our intuitive conviction of the primacy of justice. No doubt they are expressed too strongly. In any event I wish to inquire whether these contentions or others similar to them are sound, and if so how they can be accounted for. To this end it is necessary to work out a theory of justice in the light of which these assertions can be interpreted and assessed. I shall begin by considering the role of the principles of justice. Let us assume, to fix ideas, that a society is a more or less self-sufficient association of person who in their relations to one another recognize certain rules of conduct as binding and who for the most part act in accordance with them. Suppose further that these rules specify a system of cooperation designed to advance the good of those taking part in it. Then, although a society is a cooperative venture for mutual advantage, it is typically marked by a conflict as well as by an identity of interests. There is an identity of interests since social cooperation makes possible a better life for all than any would have if each were to live solely by his own efforts. There is a conflict of interests since persons are not indifferent as to how the greater benefits produced by their collaboration are distributed, for in order to pursue their ends they each prefer a larger to a lesser share. A set of principles is required for choosing among the various social arrangements which determine this division of advantages and for underwriting an agreement on the proper distributive shares. These principles are the principles of social justice: they provide a way of assigning rights and duties in the basic institutions of society and they define the appropriate distribution of the benefits and burdens of social cooperation.

---

* Edited by Lawrence Lengbeyer, Mar. 2010, with assistance from www.econ.iastate.edu/classes/econ362/Hallam/Readings/Rawl_Justice.pdf and http://philosophyfaculty.ucsd.edu/faculty/rarneson/Courses/13-2008.html.

If men's inclination to self-interest makes their vigilance against one another necessary, their public sense of justice makes their secure association together possible. Among individuals with disparate aims and purposes a shared conception of justice establishes the bonds of civic friendship; the general desire for justice limits the pursuit of other ends.

The Subject of Justice

Many different kinds of things are said to be just and unjust: not only laws, institutions, and social systems, but also particular actions of many kinds, including decisions, judgments, and imputations. We also call the attitudes and dispositions of persons, and persons themselves, just and unjust. Our topic, however, is that of social justice. For us the primary subject of justice is the basic structure of society, or more exactly, the way in which the major social institutions distribute fundamental rights and duties and determine the division of advantages from social cooperation. By major institutions I understand the political constitution and the principal economic and social arrangements. Thus the legal protection of freedom of thought and liberty of conscience, competitive markets, private property in the means of production, and the monogamous family are examples of major social institutions. Taken together as one scheme, the major institutions define men's rights and duties and influence their life-prospects, what they can expect to be and how well they can hope to do. The basic structure is the primary subject of justice because its effects are so profound and present from the start. The intuitive notion here is that this structure contains various social positions and that men born into different positions have different expectations of life determined, in part, by the political system as well as by economic and social circumstances. In this way the institutions of society favor certain starting places over others. These are especially deep inequalities. Not only are they pervasive, but they affect men's initial chances in life; yet they cannot possibly be justified by an appeal to the notions of merit or desert. It is these inequalities, presumably inevitable in the basic structure of any society to which the principles of social justice must in the first instance apply. These principles, then, regulate the choice of a political constitution and the main elements of the economic and social system. The justice of a social scheme depends essentially on how fundamental rights and duties are assigned and on the economic opportunities and social conditions in the various sectors of society.

A conception of social justice, then, is to be regarded as providing in the first instance a standard whereby the distributive aspects of the basic structure of society are to be assessed.

The Main Idea of the Theory of Justice

My aim is to present a conception of justice which generalizes and carries to a higher level of abstraction the familiar theory of the **social contract** as found, say, in Locke, Rousseau, and Kant. In order to do this we are not to think of the original contract as one to enter a particular society or to set up a particular form of government. Rather, the guiding idea is that the principles of justice for the basic structure of society are the object of the original agreement. They are the principles that free and rational persons concerned to further their own interests would accept in an initial position of equality as defining the fundamental terms of their association. These principles are to regulate all further agreements; they specify the kinds of social cooperation that can be entered into and the forms of government that can be established. This way of regarding the principles of justice I shall call **justice as fairness**.

Thus we are to imagine that those who engage in social cooperation choose together, in one joint act, the principles which are to assign basic rights and duties and to determine the division of social benefits. Men are to decide in advance how they are to regulate their claims against one another and what is to be the foundation charter of their society. Just as each person must decide by rational reflection what constitutes his good, that is, the system of ends which it is rational for him to pursue, so a group of persons must decide once and for all what is to count among them as just and unjust. The choice which rational men would make in this hypothetical situation of equal liberty, assuming for the present that this choice problem has a solution, determines the principles of justice.

In justice as fairness **the original position** of equality corresponds to the state of nature in the traditional theory of the social contract. This original position is not, of course, thought of as an actual historical state of affairs, much less as a primitive condition of culture. It is understood as a purely hypothetical situation characterized so as to lead to a certain conception of justice. Among the essential features of this situation is that no one knows his place in society, his class position or social status, nor does anyone know his fortune in the distribution of natural assets and abilities, his intelligence, strength, and the like. I shall even assume that the parties do not know their conceptions of the good or their special psychological propensities. The principles of justice are chosen behind a **veil of ignorance**. This ensures that no one is advantaged or disadvantaged in the choice of principles by the outcome of natural chance or the contingency of social circumstances. Since all are similarly situated and no one is able to design principles to favor his particular condition, the principles of justice are the result of a fair agreement or bargain. For given the circumstances of the original position, the symmetry of everyone's relation to each other, this initial situation is fair between individuals as moral persons, that is, as rational beings with their own ends and capable, I shall assume, of a sense of justice. The original position is, one might say, the appropriate initial status quo, and the fundamental agreements reached in it are fair. This explains the propriety of the name "justice as fairness": it conveys the idea that the principles of justice are agreed to in an initial situation that is fair. The name does not mean that the concepts of justice and fairness are the same, any more that the phrase "poetry as metaphor" means that the concepts of poetry and metaphor are the same.

Justice as fairness begins, as I have said, with one of the most general of all choices which persons might make together, namely, with the choice of the first principles of a conception of justice which is to regulate all subsequent criticism and reform of institutions. Then, having chosen a conception of justice, we can suppose that they are to choose a constitution and a legislature to enact laws, and so on, all in accordance with the principles of justice initially agreed upon. Our social situation is just if it is such that by this sequence of hypothetical agreements we would have contracted into the general system of rules which defines it. Moreover, assuming that the original position does determine a set of principles (that is, that a particular conception of justice would be chosen), it will then be true that whenever social institutions satisfy these principles those engaged in them can say to one another that they are cooperating on terms to which they would agree if they were free and equal persons whose relations with respect to one another were fair. They could all view their arrangements as meeting the stipulations which they would acknowledge in an initial situation that embodies widely accepted and reasonable constraints on the choice of principles. The general recognition of this fact would provide the basis for a public acceptance of the corresponding principles of justice. No society can, of course, be a scheme of cooperation which men enter voluntarily in a literal sense; each person finds himself placed at birth in some particular position in some particular society, and the nature of this position materially affects his life prospects. Yet a society satisfying the principles of justice as fairness comes as close as a society can to being a voluntary scheme, for it meets the principles which free and equal persons would assent to under circumstances that are fair. In this sense its members are autonomous and the obligations they recognize self-imposed.

One feature of justice as fairness is to think of the parties in the initial situation as rational and mutually disinterested. This does not mean that the parties are egoists, that is, individuals with only certain kinds of interests, say in wealth, prestige, and domination. But they are conceived as not taking an interest in one another's interests. They are to presume that even their spiritual aims may be opposed, in the way that the aims of those of different religions maybe opposed. Moreover, the concept of rationality must be interpreted as far as possible in the narrow sense, standard in economic theory, of taking the most effective means to given ends.

In working out the conception of justice as fairness one main task clearly is to determine which principles of justice would be chosen in the original position. It may be observed that once the principles of justice are thought of as arising from an original agreement in a situation of equality, it is an open question whether the principle of utility would be acknowledged. Offhand it hardly seems likely that persons who view themselves as equals, entitled to press their claims upon one another, would agree to a principle which may require lesser life prospects for some simply for the sake of a greater sum of advantages enjoyed by others. Since each desires to protect his interests, his capacity to advance his conception of the good, no one has a reason to acquiesce in an enduring loss for himself in order to bring about a greater net balance of satisfaction. In the absence of strong and lasting benevolent impulses, a rational man would not accept a basic structure merely because it maximized the algebraic sum of advantages irrespective of its permanent effects on his own basic rights and interests. Thus it seems that the principle of utility is incompatible with the conception of social cooperation among equals for mutual advantage. It appears to be inconsistent with the idea or reciprocity implicit in the notion of a well-ordered society. Or, at any rate, so I shall argue.

I shall maintain instead that **the persons in the initial situation would choose two** rather different **principles: the first requires equality in the assignment of basic rights and duties, while the second holds that social and economic inequalities, for example inequalities of wealth and authority, are just only if they result in compensating benefits for everyone, and in particular for the least advantaged members of society.** These principles rule out justifying institutions on the grounds that the hardships of some are offset by a greater good in the aggregate. It may be expedient but it is not just that some should have less in order that others may prosper. But there is no injustice in the greater benefits earned by a few provided that the situation of persons not so fortunate is thereby improved. The intuitive idea is that since everyone's well-being depends upon a scheme of cooperation without which no one could have a satisfactory life, the division of advantages should be such as to draw forth the willing cooperation of everyone taking part in it, including those less well situated. Yet this can be expected only if reasonable terms are proposed. The two principles mentioned seem to be a fair agreement on the basis of which those better endowed, or more fortunate in their social position, neither of which we can be said to deserve, could expect the willing cooperation of others when some workable scheme is a necessary condition of the welfare of all. Once we decide to look for a conception of justice that nullifies the accidents of natural endowment and the contingencies of social circumstance as counters in quest for political and economic advantage, we are led to these principles. They express the result of leaving aside those aspects of the social world that seem arbitrary from a moral point of view.

Justice as fairness is not a complete contract theory. For it is clear that the contractarian idea can be extended to the choice of more or less an entire ethical system. Obviously if justice as fairness succeeds reasonably well, a next step would be to study the more general view suggested by the name "rightness as fairness."

The Original Position and Justification

I have said that the original position is the appropriate initial status quo which insures that the fundamental agreements reached in it are fair. This fact yields the name "justice as fairness." One conception of justice is more reasonable than another, or justifiable with respect to it, if rational persons in the initial situation would choose its principles over those of the other for the role of justice. Conceptions of justice are to be ranked by their acceptability to persons so circumstanced.

One should not be misled by the somewhat unusual conditions which characterize the original position. The idea here is simply to make vivid to ourselves the restrictions that it seems reasonable to impose on arguments for principles of justice, and therefore on these principles themselves. Thus it seems reasonable and generally acceptable that no one should be advantaged or disadvantaged by natural fortune or social circumstances in the choice of principles. It also seems widely agreed that it should be impossible to tailor principles to the circumstances of one's own case. We should insure further that particular inclinations and aspirations, and persons' conceptions of their good do not affect the principles adopted. The aim is to rule out those principles that it would be rational to propose for acceptance, however little the chance of success, only if one knew certain things that are irrelevant from the standpoint of justice. For example, if a man knew that he was wealthy, he might find it rational to advance the principle that various taxes for welfare measures be counted unjust; if he knew that he was poor, he would most likely propose the contrary principle. To represent the desired restrictions one imagines a situation in which everyone is deprived of this sort of information. One excludes the knowledge of those contingencies which sets men at odds and allows them to be guided by their prejudices. In this manner the veil of ignorance is arrived at in a natural way. This concept should cause no difficulty if we keep in mind the constraints on arguments that it is meant to express. At any time we can enter the original position, so to speak, simply by following a certain procedure, namely, by arguing for principles of justice in accordance with these restrictions.

It seems reasonable to suppose that the parties in the original position are equal. That is, all have the same rights in the procedure for choosing principles; each can make proposals, submit reasons for their acceptance, and so on. Obviously the purpose of these conditions is to represent equality between human beings as moral persons, as creatures having a conception of their good and capable of a sense of justice. The basis of equality is taken to be similarity in these two respects. Systems of ends are not ranked in value; and each man is presumed to have the requisite ability to understand and to act upon whatever principles are adopted. Together with the veil of ignorance, these conditions define the principles of justice as those which rational persons concerned to advance their interests would consent to as equals when none are known to be advantaged or disadvantaged by social and natural contingencies.

There is, however, another side to justifying a particular description of the original position. This is to see if the principles which would be chosen match our considered convictions of justice or extend them in an acceptable way. We can note whether applying these principles would lead us to make the same judgments about the basic structure of society which we now make intuitively and in which we have the greatest confidence; or whether, in cases where our present judgments are in doubt and given with hesitation, these principles offer a resolution which we can affirm on reflection. There are questions which we feel sure must be answered in a certain way. For example, we are confident that religious intolerance and racial discrimination are unjust. We think that we have examined these things with care and have reached what we believe is an impartial judgment not likely to be distorted by an excessive attention to our own interests. These convictions are provisional fixed points which we presume any conception of justice must fit. But we have much less assurance as to what is the correct distribution of wealth and authority. Here we may be looking for a way to remove our doubts. We can check an interpretation of the initial situation, then, by the capacity of its principles to accommodate our firmest convictions and to provide guidance where guidance is needed.

A final comment. We shall want to say that certain principles of justice are justified because they would be agreed to in an initial situation of equality. I have emphasized that this original position is purely hypothetical. It is natural to ask why, if this agreement is never actually entered into, we should take any interest in these principles, moral or otherwise. The answer is that the conditions embodied in the description of the original position are ones that we do in fact accept. Or if we do not, then perhaps we can be persuaded to do so by philosophical reflection. Each aspect of the contractual situation can be given supporting grounds. Thus what we shall do is to collect together into one conception a number of conditions on principles that we are ready upon due consideration to recognize as reasonable. These constraints express what we are prepared to regard as limits on fair terms of social cooperation. One way to look at the idea or the original position, therefore, is to see it as an expository device which sums up the meaning of these conditions and helps us to extract their consequences. Led on by it we define more clearly the standpoint from which we can best interpret moral relationships.

Classical Utilitarianism

There are many forms of utilitarianism, and my aim is to work out a theory of justice that represents an alternative to utilitarian thought generally and so to all of these different versions of it. With this end in mind, the kind of utilitarianism I shall describe here is the strict classical doctrine. The main idea is that society is rightly ordered, and therefore just, when its major institutions are arranged so as to achieve the greatest net balance of satisfaction summed over all the individuals belonging to it.

We may note first that there is, indeed, a way of thinking of society which makes it easy to suppose that the most rational conception of justice is utilitarian. For consider: each man in realizing his own interests is certainly free to balance his own losses against his own gains. We may impose a sacrifice on ourselves now for the sake of a greater advantage later. A person quite properly acts, at least when others are not affected, to achieve his own greatest good, to advance his rational ends as far as possible. Now why should not a society act on precisely the same principle applied to the group and therefore regard that which is rational for one man as right for an association of men? Just as the well-being of a person is constructed from the series of satisfactions that are experienced at different moments in the course of his life, so in very much the same way the well-being of society is to be constructed from the fulfillment of the systems of desires of the many individuals who belong to it. Since the principle for an individual is to advance as far as possible his own welfare, his own system of desires, the principle for society is to advance as far as possible the welfare of the group, to realize to the greatest extent the comprehensive system of desire arrived at from the desires of its members. Just as an individual balances present and future gains against present and future losses, so a society may balance satisfactions and dissatisfactions between different individuals. And so by these reflections one reaches the principle of utility in a natural way: a society is properly arranged when its institutions maximize the net balance of satisfaction. The principle of choice for an association of men is interpreted as an extension of the principle of choice for one man. Social justice is the principle of rational prudence applied to an aggregative conception of the welfare of the group. The appropriate terms of social cooperation are settled by whatever in the circumstances will achieve the greatest sum of satisfaction of the rational desires of individuals. It is impossible to deny the initial plausibility and attractiveness of this conception.

The striking feature of the utilitarian view of justice is that it does not matter, except indirectly, how this sum of satisfactions is distributed among individuals any more than it matters, except indirectly, how one man distributes his satisfactions over time. The correct distribution in either case is that which yields the maximum fulfillment. Society must allocate its means of satisfaction whatever these are, rights and duties, opportunities and privileges, and various forms of wealth, so as to achieve this maximum if it can. But in itself no distribution of satisfaction is better than another. Thus there is no reason in principle why the greater gains of some should not compensate for the lesser losses of others; or more importantly, why the violation of the liberty of a few might not be made right by the greater good shared by many. It simply happens that under most conditions, at least in a reasonably advanced stage of civilization, the greatest sum of advantages is not attained in this way.

The most natural way, then, of arriving at utilitarianism is to adopt for society as a whole the principle of rational choice for one man. The nature of the decision made by the ideal legislator is not, therefore, materially different from that of a consumer deciding how to maximize his satisfaction by the purchase of this or that collection of goods. The correct decision is essentially a question of efficient administration. This view of social cooperation is the consequence of extending to society the principle of choice for one man, and then, to make this extension work, conflating all persons into one through the imaginative acts of the impartial sympathetic spectator. Utilitarianism does not take seriously the distinction between persons. [Neither does it] recognize as the basis of justice that to which men would consent.

THE PRINCIPLES OF JUSTICE

Two Principles of Justice

I shall now in a provisional form the two principles of justice that I believe would be chosen in the original position. The first statement of the two principles reads as follows.

- First: each person is to have an equal right to the most extensive basic liberty compatible with a similar liberty for others.
- Second: social and economic inequalities are to be arranged so that they are both (a) reasonably expected to be to everyone's advantage, and (b) attached to positions and offices open to all.

By way of general comment, these principles are to govern the assignment of rights and duties and to regulate the distribution of social and economic advantages. They distinguish between those aspects of the social system that define and secure the equal liberties of citizenship and those that specify and establish social and economic inequalities. The basic liberties of citizens are, roughly speaking, political liberty (the right to vote and to be eligible for public office) together with freedom of speech and assembly; liberty of conscience and freedom of thought; freedom of the person along with the right to hold (personal) property; and freedom from arbitrary arrest and seizure as defined by the concept of the rule of law. These liberties are all required to be equal by the first principle, since citizens of a just society are to have the same basic rights.

The second principle applies, in the first approximation, to the distribution of income and wealth and to the design of organizations that make use of differences in authority and responsibility, or chains of command. While the distribution of wealth and income need not be equal, it must be to everyone's advantage, and at the same time, positions of authority and offices of command must be accessible to all. One applies the second principle by holding positions open, and then, subject to this constraint, arranges social and economic inequalities so that everyone benefits.

These principles are to be arranged in a serial order with the first principle prior to the second. This ordering means that a departure from the institutions of equal liberty required by the first principle cannot be justified by, or compensated for, by greater social and economic advantages. The distribution of wealth and income, and the hierarchies of authority, must be consistent with both the liberties of equal citizenship and equality of opportunity.

The two principles are a special case of a more general conception of justice that can be expressed as follows.

> All social values—liberty and opportunity, income and wealth, and the bases of self-respect—are to be distributed equally unless an unequal distribution of any, or all, of these values is to everyone's advantage.

Injustice, then, is simply inequalities that are not to the benefit of all. Of course, this conception is extremely vague and requires interpretation.

As a first step, suppose that the basic structure of society distributes certain **primary goods**, that is, things that every rational man is presumed to want. These goods normally have a use whatever a person's rational plan of life. For simplicity, assume that the chief primary goods at the disposition of society are rights and liberties, powers and opportunities, income and wealth. (Later on the primary good of self-respect has a central place.) These are the social primary goods. Other primary goods such as health and vigor, intelligence and imagination, are natural goods; although their possession is influenced by the basic structure, they are not so directly under its control. Imagine, then, a hypothetical initial arrangement in which all the social primary goods are equally distributed: everyone has similar rights and duties, and income and wealth are evenly shared. This state of affairs provides a benchmark for judging improvements. If certain inequalities of wealth and organizational powers would make everyone better off than in this hypothetical starting situation, then they accord with the general conception.

Now it is possible, at least theoretically, that by giving up some of their fundamental liberties men are sufficiently compensated by the resulting social and economic gains. The general conception of justice imposes no restrictions on what sort of inequalities are permissible; it only requires that everyone's position be improved. We need not suppose anything so drastic as consenting to a condition of slavery. Imagine instead that men forego certain political rights when the economic returns are significant and their capacity to influence the course of policy by the exercise of these rights would be marginal in any case. It is this kind of exchange which the two principles as stated rule out; being arranged in serial order they do not permit exchanges between basic liberties and economic and social gains. The serial ordering of principles expresses an underlying preference among primary social goods. The only reason for circumscribing the rights defining liberty and making men's freedom less extensive than it might otherwise be is that these equal rights as institutionally defined would interfere with one another.

Now the second principle insists that each person benefit from permissible [economic and social] inequalities in the basic structure. This means that it must be reasonable for each relevant representative man defined by this structure, when he views it as a going concern, to prefer his prospects with the inequality to his prospects without it. One is not allowed to justify differences in income or organizational powers on the ground that the disadvantages of those in one position are outweighed by the greater advantages of those in another. Much less can infringements of liberty be counterbalanced in this way.

[After subsequent refinements, the two principles of justice can be restated as follows:

- **First Principle**: Each person has the same irrevocable claim to the most extensive system of equal basic liberties compatible with the same system of liberties for all.
- **Second Principle**: Social and economic inequalities are to be arranged so that they are both:
    a. attached to offices and positions open to all under conditions of *fair equality of opportunity*; and
    b. to the greatest benefit of the least advantaged [**the "difference principle"**].]

The Tendency to Equality

Since inequalities of birth and natural endowment are undeserved, these inequalities are to be somehow compensated for. Thus the [difference] principle holds that in order to treat all persons equally, to provide genuine equality of opportunity, society must give more attention to those with fewer native assets and to those born into the less favorable social positions. The idea is to redress the bias of contingencies in the direction of equality. In pursuit of this principle greater resources might be spent on the education of the less rather than the more intelligent, at least over a certain time of life, say the earlier years of school.

Now the difference principle does not require society to try to even out handicaps as if all were expected to compete on fair basis in the same race. But the difference principle would allocate resources in education, say, so as to improve the long-term expectation of the least favored. If this end is attained by giving more attention to the better endowed, it is permissible; otherwise not. And in making this decision, the value of education should not be assessed solely in terms of economic efficiency and social welfare. Equally if not more important is the role of education in enabling a person to enjoy the culture of his society and to take part in its affairs, and in this way to provide for each individual a secure sense of his own worth.

We see then that the difference principle represents, in effect, an agreement to regard the distribution of natural talents as a common asset and to share in the benefits of this distribution whatever it turns out to be. Those who have been favored by nature, whoever they are, may gain from their good fortune only on terms that improve the situation of those who have lost out. The naturally advantaged are not to gain merely because they are more gifted, but only to cover the costs of training and education and for using their endowments in ways that help the less fortunate as well. No one deserves his greater natural capacity nor merits a more favorable starting place in society. But it does not follow that one should eliminate these distinctions. There is another way to deal with them. The basic structure can be arranged so that these contingencies work for the good of the least fortunate. Thus we are led to the difference principle if we wish to set up the social system so that no one gains or loses from his arbitrary place in the distribution of natural assets or his initial position in society without giving or receiving compensating advantages in return.

The natural distribution is neither just nor unjust; nor is it unjust that persons are born into society at some particular position. These are simply natural facts. What is just and unjust is the way that institutions deal with these facts. Aristocratic and caste societies are unjust because they make these contingencies the ascriptive basis for belonging to more or less enclosed and privileged social classes. The basic structure of these societies incorporates the arbitrariness found in nature. But there is no necessity for men to resign themselves to these contingencies. The social system is not an unchangeable order beyond human control but a pattern of human action. In justice as fairness men agree to share one another's fate. In designing institutions they undertake to avail themselves of the accidents of nature and social circumstance only when doing so is for the common benefit. The two principles are a fair way of meeting the arbitrariness of fortune; and while no doubt imperfect in other ways, the institutions which satisfy these principles are just.

A further point is that the difference principle expresses a conception of reciprocity. It is a principle of mutual benefit. Each representative man can accept the basic structure as designed to advance his interests. The social order can be justified to everyone, and in particular to those who are least favored. Consider any two representative men A and B, and let B be the one who is the least favored man. Now B can accept A's being better off since A's advantages have been gained in ways that improve B's prospects. If A were not allowed his better position, B would be even worse off than he is. The difficulty is to show that A has no grounds for complaint. Perhaps he is required to have less than he might since his having more would result in some loss to B. Now what can be said to the more favored man? To begin with, it is clear that the well-being of each depends on a scheme of social cooperation without which no one could have a satisfactory life. Secondly, we can ask for the willing cooperation of everyone only if the terms of the scheme are reasonable. The difference principle, then, seems to be a fair basis on which those better endowed, or more fortunate in their social circumstances could expect others to collaborate with them when some workable arrangement is a necessary condition of the good of all.

There is a natural inclination to object that those better situated deserved their greater advantages whether or not they are to the benefit of others. At this point it is necessary to be clear about the notion of desert. It is perfectly true that given a just system of cooperation, those who have done what the system announces that it will reward are entitled to their advantages. In this sense the more fortunate have a claim to their better situation; their claims are legitimate expectations established by social institutions, and the community is obligated to meet them. But this sense of desert presupposes the existence of the cooperative scheme; it is irrelevant to the question whether in the first place the scheme is to be designed in accordance with the difference principle or some other criterion.

Perhaps some will think that the person with greater natural endowments deserves those assets and the superior character that made their development possible. Because he is more worthy in this sense, he deserves the greater advantages that he could achieve with them. This view, however, is surely incorrect. It seems to be one of the fixed points of our considered judgments that no one deserves his place in the distribution of native endowments, any more than one deserves one's initial starting place in society. The assertion that a man deserves the superior character that enables him to make the effort to cultivate his abilities is equally problematic; for his character depends in large part upon fortunate family and social circumstances for which he can claim no credit. Thus the more advantaged representative man cannot say that he deserves and therefore has a right to a scheme of cooperation in which he is permitted to acquire benefits in ways that do not contribute to the welfare of others. There is no basis for his making this claim. From the standpoint of common sense, then, the difference principle appears to be acceptable both to the more advantaged and to the less advantaged individual.

Distributive Justice*
Julian Lamont & Christi Favor

Principles of distributive justice are designed to guide the allocation of the benefits and burdens of economic activity.

1. Scope and Role of Distributive Principles

Distributive principles can vary in what is subject to distribution (income, wealth, opportunities, jobs, welfare, utility, etc.) and on what basis distribution should be made (equality, maximization, according to individual characteristics, according to free transactions, etc.). This entry will focus on the distribution of the benefits and burdens of economic activity among individuals in a society. Although principles of this kind have been the dominant source of Anglo-American debate about distributive justice over the last four decades, there are other important distributive justice questions, some of which are distributive justice at the global level rather than just at the national level, [and] distributive justice across generations.

Throughout most of history, people were born into, and largely stayed in, a fairly rigid economic position. The distribution of economic benefits and burdens was seen as fixed, either by nature or by God. Only when people realized that the distribution of economic benefits and burdens could be affected by government did distributive justice become a live topic. Now the topic is unavoidable. Governments continuously make and change laws affecting the distribution of economic benefits and burdens in their societies. Almost all changes, from the standard tax and industry laws through to divorce laws, have some distributive effect, and, as a result, different societies have different distributions. Every society then is always faced with a choice about whether to stay with the current laws and policies or to modify them. Distributive justice theory contributes practically by providing guidance for these unavoidable and constant choices. Sometimes a number of the theories will recommend the same change in policy; other times they will diverge.

2. Strict Egalitarianism

One of the simplest principles of distributive justice is that of strict or radical equality. The principle says that every person should have the same level of material goods and services. The principle is most commonly justified on the grounds that people are owed equal respect and that equality in material goods and services is the best way to give effect to this ideal.

The strict equality principle stated above says that there should be 'the same *level* of material goods and services'. The problem is how to specify and measure levels. One way is to specify that everyone should have the same *bundle* of material goods and services (so everyone would have 4 oranges, 6 apples, 1 bike, etc.). The main problem with this solution is that there will be many other allocations

---

* Lamont, Julian, Favor, Christi, "Distributive Justice", *The Stanford Encyclopedia of Philosophy (Fall 2008 Edition)*, Edward N. Zalta (ed.), URL = <http://plato.stanford.edu/archives/fall2008/entries/justice-distributive/>. Edited (material excised, emphases and formatting altered) by Lawrence Lengbeyer.

of material goods and services which will make some people better off without making anybody else worse off. For instance, a person preferring apples to oranges will be better off if she swaps some of the oranges from her bundle for some of the apples belonging to a person preferring oranges to apples. Indeed, it is likely that everybody will have something they would wish to trade in order to make themselves better off. As a consequence, requiring identical bundles will make virtually everybody materially worse off than they would be under an alternative allocation. So specifying that everybody must have the same *bundle* of goods does not seem to be satisfactory.

[U]sing *money* as [an] index for the value of material goods and services is the most practical response and is widely used in the specification and implementation of distributive principles.

The most common form of strict equality principle specifies that *income* (measured in terms of money) should be equal in each time-frame, though even this may lead to significant disparities in *wealth* if variations in savings are permitted.

There are a number of direct moral criticisms made of strict equality principles: that they unduly restrict freedom, that they do not give best effect to equal respect for persons, that they conflict with what people deserve, etc. But the most common criticism is a welfare-based one: that everyone can be materially better off if incomes are not strictly equal. It is this fact which partly inspired the Difference Principle.

3. The Difference Principle

The wealth of an economy is not a fixed amount. More wealth can be produced. The most common way of producing more wealth is to have a system where those who are more productive earn greater incomes. This partly inspired the formulation of the Difference Principle.

The most widely discussed theory of distributive justice in the past three decades has been that proposed by John Rawls. Rawls proposes the following two principles of justice:

1. Each person has an equal claim to a fully adequate scheme of equal basic rights and liberties, compatible with the same scheme for all.
2. Social and economic inequalities are to satisfy two conditions: (a) They are to be attached to positions and offices open to all under conditions of fair equality of opportunity; and (b), they are to be to the greatest benefit of the least advantaged members of society.

Under Rawls' proposed system Principle (1) has priority over Principle (2). [I]t is possible to think of Principles (1) and (2a) as principles of distributive justice: to govern the distribution of liberties, and the distribution of opportunities. Looking at the principles of justice in this way makes all principles of justice, principles of distributive justice (even principles of retributive justice will be included on the basis that they distribute negative goods). [But] let us concentrate on (2b), known as the Difference Principle.

The main moral motivation for the Difference Principle is similar to that for strict equality: equal respect for persons.

If it is possible to raise the absolute position of the least advantaged further by having some inequalities of income and wealth, then the Difference Principle prescribes inequality up to that point where the absolute position of the least advantaged can no longer be raised. The inequalities consistent with the Difference Principle are only permitted so long as they do not compromise the political liberties. So, for instance, very large wealth differentials may make it practically impossible for poor people to be elected to political office or to have their political views represented. These inequalities of wealth, even if they increase the material position of the least advantaged group, may need to be reduced in order for the first principle to be implemented.

Briefly, the main criticisms [of the Difference Principle] are as follows. Advocates of strict equality argue that inequalities permitted by the Difference Principle are unacceptable even if they do benefit the least advantaged. The most common explanation appeals to *solidarity*: that being materially equal is an important expression of the equality of persons. The Utilitarian objection to the Difference Principle is that it does not maximize utility. Libertarians object that the Difference Principle involves unacceptable infringements on liberty[,] [f]or instance redistributive taxation [that] involves the immoral taking of just holdings. Advocates of Desert-Based Principles argue that some may deserve a higher level of material goods because of their hard work or contributions even if their unequal rewards do not also function to improve the position of the least advantaged. They also argue that the explanations of how people come to be in more or less advantaged positions is relevant to their fairness, yet the Difference Principle wrongly ignores these explanations.

4. Resource-Based Principles

Resource-based principles (also called Resource Egalitarianism) prescribe equality of resources. Interestingly, resource-based principles do not normally prescribe a patterned outcome — the idea being that the outcomes are determined by people's free use of their resources. Resource-theorists claim that provided people have equal resources they should live with the consequences of their choices. They argue, for instance, that people who choose to work hard to earn more income should not be required to subsidize those choosing more leisure and hence less income.

While part of Rawls' motivation for the Difference principle is that people have unequal [natural and social] endowments, resource-theorists explicitly emphasize this feature of their theory. They agree that, ideally, social circumstances over which people have no control should not adversely affect life prospects or earning capacities. For instance, people born with handicaps, ill-health, or low levels of natural talents have not brought these circumstances upon themselves and hence, should not be disadvantaged in their life prospects.

Because the Resource-based theory has a similar motivation to the Difference Principle the moral criticisms of it tend to be variations on those leveled against the Difference Principle. However, it is not at all clear what would constitute an implementation of Resource-based theories and their variants in a real economy. It seems impossible to measure differences in people's natural talents—unfortunately, people's talents do not neatly divide into the natural and developed categories. A system of special assistance to the physically and mentally handicapped and to the ill would be a partial implementation, but most natural inequalities would be left untouched by such assistance.

5. Welfare-Based Principles

Welfare-based principles are motivated by the idea that what is of primary moral importance is the level of welfare of people. Resources, equality, desert-claims, or liberty are only valuable in so far as they increase welfare, so that all distributive questions should be settled according to which distribution maximizes welfare. However, 'maximizes welfare' is imprecise.

[M]ost philosophical activity has concentrated on a variant known as Utilitarianism. This theory can be used to illustrate most of the main characteristics of Welfare-based principles.

Historically, Utilitarians have used the term 'utility' rather than 'welfare' and utility has been defined variously as pleasure, happiness, or preference-satisfaction. So, for instance, the principle for distributing economic benefits for Preference Utilitarians is to distribute them so as to maximize the arithmetic sum of all satisfied preferences (unsatisfied preferences being negative), weighted for the intensity of those preferences.

The basic theory of Utilitarianism is one of the simplest to state and understand. Much of the work on the theory therefore has been directed towards defending it against moral criticisms. The [most fundamental] is that Utilitarianism fails to take the distinctness of persons seriously. Maximization of preference-satisfaction is often taken as prudent in the case of *individuals* — Utilitarianism uses it on an entity, *society*, unlike individuals in important ways. While it may be acceptable for a person to choose to suffer at some period in her life (be it a day, or a number of years) so that her overall life is better, it is often argued against Utilitarianism that it is immoral to make some people suffer so that there is a net gain for other people. In the individual case, there is a single entity experiencing both the sacrifice and the gain. Also, the individuals, who suffer or make the sacrifices, *choose* to do so in order to gain some benefit they deem worth their sacrifice. [U]nder Utilitarianism, there is no requirement for people to consent to the suffering or sacrifice.

6. Desert-Based Principles

Aristotle argued that virtue should be a basis for distributing rewards, but most contemporary principles owe a larger debt to John Locke. Locke argued [that] people deserve to have those items [that have been] produced by their toil and industry, the products (or the value thereof) being a fitting reward for their effort. His underlying idea was to guarantee to individuals the fruits of their own labor and abstinence. According to the contemporary desert theorist, people freely apply their abilities and talents, in varying degrees, to socially productive work. People come to deserve varying levels of income by providing goods and services desired by others. Distributive systems are just insofar as they distribute incomes according to the different levels earned or deserved by the individuals in the society for their productive labors, efforts, or contributions. [O]nly activity directed at raising the social product [i.e., the collective standard of living] will serve as a basis for deserving income.

[I]t is difficult to identify what is to count as a contribution, an effort or a cost, and it is even more difficult to measure these in a complex modern economy.

The main moral objection to desert-based principles is that they make economic benefits depend on factors over which people have little control. John Rawls has made one of the most widely discussed arguments to this effect, and it remains a problem for desert-based principles. The problem is most pronounced in the case of productivity-based principles — a person's productivity seems clearly to be influenced by many factors over which the person has little control.

7. Libertarian Principles

[For Libertarians,] just outcomes are those arrived at by the separate just actions of individuals; a particular distributive pattern is not required for justice. Robert Nozick has advanced this version of Libertarianism, and is its most well-known contemporary advocate.

Nozick proposes a 3-part "Entitlement Theory".

a. A person who acquires a holding in accordance with the principle of justice in acquisition is entitled to that holding.

b. A person who acquires a holding in accordance with the principle of justice in transfer, from someone else entitled to the holding, is entitled to the holding.

c. No one is entitled to a holding except by (repeated) applications of (a) and (b).

The distribution is just if everyone is entitled to the holdings they possess under the distribution.

The principle of *justice in transfer* is the least controversial and is designed to specify fair contracts while ruling out stealing, fraud, etc. The principle of *justice in acquisition* is more complicated and more controversial. The principle is meant to govern the gaining of exclusive property rights over the material world. Nozick takes his inspiration from John Locke's idea that everyone 'owns' themselves and, by mixing one's labors with the world, self-ownership can generate ownership of some part of the material world. However, of Locke's mixing metaphor, Nozick legitimately asks: '. . . why isn't mixing what I own with what I don't own a way of losing what I own rather than a way of gaining what I don't? If I own a can of tomato juice and spill it in the sea so its molecules. . . mingle evenly throughout the sea, do I thereby come to own the sea, or have I foolishly dissipated my tomato juice?' Nozick concludes that what is significant about mixing our labor with the material world is that in doing so, we tend to increase the value of it, so that self-ownership can lead to ownership of the external world in such cases.

The obvious objection is that it is not clear why the *first* people to acquire some part of the material world should be able to exclude others from it (and, for instance, be the land owners while the later ones become the wage laborers). In response to this objection, Nozick puts a qualification on just acquisition, called the *Lockean Proviso*, whereby an exclusive acquisition of the external world is just, if, after the acquisition, there is 'enough and as good left in common for others'. One of the main challenges for Libertarians has been to formulate a morally plausible interpretation of this proviso. According to Nozick's interpretation, an acquisition is just if and only if the position of others after the acquisition is no worse than their position was when the acquisition was unowned or 'held in common'. For

Nozick's critics, his proviso is unacceptably weak because it fails to consider the position others may have achieved under *alternative distributions* and thereby instantiates the morally dubious criterion of whoever is first gets the exclusive spoils.

Of course, many existing holdings are the result of acquisitions or transfers which at some point did not satisfy principles (a) and (b) above. Hence, Nozick must supplement those principles with a *principle of rectification* for past injustice.

> The principle of rectification will make use of its best estimate of information about what *would have* occurred... if the injustice had not taken place. [O]ne of the descriptions yielded must be realized [through appropriate transfers, if necessary].

Past injustices systematically undermine the justice of every subsequent distribution in historical theories. Nozick is clear that his historical theory is of no use in evaluating the justice of actual societies until such a theory of rectification is given:

> In the absence of [a full treatment of the principle of rectification] applied to a particular society, one *cannot* use the analysis and theory presented here to condemn any particular scheme of transfer payments, unless it is clear that no considerations of rectification of injustice could apply to justify it.

Unfortunately for the theory, no such treatment will ever be forthcoming because the task is, for all practical purposes, impossible. The numbers of injustices perpetrated throughout history, both within nations and between them, are enormous and the necessary details of the vast majority of injustices are unavailable. As a consequence, Nozick's entitlement theory will never provide any guidance as to what the current distribution of material holdings should be nor what distributions or redistributions are legitimate or illegitimate. (Indeed Nozick suggests, for instance, [that] the Difference Principle may be the best implementation of the principle of rectification.)

Libertarians inspired by Nozick usually advocate a system in which there are exclusive property rights, with the role of the government restricted to the protection of these property rights. The property rights commonly rule out taxation for purposes other than raising the funds necessary to protect property rights. The strongest critique of any attempt to institute such a system of legally protected property rights comes, as we have seen, from Nozick's theory itself — there seems no obvious reason to give strong legal protection to property rights which have arisen through violations of the just principles of acquisition and transfer. But putting this critique to one side, what other arguments are made in favor of exclusionary property rights?

Nozick argues that because people own themselves and hence their talents, they own whatever they can produce with these talents. Moreover, it is possible in a free market to sell the products of exercising one's talents. Any taxation of the income from such selling, according to Nozick, 'institute[s] (partial) ownership by others of people and their actions and labor'. People, according to this argument, have these exclusive rights of ownership. Taxation then, simply involves violating these rights and allowing some people to own (partially) other people. Moreover, it is argued, any system not legally recognizing these rights violates Kant's maxim to treat people always as ends in themselves and never merely as a means. The two main difficulties with this argument have been: (1) to show that self-ownership is only compatible with having such strong exclusive property rights; and (2) that a system of exclusive property rights is the best system for treating people with respect, as ends in themselves.

[T]he most common other route for trying to justify exclusive property rights has been to argue that they are required for the maximization of freedom and/or liberty or the minimization of violations of these. As an empirical claim though, this appears to be false. If we compare countries with less exclusionary property rights (e.g. more taxation) with countries with more exclusionary property regimes, we see no systematic advantage in freedoms/liberties enjoyed by people in the latter countries. (Of course, we do see a difference in *distribution* of such freedoms/liberties in the latter countries[:] the richer have more and the poorer less, while in the former they are more evenly distributed.)

8. Feminist Principles

The distributive principles so far outlined, with the exception of strict egalitarianism, could be classified as liberal theories — they both inform, and are the product of, the liberal democracies which have emerged over the last two centuries.

John Stuart Mill in *The Subjection of Women* (1869) gives one of the clearest early feminist critiques of the political and distributive structures of the emerging liberal democracies. Mill argued that the principles associated with the developing liberalism of his time required equal political status for women[:] equal opportunity in education and in the marketplace, equal rights to hold property, a rejection of the man as the legal head of the household, and equal rights to political participation. Feminists who follow Mill believe that a proper recognition of the position of women in society requires that women be given equal and the same rights as men have, and that these primarily protect their liberty and their status as equal persons under the law. Thus, government regulation should not prevent women from competing on equal terms with men in educational, professional, marketplace and political institutions. From the point of view of other feminisms, the liberal feminist position is a conservative one, in the sense that it requires the proper inclusion for women of the rights, protections, and opportunities previously secured for men, rather than a fundamental change to the traditional liberal position.

One phrase or motto around which feminists have rallied, however, marks a significant break with Mill's liberalism: 'the personal is political.' Mill was crucial in developing the liberal doctrine of limiting the state's intervention in the private lives of citizens. Many contemporary feminists have argued that the resulting liberal theories of justice have fundamentally been unable to accommodate the injustices that have their origins in this 'protected' private sphere of government non-interference. Susan Moller Okin and others demonstrate, for example, that women have substantial disadvantages in competing in the market because of childrearing responsibilities which are not equally shared with men. As a consequence, any theory relying on market mechanisms, including most liberal theories, will yield systems which result in women systematically having less income and wealth than men.

9. Methodology and Empirical Beliefs about Distributive Justice

As noted above, the overarching methodological concern of the distributive justice literature must be, in the first instance, the pressing choice of how the benefits and burdens of economic activity should be distributed, rather than the mere uncovering of abstract truth. Principles are to be implemented in real societies with the problems and constraints inherent in such application. Given this, pointing out that the application of any particular principle will have some, perhaps many, immoral results will not by itself constitute a fatal counterexample to any distributive theory. Such counter-evidence to a theory would only be fatal if there were an alternative theory, which, if fully implemented, would yield a morally preferable society overall. So, it is at least possible that the best distributive theory, when implemented, might yield a system which still has many injustices and/or negative consequences.

Distributive justice is not an area where we can say an idea is good in theory but not in practice. If it is not good in practice, then it is not good in theory either.

Crime and Punishment
R.A. Duff*

1 Punishment, the State and the Criminal Law

... Punishment can be initially defined as the deliberate infliction of something meant to be burdensome, by an authority, on an alleged offender, for an alleged offence. It needs justification because it involves doing things (depriving people of life, liberty or money) which are normally wrong. Different moral perspectives, however, generate different accounts of why punishment is morally problematic and thus of what could justify it. Is what matters the infliction of pain, for instance; or the apparent coercive infringement of rights which punishment involves?

...

... [N]ot all breaches of socially (or legally) authoritative norms ... merit punishment: we must ask what kinds of response are appropriate to different kinds of wrongdoing.

Censure [expression of disapproval or condemnation] is one proper response to breaches of authoritative norms.... But censure can be expressed by formal declarations, or by symbolic punishments which are painful only in virtue of their expressive meaning, whereas criminal punishments typically inflict 'hard treatment' (the loss of liberty, money or life) which is painful independently of its expressive meaning. Why should such hard treatment be an appropriate response to socially proscribed wrongdoing?

...

2 Consequentialism and Retributivism

Penal theory has long been a battleground between consequentialists [such as Utilitarians] and retributivists [such as Kantians]....

Consequentialists justify punishment by its instrumental contribution to certain goods: most obviously, the good of crime-prevention. A penal system is justified if its crime-preventive and other benefits outweigh its costs, and no alternative practice could achieve such goods more cost-effectively. Punishment prevents crime by deterring, incapacitating or reforming potential offenders: by giving them prudential disincentives to crime, by subjecting them to restraints which make it harder for them to break the law, or by so modifying their attitudes that they will obey the law willingly.

...

One objection to ... consequentialist theory concerns the moral standing of those who are punished or threatened with punishment: that a consequentialist system fails to respect its citizens (criminals and non-criminals) as responsible agents. A system of deterrent punishments, Hegel argued, treats all those whom it threatens with punishment like 'dogs': rather than seeking their allegiance to the law by appeal to the moral reasons which justify its demands, it coerces their obedience by threats.

---

*From *The Routledge Encyclopedia of Philosophy*, vol. 2 pp. 704–11 (New York: Routledge, 1998). Edited by Lawrence Lengbeyer, June 2003.

A consequentialist system of reform similarly treats those subjected to it as objects to be remoulded, rather than as responsible agents who must determine their own conduct.

Against such objections, some argue that a . . . system of deterrent punishments can respect the moral standing of those it threatens and punishes; or that 'rehabilitation' and 'reform' need not be improperly manipulative or coercive. But one stimulus to the retributivist revival in the 1970s was the claim that only retributivism respects the moral standing of criminals: their right to receive 'fair and certain punishment', rather than being 'used merely as means' to the deterrence of others, or being subjected to indefinite terms of reformative 'treatment'.

The central retributivist slogan is that (only) the guilty deserve punishment, and deserve punishments proportionate to the seriousness of their crimes. This demand for 'just deserts' may be interpreted negatively, as forbidding the punishment of the innocent (or the excessive punishment of the guilty); or positively, as requiring that the guilty be punished as they deserve. The negative reading . . . suggests a 'mixed' account, . . . requiring that punishment be both deserved and consequentially beneficial. The positive reading makes guilt a necessary and sufficient condition of justified punishment: the guilty should be punished because they deserve it, whether or not their punishment achieves any consequential good.

The central task for any retributivist is to explain this supposed justificatory relation between guilt and punishment: what is it about crime that makes punishment an appropriate response to it? The central objection to all retributivist theories is that they fail to discharge this task: they either fail to explain this notion of penal desert, falling back on unexplained intuition or metaphysical mystery-mongering, or offer covertly consequentialist explanations.

The 'new retributivism' . . . offered various accounts of the idea of penal desert. One was that criminals gained by their crimes an unfair advantage over the law-abiding, since they accepted the benefits of the law-abiding self-restraint of others, but evaded that burden of self-restraint themselves: their punishment removed that unfair advantage, thus restoring the fair balance of benefits and burdens which the law should preserve. One objection to this account is that it distorts the nature of crime: what makes rape punishable as a crime is surely the wrong done to its victim, not the unfair advantage the rapist supposedly takes over all those who obey the law.

Another trend in recent retributivist thought has rather built on the idea of punishment as an expressive or communicative practice.

3 Punishment and Communication

. . . [C]onsequentialists can advocate expressive punishments. But the expressive or communicative aspect of punishment can explain the retributivist's slogan that the guilty deserve punishment: if they have broken a law which justifiably claimed their obedience, they deserve censure; and it is a proper task for the state, speaking on behalf of the community, to communicate that censure to them . . . as rational and responsible agents.

. . .

. . . But why, if not for the consequentialist reason that this will make the punishment a more effective deterrent, is effective communication so crucial that we must inflict hard-treatment punishments to achieve it?

. . . The communication of censure can itself be the central justifying aim of punishment, . . . [b]ut recognizing that, as fallible human beings, we will not always be adequately motivated by . . . moral reasons for obeying the law, we communicate that censure through hard treatment in order to provide an additional prudential [self-interest] incentive for obedience. . . .

More ambitiously communicative accounts of punishment portray the hard treatment as a mode of moral communication which aims to reform or educate. Punishment aims to bring wrongdoers to understand and to repent their crimes, and thus to reform their future conduct. Hard treatment assists this purpose by helping to bring home to them the meaning and implications of what they have done; it can also, if it is willingly undergone, enable them to express their repentance and thus reconcile themselves with their victims and the community. Such accounts are retributivist, since punishment must be focused on the past crime as an appropriate censuring response to it, but they also give punishment a forward-looking purpose: the offender's reform or rehabilitation, the restoration of the relationships which the crime damaged, the making of symbolic (and perhaps material) reparation to the victim and the community....

We must ask, however, whether hard-treatment punishment... seems incompatible with ... insistence on the need to protect individual rights and privacy against intrusive state or community power. [One] ... can argue that punishment's primary purpose should be the communication of appropriate censure, but may deny that the state should try, by such coercive means, to secure repentance and reform; in which case hard-treatment punishments could be justified only as prudential deterrents which do not seek to invade the criminal's soul.

4 Penal Theory and Sentencing

...

... What makes a particular kind of punishment appropriate, either generally or for a particular crime? How should sentencers determine the severity of punishment to be imposed on particular crimes or criminals?

Discussion of the last question often focuses on the principle of proportionality: the severity of punishment should be proportionate to the seriousness of the crime. Some such principle is integral to any retributivist theory, including communicative theories.... [But]while such a principle can help to determine the relative severity of sentences, requiring that more serious crimes be punished more severely, and so on, it is not clear whether it can help to fix absolute levels of punishment.

...

... If punishment is given an educative, reformative, or penitential aim, courts should seek punishments ... which will appropriately address the particular criminal. But this would require the courts to be given a more flexible and creative discretion in sentencing, to find or construct sentences appropriate to the particular case: a discretion which might undermine demands for strict and formal proportionality.

...

from Billy Budd
*Herman Melville**

In this excerpt from Melville's posthumously-published masterpiece, fundamental questions regarding the relations between subjective motives and objective effects of one's actions are raised in a compelling manner. Billy Budd, an admirable character, accidentally kills a malicious superior who has falsely accused him of mutiny. This is a capital offense. Billy seems subjectively innocent, yet objectively guilty. Vere's decision and his reasoning have been hotly debated many times.

Billy Budd takes place in 1797 on the British naval ship Bellipotent, just following two notorious mutinies at Spithead and Nore. Billy Budd, a sailor on the Bellipotent, is gentle and trusting and well loved by the crew. He is also uneducated and has difficulty speaking when he is upset. John Claggart, Billy's superior officer is a malicious and cruel man who deeply resents Billy's kindly nature and popularity among the men. Billy is unaware of Claggart's hatred until the moment he brings Billy before the ship's master Captain Vere, and falsely accuses Billy of plotting a mutiny. Billy, stunned by Claggart's vicious lies and unable to speak, strikes out at him, accidentally killing him by the blow.

Everyone sympathizes with Billy. But Captain Vere (a good man who has been acting strangely of late) sets up a military tribunal and, to everyone's surprise, testifies against Billy. In his testimony, Captain Vere acknowledges that Claggart was an evil man, but reminds the tribunal that they are a military court empowered only to judge Billy's deed—not his motives. According to military law, the punishment for striking a superior officer is death by hanging. Just as sailors must obey their superiors and not take the law into their own hands, so the tribunal has an absolute duty to obey the law. Moreover, because there had been several mutinies recently, it was all the more important that military law be enforced. Captain Vere says to the court: "Let not warm hearts betray heads that should be cool." Billy is convicted and hanged.

Critics disagree about the moral implications of Billy Budd. Some see Captain Vere as an evil man whose abstract notion of duty blinded him to true justice and compassion. For others, Vere is a moral hero who rises above sentiment to meet the need for order, authority, and law in human affairs.

Who in the rainbow can draw the line where the violet tint ends and the orange tint begins? Distinctly we see the difference of the colors, but where exactly does the one first blendingly enter into the other? So with sanity and insanity. In pronounced cases there is no question about them. But in some supposed cases, in various degrees supposedly less pronounced, to draw the exact line of demarcation few will undertake, though for a fee becoming considerate some professional experts will. There is nothing namable but that some men will, or undertake to, do it for pay.

---

*"Billy Budd" from *Billy Budd: Sailor* by Herman Melville, edited by Harrison Hayford and Merton Sealts. Copyright © 1972. University of Chicago Press. Introduction by Christina and Fred Sommers, *Vice and Virtue in Everyday Life* (NY, Harcourt, 1985) p. 54.

Whether Captain Vere, as the surgeon professionally and privately surmised, was really the sudden victim of any degree of aberration, every one must determine for himself by such light as this narrative may afford.

That the unhappy event which has been narrated could not have happened at a worse juncture was but too true. For it was close on the heel of the suppressed insurrections, an aftertime very critical to naval authority, demanding from every English sea commander two qualities not readily interfusable—prudence and rigor. Moreover, there was something crucial in the case.

In the jugglery of circumstances preceding and attending the event on board the *Bellipotent*, and in the light of that martial code whereby it was formally to be judged, innocence and guilt personified in Claggart and Budd in effect changed places. In a legal view the apparent victim of the tragedy was he who had sought to victimize a man blameless; and the indisputable deed of the latter, navally regarded, constituted the most heinous of military crimes. Yet more. The essential right and wrong involved in the matter, the clearer that might be, so much the worse for the responsibility of a loyal sea commander, inasmuch as he was not authorized to determine the matter on that primitive basis.

Small wonder then that the *Bellipotent's* captain, though in general a man of rapid decision, felt that circumspectness not less than promptitude was necessary. Until he could decide upon his course, and in each detail; and not only so, but until the concluding measure was upon the point of being enacted, he deemed it advisable, in view of all the circumstances, to guard as much as possible against publicity. Here he may or may not have erred. Certain it is, however, that subsequently in the confidential talk of more than one or two gun rooms and cabins he was not a little criticized by some officers, a fact imputed by his friends and vehemently by his cousin Jack Denton to professional jealousy of Starry Vere. Some imaginative ground for invidious comment there was. The maintenance of secrecy in the matter, the confining all knowledge of it for a time to the place where the homicide occurred, the quarterdeck cabin; in these particulars lurked some resemblance to the policy adopted in those tragedies of the palace which have occurred more than once in the capital founded by Peter the Barbarian.

The case indeed was such that fain would the *Bellipotent's* captain have deferred taking any action whatever respecting it further than to keep the foretopman a close prisoner till the ship rejoined the squadron and then submitting the matter to the judgment of his admiral.

But a true military officer is in one particular like a true monk. Not with more of self-abnegation will the latter keep his vows of monastic obedience than the former his vows of allegiance to martial duty.

Feeling that unless quick action was taken on it, the deed of the foretopman, so soon as it should be known on the gun decks, would tend to awaken any slumbering embers of the *Nore* among the crew, a sense of the urgency of the case overruled in Captain Vere every other consideration. But though a conscientious disciplinarian, he was no lover of authority for mere authority's sake. Very far was he from embracing opportunities for monopolizing to himself the perils of moral responsibility, none at least that could properly be referred to an official superior or shared with him by his official equals or even subordinates. So thinking, he was glad it would not be at variance with usage to turn the matter over to a summary court of his own officers, reserving to himself, as the one on whom the ultimate accountability would rest, the right of maintaining a supervision of it, or formally or informally interposing at need. Accordingly a drumhead court was summarily convened, he electing the individuals composing it: the first lieutenant, the captain of marines and the sailing master.

In associating an officer of marines with the sea lieutenant and the sailing master in a case having to do with a sailor, the commander perhaps deviated from general custom. He was prompted thereto by the circumstance that he took that soldier to be a judicious person, thoughtful, and not altogether incapable of grappling with a difficult case unprecedented in his prior experience. Yet even as to him he was not without some latent misgiving, for withal he was an extremely good-natured man, an enjoyer of his dinner, a sound sleeper, and inclined to obesity—a man who though he would always maintain his manhood in battle might not prove altogether reliable in a moral dilemma involving aught of the tragic. As to the first lieutenant and the sailing master, Captain Vere could not but be aware that though honest natures, of approved gallantry upon occasion, their intelligence was mostly confined to the matter of active seamanship and the fighting demands of their profession.

The court was held in the same cabin where the unfortunate affair had taken place. This cabin, the commander's, embraced the entire area under the poop deck. Aft, and on either side, was a small stateroom, the one now temporarily a jail and the other a dead-house, and a yet smaller compartment, leaving a space between expanding forward into a goodly oblong of length coinciding with the ship's beam. A skylight of moderate dimension was overhead, and at each cut of the oblong space were two sashed porthole windows easily convertible back into embrasures for short carronades.

All being quickly in readiness, Billy Budd was arraigned, Captain Vere necessarily appearing as the sole witness in the case, and as such temporarily sinking his rank, though singularly maintaining it in a matter apparently trivial, namely that he testified from the ship's weather side, with that object having caused the court to sit on the lee side. Concisely he narrated all that had led up to the catastrophe, omitting nothing in Claggart's accusation and deposing as to the manner in which the prisoner had received it. At this testimony the three officers glanced with no little surprise at Billy Budd, the last man they would have suspected either of the mutinous design alleged by Claggart or the undeniable deed he himself had done. The first lieutenant, taking judicial primacy and turning toward the prisoner, said, "Captain Vere has spoken. Is it or is it not as Captain Vere says?"

In response came syllables not so much impeded in the utterance as might have been anticipated. They were these: "Captain Vere tells the truth. It is just as Captain Vere says, but it is not as the master-at-arms said. I have eaten the King's bread and I am true to the King."

"I believe you, my man," said the witness, his voice indicating a suppressed emotion not otherwise betrayed.

"God will bless you for that, your honor!" not without stammering said Billy, and all but broke down. But immediately he was recalled to self-control by another question, to which with the same emotional difficulty of utterance he said, "No, there was no malice between us. I never bore malice against the master-at-arms. I am sorry that he is dead. I did not mean to kill him. Could I have used my tongue I would not have struck him. But he foully lied to my face and in presence of my captain, and I had to say something, and I could only say it with a blow, God help me!"

In the impulsive aboveboard manner of the frank one the court saw confirmed all that was implied in words that just previously had perplexed them, coming as they did from the testifier to the tragedy and promptly following Billy's impassioned disclaimer of mutinous intent—Captain Vere's words, "I believe you, my man."

Next it was asked of him whether he knew of or suspected aught savoring of incipient trouble (meaning mutiny, though the explicit term was avoided) going on in any section of the ship's company.

The reply lingered. This was naturally imputed by the court to the same vocal embarrassment which had retarded or obstructed previous answers. But in main it was otherwise here, the question immediately recalling to Billy's mind the interview with the afterguardsman in the forechains. But an innate repugnance to playing a part at all approaching that of an informer against one's own shipmates—the same erring sense of uninstructed honor which had stood in the way of his reporting the matter at the time, though as a loyal man-of-war's man it was incumbent on him, and failure so to do, if charged against him and proven, would have subjected him to the heaviest of penalties; this, with the blind feeling now his that nothing really was being hatched, prevailed with him. When the answer came it was a negative.

"One question more," said the officer of marines, now first speaking and with a troubled earnestness. "You tell us that what the master-at-arms said against you was a lie. Now why should he have so lied, so maliciously lied, since you declare there was no malice between you?"

At that question, unintentionally touching on a spiritual sphere wholly obscure to Billy's thoughts, he was nonplused, evincing a confusion indeed that some observers, such as can readily be imagined, would have construed into involuntary evidence of hidden guilt. Nevertheless, he strove some way to answer, but all at once relinquished the vain endeavor, at the same time turning an appealing glance toward Captain Vere as deeming him his best helper and friend. Captain Vere, who had been seated for a time, rose to his feet, addressing the interrogator. "The question you put to him comes naturally enough. But how can he rightly answer it?—or anybody else, unless indeed it be he who lies within there," designating the compartment where lay the corpse. "But the prone one there will not rise to our summons. In effect though, as it seems to me, the point you make is hardly material. Quite aside from any conceivable motive actuating the master-at-arms, and irrespective of the provocation to the blow, a martial court must needs in the present case confine its attention to the blow's consequence, which consequence justly is to be deemed not otherwise than as the striker's deed."

This utterance, the full significance of which it was not at all likely that Billy took in, nevertheless caused him to turn a wistful interrogative look toward the speaker, a look in its dumb expressiveness not unlike that which a dog of generous breed might turn upon his master, seeking in his face some elucidation of a previous gesture ambiguous to the canine intelligence. Nor was the same utterance without marked effect upon the three officers, more especially the soldier. Couched in it seemed to them a meaning unanticipated, involving a prejudgment on the speaker's part. It served to augment a mental disturbance previously evident enough.

The soldier once more spoke, in a tone of suggestive dubiety addressing at once his associates and Captain Vere: "Nobody is present—none of the ship's company, I mean—who might shed lateral light, if any is to be had, upon what remains mysterious in this matter."

"That is thoughtfully put," said Captain Vere; "I see your drift. Ay, there is a mystery; but, to use a scriptural phrase, it is a 'mystery of iniquity,' a matter for psychologic theologians to discuss. But what has a military court to do with it? Not to add that for us any possible investigation of it is cut off by the lasting tongue-tie of him-in-yonder," again designating the mortuary stateroom. "The prisoner's deed—with that alone we have to do."

To this, and particularly the closing reiteration, the marine soldier, knowing not how aptly to reply, sadly abstained from saying aught. The first lieutenant, who at the outset had not unnaturally assumed primacy in the court, now overrulingly instructed by a glance from Captain Vere, a glance more effective than words, resumed that primacy. Turning to the prisoner "Budd," he said, and scarce in equable tones, "Budd, if you have aught further to say for yourself, say it now."

Upon this the young sailor turned another quick glance toward Captain Vere; then, as taking a hint from that aspect, a hint confirming his own instinct that silence was now best, replied to the lieutenant, "I have said all, sir."

The marine—the same who had been the sentinel without the cabin door at the time that the foretopman, followed by the master-at-arms, entered it—he, standing by the sailor throughout these judicial proceedings, was now directed to take him back to the after compartment originally assigned to the prisoner and his custodian. As the twain disappeared from view, the three officers, as partially liberated from some inward constraint associated with Billy's mere presence, simultaneously stirred in their seats. They exchanged looks of troubled indecision, yet feeling that decide they must and without long delay. For Captain Vere, he for the time stood—unconsciously with his back toward them, apparently in one of his absent fits—gazing out from a sashed porthole to windward upon the monotonous blank of the twilight sea. But the court's silence continuing broken only at moments by brief consultations, in low earnest tones this served to arouse him and energize him. Turning, he to-and-fro paced the cabin athwart; in the returning ascent to windward climbing the slant deck in the ship's lee roll, without knowing it symbolizing thus in his action a mind resolute to surmount difficulties even if against primitive instincts strong as the wind and the sea. Presently he came to a stand before the three. After scanning their faces he stood less as mustering his thoughts for expression than as one only deliberating how best to put them to well-meaning men not intellectually mature, men with whom it was necessary to demonstrate certain principles that were axioms to himself. Similar impatience as to talking is perhaps one reason that deters some minds from addressing any popular assemblies.

When speak he did, something, both in the substance of what he said and his manner of saying it, showed the influence of unshared studies modifying and tempering the practical training of an active career. This, along with his phraseology, now and then was suggestive of the grounds whereon rested that imputation of a certain pedantry socially alleged against him by certain naval men of wholly practical cast, captains who nevertheless would frankly concede that His Majesty's navy mustered no more efficient officer of their grade than Starry Vere.

What he said was to this effect: "Hitherto I have been but the witness, little more; and I should hardly think now to take another tone, that of your coadjutor for the time, did I not perceive in you—at the crisis too—a troubled hesitancy, proceeding, I doubt not, from the clash of military duty with moral scruple—scruple vitalized by compassion. For the compassion, how can I otherwise than share it? But, mindful of paramount obligations, I strive against scruples that may tend to enervate decision. Not, gentlemen, that I hide from myself that the case is an exceptional one. Speculatively regarded, it well might be referred to a jury of casuists. But for us here, acting not as casuists or moralists, it is a case practical, and under martial law practically to be dealt with.

"But your scruples: do they move as in a dusk? Challenge them. Make them advance and declare themselves. Come now; do they import something like this: If, mindless of palliating circumstances, we are bound to regard the death of the master-at-arms as the prisoner's deed, shell does that deed constitute a capital crime whereof the penalty is a mortal one. But in natural justice is nothing but the prisoner's overt act to be considered? How can we adjudge to summary and shameful death a fellow creature innocent before God, and whom we feel to be so?—Does that state it aright? You sign sad assent. Well, I too feel that, the full force of that. It is Nature. But do these buttons that we wear attest that our allegiance is to Nature? No, to the King. Though the ocean, which is inviolate Nature primeval, though this be the element where we move and have our being as sailors, yet as the King's officers lies our duty in a sphere correspondingly natural? So little is that true, that in receiving our commissions we in the most important regards ceased to be natural free agents. When war is declared are we the commissioned fighters previously consulted? We fight at command. If our judgments approve the war, that is but coincidence. So in other particulars. So now. For suppose condemnation to follow these present proceedings. Would it be so much we ourselves that would condemn as it would be martial law operating through us? For that law and the rigor of it, we are not responsible. Our vowed responsibility is in this: That however pitilessly that law may operate in any instances, we nevertheless adhere to it and administer it.

"But the exceptional in the matter moves the hearts within you. Even so too is mine moved. But let not warm hearts betray heads that should be cool. Ashore in a criminal case, will an upright judge allow himself off the bench to be waylaid by some tender kinswoman of the accused seeking to touch him with her tearful plea? Well, the heart here, sometimes the feminine in man, is as that piteous woman, and hard though it be, she must here be ruled out."

He paused, earnestly studying them for a moment; then resumed.

"But something in your aspect seems to urge that it is not solely the heart that moves in you, but also the conscience, the private conscience. But tell me whether or not, occupying the position we do, private conscience should not yield to that imperial one formulated in the mode under which alone we officially proceed?"

Here the three men moved in their seats, less convinced than agitated by the course of an argument troubling but the more the spontaneous conflict within.

Perceiving which, the speaker paused for a moment; then abruptly changing his tone, went on.

"To steady us a bit let us recur to the facts.—In wartime at sea a man-of-war's man strikes his superior in grade, and the blow kills. Apart from its effect the blow itself is, according to the Articles of War, a capital crime, Furthermore—"

"Ay, sir," emotionally broke in the officer of marines, "in one sense it was. But surely Budd purposed neither mutiny nor homicide."

"Surely not, my good man. And before a court less arbitrary and more merciful than a martial one, that plea would largely extenuate. At the Last Assizes it shall acquit. But how here? We proceed under the law of the Mutiny Act. In feature no child can resemble his father more than that Act resembles in spirit the thing from which it derives—War. In His Majesty's service—in this ship, indeed—there are Englishmen forced to fight for the King against their will. Against their conscience, for aught we know. Though as their fellow creatures some of us may appreciate their position, yet as navy officers what reck we of it? Still less recks the enemy. Our impressed men he would fain cut down in the same swath with our volunteers. As regards the enemy's naval conscripts, some of whom may even share our own abhorrence of the regicidal French Directory, it is the same on our side. War looks but to the frontage, the appearance. And the Mutiny Act War's child, takes after the father. Budd's intent or non-intent is nothing to the purpose.

"But while, put to it by those anxieties in you which I cannot but respect, I only repeat myself—while thus strangely we prolong proceedings that should be summary—the enemy may be sighted and an engagement result. We must do; and one of two things must we do—condemn or let go."

"Can we not convict and yet mitigate the penalty?" asked the sailing master, here speaking, and falteringly, for the first.

"Gentlemen, were that clearly lawful for us under the circumstances, consider the consequences of such clemency. The people" (meaning the ship's company) "have native sense; most of them are familiar with our naval usage and tradition; and how would they take it? Even could you explain to them—which our official position forbids—they, long molded by arbitrary discipline, have not that kind of intelligent responsiveness that might qualify them to comprehend and discriminate. No, to the people the foretopman's deed, however it be worded in the announcement will be plain homicide committed in a flagrant act of mutiny. What penalty for that should follow, they know. But it does not follow. *Why?* they will ruminate. You know what sailors are. Will they not revert to the recent outbreak at the *Nore?* Ay. They know the well-founded alarm—the panic it struck throughout England. Your clement sentence they would account pusillanimous. They would think that we flinch, that we are afraid of them—afraid of practicing a lawful rigor singularly demanded at this juncture, lest it should provoke new troubles. What shame to us such a conjecture on their part, and how deadly to discipline. You see then, whither, prompted by duty and the law, I steadfastly drive. But I beseech you, my friends, do not take me amiss. I feel as you do for this unfortunate boy. But did he know our hearts, I take him to be of that generous nature that he would feel even for us on whom this military necessity so heavy a compulsion is laid."

With that, crossing the deck he resumed his place by the sashed porthole, tacitly leaving the three to come to a decision. On the cabin's opposite side the troubled court sat silent. Loyal lieges, plain and practical, though at bottom they dissented from some points Captain Vere had put to them, they were without the faculty, hardly had the inclination, to gainsay one whom they felt to be an earnest man, one too not less their superior in mind than in naval rank. But it is not improbable that even such of his words as were not without influence over them, less came home to them than his closing appeal to their instinct as sea officers: in the forethought he threw out as to the practical consequences to discipline, considering the unconfirmed tone of the fleet at the time, should a mall-of-war's mall's violent killing at sea of a superior in grade be allowed to pass for aught else than a capital crime demanding prompt infliction of the penalty.

Not unlikely they were brought to something more or less akin to that harassed frame of mind which in the year 1842 actuated the commander of the U.S. brig-of-war *Somers* to resolve, under the so-called Articles of War, Articles modeled upon the English Mutiny Act, to resolve upon the execution at sea of a midshipman and two sailors as mutineers designing the seizure of the brig. Which resolution was carried out though in a time of peace and within not many days' sail of home. An act vindicated by a naval court of inquiry subsequently convened ashore. History, and here cited without comment. True, the circumstances on board the *Somers* were different from those on board the *Bellipotent*. But the urgency felt, well-warranted or otherwise, was much the same.

Says a writer whom few know, "Forty years after a battle it is easy for a noncombatant to reason about how it ought to have been fought. It is another thing personally and under fire to have to direct the fighting while involved in the obscuring smoke of it. Much so with respect to other emergencies involving considerations both practical and moral, and when it is imperative promptly to act. The greater the fog the more it imperils the steamer, and speed is put on though at the hazard of running somebody down. Little ween the snug card players in the cabin of the responsibilities of the sleepless man on the bridge."

In brief, Billy Budd was formally convicted and sentenced to be hung at the yardarm in the early morning watch, it being now night. Otherwise, as is customary in such cases, the sentence would forthwith have been carried out. In wartime on the field or in the fleet, a mortal punishment decreed by a drumhead court—on the field sometimes decreed by but a nod from the general—follows without delay on the heel of conviction, without appeal.

C. Upholding the Truth

The need for secrecy, and for protecting the confidentiality of information about vital military operations during wartime, seems obvious. Few would deny, in addition, the utility of deception in military tactics, including use of covert agents and informants to gather vital information on threats to national security (intelligence), and in carrying out operations designed to thwart those who would do us, or others, unjustifiable harm (as in the famous ploy by General Dwight D. Eisenhower, Supreme Allied Commander, to confuse the German army about the details of his plan to invade Europe during the Second World War).

It may seem paradoxical, and even hypocritical to some, therefore, to witness the manner in which honesty and truthfulness are held forth as essential virtues for a military officer. In what sense can it be said that a military officer has a unique duty to "uphold the truth?"

In response, many senior military leaders would reply, notwithstanding the examples above, that honesty is the best policy, not only throughout life in general, but especially in the military. This is because *trust* among commanders and subordinates, and between shipmates and comrades-in-arms, is absolutely essential for organizational effectiveness. Falsifying information on a report that certified a ship or unit as fully operational and combat-ready when this was untrue, for example, could (and often has) led to disaster, and placed lives needlessly at risk. Covering up or otherwise lying about serious known engineering deficiencies in a vital weapons system, such as the Patriot missile system, or an aircraft design, such as the V-22 Osprey, for example, can lead (and has led) to the needless and avoidable deaths of good soldiers, sailors, and pilots, and also to national scandal and disgrace to the respective services implicated in these deceptions.

An officer's duty to be truthful, however, transcends its apparent utility within a military organization. The military services of this nation do not permit an officer to lie whenever he or she has sufficient reason to believe that lying will result in more good or less harm than being truthful. No sensible officer, for example, would grant his or her own superiors or subordinates permission and license to lie without his or her knowledge, whenever those other persons simply believed they had a good reason to think their actions would do more good than harm. Why, then, would officers grant such license to themselves? Finally, of course, there are some circumstances (such as in providing testimony before Congress in carrying out its Constitutionally-mandated mission to oversee the armed forces) in which the prohibition against an officer's lying is absolute.

These are among the reasons why the presumptions against lying in the military are so strong. The military services, however, are not alone in their formal abhorrence of the practice of lying. In fact, their stern attitude mirrors a longstanding and strong presumption within our western culture that, of all sins or moral lapses, including even homicide, lying is held to be the most reprehensible and inexcusable. As our reading selection suggests, there are some philosophers, such as Augustine and Kant, who hold the remarkably strong conviction that it can never, under any circumstances, be permissible or excusable to tell a lie. Dante famously relegated liars to one of the lowest, the "eighth" circle of hell.

What is awkward about this absolute prohibition against lying, of course, is that few if any individuals—certainly not even General George Washington, despite that famous myth—can be said to have navigated through life successfully without having withheld, distorted, or misrepresented the truth. Dante's "eighth circle" will, when we each arrive, very likely be found overpopulated!

Thus, we might also wish to investigate whether (and if so, why) it might seem unreasonable to maintain that there can be *no circumstance whatsoever* in which lying could be justified. If the absolute and unconditional prohibition against lying cannot be maintained, what we would be left with is *a very strong presumption against* ever lying. That is only a slightly less stern position, but it adds a new dimension of responsibility to each moral agent. If this second position, the "strong presumption against lying," is in fact what we mean to uphold, then it falls to each of us to consider what may justify an exception, or a release from that presumption. The cases of tactical deception and intelligence gathering cited above constitute two examples. Surely we would not mean to condemn General Eisenhower as a "liar," or argue that he was somehow wrong to deceive the enemy by using General George Patton and his "phantom army" as a ruse. Neither would we wish to condemn as a liar the intelligence agent attempting, under false pretenses, to gather information about a pending terrorist attack in order to prevent it.

Clearly, however, whatever it is about those instances that may make them an exception to the general presumption against lying, that property of being an exception does not carry over from those cases to include also a Marine Lieutenant Colonel lying to his own elected legislators about his actions in defiance of laws they have properly passed, even if he believes those laws to be in error. Neither he, nor we, have the authority to grant ourselves exemptions from the prohibition or presumption against lying, certainly to our own fellow officers and fellow citizens.

Nothing in the putative wartime or intelligence exception, likewise, carries over to justify the actions of lying to commanding officers about battle fitness or preparedness, or to deceiving shipmates, comrades-in-arms, let alone the general public, about the reliability of expensive weapons systems or aircraft that do not function as they were intended. Yet these kinds of things occur, daily, inside military organizations, as well as in civilian organizations like the National Aeronautics and Space Administration. In this, moreover, the behavior of individuals in these organizations mimics widespread practice in business, government, and the private sector. Can it be that, the admonitions of centuries of philosophers and theologians notwithstanding, we have eagerly cultivated a culture of cynicism and deception, in which lies are deemed preferable to the truth?

Individuals lie, on occasion, because they feel themselves under pressure to perform in certain ways, and because they believe they discern a climate of leadership more interested in the appearance of success than in knowing the cold, hard facts. They discern that the "reward system" in such organizations does not recognize and reward honesty, but encourages duplicity and mendacity instead. Individuals lie to protect themselves from punishment, or from other shameful effects of their own weakness and failure. They lie because they see others doing it, and getting away with it, even seeming to be honored and rewarded for it. They lie because they lack the moral courage to do otherwise. They lie, because they come to perceive that it is acceptable to do so, that no real blame attaches to it. They lie, because they come to believe "the system" to which they belong is itself corrupt, and that lying helps them maintain a certain degree of control over their own circumstances. When such perceptions become widespread, however, all is lost, including, most tragically, the innocent lives of unsuspecting comrades who mistakenly place their trust in us and in our organization.

The Reading

Our reading elaborates a process of thought with which one may test proposed justifications for lying. Readers should notice that this process is applicable not just to the case of lying, but to ethical decision making in general. It invokes a procedure of justification reminiscent of Martin Luther King, Jr.'s criteria for civil disobedience, or for reflecting upon orders and obedience according to the "Constitutional paradigm." Indeed, a great many of the provisions suggested by Sissela Bok, when closely examined, resemble those that emerged in the philosophical reflection upon "just war:" the reasons for lying must be compelling ("just cause"), one must exhaust all alternatives before resorting to a lie ("last resort"), and one must acknowledge a "Principle of Publicity" ("legitimate authority") according to which we are prepared to place our case before a tribunal of peers, if necessary, to gain concurrence (or to accept blame for wrong-doing).

In all these instances, the similar structure of justification mirrors the gravity of the actions being contemplated: taking a life, disobeying a law, prosecuting a war, or telling a lie, all involve the moral agent in deciding *to make himself or herself an exception to rules and principles that we believe should otherwise prevail*. These principles, in turn, uphold honor and integrity, protect us from grave harm, promote the general welfare, and call upon the decency and humanity of those whose behavior they govern. They are thus not to be lightly set aside. Indeed, in the example of lying and truth-telling, Bok enjoins us to consider, not only the specific reasons for requesting an exemption from, or making ourselves an exception to the general rule, but also to consider the long-term damage each action of lying does to the essential institution of trust, without which society as we know it is irreparably degraded.

Lying:
Moral Choice in Public & Private Life
Sissela Bok*

Is the "Whole Truth" Attainable?

The moral question of whether you are lying or not is not settled by establishing the truth or falsity of what you say. In order to settle this question, we must know whether you *intend your statement to mislead*.

Any number of appearances and words can mislead us; but only a fraction of them are *intended* to do so. A mirage may deceive us, through no one's fault. Our eyes deceive us all the time. Yet we often know when we mean to be honest or dishonest.

We must single out, therefore, from the countless ways in which we blunder misunderstood through life, that which is done with the *intention to mislead*.

When we undertake to deceive others intentionally, we communicate messages meant to mislead them, meant to make them believe what we ourselves do not believe. We can do so through gesture, through disguise, by means of action or inaction, even through silence. Which of these innumerable deceptive messages are also lies? I shall define as a lie any intentionally deceptive message which is stated. Such statements are most often made verbally or in writing, but can of course also be conveyed via smoke signals, Morse code, sign language, and the like. Deception, then, is the larger category, and lying forms part of it.

This definition resembles some of those given by philosophers and theologians, but not all. For it turns out that the very choice of definition has often presented a moral dilemma all its own. Certain religious and moral traditions were rigorously opposed to all lying. Yet many adherents wanted to recognize at least a few circumstances when intentionally misleading statements could be allowed. The only way out for them was, then, to define lies in such a way that some falsehoods did not count as lies. Thus Grotius, followed by a long line of primarily Protestant thinkers, argued that speaking falsely to those—like thieves—to whom truthfulness is not owed cannot be called lying. Sometimes the rigorous tradition was felt to be so confining that a large opening to allowable misstatements was needed. In this way, casuist thinkers developed the notion of the "mental reservation," which, in some extreme formulations, can allow you to make a completely misleading statement, so long as you add something in your own mind to make it true. Thus, if you are asked whether you broke somebody's vase, you could answer "No," adding in your own mind the mental reservation "not last year" to make the statement a true one.

---

*Excerpts from Sissela Bok, *Lying: Moral Choice in Public and Private Life* (New York: Vintage Books, 1978). Editing by Lawrence Lengbeyer includes omission of citations and footnotes, introductory chapter quotations, and subheading within chapters.

Such definitions serve the special purpose of allowing persons to subscribe to a strict tradition yet have the leeway in actual practice which they desire. When the strict traditions were at their strongest, as with certain forms of Catholicism and Calvinism, such "definitional" ways out often flourished. Whenever a law or rule is so strict that most people cannot live by it, efforts to find loopholes will usually ensue; the rules about lying are no exception.

I propose that we look primarily at clear-cut lies—lies where the intention to mislead is obvious, where the liar knows that what he is communicating is not what he believes, and where he has not deluded himself into believing his own deceits.

Truthfulness, Deceit, and Trust

Deceit and violence—these are the two forms of deliberate assault on human beings.

Of course, deception—again like violence—can be used also in self-defense, even for sheer survival. Its use can also be quite trivial, as in white lies. Yet its potential for coercion and for destruction is such that society could scarcely function without some degree of truthfulness in speech and action.

There must be a minimal degree of trust in communication for language and action to be more than stabs in the dark. This is why some level of truthfulness has always been seen as essential to human society, no matter how deficient the observance of other moral principles.

All our choices depend on our estimates of what is the case; these estimates must in turn often rely on information from others. Lies distort this information. A lie, in Hartmann's words, "injures the deceived person in his life; it leads him astray."

To the extent that knowledge gives power, to that extent do lies affect the distribution of power; they add to that of the liar, and diminish that of the deceived, altering his choices.

The Perspective of the Deceived

Those who learn that they have been lied to in an important matter are resentful, disappointed, and suspicious. They feel wronged; they are wary of new overtures. And they look back on their past beliefs and actions in the new light of the discovered lies. They see that they were manipulated, that the deceit made them unable to make choices for themselves according to the most adequate information available, unable to act as they would have wanted to act had they known all along.

Of course, we know that many lies are trivial. But since we, when lied to, have no way to judge which lies are the trivial ones, and since we have no confidence that liars will restrict themselves to just such trivial lies, the perspective of the deceived leads us to be wary of *all* deception.

Nor is this perspective restricted to those who are actually deceived in any given situation. Though only a single person may be deceived, many others may be harmed as a result. If a mayor is deceived about the need for new taxes, the entire city will bear the consequences. Accordingly, the perspective of the deceived is shared by all those who feel the consequences of a lie, whether or not they are themselves lied to.

Deception, then, can be coercive. When it succeeds, it can give power to the deceiver—power that all who suffer the consequences of lies would not wish to abdicate. This is especially true because lying so often accompanies every *other* form of wrongdoing.

For this reason, the perspective of the deceived supports the statement by Aristotle:

Falsehood is in itself mean and culpable, and truth noble and full of praise.

There is an initial imbalance in the evaluation of truth-telling and lying. Lying requires a *reason*, while truth-telling does not. It must be excused; reasons must be produced, in any one case, to show why a particular lie is not "mean and culpable."

The Perspective of the Liar

Liars share with those they deceive the desire not to be deceived. As a result, their choice to lie is one which they would like to reserve for themselves while insisting that others be honest. They would prefer, in other words, a "free-rider" status, giving them the benefits of lying without the risks of being lied to. The free rider trades upon being an exception, and could not exist in a world where everybody chose to exercise the same prerogatives.

[I]n th[e] benevolent self-evaluation by the liar of the lies he might tell, certain kinds of disadvantage and harm are almost always overlooked. Liars usually weigh only the immediate harm to others from the lie against the benefits they want to achieve. The flaw in such an outlook is that it ignores or underestimates two additional kinds of harm—the harm that lying does to the liars themselves and the harm done to the general level of trust and social cooperation. Both are cumulative; both are hard to reverse.

How is the liar affected by his own lies? The very fact that he *knows* he has lied, first of all, affects him. He may regard the lie as an inroad on his integrity; he certainly looks at those he has lied to with a new caution. And if they find out that he has lied, he knows that his credibility and the respect for his word have been damaged.

No one trivial lie undermines the liar's integrity. But the problem for liars is that they tend to see *most* of their lies in this benevolent light and thus vastly underestimate the risks they run. While no one lie always carries harm for the liar, there is *risk* of such harm in most.

These risks are increased by the fact that so few lies are solitary ones. It is easy to tell a lie, but hard to tell only one. The first lie "must be thatched with another or it will rain through." More and more lies may come to be needed; the liar always has more mending to do.

After the first lies, moreover, others can come more easily. Psychological barriers wear down; lies seem more necessary, less reprehensible; the ability to make moral distinctions can coarsen; the liar's perception of his chances of being caught may warp. Paradoxically, once his word is no longer trusted, he will be left with greatly *decreased* power—even though a lie often does bring at least a short-term gain in power over those deceived.

Yet such risks rarely enter his calculations. Bias skews all judgment, but never more so that in the search for good reasons to deceive.

Bias causes liars often to ignore the second type of harm as well. For even if they make the effort to estimate the consequences to *individuals* of their lies, they often fail to consider the many ways in which deception can spread and give rise to practices very damaging to human communities. The veneer of social trust is often thin. As lies spread—by imitation, or in retaliation, or to forestall suspected deception—trust is damaged. Yet trust is a social good to be protected just as much as the air we breathe or the water we drink. When it is damaged the community as a whole suffers; and when it is destroyed, societies falter and collapse.

Take the example of a government official hoping to see Congress enact a crucial piece of antipoverty legislation. Should he lie to a Congressman he believes unable to understand the importance and urgency of the legislation, yet powerful enough to block its passage? Should he tell him that, unless the proposed bill is enacted, the government will push for a much more extensive measure?

In answering, shift the focus from this case taken in isolation to the vast practices of which it forms a part. What is the effect on colleagues and subordinates who witness the deception so often resulting from such a choice? What is the effect on the members of Congress as they inevitably learn of a proportion of these lies? And what is the effect on the electorate as it learns of these and similar practices? Then shift back to the narrower world of the official troubled about the legislation he believes in, and hoping by a small deception to change a crucial vote.

The Principle of Veracity

For these reasons, I believe that we must accept as an initial premise Aristotle's view that lying is "mean and culpable" and that truthful statements are preferable to lies in the absence of special considerations. This premise gives an initial negative weight to lies. It holds that lying requires explanation, whereas truth ordinarily does not.. It provides a counterbalance to the crude evaluation by liars of their own motives and of the consequences of their lies. And it places the burden of proof squarely on those who assume the liar's perspective.

This presumption against lying can also be stated so as to stress the positive worth of truthfulness or veracity. I would like to refer to the "principle of veracity" as an expression of this initial imbalance in our weighing of truthfulness and lying.

[T]rust in some degree of veracity functions as a *foundation* of relations among human beings; when this trust shatters or wears away, institutions collapse.

Such a principle need not indicate that all lies should be ruled out by the initial negative weight given to them. But it does make at least one immediate limitation on lying: in any situation where a lie is a possible choice, one must first seek truthful alternatives. If lies and truthful statements appear to achieve the same result or appear to be as desirable the lies should be ruled out. And only where a lie is a *last resort* can one even begin to consider whether or not it is morally justified.

Never to Lie?

> But every liar says the opposite of what he thinks in his heart, with purpose to deceive. Now it is evident that speech was given to man, not that men might therewith deceive one another, but that one man might make known his thoughts to another. To use speech, then, for the purpose of deception, and not for its appointed end, is a sin. Nor are we to suppose that there is any lie that is not a sin, because it is sometimes possible, by telling a lie, to do service to another.
> - St. Augustine, *The Enchiridion*

> By a lie a man throws away and, as it were, annihilates his dignity as a man.
> - Immanuel Kant, *Doctrine of Virtue*

The simplest answer to the problems of lying, at least in principle, is to rule out all lies. Many theologians have chosen such a position; foremost among them is St. Augustine. He cut a clear swath through all the earlier opinions holding that some lies might be justified. He claimed that God forbids all lies and that liars therefore endanger their immortal souls.

His definition left no room at all for justifiable falsehood.

But such a doctrine turned out to be very difficult to live by. Many ways were tried to soften the prohibition, to work around it and to allow at least a few lies. Three different paths were taken: to allow for pardoning of some lies; to claim that some deceptive statements are not falsehoods, merely misinterpreted by the listener; and finally to claim that certain falsehoods do not *count* as lies.

Aquinas [subsequently] set a pattern which is still followed by Catholic theologians. He distinguished three kinds of lies: the officious, or helpful, lies; the jocose lies, told in jest; and the mischievous, or malicious, lies, told to harm someone. Only the latter constitute mortal sins for Aquinas. He agreed with Augustine that all lies are sins, but regarded the officious and jocose lies as less serious. The pardoning function came to grow more and more important and ultimately created great discord within the Church. Should one be able to tell lies and then have them wiped from one's conscience? Ought it be possible to do so repeatedly, perhaps even to plan the lies with the pardon in mind? And by what means should the pardon be sought?

The two other paths around Augustine's strict prohibition occasioned similar disputes. They assumed, in effect, that certain intentionally deceptive statements are not lies in the first place. They might then be used in good conscience.

One such was the "mental reservation" or "mental restraint." If you say something misleading to another and merely add a qualification to it in your mind so as to make it true, you cannot be responsible for the "misinterpretation" made by the listener.

Nor is the mental reservation altogether a thing of the past. We still swear to omit it in many official oaths of citizenship and public office. And some still recommend it. A well-known Catholic textbook advises doctors and nurses to deceive patients by this method when they see fit to do so.

The final way to avoid Augustine's across-the-board prohibition of all lies seeks to argue that not all intentionally false statements ought to count as lies from a moral point of view. This view found powerful expression in Grotius. He argued that a falsehood is a lie in the strict sense of the word only if it conflicts with a right of the person to whom it is addressed. A robber, for instance, has no right to the information he tries to extort; to speak falsely to him is therefore not to lie in the strict sense of the word. The right in question is that of liberty of judgment, which is implied in all speech; but it can be lost if the listener has evil intentions; or not yet acquired, as in the case of children; or else freely given up, as when two persons agree to deceive one another. Grotius helped to bring back into the discourse on lying the notion that falsehood is at times justifiable.

Among those who discussed such doctrines with their students in ethics was Kant.

It is all the more striking, then, that Kant set forth the strongest arguments we have against all lying.

Kant takes issue, first with the idea that any generous motive, any threat to life, could excuse a lie. He argues that:

> Truthfulness in statements which cannot be avoided is the formal duty of an individual to everyone, however great may be the disadvantage accruing to himself or to another.

This is the absolutist position, prohibiting all lies, even those told for the best of purposes or to avoid the most horrible of fates.

But can we agree with Kant? His position has seemed too sweeping to nearly all his readers, even obsessive to some. For although veracity is undoubtedly an important duty, most assume that it leaves room for exceptions. It can clash with other duties, such as that of averting harm to innocent persons.

Most have held that there are times when truthfulness causes or fails to avert such great harm that a lie is clearly justifiable. One such time is where a life is threatened and where a lie might avert the danger. The traditional testing case advanced against the absolutist position is that discussed by Kant himself, where a would-be murderer inquires whether "our friend who is pursued by him had taken refuge in our house." Should one lie in order to save one's friend? Or should one tell the truth?

This is a standard case, familiar from Biblical times, used by the Scholastics in many variations, and taken up by most commentators on deception. It assumes, of course, that mere silence or evasion will not satisfy the assailant. If this case does not weaken your resistance to all lies, it is hard to think of another that will.

Most others have argued that, in such cases, where innocent lives are at stake, lies are morally justified, if indeed they are lies in the first place. Kant believes that to lie is to annihilate one's human dignity; yet for these others, to reply honestly, and thereby betray one's friend, would in itself constitute a compromise of that dignity. In such an isolated case, they would argue, the costs of lying are small and those of telling the truth catastrophic.

Similarly, a captain of a ship transporting fugitives from Nazi Germany, if asked by a patrolling vessel whether there were any Jews on board would, for Kant's critics, have been justified in answering No. His duty to the fugitives, they claim, would then have conflicted with the duty to speak the truth and would have far outweighed it. In fact, in times of such crisis, those who share Kant's opposition to lying clearly put innocent persons at the mercy of wrongdoers.

Furthermore, force has been thought justifiable in all such cases of wrongful threat to life. If to use force in self-defense or in defending those at risk of murder is right, why then should a lie in self-defense be ruled out? Surely if force is allowed, a lie should be equally, perhaps at times more, permissible. Kant's single-minded upholding of truthfulness above all else would clearly create guilt for many: guilt at having allowed the killing of a fellow human rather than lie to a murderer. Kant attempts to assuage this guilt by arguing as follows: If one stays close to the truth, one cannot strictly speaking, be responsible for the murderous acts another commits. The murderer will have to take the whole blame for his act. In speaking to him truthfully, one has done nothing blameworthy. If, on the other hand, one tells him a lie, Kant argues, one becomes responsible for all the bad consequences which might befall the victim and anyone else. One may, for instance, point the murderer in what one believes to be the wrong direction, only to discover with horror that that is exactly where the victim has gone to hide.

But it is a very narrow view of responsibility which does not also take some blame for a disaster one could easily have averted, no matter how much others are also to blame. A world where it is improper even to tell a lie to a murderer pursuing an innocent victim is not a world that many would find safe to inhabit.

Beneath the belief in the divine command to forgo all lying at all costs is yet another belief: that some grievous punishment will come to those who disobey such commands. Augustine stated the matter starkly: Death kills but the body, but a lie loses eternal life for the soul. To lie to save the life of another, then, is a foolish bargain. To the degree that one believes in the immortality of the soul and in its "death" through lying, to that degree it does make sense to eschew lies, even when one might have saved a life by lying. Any complete prohibition of lying, even in circumstances of threats to innocent lives, must, in order for it to be reasonable, rely on some belief that the lie is associated with a fate "worse than death."

And even among nonbelievers, one can imagine some terror which might make any lie seem "worse than death": the fear, perhaps, of some authority figure who has outlawed lying and seems to have knowledge of any breach of the rules; or an exalted view of the injury to one's integrity which a lie might bring.

To sum up, two beliefs often support the rigid rejection of all lies: that God rules out all lies and that He will punish those who lie. These beliefs cannot be proved or disproved. Many, including many Christians, refuse to accept one or both. Other religions, while condemning lying, rarely do so without exceptions.

I have to agree that there are at least *some* circumstances which warrant a lie. And foremost among them are those where innocent lives are at stake, and where only a lie can deflect the danger.

But, in taking such a position, it would be wrong to lose the profound concern which the absolutist theologians and philosophers express—the concern for the harm to trust and to oneself from lying, quite apart from any immediate effects from any one lie.

Similarly, Kant stressed the injury to humankind from lying and dramatized this to the culprit by stating that "by a lie a man throws away and, as it were, annihilates his dignity as a man." It may seem exaggerated to apply this statement to any one small lie, but if one sees it instead as a warning against *practices* of lying, against biased calculations of pros and cons, and against assuming the character of a liar, it may be closer to the mark. For a liar often *does* diminish himself by lying, and the loss is precisely to his dignity, his integrity.

Weighing the Consequences

[T]he utilitarian philosophers did not accept the premise that God has ruled out all lies. They brought a great sense of freedom to those whom they could convince that what ought to be done was not necessarily what the soothsayer or the ruler or the priests required, but rather, quite simply, what brought about the greatest balance of good over evil.

Utilitarians also differ from Kant (thought not from Augustine) in stressing the differences in seriousness between one lie and another. They are therefore much closer to our actual moral deliberation in many cases where we are perplexed. In choosing whether or not to lie, we *do* weigh benefits against harm and happiness against unhappiness.

[M]ost lies do have negative consequences for liars, dupes, all those affected, and for social trust. And when liars evaluate these consequences, they are peculiarly likely to be biased; their calculations frequently go astray. Therefore, even strict utilitarians might be willing to grant the premise that in making moral choices, we should allow an initial presumption against lies. There would be no need to see this presumption as something mysterious or abstract, nor to say that lies are somehow bad "in themselves." Utilitarians could view the negative weight instead as a correction, endorsed by experience, of the inaccurate and biased calculations of consequences made by any one liar.

The common assumption that lies can be evaluated on a risk-benefit scale *determined by the liar* can therefore be set aside on utilitarian grounds. The long-range results of an acceptance of such facile calculations, made by those most biased to favor their own interests and to disregard risks to others, would be severe.

White Lies

Harmless Lying

[A] white lie is a falsehood not meant to injure anyone, and of little moral import. I want to ask whether there *are* such lies; and if there are, whether their cumulative consequences are still without harm; and, finally, whether many lies are not defended as "white" which are in fact harmful in their own right.

Many small subterfuges may not even be intended to mislead. Take, for example, the many social exchanges: "How nice to see you!" or "Cordially yours." The justification for continuing to use such accepted formulations is that they deceive no one, except possibly those unfamiliar with the language.

A social practice more clearly deceptive is that of giving a false excuse so as not to hurt the feelings of someone making an invitation or request: to say one "can't" do what in reality one may not *want* to do.

Still other white lies are told in an effort to flatter, to throw a cheerful interpretation on depressing circumstances, or to show gratitude for unwanted gifts. In the eyes of many, such white lies do no harm, provide needed support and cheer, and help dispel gloom and boredom. They preserve the equilibrium and often the humaneness of social relationships.

A white lie, [utilitarians] hold, is trivial; it is either completely harmless, or so marginally harmful that the cost of detecting and evaluating the harm is much greater than the minute harm itself. In addition, the while lie can often actually be beneficial, thus further tipping the scales of utility.

But the harmlessness of lies is notoriously disputable. What the liar perceives as harmless or even beneficial may not be so in the eyes of the deceived. Second, the failure to look at an entire practice rather than at their own isolated case often blinds liars to cumulative harm for liars, those deceived, and honesty and trust more generally.

[L]ies tend to spread. Disagreeable facts come to be sugar-coated, and sad news softened or denied altogether. Many lie to children and to those who are ill about matters no longer peripheral but quite central, such as birth, adoption, divorce, and death. Deceptive propaganda and misleading advertising abound. All these lies are often dismissed on the same grounds of harmlessness and triviality used for white lies in general.

Letters of Recommendation

It is worth taking a closer look at practices where lies believed trivial are common. Triviality in an isolated lie can then be more clearly seen to differ from the costs of an entire practice. One such practice which has high accumulated costs is that of the inflated recommendation. In the harsh competition for employment and advancement, such a gesture is natural. It helps someone, while injuring no one in particular, and balances out similar gestures on the part of many others. Yet the practice obviously injures those who do not benefit from this kind of assistance.

Consider officer evaluation reports in the U.S. Army. Those who rate officers are asked to give them scores of "outstanding," "superior," "excellent," "effective," "marginal," and "inadequate." Raters know, however, that those who are ranked anything less than "outstanding" are then at a great disadvantage.

In such cases, honesty might victimize innocent persons. At the same time, the blurring of the meaning of words in these circumstances can make it easier not to be straightforward in others.

It is difficult for raters to know what to do in such cases. Some feel forced to say what they do not mean. Others adhere to a high standard of accuracy and thereby perhaps injure those who must have their recommendations.

To make choices on the basis of such inflated recommendations is equally difficult.

The entire practice, then, is unjust for those rated and bewildering for those who give and make use of ratings. [T]he costs to deceivers and deceived alike are great.

For this reason, those who give ratings should make every effort to come closer to the standard of accuracy. But one does have the responsibility of indicating that one is doing so, in order to minimize the effect on those rated. To do so requires time, power, and consistency. It is hard to resist [the inflated practices] singlehandedly.

Institutions, on the other hand, do have more leverage. Some can seek to minimize the reliance on such reports altogether. Others can try to work at the verbal inflation itself. But it is very difficult to do so, especially for large organizations.

Truthfulness at What Price?

[M]ost lies believed to be "white" are unnecessary if not downright undesirable. Many are not as harmless as liars take them to be. And even those lies which would generally be accepted as harmless are not needed whenever their goals can be achieved through completely honest means. Why tell a flattering lie about someone's hat rather than a flattering truth about their flowers? Why tell a general white lie about a gift, a kind act, a newborn baby, rather than a more specific truthful statement? If the purpose is understood by both speaker and listener to be one of civility and support, the *full* truth in such cases is not called for.

I would not wish to argue that all white lies should be ruled out.

But these are very few. And it is fallacious to argue that all white lies are right because a few are. As a result, those who undertake to tell white lies should look hard for alternatives.

A word of caution is needed here. To say that white lies should be kept at a minimum is *not* to endorse the telling of truths to all comers. Silence and discretion, respect for the privacy and for the feelings of others must naturally govern what is spoken. And the truth told in such a way as to wound may be unforgivably cruel.

Excuses

People look for moral reasons when they need to persuade themselves or others that the usual presumption against lying is outweighed in their particular case. I shall set forth the reasons most commonly used to defend lies as appealing to four principles: avoiding harm, producing benefits, fairness, and veracity.

Avoiding harm and producing benefit go together, yet it will be helpful to consider them separately. We can sense the difference in urgency between them.

Just as lies intended to avoid serious harm have often been thought more clearly excusable than others, so lies meant to *do* harm are often thought least excusable.

Among lies which do harm, those which do the greatest harm are judged the worst. Lies which are planned are judged more harshly than those told without forethought; single lies less severely than repeated ones. Planned practices of deception are therefore especially suspect, no matter how repentant the liar claims to be between lies.

Many lies invoke self-defense—the avoidance of harm to oneself—as an excuse. Lies told in court by those dreading a sentence, or lies by those caught stealing or cheating; lies by those threatened with violence; lies to get out of trouble of all kinds, to save face, to avoid losing work—all claim the overriding importance of avoiding harm to oneself.

Lies to bring about some benefit are often harder to excuse than those which prevent harm. [So] those who [pursue] lies to benefit others or themselves very often place great stress on the harm those lies will prevent.

Appeals to fairness claim to correct or forestall an injustice, or to help provide a fair distribution. [Some, for instance,] are held excusable by liars because they retrieve or protect what the liars think rightfully their own—their property, their liberty, or even their children, as in lies told to kidnappers. Still other lies appeal to fairness in a tit-for-tat way: "He lied to me, now it is all right for me to lie to him." Another way of appealing to fairness is to use a pseudonym [or disguise] when one is afraid that the use of one's real name might confer unfair advantages [or disadvantages].

Still another large category of excuses rely on confidentiality. One may feel forced to lie to some persons in order to protect the privacy or the confidences of others.

A final category where fairness is invoked is that where the dupes have agreed in advance to a practice involving deception[, such as] a game of poker, [or] medical experimentation.

Some claim they lie so as to *protect* the truth.

The lie to undo the effect of another lie; the lie to further some larger or more important truth; the lie to preserve confidence in one's own truthfulness—all make appeal to truth in some sense.

[W]hile excuses abound, justification is hard to come by.

Justification

Justification and Publicity

How can we single out, then, justifiable lies from all those that their perpetrators regard as so highly excusable? Assume, as before, that we are dealing with clearcut lies, deliberate efforts to mislead. We can examine the alternatives confronting the liar, and the excuses he gives. Which excuses not only mitigate and extenuate, but remove moral blame? And if we accept the excuses for some lies, do we thereby merely remove blame from the liar retroactively? Or are we willing to allow those lies ahead of time under certain circumstances? Could we, finally, recommend a *practice* of telling such lies whenever those circumstances arise—whenever, for instance, an innocent life is otherwise threatened?

We have already seen how often the liar is caught in a distorting perspective; his efforts to answer questions of justification can then show a systematic bias. His appeals to principle may be hollow, his evaluation flimsy. The result is that he can arrive at diametrically opposed weighings of alternatives and reasons, depending upon what he puts into the weighing process in the first place.

Justification must involve more than such untested personal steps of reasoning. To justify is to defend as just, right, or proper, by providing adequate reasons. It means to hold up to some standard. Such justification requires an audience: in ethics it is most appropriately aimed, not at any one individual or audience, but rather at "reasonable persons" in general.

Moral justification, therefore, cannot be exclusive or hidden; it has to be capable of being made public. In going beyond the purely private, it attempts to transcend also what is merely subjective. Many moral philosophers have assumed that such an appeal is of the very essence in reasoning about moral choice. According to such a [*publicity*] constraint, a moral principle must be capable of public statement and defense. A secret moral principle, or one which could be disclosed only to a sect or a guild, could not satisfy such a condition.

Such publicity is crucial to the justification of all moral choice. But it is, perhaps, particularly indispensable to the justification of lies and other deceptive practices. For publicity is connected more directly to veracity than to other moral principles. In addition, openly performed problematic acts are more likely to arouse controversy eventually, whereas lies, if they succeed, may never do so.

I would like to combine this concept of *publicity* with the view of justification in ethics as being *directed to reasonable persons,* in order to formulate a workable test to weigh the various excuses advanced for lies. Such a test counters the self-deception and bias inherent in the liar's perspective. It challenges privately held assumptions and hasty calculations. It requires clear and understandable formulation of the arguments used to defend the lie. Its advantages, moreover, are cumulative: through [its] exercise a more finely tuned moral sense will develop.

The test of publicity asks which lies, if any, would survive the appeal for justification to reasonable persons. It requires us to seek concrete and open performance of an exercise crucial to ethics: the Golden Rule, basic to so many religious and moral traditions.[1] We must share the perspective of those affected by our choices, and ask how we would react if the lies we are contemplating were told to us. We must, then, adopt the perspective not only of liars but of those lied to; and not only of particular persons but of all those affected by lies—the collective perspective of reasonable persons seen as potentially deceived. We must formulate the excuses and the moral arguments used to defend the lies and ask how they would stand up under the public scrutiny of these reasonable persons.

But exactly how is such a test best undertaken? Is the traditional appeal to conscience sufficient? Or, if there is to be more of a "public" involved, can it consist of just a few persons or need there be many? Need they be real or can they be merely imagined? And what are the limitations of such a test?

Levels of Justification

The initial and indispensable first effort at weighing moral choice from a reflective point of view that is already somewhat "public" is familiar: it is to have recourse to one's own conscience. The [inner] judge may be an ideal one, perhaps even held divine; at other times, simply a commentator on one's acts to whom one tries to justify one's actions and beliefs. Seneca describes the appeal of such an onlooker:

> It is nobler to live as you would live under the eyes of some good person always at your side; but nevertheless I am content if you only act, in whatever you do, as you would act if anyone at all were looking on; because solitude prompts us to all kinds of evil.

Some such method of personal soul-searching is undoubtedly necessary for even the most rudimentary moral choices. It is often all one can do. But it cannot be a *sufficient* guarantee that the conditions of publicity have been met. For while consciences can ravage, they can also be very accommodating and malleable. Most often, those who lie have a much easier time in justifying their behavior so long as their only audience is their own conscience or their self-appointed imaginary onlooker. And even for the well-intentioned, a conscience is unreliable as soon as matters of complexity and intensity must be weighed. Arguments may not be well formulated; implicit assumptions may go unchallenged; blurred analogies or faulty reasoning may continue unchecked. While the appeal to one's conscience is certainly indispensable, therefore, it does not provide for sufficient publicity. It can be quite unable to counter bias, and never more so than for those locked into the biased perspective of the liar.

For all the same reasons, appealing in one's mind to what others might say can come similarly to grief, no matter how numerous or how judicious one takes these others to be. It is doubtless helpful to imagine that one is justifying one's lie before a public assembly or a jury or even a television audience. And it certainly helps to try to formulate the maxim on which one is acting in order to see if it can serve as a maxim others could accept and live by. But so long as this process is purely an imagined one, so long as one is both actor and audience, both defender and jury, both law-giver and citizen, the risk of bias remains very high.

The next "level" of public justification, then, goes beyond one's internal thought-experiment. Asking friends, elders, or colleagues for advice, looking up precedents, consulting with those who have a special knowledge in questions of religion or ethics—these are well-trodden paths which can bring objectivity, sometimes wisdom, to moral choices and lead to the demise of many an ill-conceived scheme.

Unfortunately, in the more difficult cases, where the stakes are high, such consultation is still insufficiently "public." It does not eliminate bias; nor does it question shared assumptions and fallacious reasoning. This is especially often the case, once more, in professional and powerful circles, where those who might object are not given a voice, and where those considered "wise" can be those most likely to agree with the questionable scheme. There was collegial consultation, for example, in the decision to deny falsely that the United States of America was bombing Cambodia. And there was consultation before the adoption of a deceptive cover story for the Bay of Pigs invasion. Irving Janis has described the failures of such systems of consultation among the like-minded in foreign policy decisions:

> "Since our group's objectives are good," the members feel, "any means we decide to use must be good." This shared assumption helps the members avoid feelings of shame or guilt about decisions that may violate their personal code of ethical behavior. Negative stereotypes of the enemy enhance their sense of moral righteousness as well as their pride in the lofty mission of the ingroup.

More than consultation with chosen peers is needed whenever crucial interests are thus at stake.

A third "level" of public justification is required. At this level, persons of all allegiances must be consulted, or at least not excluded or bypassed. "Publicity" in this sense rules out the hand-picking of those who should be consulted. It is not so much a matter of whether many or a few have access to the public justification, as that no one should be denied access. Naturally, the more complex and momentous the decision, the more consultation will be judged necessary.

If possible, such open discussion should take place before the initiation of the deceptive scheme, giving those to be deceived an opportunity to be heard. To do so is the only sure way of having the perspective of the deceived represented.

---

[1] The Golden Rule has a very powerful negative form, as in the *Analects* of Confucius: Tzu Kung asked: "Is there any one word that can serve as a principle for the conduct of life?" Confucius said: "Perhaps... 'reciprocity': Do not do to others what you would not want others to do to you." See also Rabbi Hillel's saying: "What is hateful to you do not do to your neighbour; that is the whole Torah, while the rest is commentary thereof."

But is it not illogical to expect that those very persons lied to might be thus forewarned?

Here we must distinguish, again, between [individual] cases and [general] practices. It would certainly be self-defeating to preface any one lie by consultation with the dupe. But it is not at all self-defeating to discuss deceptive *policies* beforehand, nor to warn the deceived themselves. For instance, in deceptive games, players obviously choose whether or not to participate. The same is true in those deceptive medical experiments where consent is required as a preliminary. Similarly, in the conduct of foreign policy, a national discussion of the purposes and limits of deception could set standards for allowable deception in times of emergency. Examples of past deceptions held necessary for national defense could be debated and procedures set up for coping with similar choices in the future.

Take the use of unmarked police cars. If a society has openly debated their use and chosen to allow it in order to lull speeders and others into false confidence, then those who still choose to break the speed laws will be aware of the deceptive practice and can decide whether to take their chances or not. Once again, while each deceptive *act* does not lend itself easily to public justification, nothing stands in the way of a public scrutiny of the *practice*.

The questions about the nature of the publicity required for justifying lies can now be answered as follows. First, the "public" required for the justification of deceptive practices should ideally be wider than our conscience and more critical than the imagined audience, important though these are in their own right. If the choice is one of importance for others, or if, even though it seems trivial in itself, it forms part of a practice of deceit, then greater accountability should be required. Can the lie or the entire practice be defended in the press or on television? Can they be justified in advance in classrooms, workshops, or public meetings?

Second, there can be many or few in the public so addressed; but no one should be excluded from it on principle, least of all those representing the deceived or others affected by the lie.

[T]hose who plan to enter professions where deceptive practices are common would have the opportunity in professional schools to consider how to respond before becoming enmeshed in situations which seem to require lying. They could confront hypothetical cases similar to many they will later encounter; articulate and weigh the reasons supporting the conflicting choices; and debate their strengths and weaknesses. A public test of this kind would limit lies by professionals who believe that, as a group, they share a concern for the well-being of mankind which puts them beyond scrutiny.

The last question concerns the limitations on this test of publicity. They are substantial. While the test is a useful check on bias and rationalization, it is no more than a check. It is obviously of no avail in situations where the opportunity to reflect and to discuss is absent, as where immediate action is required. Nor does the test work well in moral quandaries which have no good answer, given our limited information, powers of reasoning, and foreknowledge.

These two limitations can be reduced in scope: the test can be used in advance to consider what to do in situations where there will be no opportunity to reflect or to discuss; and it can help us to work out modes of response even to those circumstances where uncertainty prevents a clear choice—who is to decide at such times, for instance, and how.

Caution and Risk-Taking

What steps does public justification require? What might reasonable persons do, when presented with someone's excuses for a particular lie? If they were asked to judge the degree of justification, how would they go about seeking an answer?

They would, first of all, look carefully for any alternatives of a non-deceptive nature available to the liar. Assuming that lies always carry a negative value, they would only begin to consider possible excuses after ascertaining that no statement devoid of that negative value would do.

In the second place, they would proceed to the weighing of the moral reasons for and against the lie. In so doing, they would share the perspective of the deceived and those affected by lies. They would, therefore, tend to be much more cautious than those with the optimistic perspective of the liar. They would value veracity and accountability more highly than would individual liars or their apologists.

In weighing the moral reasons, the excuses advanced, and the principles invoked, these persons would keep in mind the analogy between the use of force and the use of deception. Both, in their view, would be acceptable when *consented* to, given certain restrictions. The consent would have to be based on adequate information and ability to make a choice; and there would have to be freedom to opt out of the violent or deceptive situation. Where such informed and voluntary consent obtains, there is no longer a discrepancy of perspectives between liar and dupe, agent and victim. Deceptive bargaining in a bazaar, for instance, where buyer and seller try to outwit one another, would present few problems to these reasonable persons. The same is true of professional boxing matches. But deception can be justified in such situations only if they are knowingly and freely entered into, with complete freedom to leave. The naive newcomer may not be informed; many others have little genuine free choice; and if the practice is known to be deceptive yet entered into for survival, as in the case of a widespread black market, there is no longer freedom to leave.

Both violence and deception, moreover, would be more acceptable to these reasonable persons when used for the purposes of self-defense or life-saving. Finally, both would be more excusable the more trivial their effect on others.

But under all circumstances, these reasonable persons would need to be very wary because of the great susceptibility of deception to spread, to be abused, and to give rise to even more undesirable practices. The third step in their debate, therefore, would have to look beyond the individual excuses brought forth by liars and the individual counter-arguments on behalf of dupes. Here, the importance of practices would be stressed, and the harm to persons quite outside the deceptive situation considered. Spread multiplies the harm resulting from lies; abuse increases the damage for each and every instance. Both spread and abuse result in part from the lack of clear-cut standards as to what is acceptable. In the absence of such standards, instances of deception can and will increase, bringing distrust and thus more deception, loss of personal standards on the part of liars and so yet more deception, imitation by those who witness deception and the rewards it can bring, and once again more deception.

Reasonable persons might be especially eager to circumscribe the lies told by all those whose power renders their impact on human lives greater than usual.

Such, then, are the general principles which I believe govern the justification of lies. When we consider different kinds of lies, we must ask, first, whether there are alternative forms of action which will resolve the difficulty without the use of a lie; second, what might be the moral reasons brought forward to excuse the lie, and what reasons can be raised as counter-arguments. Third, as a test of these two steps, we must ask what a public of reasonable persons might say about such lies.

Most lies will clearly fail to satisfy these questions of justification.

Part VI

Moral Leaders and Moral Warriors: Vice Admiral James B. Stockdale and Stoic Philosophy

Consider the situation of the warrior and the military leader who has done exactly what his country has ordered him to do. Behaving honorably, abiding by the code of his profession, he performs well in the line of duty, but is captured, wounded, and imprisoned by his enemies. There he is daily subject to constant hardship and deprivation, and frequently to cruel and painful torture. His enemies do not abide by the provisions of that professional code, nor do they think themselves bound by it. Thus, although there are others like him in his prison, neither he nor they are allowed to visit, or otherwise to communicate or enjoy even the most rudimentary companionship. Instead, he is cut off from the outside world, isolated, alone, and as far as he knows, forgotten. Outside the very walls of his prison, moreover, he becomes aware that citizens from his own nation have come to denounce both he and his fellow prisoners as criminals and traitors, and disavow any connection between themselves, their country, and his actions.

Caught up in just these circumstances, Commander James B. Stockdale, U. S. Navy, turned within, to his own inner resources, and to the thoughts and writings of a one-time Roman slave named Epictetus, to find comfort and solace in his terrible circumstances, and to seek both to understand and to find the strength to persevere in his loyalty and his faith, despite those circumstances. Reflecting on the sayings of that ancient Stoic philosopher, Stockdale himself resolved, like Epictetus, never to surrender, never to falter, never to weaken, and never, ever to wallow in self-pity, recrimination, bitterness or hatred. Instead, Epictetus comforted Stockdale, as his simple writings had comforted countless numbers of Roman legionnaires centuries before who had carried the small "Enchiridion" or "handbook" of his teachings. Those teachings, reflecting the central thrust of Stoic philosophy, explained how to achieve inner peace and serenity in a world of nightmares, what Stockdale himself described as "The World of Epictetus," a world which is devoid of justice, and characterized by arbitrary and brutal principles over which one has little control.

Stoicism, as we have noted in these pages, is the school of thought that helped give rise to concepts like natural law, natural rights, and the moral equality of all human beings. It teaches that "freedom" is a state of mind, indomitable, even when the body is in chains. It teaches that inner will can predominate over external circumstances, and give to each of us the strength and the courage to "carry on," to remain steadfast, like Socrates, in our station, doing our duty, fearing neither death nor any other evil circumstance as much as the dishonor of abandoning our post. It teaches that there are things that are in our power: our emotions, our will, our desires, our inner states of being, and urges us to attend to, and to govern these things. It teaches that there are other things, not in our power, such as health, and fortune, and reputation, the attitudes and actions of others, the workings of nature, and the moments of our living and the moment of our dying, over which we have little or no control, and urges us to accept with serenity these things.

Stoicism teaches, in the words of Cicero, that Reason, divine *Logos*, permeates the universe, and brings ultimate order and justice and the rule of law out of chaos, anarchy, corruption, and treachery. It teaches us about the dignity and equality of emperor and subject, Jew and Greek, male and female, slave or freedman, who are all equal in the clear light of Reason, and who are regarded with equal tenderness and mercy by Reason's God. It teaches that immorality, duplicity, and betrayal are not the last word in the cosmos, and that nothing, and no one, is ever beyond the reach of justice, or of what Kant would later term, in this same spirit, the dictates of the Moral Law. Stoicism also teaches, in the words of another famous prisoner from the ancient world, that neither life, nor death, nor angels, nor principalities, nor things that are, nor things that are to come, nor powers, nor height, nor depth, nor anything else in all Creation shall ever separate us from the love of friends and family, from the faith and loyalty of comrades, or from the love of God, and that in all these things we are, as Stockdale demonstrated, "more than conquerors."

Stockdale did not simply draw comfort for himself from such reflections. He assumed command of his "outpost," united his fellow prisoners in faith and discipline, established rules, dispensed justice, and offered mercy and comfort to those, like himself, caught up in these terrible circumstances beyond their control. He did not simply survive. He triumphed. He did not waiver in his faith in friends, or in his loyalty to his service, or in his duty to his country. He helped his fellow prisoners to do likewise, and in so doing, to be "more than themselves." Upon his release—and, like his mentor, Epictetus, bearing the permanent and crippling scars of his physical abuse—he continued to serve, to lead, to study, to teach, and to write.

No one could more completely exemplify the moral concepts we have studied in this book, nor better illustrate their applicability to the life of the military officer. His account of these experiences, "Courage under Fire"—a speech at Jeremy Bentham's university in London in 1993—and several of his other writings and teachings, together with those of his own spiritual teacher, Epictetus, conclude our study of the moral resources of western thought, and beautifully summarize the moral foundations of military leadership for the years that lie ahead.

Courage Under Fire
*VADM James B. Stockdale, USN**

1. Testing Epictetus' Doctrines in a Laboratory of Human Behavior[1]

I came to the philosophic life as a thirty-eight-year-old naval pilot in grad school at Stanford University. I had been in the navy for twenty years and scarcely ever out of a cockpit. In 1962, I began my second year of studying international relations so I could become a strategic planner in the Pentagon. But my heart wasn't in it. I had yet to be inspired at Stanford and saw myself as just processing tedious material about how nations organized and governed themselves. I was too old for that. I knew how political systems operated; I had been beating systems for years.

Then, in what we call a "feel out pass" in stunt flying, I cruised into Stanford's philosophy corner one winter morning. I was gray-haired and in civilian clothes. A voice boomed out of an office, "Can I help you?" The speaker was Philip Rhinelander, dean of Humanities and Sciences, who taught Philosophy 6: The Problems of Good and Evil.

At first he thought I was a professor, but we soon found common ground in the navy because he'd served in World War II. Within fifteen minutes we'd agreed that I would enter his two-term course in the middle, and to make up for my lack of background, I would meet him for an hour a week for a private tutorial in the study of his campus home.

Phil Rhinelander opened my eyes. In that study it all happened for me—my inspiration, my dedication to the philosophic life. From then on, I was out of international relations—I already had enough credits for the master's—and into philosophy. We went from Job to Socrates to Aristotle to Descartes. And then on to Kant, Hume, Dostoyevsky, Camus. All the while, Rhinelander was psyching me out trying to figure out what I was seeking. He thought my interest in Hume's *Dialogues Concerning Natural Religion* was quite interesting. On my last session, he reached high in his wall of books and brought down a copy of *The Enchiridion*. He said, "I think you'll be interested in this."

Enchiridion means "ready at hand." In other words, it's a handbook. Rhinelander explained that its author, Epictetus, was a very unusual man of intelligence and sensitivity, who gleaned wisdom rather than bitterness from his early firsthand exposure to extreme cruelty and firsthand observations of the abuse of power and self-indulgent debauchery.

Epictetus was born a slave in about A.D. 50 and grew up in Asia Minor speaking the Greek language of his slave mother. At the age of fifteen or so, he was loaded off to Rome in chains in a slave caravan. He was treated savagely for months while en route. He went on the Rome auction block as a permanent cripple, his knee having been shattered and left untreated. He was "bought cheap" by a freedman named Epaphroditus, a secretary to Emperor Nero. He was taken to live at the Nero White

---

*Excerpts from *Courage Under Fire: Testing Epictetus' Doctrines in a Lab of Human Behavior* by James Bond Stockdale are reprinted with the permission of the publisher, Hoover Institution Press. Copyright © 1994 the Board of Trustees of the Leland Stanford Junior University.

[1]Speech delivered at the Great Hall, King's College, London, Monday, November 15, 1993.

House at a time when the emperor was neglecting the empire as he frequently toured Greece as actor, musician, and chariot race driver. When home in Rome in his personal quarters, Nero was busy having his half-brother killed, his wife killed, his mother killed, his second wife killed. Finally, it was Epictetus' master Epaphroditus who cut Nero's throat when he fumbled his own suicide as the soldiers were breaking down his door to arrest him.

That put Epaphroditus under a cloud, and, fortuitously, the now cagey slave Epictetus realized he had the run of Rome. And being a serious and doubtless disgusted young man, he gravitated to the high-minded public lectures of the Stoic teachers who *were* the philosophers of Rome in those days. Epictetus eventually became apprenticed to the very best Stoic teacher in the empire, Musonius Rufus, and, after ten or more years of study, achieved the status of philosopher in his own right. With that came true freedom in Rome, and the preciousness of that was duly celebrated by the former slave. Scholars have calculated that in his works individual freedom is praised six times more frequently than it is in the New Testament. The Stoics held that all human beings were equal in the eyes of God: male/female, black/white, slave and free.

I read every one of Epictetus' extant writings twice, through two translators. Even with the most conservative translators, Epictetus comes across speaking like a modern person. It is "living speech," not the *literary* Attic Greek we're used to in men of that tongue. *The Enchiridion* was actually penned not by Epictetus, who was above all else a determined teacher and man of modesty who would never take the time to transcribe his own lectures, but by one of his most meticulous and determined students. The student's name was Arrian, a very smart, aristocratic Greek in his twenties. After hearing his first few lectures, he is reported to have exclaimed something like, "Son of a gun! We've got to get this guy down on parchment!"

With Epictetus' consent, Arrian took down his words verbatim in some kind of frantic shorthand he devised. He bound the lectures into books; in the two years he was enrolled in Epictetus' school, he filled eight books. Four of them disappeared sometime before the Middle Ages. It was then that the remaining four got bound together under the title *Discourses of Epictetus.* Arrian put *The Enchiridion* together after he had finished the eight. It is just highlights from them "for the busy man." Rhinelander told me that last morning, "As a military man, I think you'll have a special interest in this. Frederick the Great never went on a campaign without a copy of this handbook in his kit."

I'll never forget that day, and the essence of what that great man had to say as we said good-bye was burned into my brain. It went very much like this: Stoicism is a noble philosophy that proved more practicable than a modern cynic would expect. The Stoic viewpoint is often misunderstood because the casual reader misses the point that all talk is in reference to the "inner life" of man. Stoics belittle physical harm, but this is not braggadocio. They are speaking of it in comparison to the devastating agony of shame they fancied good men generating when they knew in their hearts that they had *failed* to do their duty vis-a-vis their fellow men or God. Although pagan, the Stoics had a monotheistic, natural religion and were great contributors to Christian thought. The fatherhood of God and the brotherhood of man were Stoic concepts before Christianity. In fact, one of their early theoreticians, named Chrysippus, made the analogy of what might be called the *soul* of the universe to the *breath* of a human, *pneuma* in Greek. This Stoic conception of a celestial pneuma is said to be the great-grandfather of the Christian Holy Ghost. Saint Paul, a Hellenized Jew brought up in Tarsus, a Stoic town in Asia Minor, always used the Greek word *pneuma,* or breath, for "soul."

Rhinelander told me that the Stoic demand for disciplined thought naturally won only a small minority to its standard, but that those few were everywhere the best. Like its Christian counterparts, Calvinism and Puritanism, it produced the strongest characters of its time. In theory, a doctrine of pitiless perfection, it actually created men of courage, saintliness, and goodwill. Rhinelander singled out three examples: Cato the Younger, Emperor Marcus Aurelius, and Epictetus. Cato was the great Roman republican who pitted himself against Julius Caesar. He was the unmistakable hero of George Washington; scholars find quotations of this man in Washington's farewell address—without quotation marks. Emperor Marcus Aurelius took the Roman Empire to the pinnacle of its power and influence. And Epictetus, the great teacher, played his part in changing the leadership of Rome from the swill he had known in the Nero White House to the power and decency it knew under Marcus Aurelius.

Marcus Aurelius was the last of the five emperors (all with Stoic connections) who successively ruled throughout that period Edward Gibbon described in his *Decline and Fall of the Roman Empire* as follows: "If a man were called upon to fix the period in the history of the world during which the condition of the human race was most happy and prosperous, he would without hesitation name that which elapsed from the accession of Nerva (A.D. 96) to the death of Marcus Aurelius (A.D. 180). The united reigns of the five emperors of the era are possibly the only period of history in which the happiness of a great people was the sole object of government."

Epictetus drew the same sort of audience Socrates had drawn five hundred years earlier—young aristocrats destined for careers in finance, the arts, public service. The best families sent him their best sons in their middle twenties—to be told what the good life consisted of, to be disabused of the idea that they deserved to become playboys, the point made clear that their job was to serve their fellow men.

In his inimitable, frank language, Epictetus explained that his curriculum was *not* about "revenues or income, or peace or war, but about happiness and unhappiness, success and failure, slavery and freedom." His model graduate was not a person "able to speak fluently about philosophic principles as an idle babbler, but about things that will do you good if your child dies, or your brother dies, or if you must die or be tortured." "Let others practice lawsuits, others study problems, others syllogisms; here you practice how to die, how to be enchained, how to be racked, how to be exiled." A man is responsible for his own "judgments, even in dreams, in drunkenness, and in melancholy madness." Each individual brings about his own good and his own evil, his good fortune, his ill fortune, his happiness, and his wretchedness. And to top all this off, he held that it is *unthinkable* that one man's error could cause another's suffering. Suffering, like everything else in Stoicism, was all down here—remorse at destroying *yourself*.

So what Epictetus was telling his students was that there can be no such thing as being the "victim" of another. You can only be a "victim" of *yourself*. It's all how you discipline your mind. Who is your master? "He who has authority over any of the things on which you have set your heart." "What is the result at which all virtue aims? *Serenity*." "Show me a man who though sick is happy, who though in danger is happy, who though in prison is happy, and I'll show you a Stoic."

When I got my degree, Sybil and I packed up our four sons and family belongings and headed to Southern California. I was to take command of Fighter Squadron 51, flying supersonic F-8 Crusaders, first at the Miramar Naval Air Station, near San Diego, and later, of course, at sea aboard various aircraft carriers in the western Pacific. Exactly three years after we drove up to our new home near San Diego, I was shot down and captured in North Vietnam.

During those three years, I had launched on three seven-month cruises to the waters off Vietnam. On the first we were occupied with general surveillance of the fighting erupting in the South; on the second I led the first-ever American bombing raid against North Vietnam; and on the third, I was flying in combat almost daily as the air wing commander of the *USS Oriskany*. But on my bedside table, no matter what carrier I was aboard, were my Epictetus books: *Enchiridion, Discourses,* Xenophon's *Memorabilia* of Socrates, and *The Iliad* and *The Odyssey* (Epictetus expected his students to be familiar with Homer's plots.) I didn't have time to be a bookworm, but I spent several hours each week buried in them.

I think it was obvious to my close friends, and certainly to me, that I was a changed man and, I have to say a better man for my introduction to philosophy and especially to Epictetus. I was on a different track—certainly not an antimilitary track but to some extent an anti-organization track. Against the backdrop of all the posturing and fumbling around peacetime military organizations seem to have to go through, to accept the need for graceful and unself-conscious improvisation under pressure, to break away from set procedures forces you to be reflective, reflective as you put a new mode of operation together. I had become a man detached—not aloof but detached—able to throw out the book without the slightest hesitation when it no longer matched the external circumstances. I was able to put juniors over seniors without embarrassment when their wartime instincts were more reliable. This new abandon, this new built-in flexibility I had gained, was to pay off later in prison.

But undergirding my new confidence was the realization that I had found the proper philosophy for the military arts as I practiced them. The Roman Stoics coined the formula *Vivere militare!*—"Life is being a soldier." Epictetus in *Discourses*: "Do you not know that life is a soldier's service? One must keep guard, another go out to reconnoitre, another take the field. If you neglect your responsibilities when some severe order is laid upon you, do you not understand to what a pitiful state you bring the army in so far as in you lies?" *Enchiridion*: "Remember, you are an actor in a drama of such sort as the

Author chooses—if short, then in a short one; if long, then in a long one. If it be his pleasure that you should enact a poor man, or a cripple, or a ruler, see that you act it well. For this is your business—to act well the given part, but to choose it belongs to Another." "Every one of us, slave or free, has come into this world with *innate* conceptions as to good and bad, noble and shameful, becoming and unbecoming, happiness and unhappiness, *fitting and inappropriate.*" "If you regard yourself as a man and as a part of some whole, it is fitting for you now to be sick and now to make a voyage and run risks, and now to be in want, and on occasion to die before your time. Why, then are you vexed? Would you have someone else be sick of a fever now, someone else go on a voyage, someone else die? For it is impossible in such a body as ours, that is, in this universe that envelops us, among these fellow-creatures of ours, that such things should not happen, some to one man, some to another."

On September 9, 1965, I flew at 500 knots right into a flak trap, at treetop level, in a little A-4 airplane—the cockpit walls not even three feet apart—which I couldn't steer after it was on fire, its control system shot out. After ejection I had about thirty seconds to make my last statement in freedom before I landed in the main street of a little village right ahead. And so help me, I whispered to myself: "Five years down there, at least. I'm leaving the world of technology and entering the world of Epictetus."

"Ready at hand" from the *Enchiridion* as I ejected from that airplane was the understanding that a Stoic always kept *separate* files in his mind for (A) those things that are "up to him" and (B) those things that are "not up to him." Another way of saying it is (A) those things that are "within his power" and (B) those things that are "beyond his power." Still another way of saying it is (A) those things that are within the grasp of "his Will, his Free Will" and (B) those things that are beyond it. All in category B are "external," beyond my control, ultimately dooming me to fear and anxiety if I covet them. All in category A are up to me, within my power, within my will, and properly subjects for my total concern and involvement. They include my opinions, my aims, my aversions, my own grief, my own joy, my judgments, my attitude about what is going on, my own good, and my own evil.

To explain why "your own good and your own evil" is on that list, I want to quote Alexander Solzhenitsyn from his Gulag book. He writes about that point in prison when he realizes the strength of his residual powers, and starts what I called to myself "gaining moral leverage," riding the updrafts of occasional euphoria as you realize you are getting to know yourself and the world for the first time. He calls it "ascending" and names the chapter in which this appears "The Ascent":

> It was only when I lay there on the rotting prison straw that I sensed within myself the first stirrings of *good*. Gradually it was disclosed to me that the line separating good and evil passes not between states nor between classes nor between political parties, but right through every human heart, through all human hearts. And that is why I turn back to the years of my imprisonment and say, sometimes to the astonishment of those about me, "Bless you, prison, for having been a part of my life."

I came to understand that long before I read it. Solzhenitsyn learned, as I and others have learned, that good and evil are not just abstractions you kick around and give lectures about and attribute to this person and that. The only good and evil that means anything is right in your own heart within your will, within your power, where it's up to you. *Enchiridion 32:* "Things that are not within our own power, not without our Will, can by no means be either good or evil." *Discourses:* "Evil lies in the evil use of moral purpose, and good the opposite. The course of the Will determines good or bad fortune, and one's balance of misery and happiness." In short, what the Stoics say is "Work with what you have control of and you'll have your hands full."

What is not up to you? beyond your power? not subject to your will in the last analysis? For starters, let's take "your station in life." As I glide down toward that little town on my short parachute ride, I'm just about to learn how negligible is my control over my station in life. It's not at all up to me. I'm going right now from being the leader of a hundred-plus pilots and a thousand men and, goodness knows, all sorts of symbolic status and goodwill, to being an *object of contempt*. I'll be known as a "criminal." But that's not *half* the revelation that is the realization of your own *fragility*—that you can be reduced by wind and rain and ice and seawater or *men* to a helpless, sobbing wreck—unable to control even your own bowels—in a matter of *minutes*. And, more than even that, you're going to face fragilities you never before let yourself believe you could have—like after mere minutes, in a flurry of action while being bound with tourniquet-tight ropes, with care, by a professional, hands behind, jackknifed forward and down toward your ankles held secure in lugs attached to an iron bar, that with the onrush of anxiety knowing your upper body's circulation has been stopped and feeling the ever-growing induced pain and the ever-closing-in of claustrophobia, you can be made to blurt out answers, sometimes correct answers, to questions about anything they know you know. (Hereafter, I'll just call that situation "taking the ropes.")

"Station in life," then, can be changed from that of a dignified and competent gentleman of culture to that of a panic-stricken, sobbing, self-loathing wreck in a matter of minutes. So what? To live under the false pretense that you will forever have control of your station in life is to ride for a fall; you're asking for disappointment. So make sure in your heart of hearts, in your inner self, that you treat your station in life with *indifference*, not with contempt, only with *indifference.*

And so also with a long long list of things that some unreflective people assume they're assured of controlling to the last instance: your body, property, wealth, health, life, death, pleasure, pain, reputation. Consider "reputation," for example. Do what you will, reputation is at least as fickle as your station in life. *Others* decide what your reputation is. Try to make it as good as possible, but don't get hooked on it. Don't be ravenous for it and start chasing it in tighter and tighter circles. As Epictetus says, "For what are tragedies but the portrayal in tragic verse of the sufferings of men who have admired things external?" In your heart of hearts, when you get out the key and open up that old roll-top desk where you really keep your stuff, don't let "reputation" get mixed up with your *moral purpose* or your *will power;* they *are* important. Make sure "reputation" is in that box in the bottom drawer marked "matters of indifference." As Epictetus says, "He who craves or shuns things not under his control can neither be faithful nor free, but must himself be changed and tossed to and fro and must end by subordinating himself to others."

I know the difficulties of gulping this down right away. You keep thinking of practical problems. Everybody has to play the game of life. You can't just walk around saying, "I don't give a damn about health or wealth or whether I'm sent to prison or not." Epictetus took time to explain better what he meant. He says everybody should play the game of life—that the best play it with "skill, form, speed, and grace." But like most games, you play it with a ball. Your team devotes all its energies to getting the ball across the line. But after the game, what do you do with the ball? Nobody much cares. It's not worth anything. The competition, the game, was the thing. The ball was "used" to make the game possible, but it in itself is not of any value that would justify falling on your sword for it.

Once the game is over, the ball is properly a matter of indifference. Epictetus on another occasion used the example of shooting dice—the dice being matters of indifference, once their numbers had turned up. To exercise *judgment* about whether to accept the numbers or roll again is a *willful* act, and thus *not* a matter of indifference. Epictetus' point is that our *use* of externals is not a matter of indifference because our actions are products of our will and we totally control that, but that the dice themselves, like the ball, are material over which we have no control. They are externals that we cannot afford to covet or be earnest about, else we might set our hearts on them and become slaves of such others as control them.

These explanations of this concept seem so modern, yet I have just given you practically verbatim quotes of Epictetus' remarks to his students in Nicopolis, colonial Greece, two thousand years ago.

So I took those core thoughts into prison; I also remembered a lot of attitude-shaping remarks. Here's Epictetus on how to stay off the hook: "A man's master is he who is able to confer or remove whatever that man seeks or shuns. Whoever then would be free, let him wish nothing, let him decline nothing, which depends on others; else he must necessarily be slave." And here's why never to beg: "For it is better to die of hunger, exempt from fear and guilt, than to live in affluence with perturbation." Begging sets up a demand for quid pro quos, deals, agreements, reprisals, the pits.

If you want to protect yourself from "fear and guilt," and those are the crucial pincers, the real long-term destroyers of will, you have to get rid of all your instincts to compromise, to meet people halfway. You have to learn to stand aloof, never give openings for deals, never level with your adversaries. You have to become what Ivan Denisovich called a "slow movin' cagey prisoner."

All that, over the previous three years, I had *unknowingly* put away for the future. So, to return to my bailing out of my A-4, I can hear the noontime shouting and pistol shots and whining bullets ripping my parachute canopy and see the fists waving in the street below as my chute hooks a tree but deposits me on the ground in good shape. With two quick-release fastener flips, I'm free of the parachute, and immediately gang tackled by the ten or fifteen town roughnecks I had seen in my peripheral vision, pounding up the road from my right.

I don't want to exaggerate this or indicate that I was surprised at my reception. It was just that when the gang tackling and pummeling was all over, and it lasted for two or three minutes before a man with a pith helmet got there to blow his police whistle, I had a very badly broken leg that I felt sure would be with me for life. My hunch turned out to be right. Later, I felt some relief—but only minor—from another Epictetus admonition I remembered: "Lameness is an impediment to the leg, but not to the Will; and say this to yourself with regard to everything that happens. For you will find such things to be an impediment to something else, but not truly to yourself."

But during the time interval between pulling the ejection handle and coming to rest on the street, I had become a man with a mission. I can't explain this without unloading a little emotional baggage that was part of my military generation's legacy in 1965.

In the aftermath of the Korean War, just over ten years before, we all had memories of reading about, and seeing early television news accounts of, U.S. government investigations into the behavior of some American prisoners of war in North Korea and mainland China. There was a famous series of articles in the *New Yorker* magazine that later became a book entitled *In Every War but One*. The gist of it was that in prison camps for Americans, it was every man for himself. Since those days, I've come to know officers who were prisoners of war there, and I now see much of that as selective reporting and as a bum rap. However, there were cases of young soldiers who were confused by the times, scared to death, in cold weather, treating each other like dogs fighting over scraps, throwing each other out in the snow to die, and nobody doing anything about it.

This could not go on, and President Eisenhower commissioned the writing of the American Fighting Man's Code of Conduct. It is written in the form of a personal pledge. Article 4: "If I become a prisoner of war, I will keep faith with my fellow prisoners. I will give no information or take part in any action which might be harmful to my comrades. If I am senior, I will take command. If not, I will obey the lawful orders of those appointed over me and will back them up in every way." In other words, as of the moment Eisenhower signed the document, American prisoners of war were never to escape the chain of command; the war goes on behind bars. As an insider, I knew the whole setup—that the North Vietnamese already held about twenty-five prisoners, probably in Hanoi, that I was the only wing commander to survive an ejection, and that I would be their senior, their commanding officer, and would remain so, very likely, throughout this war that I felt sure would last at least another five years. And here I was starting off crippled and flat on my back.

Epictetus turned out to be right. After a very crude operation, I was on crutches within a couple of months, and the crooked leg, healing itself, was strong enough to hold me up without the crutches in about a year. All told, it was only a temporary setback from things that were important to me, and being cast in the role as the sovereign head of an American expatriate colony that was destined to remain autonomous, out of communication with Washington, for years on end was very important to me. I was forty-two years old—still on crutches, dragging a leg, at considerably less than my normal weight, with hair down near my shoulders, my body unbathed since I had been catapulted from the *Oriskany*, a beard that had not seen a razor since I arrived—when I took command (clandestinely, of course, the North Vietnamese would never acknowledge our rank) of about fifty Americans. That expatriate colony would grow to over four hundred—all officers, all college graduates, all pilots or backseat electronic wizards. I was determined to "play well the given part."

The key word for all of us at first was "fragility." Each of us, before we were ever in shouting distance of another American, was made to "take the ropes." That was a real shock to our systems—and, as with all shocks, its impact on our inner selves was a lot more impressive and lasting and important than to our limbs and torsos. These were the sessions where we were taken down to submission and made to blurt out distasteful confessions of guilt and American complicity into antique tape recorders, and then to be put in what I call "cold soak," a month or so of total isolation to "contemplate our crimes." What we actually contemplated was what even the most laid-back American saw as his betrayal of himself and everything he stood for. It was there that I learned what "Stoic Harm" meant. A shoulder broken, a bone in my back broken, a leg broken twice were *peanuts* by comparison. Epictetus: "Look not for any greater harm than this: destroying the trustworthy, self-respecting, well-behaved man within you."

When put into a regular cell block, hardly an American came out of that experience without responding something like this when first whispered to by a fellow prisoner next door: "You don't want to talk to me; I am a traitor." And because we were equally fragile, it seemed to catch on that we all replied something like this: "Listen, pal, there are no virgins in here. You should have heard the kind of statement I made. Snap out of it. We're all in this together. What's your name? Tell me about yourself." To hear that last was, for most new prisoners just out of initial shakedown and cold soak, a turning point in their lives.

But the new prisoner's learning process was just beginning. Soon enough he would realize that things were not at all like some had told him in survival training—that if you made a good stiff showing of resistance in the opening chapters, the interrogators would lose interest in you and you would find yourself merely relegated to boredom, to "sitting out the war," to "languishing in your cell," as the uninitiated novelists love to describe the predicament. No, the war went on behind bars—there was no such thing as the jailers giving up on you as a hopeless case. Their political beliefs *made* them believe you could be made to see things their way; it was just a matter of time. And so you were marched to the interrogation room endlessly, particularly on the occasions of your being apprehended breaking one of the myriad rules that were posted on your cell wall—"trip wire" rules, which paid dividends for the commissar if his interrogator could get you to fall prey to his wedge of *shame.* The currency at the game table, where you and the interrogator faced one another in a duel of wits, was *shame,* and I learned that unless he could impose shame on me, or unless I imposed it on myself, he had nothing going for him. (Force was available, but that required the commissar's okay.)

For Epictetus, emotions were acts of will. Fear was not something that came out of the shadows of the night and enveloped you; he charged you with the total responsibility of starting it, stopping it, controlling it. This was one of Stoicism's biggest demands on a person. Stoics can be made to sound like lazy brutes when they are described merely as people indifferent to most everything but good and evil, people who make stingy use of emotions like pity and sympathy. But add this requirement of total personal responsibility for each and every one of your emotions, and you're talking about a person with his hands full. I whispered a "chant" to myself as I was marched at gunpoint to my daily interrogation: "control fear, control guilt, control fear, control guilt." And I devised methods of deflecting my gaze to obscure such fear or guilt as doubtless emerged in my eyes when I temporarily lost control under questioning. You could be bashed for failure to look at the face of your interrogator; I concentrated on his left earlobe, and he seemed to get used to it—thought I was a little cockeyed, probably. Controlling your emotions is difficult but can be *empowering.* Epictetus: "For it is *within you,* that both your destruction and deliverance lie." Epictetus: "The judgment seat and a prison is each a place, the one high, the other low; but the *attitude of your will* can be kept the same, if you *want* to keep it the same, in either place."

We organized a clandestine society via our wall tap code—a society with our own laws, traditions, customs, even heroes. To explain how it could be that we would order each other into more torture, order each other to refuse to comply with specific demands, intentionally call the bluff of our jailers and in a real sense force them to repeat the full ropes process to another submission, I'll quote a statement that could have come from at least half of those wonderful competitive fly-boys I found myself locked up with: "We are in a spot like we've never been in before. But we deserve to maintain our self-respect, to have the feeling we are fighting back. We can't refuse to do every degrading thing they demand of us, but it's up to you, boss, to pick out things we must all refuse to do unless and until they put us through the ropes again. We deserve to sleep at night. We at least deserve to have the satisfaction that we are hewing to our leader's orders. Give us the list; what are we to take torture for?"

I know this sounds like strange logic, but in a sense it was a first step in claiming what was rightfully *ours.* Epictetus said, "The judge will do some things to you which are thought to be terrifying; but how can he stop you from taking the punishment *he threatened?*" That's *my* kind of Stoicism. You have a right to make them hurt you, and they don't like to do that. When my fellow prisoner Ev Alvarez, the very first pilot they captured, was released with the rest of us, the prison commissar told him: "You Americans were nothing like the French; we could count on them to be reasonable." Ha.

I put a lot of thought into what those first orders should be. They would be orders that *could be obeyed,* not a "cover your ass" move of reiterating some U.S government policy like "name, rank, serial number, and date of birth," which had no chance of standing up in the torture room. My mind-set was "we here under the gun are the experts, we are the masters of our fate, ignore guilt-inducing echoes of hollow edicts, throw out the book and write your own." My orders came out as easy-to-remember acronyms. The principal one was BACK US: Don't Bow in public; stay off the Air; admit no Crimes, never Kiss them goodbye. "US" could be interpreted as United States, but it *really* meant, "Unity over Self." Loners make out in an enemy's prison, so my first rule of togetherness in there was that each of us had to work at the lowest common denominator, never negotiating for himself but only for *all*.

Prison life became a crazy mixture of an old regime and a new one. The old was the political prison routine, mainly for dissenters and domestic enemies of the state. It was designed and run by old-fashioned Third World Communists of the Ho Chi Minh cut. It revolved around the idea of "repentance" for your "crimes" of anti-social behavior. American prisoners, street criminals, and domestic political enemies of the state were all in the same prison. We never saw a "POW camp" like the movies show. The communist jail was part psychiatric clinic and part reform school. North Vietnam protocol called for making *all* their inmates demonstrate shame—bowing to all guards, heads low, never looking at the sky, frequent sessions with your interrogator if, for no other reason, to check your *attitude* and, if judged "wrong," then maybe down the torture chute of confession of guilt, of apology and then the inevitable payoff of atonement.

The new regime, superimposed on the above, was for Americans only. It was a propaganda factory, supervised by English-speaking young bureaucratic army officers with quotas to fill, quotas set by the political arm of the government: press interviews with visiting left-wing Americans, propaganda films to shoot (starring intimidated "American air pirates"), and so on.

An encapsulated history of how this bifurcated prison philosophy fared is that the propaganda footage and interviews started to backfire. Smart American college men were salting their acts with sentences with double-meanings, gestures read as funny-obscene by Western audiences, and practical jokes. One of my best friends, tortured to give names of pilots he knew who had turned in their wings in opposition to the war, said there were only two: Lieutenants Clark Kent and Ben Casey (then-popular fictional characters in America). That joke was headlined on the front page of the *San Diego Union,* and somebody sent a copy back to the government in Hanoi. As a result of that friendly gesture from a fellow American, Nels Tanner went into three successive days of rope torture, followed by 123 days in leg stocks, all while isolated of course.

So after several of these stunts, which cost the Vietnamese much loss of face, North Vietnamese resorted to getting their propaganda only from the relatively *few* (less than 5 percent) of the Americans they could trust not to act up: real loners who, for different reasons, never joined the prisoner organization, never wanted to get into the tap code network, well-known sleaze balls we came to call *finks*. The vast majority of my constituents were enraged by their actions and took it upon themselves to diligently memorize data that would convict them in an American court-martial. But when we got home our government ruled against my bringing charges.

The great mass of all other Americans in Hanoi were by all standards "honorable prisoners," but that is not to say there was anything like a homogeneous prison regime we all shared. People like to think that because we were all in the Hanoi prison system, we had all these common experiences. It's not so. These *differing* regimes became marked when our prison organization stultified the propaganda efforts of this two-headed monster they called the "Prison Authority." They turned to vengeance against the leadership of my organization and to an effort to break down the morale of the others by baiting them with an amnesty program in which they would compete for early release by being compliant with North Vietnam's wishes.

In May 1967, the public address system blared out: "Those of you who repent, truly repent, will be able to go home before the war is over. Those few diehards who insist on inciting the other criminals to oppose the camp authority will be sent to a special dark place." I immediately put out an order forbidding any American to accept early release, but that is not to say I was a lone man on a white horse. I didn't have to sell that one; it was accepted with obvious relief and spontaneous jubilation by the overwhelming majority.

Guess who went to the "dark place." They isolated my leadership team—me and my cohort of ten top men—and sent us into exile. The Vietnamese worked very hard to learn our habits, and they knew who were the troublemakers and who were "not making any waves." They isolated those I trusted most; everybody had a long record of solitary and rope-mark pedigrees. Not all were seniors; we had seniors in prison who would not even communicate with the man next door. One of my ten was only twenty-four years old—born after I was in the navy. He was a product of my recent shipboard tendencies: "When instincts and rank are out of phase, take the guy with the instincts." All of us stayed in solitary throughout, starting with two years in leg irons in a little high-security prison right beside North Vietnam's "Pentagon"—their Ministry of Defense, a typical old French building. There are chapters upon chapters after that, but what they came down to in my case was a strung-out vengeance fight between the "Prison Authority" and those of us who refused to quit trying to be our brothers' keepers. The stakes grew to *nervous breakdown* proportions. One of the eleven of us died in that little prison we called Alcatraz, but even including him, there was not a man who wound up with less than three and a half years of solitary and four of us had more than four years. To give you a sense of proportion on how the total four hundred fared on solo, one hundred had none, more than half of the other three hundred had less than a year, and half of those with less than a year had less than a month. So the average for the four hundred was considerably less than six months.

Howie Rutledge, one of the four of us with more than four years, went back to school and got a master's degree after we got home, and his thesis concentrated on the question of whether long-term erosion of human purpose was more effectively achieved by torture or isolation. He mailed out questionnaires to us (who had also all taken the ropes at least ten times) and others with records of extreme prison abuse. He found that those who had less than two years' isolation and plenty of torture said torture was the trump card; those with more than two years' isolation and plenty of torture said that for long-term modification of behavior, isolation was the way to go. From my viewpoint, you can get used to repeated rope torture—there are some tricks for minimizing your losses in that game. But keep a man, even a very strong-willed man, in isolation for three or more years, and he starts looking for a friend—*any* friend, regardless of nationality or ideology.

Epictetus once gave a lecture to his faculty complaining about the common tendency of new teachers to slight the stark realism of Stoicism's challenges in favor of giving the students an uplifting, rosy picture of how they could meet the harsh requirements of the good life painlessly. Epictetus said: "Men, the lecture-room of the philosopher is a hospital; students ought not to walk out of it in pleasure, but in pain." If Epictetus' lecture room was a hospital, my prison was a laboratory—a laboratory of human behavior. I chose to test his postulates against the demanding real-life challenges of my laboratory. And as you can tell, I think he passed with flying colors.

It's hard to discuss in public the real-life challenges of that laboratory because people ask all the wrong questions: How was the food? That's always the first one, and in a place like I've been, that's so far down the scale you want to cry. Did they harm you physically? What was the nature of the *device* they used to harm you? Always the device or the truth serum or the electric shock treatment—all of which would totally defeat the purpose of a person seriously trying to break down your will. All those things would give *you* a feeling of moral superiority which is the last thing he would want to have happen. I'm not talking about brainwashing; there is no such thing. I'm talking about having looked over the brink and seen the bottom of the pit and realized the truth of that linchpin of Stoic thought: that the thing that brings down a man is not *pain* but *shame!*

Why did those men in "cold soak" after their first rope trip eat their hearts out and feel so unworthy when the first American contacted them? Epictetus knew human nature well. In that prison laboratory, I do not know of a single case where a man was able to erase his conscience pangs with some laid-back pop psychology theory of cause and effect. Epictetus emphasizes time and again that a man who lays off the causes of his actions to third parties or forces is not leveling with himself. He must live with his own judgments if he is to be honest with himself. (And the "cold soak" tends to make you honest.) "But if a person subjects me to fear of death, he compels me," says a student. "No," says Epictetus, "It is neither death, nor exile, nor toil, nor any such things that is the cause of your doing, or not doing, *anything,* but only your opinions and the decisions of your Will." "What is the fruit of your doctrines?" someone asked Epictetus. "Tranquility, fearlessness, and freedom," he answered. You can have these only if you are honest and take responsibility for your own actions. You've got to get it *straight! You* are in charge of *you.*

Did I preach these things in prison? Certainly not. You soon learned that if the guy next door was doing okay that meant that he had all his philosophical ducks lined up in his own way. You soon realized that when you dared to spout high-minded philosophical suggestions through the wall, you always got a very reluctant response.

No, I never tapped or mentioned Stoicism once. But some sharp guys read the signs in my actions. After one of my long isolations outside the cell blocks of the prison, I was brought back into signaling range of the fold, and my point of contact was a man named Dave Hatcher. As was standard operating procedure on a first contact after a long separation, we started off not with gushes of news but with first, an agreed-upon danger signal, second, a cover story for each of us if we were caught, and third, a backup communications system if this link was compromised—"slow movin' cagey prisoner" precautions. Hatcher's backup communication for me was a note drop by an old sink near a place we called the Mint, the isolation cell block of Hatcher's "Las Vegas" wing of the prison—a place he rightly guessed I would soon enough be in. Every day we would signal for fifteen minutes over a wall between his cell block and my "no man's land."

Then I got back into trouble. At that time the commissar of prisons had me isolated and under almost constant surveillance for the year since I had staged a riot in Alcatraz to get us out of leg irons. I was barred from all prisoner cell blocks. I had special handlers, and they caught me with an outbound note that gave leads I knew the interrogators could develop through torture. The result would be to implicate my friends in "black activities" (as the North Vietnamese called them). I had been through those ropes more than a dozen times, and I knew I could *contain* material—*so long as they didn't know I knew it*. But this note would open doors that could lead to more people getting killed in there. We had lost a few in big purges—I think in torture overshoots—and I was getting tired of it. It was the fall of 1969, and I had been in this role for four years and saw nothing left for me to do but check out. I was solo in the main torture room in an isolated part of the prison, the night before what they told me would be my day to spill my guts. There was an eerie mood in the prison. Ho Chi Minh had just died, and his special dirge music was in the air. I was to sit up all night in traveling irons. My chair was near the only paned glass window in the prison. I was able to waddle over and break the window stealthily. I went after my wrist arteries with the big shards. I had knocked the light out, but the patrol guard happened to find me passed out in a pool of blood but still breathing. The Vietnamese sounded the alert, got their doctor, and saved me.

Why? It was not until after I was released years later that I learned that that very week, Sybil had been in Paris demanding humane treatment for prisoners. She was on world news, a public figure, and the last thing the North Vietnamese needed was me dead. There had been a very solemn crowd of senior North Vietnamese officers in that room as I was revived.

Prison torture, *as we had known it in Hanoi*, ended for everybody that night.

Of course it was months before we could be sure that was so. All I knew at the time was that in the morning, after my arms had been dressed and bandaged, the commissar himself brought in a hot cup of sweet tea, told my surveillance guard to take off my leg irons, and asked me to sit at the table with him. "Why did you do this, Stockdale? You know I sit with the army's General Staff, they've asked for a full report this morning." (It was not unusual for us to talk like that by that time.) But he never once mentioned the note, nor did anybody else thereafter. *That* was unprecedented. After a couple of months in a tiny isolated cell we called Calcutta to let my arms heal, they blindfolded me and walked me right into the Las Vegas cell block. The isolation and special surveillance were over. I was put solo, of course, in the Mint.

Dave Hatcher knew I was back because I was walked under his window, and though he could not peek out, he could listen and over the years had attuned his ear to my walking "signature," my limping gait. Soon enough, the rusty wire over the sink in the washroom was bent to the north—Dave Hatcher's signal for "note in the bottle under the sink for Stockdale." Like an old fighter pilot, I checked my six o'clock, scooped the note up fast, and concealed it in my prison pajama pants, carefully. Back in my cell, after the guard locked the door, I sat on my toilet bucket—where I could stealthily jettison the note if the peephole cover moved—and unfolded Hatcher's sheet of low-grade paper toweling on which, with a rat dropping, he had printed, without comment or signature, the last verse of Ernest Henley's poem *Invictus*:

> *It matters not how strait the gate,*
> *How charged with punishment the scroll,*
> *I am the master of my fate:*
> *I am the captain of my soul.*

The Enchiridion
Epictetus

I

There are things which are within our power, and there are things which are beyond our power. Within our power are opinion, aim, desire, aversion, and, in one word, whatever affairs are our own. Beyond our power are body, property, reputation, office, and, in one word, whatever are not properly our own affairs.

Now the things within our power are by nature free, unrestricted, unhindered; but those beyond our power are weak, dependent, restricted, alien. Remember, then, that if you attribute freedom to things by nature dependent and take what belongs to others for your own, you will be hindered, you will lament, you will be disturbed, you will find fault both with gods and men. But if you take for your own only that which is your own and view what belongs to others just as it really is, then no one will ever compel you, no one will restrict you; you will find fault with no one, you will accuse no one, you will do nothing against your will; no one will hurt you, you will not have an enemy, nor will you suffer any harm.

Aiming, therefore, at such great things, remember that you must not allow yourself any inclination, however slight, toward the attainment of the others; but that you must entirely quit some of them, and for the present postpone the rest. But if you would have these, and possess power and wealth likewise, you may miss the latter in seeking the former; and you will certainly fail of that by which alone happiness and freedom are procured.

Seek at once, therefore, to be able to say to every unpleasing semblance, "You are but a semblance and by no means the real thing." And then examine it by those rules which you have; and first and chiefly by this: whether it concerns the things which are within our own power or those which are not; and if it concerns anything beyond our power, be prepared to say that it is nothing to you.

II

Remember that desire demands the attainment of that of which you are desirous; and aversion demands the avoidance of that to which you are averse; that he who fails of the object of his desires is disappointed; and he who incurs the object of his aversion is wretched. If, then, you shun only those undesirable things which you can control, you will never incur anything which you shun; but if you shun sickness, or death, or poverty, you will run the risk of wretchedness. Remove [the habit of] aversion, then, from all things that are not within our power, and apply it to things undesirable which are within our power. But for the present, altogether restrain desire; for if you desire any of the things not within your own power, you must necessarily be disappointed; and you are not yet secure of those which are within our power, and so are legitimate objects of desire. Where it is practically necessary for you to pursue or avoid anything, do even this with discretion and gentleness and moderation.

III

With regard to whatever objects either delight the mind or contribute to use or are tenderly beloved, remind yourself of what nature they are, beginning with the merest trifles: if you have a favorite cup, that it is but a cup of which you are fond of—for thus, if it is broken, you can bear it; if you embrace your child or your wife, that you embrace a mortal—and thus, if either of them dies, you can bear it.

IV

When you set about any action, remind yourself of what nature the action is. If you are going to bathe, represent to yourself the incidents usual in the bath—some persons pouring out, others pushing in, others scolding, others pilfering. And thus you will more safely go about this action if you say to yourself, "I will now go to bathe and keep my own will in harmony with nature." And so with regard to every other action. For thus, if any impediment arises in bathing, you will be able to say, "It was not only to bathe that I desired, but to keep my will in harmony with nature; and I shall not keep it thus if I am out of humor at things that happen."

V

Men are disturbed not by things, but by the views which they take of things. Thus death is nothing terrible, else it would have appeared so to Socrates. But the terror consists in our notion of death, that it is terrible. When, therefore, we are hindered or disturbed, or grieved, let us never impute it to others, but to ourselves—that is, to our own views. It is the action of an uninstructed person to reproach others for his own misfortunes; of one entering upon instruction, to reproach himself; and one perfectly instructed, to reproach neither others nor himself.

VI

Be not elated at any excellence not your own. If a horse should be elated, and say, "I am handsome," it might be endurable. But when you are elated and say, "I have a handsome horse," know that you are elated only on the merit of the horse. What then is your own? The use of the phenomena of existence. So that when you are in harmony with nature in this respect you will be elated with some reason; for you will be elated at some good of your own.

VII

As in a voyage, when the ship is at anchor, if you go on shore to get water, you may amuse yourself with picking up a shellfish or a truffle in your way, but your thoughts ought to be bent toward the ship, and perpetually attentive, lest the captain should call, and then you must leave all these things, that you may not have to be carried on board the vessel, bound like a sheep; thus likewise in life, if, instead of a truffle or shellfish, such a thing as a wife or a child be granted you, there is no objection; but if the captain calls, run to the ship, leave all these things, and never look behind. But if you are old, never go far from the ship, lest you should be missing when called for.

VIII

Demand not that events should happen as you wish; but wish them to happen as they do happen, and you will go on well.

IX

Sickness is an impediment to the body, but not to the will unless itself pleases. Lameness is an impediment to the leg, but not to the will; and say this to yourself with regard to everything that happens. For you will find it to be an impediment to something else, but not truly to yourself.

X

Upon every accident, remember to turn toward yourself and inquire what faculty you have for its use. If you encounter a handsome person, you will find continence the faculty needed; if pain, then fortitude; if reviling, then patience. And when thus habituated, the phenomena of existence will not overwhelm you.

XI

Never say of anything, "I have lost it," but "I have restored it." Has your child died? It is restored. Has your wife died? She is restored. Has your estate been taken away? That likewise is restored. "But it was a bad man who took it." What is it to you by whose hands he who gave it has demanded it again? While he permits you to possess it, hold it as something not your own, as do travelers at an inn.

XII

If you would improve, lay aside such reasonings as these: "If I neglect my affairs, I shall not have a maintenance; if I do not punish my servant, he will be good for nothing." For it were better to die of hunger, exempt from grief and fear, than to live in affluence with perturbation; and it is better that your servant should be bad than you unhappy.

Begin therefore with little things. Is a little oil spilled or a little wine stolen? Say to yourself, "This is the price paid for peace and tranquility; and nothing is to be had for nothing." And when you call your servant, consider that it is possible he may not come at your call; or, if he does, that he may not do what you wish. But it is not at all desirable for him, and very undesirable for you, that it should be in his power to cause you any disturbance.

XIII

If you would improve, be content to be thought foolish and dull with regard to externals. Do not desire to be thought to know anything; and though you should appear to others to be somebody, distrust yourself. For be assured, it is not easy at once to keep your will in harmony with nature and to secure externals; but while you are absorbed in the one, you must of necessity neglect the other.

XIV

If you wish your children and your wife and your friends to live forever, you are foolish, for you wish things to be in your power which are not so, and what belongs to others to be your own. So likewise, if you wish your servant to be without fault, you are foolish, for you wish vice not to be vice but something else. But if you wish not to be disappointed in your desires, that is in your own power. Exercise, therefore, what is in your power. A man's master is he who is able to confer or remove whatever that man seeks or shuns. Whoever then would be free, let him wish nothing, let him decline nothing, which depends on others; else he must necessarily be a slave.

XV

Remember that you must behave as at a banquet. Is anything brought round to you? Put out your hand and take a moderate share. Does it pass by you? Do not stop it. Is it not yet come? Do not yearn in desire toward it but wait till it reaches you. So with regard to children, wife, office, riches; and you will some time or other be worthy to feast with the gods. And if you do not so much as take the things which are set before you, but are able even to forego them, then you will not only be worthy to feast with the gods, but to rule with them also. For, by thus doing, Diogenes and Heraclitus, and others like them, deservedly became divine, and were so recognized.

XVI

When you see anyone weeping for grief, either that his son has gone abroad or that he has suffered in his affairs, take care not to be overcome by the apparent evil, but discriminate and be ready to say, "What hurts this man is not this occurrence itself—for another man might not be hurt by it—but the view he chooses to take of it." As far as conversation goes, however, do not disdain to accommodate yourself to him and, if need be, to groan with him. Take heed, however, not to groan inwardly, too.

XVII

Remember that you are an actor in a drama of such sort as the Author chooses—if short, then in a short one; if long, then in a long one. If it be his pleasure that you should enact a poor man, or a cripple, or a ruler, or a private citizen, see that you act it well. For this is your business—to act well the given part, but to choose it belongs to another.

XVIII

When a raven happens to croak unluckily, be not overcome by appearances, but discriminate and say, "Nothing is portended to me, either to my paltry body, or property, or reputation, or children, or wife. But to me all portents are lucky if I will. For whatsoever happens, it belongs to me to derive advantage therefrom."

XIX

You can be unconquerable if you enter into no combat in which it is not in your own power to conquer. When, therefore, you see anyone eminent in honors or power, or in high esteem on my other account, take heed not to be bewildered by appearances and to pronounce him happy; for if the essence of good consists in things within our own power, there will be no room for envy or emulation. But, for your part do not desire to be a general, or a senator, or a consul, but to be free; and the only way to this is a disregard of things which lie not within our own power.

XX

Remember that it is not he who gives abuse or blows, who affronts, but the view we take of these things as insulting. When, therefore, anyone provokes you, be assured that it is your own opinion which provokes you. Try, therefore, in the first place, not to be bewildered by appearances. For if you once gain time and respite, you will more easily command yourself.

XXI

Let death and exile, and all other things which appear terrible, be daily before your eyes, but death chiefly; and you will never entertain an abject thought, nor too eagerly covet anything.

XXII

If you have an earnest desire toward philosophy, prepare yourself from the very first to have the multitude laugh and sneer, and say, "He is returned to us a philosopher all at once"; and, "Whence this supercilious look?" Now, for your part do not have a supercilious look indeed, but keep steadily to those things which appear best to you, as one appointed by God to this particular station. For remember that, if you are persistent, those very persons who at first ridiculed will afterwards admire you. But if you are conquered by them, you will incur a double ridicule.

XXIII

If you ever happen to turn your attention to externals, for the pleasure of anyone, be assured that you have ruined your scheme of life. Be content, then, in everything, with being a philosopher; and if you wish to seem so likewise to anyone, appear so to yourself, and it will suffice you.

XXIV

Let not such considerations as these distress you: "I shall live in discredit and be nobody anywhere." For if discredit be an evil, you can no more be involved in evil through another than in baseness. Is it any business of yours, then, to get power or to be admitted to an entertainment? By no means. How then, after all, is this discredit? And how it is true that you will be nobody anywhere when you ought to be somebody in those things only which are within your own power, in which you may be of the greatest consequence? "But my friends will be unassisted." What do you mean by "unassisted?" They will not have money from you, nor will you make them Roman citizens. Who told you, then, that these are among the things within our own power, and not rather the affairs of others? And who can give to another the things which he himself has not? "Well, but get them, then, that we too may have a share." If I can get them with the preservation of my own honor and fidelity and self-respect, show me the way and I will get them; but if you require me to lose my own proper good, that you may gain what is no good, consider how unreasonable and foolish you are. Besides, which would you rather have, a sum of money or a faithful and honorable friend? Rather assist me, then, to gain this character than require me to do those things by which I may lose it. Well, but my country, say you, as far as depends upon me, will be unassisted. Here, again, what assistance is this you mean? It will not have porticos nor baths of your providing? And what signifies that? Why, neither does a smith provide it with shoes, nor a shoemaker with arms. It is enough if everyone fully performs his own proper business. And were you to supply it with another faithful and honorable citizen, would not he be of use to it? Yes. Therefore neither are you yourself useless to it. "What place, then," say you, "shall I hold in the state?" Whatever you can hold with the preservation of your fidelity and honor. But if, by desiring to be useful to that, you lose these, how can you serve your country when you have become faithless and shameless?

XXV

Is anyone preferred before you at an entertainment or in courtesies, or in confidential intercourse? If these things are good, you ought to rejoice that he has them; and if they are evil, do not be grieved that you have them not. And remember that you cannot be permitted to rival others in externals without using the same means to obtain them. For how can he who will not haunt the door of any man, will not attend him, will not praise him, have an equal share with him who does these things? You are unjust, then, and unreasonable if you are unwilling to pay the price for which these things are sold, and would have them for nothing. For how much are lettuces sold? An obulus, for instance. If any other, then, paying an obulus, takes the lettuces, and you, not paying it, go without them, do not imagine that he has gained any advantage over you. For as he has the lettuces, so you have the obulus which you did not give. So, in the present case, you have not been invited to such a person's entertainment because you have not paid him the price for which a supper is sold. It is sold for praise; it is sold for attendance. Give him, then, the value if it be for your advantage. But if you would at the same time not pay the one, and yet receive the other, you are unreasonable and foolish. Have you nothing, then, in place of the supper? Yes, indeed, you have—not to praise him whom you do not like to praise; not to bear the insolence of his lackeys.

XXVI

The will of nature may be learned from things upon which we are all agreed. As when our neighbor's boy has broken a cup, or the like, we are ready at once to say, "These are casualties that will happen"; be assured, then, that when your own cup is likewise broken, you ought to be affected just as when another's cup was broken. Now apply this to greater things. Is the child or wife of another dead? There is no one who would not say, "This is an accident of mortality." But if anyone's own child happens to die, it is immediately, "Alas! how wretched am I!" It should be always remembered how we are affected on hearing the same thing concerning others.

XXVII

As a mark is not set up for the sake of missing the aim, so neither does the nature of evil exist in the world.

XXVIII

If a person had delivered up your body to some passer-by, you would certainly be angry. And do you feel no shame in delivering up your own mind to any reviler, to be disconcerted and confounded?

XXIX

In every affair consider what precedes and what follows, and then undertake it. Otherwise you will begin with spirit, indeed, careless of the consequences, and when these are developed, you will shamefully desist. "I would conquer at the Olympic Games." But consider what precedes and what follows, and then, if it be for your advantage, engage in the affair. You must conform to rules, submit to a diet, refrain from dainties; exercise your body, whether you choose it or not at a stated hour, in heat and cold; you must drink no cold water, and sometimes no wine—in a word, you must give yourself up to your trainer as to a physician. Then, in the combat, you may be thrown into a ditch, dislocate your arm, turn your ankle, swallow an abundance of dust, receive stripes [for negligence], and, after all, lose the victory. When you have reckoned up all this, if your inclination still holds, set about the combat. Otherwise, take notice, you will behave like children who sometimes play wrestlers, sometimes gladiators, sometimes blow a trumpet and sometimes act a tragedy, when they happen to have seen and admired these shows. Thus you too will be at one time a wrestler, and another a gladiator; now a philosopher, now an orator; but nothing in earnest. Like an ape you mimic all you see, and one thing after another is sure to please you, but is out of favor as soon as it becomes familiar. For you have never entered upon anything considerately; nor after having surveyed and tested the whole matter, but carelessly, and with a halfway zeal. Thus some, when they have seen a philosopher and heard a man speaking like Euphrates—though, indeed, who can speak like him?—have a mind to be philosophers, too. Consider first, man, what the matter is, and what your own nature is able to bear. If you would be a wrestler, consider your shoulders, your back, your thighs; for different persons are made for different things. Do you think that you can act as you do and be a philosopher, that you can eat, drink, be angry, be discontented, as you are now? You must watch, you must labor, you must get the better of certain appetites, must quit your acquaintances, be despised by your servant, be laughed at by those you meet; come off worse than others in everything—in offices, in honors, before tribunals. When you have fully considered all these things, approach, if you please—that is, if, by parting with them, you have a mind to purchase serenity, freedom, and tranquillity. If not, do not come hither; do not like children, be now a philosopher, then a publican, then an orator, and then one of Caesar's officers. These things are not consistent. You must be one man, either good or bad. You must cultivate either your own reason or else externals; apply yourself either to things within or without you—that is, be either a philosopher or one of the mob.

XXX

Duties are universally measured by relations. Is a certain man your father? In this are implied taking care of him, submitting to him in all things, patiently receiving his reproaches, his correction. But he is a bad father. Is your natural tie, then, to a good father? No, but to a father. Is a brother unjust? Well, preserve your own just relation toward him. Consider not what he does, but what you are to do to keep your own will in a state conformable to nature, for another cannot hurt you unless you please. You will then be hurt when you consent to be hurt. In this manner, therefore, if you accustom yourself to contemplate the relations of neighbor, citizen, commander, you can deduce from each the corresponding duties.

XXXI

Be assured that the essence of piety toward the gods lies in this—to form right opinions concerning them, as existing and as governing the universe justly and well. And fix yourself in this resolution, to obey them, and yield to them, and willingly follow them amidst all events, as being ruled by the most perfect wisdom. For thus you will never find fault with the gods, nor accuse them of neglecting you. And it is not possible for this to be affected in any other way than by withdrawing yourself from things which are not within our own power, and by making good or evil to consist only in those which are. For if you suppose any other things to be either good or evil, it is inevitable that, when you are disappointed of what you wish or incur what you would avoid, you should reproach and blame their authors. For every creature is naturally formed to flee and abhor things that appear hurtful and that which causes them; and to pursue and admire those which appear beneficial and that which causes them. It is impracticable, then, that one who supposes himself to be hurt should rejoice in the person who, as he thinks, hurts him, just as it is impossible to rejoice in the hurt itself. Hence, also, a father is reviled by his son when he does not impart the things which seem to be good; and this made Polynices and Eteocles mutually enemies—that empire seemed good to both. On this account the husbandman reviles the gods; [and so do] the sailor, the merchant or those who have lost wife or child. For where our interest is, there, too, is piety directed. So that whoever is careful to regulate his desires and aversions as he ought is thus made careful of piety likewise. But it also becomes incumbent on everyone to offer libations and sacrifices and first fruits, according to the customs of his country, purely, and not heedlessly nor negligently; not avariciously, nor yet extravagantly.

XXXII

When you have recourse to divination, remember that you know not what the event will be, and you come to learn it of the diviner; but of what nature it is you knew before coming; at least if you are of philosophic mind. For if it is among the things not within our own power, it can by no means be either good or evil. Do not, therefore, bring with you to the diviner either desire or aversion—else you will approach him trembling—but first clearly understand that every event is indifferent and nothing to you, of whatever sort it may be; for it will be in your power to make a right use of it, and this no one can hinder. Then come with confidence to the gods as your counselors; and afterwards, when any counsel is given you, remember what counselors you have assumed, and whose advice you will neglect if you disobey. Come to divination as Socrates prescribed, in cases of which the whole consideration relates to the event, and in which no opportunities are afforded by reason or any other art to discover the matter in view. When, therefore, it is our duty to share the danger of a friend or of our country, we ought not to consult the oracle as to whether we shall share it with them or not. For though the diviner should forewarn you that the auspices are unfavorable, this means no more than that either death or mutilation or exile is portended. But we have reason within us; and it directs us, even with these hazards, to stand by our friend and our country. Attend, therefore, to the greater diviner, the Pythian God, who once cast out of the temple him who neglected to save his friend.

XXXIII

Begin by prescribing to yourself some character and demeanor, such as you may preserve both alone and in company.

Be mostly silent or speak merely what is needful, and in few words. We may, however, enter sparingly into discourse sometimes, when occasion calls for it; but let it not run on any of the common subjects, as gladiators, or horse races, or athletic champions, or food, or drink—the vulgar topics of conversation—and especially not on men, so as either to blame, or praise, or make comparisons. If you are able, then, by your own conversation, bring over that of your company to proper subjects; but if you happen to find yourself among strangers, be silent.

Let not your laughter be loud, frequent, or abundant.

Avoid taking oaths, if possible, altogether; at any rate, so far as you are able.

Avoid public and vulgar entertainments; but if ever an occasion calls you to them, keep your attention upon the stretch, that you may not imperceptibly slide into vulgarity. For be assured that if a person be ever so pure himself, yet if his companion be corrupted, he who converses with him will be corrupted likewise.

Provide things relating to the body no further than absolute need requires, as meat, drink, clothing, house, retinue. But cut off everything that looks toward show and luxury. Before marriage guard yourself with all your ability from unlawful intercourse with women; yet be not uncharitable or severe to those who are led into this, nor boast frequently that you yourself do otherwise.

If anyone tells you that a certain person speaks ill of you, do not make excuses about what is said of you, but answer: "He was ignorant of my other faults, else he would not have mentioned these alone,"

It is not necessary for you to appear often at public spectacles; but if ever there is a proper occasion for you to be there, do not appear more solicitous for any other than for yourself—that is, wish things to be only just as they are, and only the best man to win; for thus nothing will go against you. But abstain entirely from acclamations and derision and violent emotions. And when you come away, do not discourse a great deal on what has passed and what contributes nothing to your own amendment. For it would appear by such discourse that you were dazzled by the show.

Be not prompt or ready to attend private recitations; but if you do attend, preserve your gravity and dignity, and yet avoid making yourself disagreeable.

When you are going to confer with anyone, and especially with one who seems your superior, represent to yourself how Socrates or Zeno would behave in such a case, and you will not be at a loss to meet properly whatever may occur.

When you are going before anyone in power, fancy to yourself that you may not find him at home, that you may be shut out, that the doors may not be opened to you, that he may not notice you. If, with all this, it be your duty to go, bear what happens and never say to yourself, "It was not worth so much"; for this is vulgar, and like a man bewildered by externals.

In company, avoid a frequent and excessive mention of your own actions and dangers. For however agreeable it may be to yourself to allude to the risks you have run, it is not equally agreeable to others to hear your adventures. Avoid likewise an endeavor to excite laughter, for this may readily slide you into vulgarity, and, besides, may be apt to lower you in the esteem of your acquaintance. Approaches to indecent discourse are likewise dangerous. Therefore, when anything of this sort happens, use the first fit opportunity to rebuke him who makes advances that way, or, at least, by silence and blushing and a serious look show yourself to be displeased by such talk.

XXXIV

If you are dazzled by the semblance of any promised pleasure, guard yourself against being bewildered by it; but let the affair wait your leisure, and procure yourself some delay. Then bring to your mind both points of time—that in which you shall enjoy the pleasure, and that in which you will repent and reproach yourself, after you have enjoyed it—and set before you, in opposition to these, how you will rejoice and applaud yourself if you abstain. And even though it should appear to you a seasonable gratification, take heed that its enticements and allurements and seductions may not subdue you, but set in opposition to this how much better it is to be conscious of having gained so great a victory.

XXXV

When you do anything from a clear judgment that it ought to be done, never shrink from being seen to do it, even though the world should misunderstand it; for if you are not acting rightly, shun the action itself; if you are, why fear those who wrongly censure you?

XXXVI

As the proposition, "either it is day or it is night," has much force in a disjunctive argument, but none at all in a conjunctive one, so, at a feast, to choose the largest share is very suitable to the bodily appetite, but utterly inconsistent with the social spirit of the entertainment. Remember, then, when you eat with another, not only the value to the body of those things which are set before you, but also the value of proper courtesy toward your host.

XXXVII

If you have assumed any character beyond your strength, you have both demeaned yourself ill in that and quitted one which you might have supported.

XXXVIII

As in walking you take care not to tread upon a nail, or turn your foot, so likewise take care not to hurt the ruling faculty of your mind. And if we were to guard against this in every action, we should enter upon action more safely.

XXXIX

The body is to everyone the proper measure of its possessions, as the foot is of the shoe. If, therefore, you stop at this, you will keep the measure; but if you move beyond it, you must necessarily be carried forward, as down a precipice; as in the case of a shoe, if you go beyond its fitness to the foot, it comes first to be gilded, then purple, and then studded with jewels. For to that which once exceeds the fit measure there is no bound.

XL

Women from fourteen years old are flattered by men with the title of mistresses. Therefore, perceiving that they are regarded only as qualified to give men pleasure, they begin to adorn themselves, and in that to place all their hopes. It is worth while, therefore, to try that they may perceive themselves honored only so far as they appear beautiful in their demeanor and modestly virtuous.

XLI

It is a mark of want of intellect to spend much time in things relating to the body, as to be immoderate in exercises, in eating and drinking, and in the discharge of other animal functions. These things should be done incidentally and our main strength be applied to our reason.

XLII

When any person does ill by you, or speaks ill of you, remember that he acts or speaks from an impression that it is right for him to do so. Now it is not possible that he should follow what appears right to you, but only what appears so to himself. Therefore, if he judges from false appearances, he is the person hurt since he, too, is the person deceived. For if anyone takes a true proposition to be false, the proposition is not hurt, but only the man is deceived. Setting out then, from these principles, you will meekly bear with a person who reviles you, for you will say upon every occasion, "It seemed so to him."

XLIII

Everything has two handles: one by which it may be borne, another by which it cannot. If your brother acts unjustly, do not lay hold on the affair by the handle of his injustice, for by that it cannot be borne, but rather by the opposite—that he is your brother, that he was brought up with you; and thus you will lay hold on it as it is to be borne.

XLIV

These reasonings have no logical connection: "I am richer than you, therefore I am your superior." "I am more eloquent than you, therefore I am your superior." The true logical connection is rather this: "I am richer than you, therefore my possessions must exceed yours." "I am more eloquent than you, therefore my style must surpass yours." But you, after all, consist neither in property nor in style.

XLV

Does anyone bathe hastily? Do not say that he does it ill, but hastily. Does anyone drink much wine? Do not say that he does ill, but that he drinks a great deal. For unless you perfectly understand his motives, how should you know if he acts ill? Thus you will not risk yielding to any appearances but such as you fully comprehend.

XLVI

Never proclaim yourself a philosopher, nor make much talk among the ignorant about your principles, but show them by actions. Thus, at an entertainment do not discourse how people ought to eat, but eat as you ought. For remember that thus Socrates also universally avoided all ostentation. And when persons came to him and desired to be introduced by him to philosophers, he took them and introduced them; so well did he bear being overlooked. So if ever there should be among the ignorant any discussion of principles, be for the most part silent. For there is great danger in hastily throwing out what is undigested. And if anyone tells you that you know nothing, and you are not nettled at it, then you may be sure that you have really entered on your work. For sheep do not hastily throw up the grass to show the shepherds how much they have eaten, but inwardly digesting their food, they produce it outwardly in wool and milk. Thus, therefore, do you not make an exhibition before the ignorant of your principles, but of the actions to which their digestion gives rise.

XLVII

When you have learned to nourish your body frugally, do not pique yourself upon it; nor, if you drink water, be saying upon every occasion, "I drink water." But first consider how much more frugal are the poor than we, and how much more patient of hardship. If at any time you would inure yourself by exercise to labor and privation, for your own sake and not for the public, do not attempt great feats; but when you are violently thirsty, just rinse your mouth with water, and tell nobody.

XLVIII

The condition and characteristic of a vulgar person is that he never looks for either help or harm from himself, but only from externals. The condition and characteristic of a philosopher is that he looks to himself for all help or harm. The marks of a proficient are that he censures no one, praises no one, blames no one, accuses no one; says nothing concerning himself as being anybody or knowing anything. When he is in any instance hindered or restrained, he accuses himself; and if he is praised, he smiles to himself at the person who praises him; and if he is censured, he makes no defense. But he goes about with the caution of a convalescent, careful of interference with anything that is doing well but not yet quite secure. He restrains desire; he transfers his aversion to those things only which thwart the proper use of our own will; he employs his energies moderately in all directions; if he appears stupid or ignorant, he does not care; and, in a word, he keeps watch over himself as over an enemy and one in ambush.

XLIX

When anyone shows himself vain on being able to understand and interpret the works of Chrysippus, say to yourself: "Unless Chrysippus had written obscurely, this person would have had nothing to be vain of. But what do I desire? To understand nature, and follow her. I ask, then, who interprets her; and hearing that Chrysippus does, I have recourse to him. I do not understand his writings. I seek, therefore, one to interpret them." So far there is nothing to value myself upon. And when I find an interpreter, what remains is to make use of his instructions. This alone is the valuable thing. But if I admire merely the interpretation, what do I become more than a grammarian, instead of a philosopher, except indeed, that instead of Homer I interpret Chrysippus? When anyone, therefore, desires me to read Chrysippus to him, I rather blush when I cannot exhibit actions that are harmonious and consonant with his discourse.

L

Whatever rules you have adopted, abide by them as laws, and as if you would be impious to transgress them; and do not regard what anyone says of you, for this, after all, is no concern of yours. How long, then, will you delay to demand of yourself the noblest improvements, and in no instance to transgress the judgments of reason? You have received the philosophic principles with which you ought to be conversant; and you have been conversant with them. For what other master, then, do you wait as an excuse for this delay in self-reformation? You are no longer a boy but a grown man. If, therefore, you will be negligent and slothful, and always add procrastination to procrastination, purpose to purpose, and fix day after day in which you will attend to yourself, you will insensibly continue to accomplish nothing and, living and dying, remain of vulgar mind. This instant, then, think yourself worthy of living as a man grown up and a proficient. Let whatever appears to be the best be to you an inviolable law. And if any instance of pain or pleasure, glory or disgrace, be set before you, remember that now is the combat, now the Olympiad comes on, nor can it be put off; and that by one failure and defeat honor may be lost or—won. Thus Socrates became perfect, improving himself by everything, following reason alone. And though you are not yet a Socrates, you ought, however, to live as one seeking to be a Socrates.

Roman Stoicism
*Shannon E. French**

The Romans took great pride in their practical achievements. They were successful and efficient architects, engineers, soldiers, administrators and statesmen. For many educated Romans, the more romantic or spiritual aspects of the accepted Greco-Roman religion, including its vision of life-after-death, were difficult to swallow. In addition, there was little moral guidance to be found in the behavior of the Olympic gods and goddesses. So for those who wished to wrestle further with questions of ethics and the meaning of human existence, the Roman philosophical community offered alternative schools of thought.

Two of these schools of thought made available to disenchanted Romans were again Greek imports: stoicism and hedonism. Both philosophies rejected the tenets of the standard Greco-Roman religion, including the belief in life-after-death. They focused instead on how to live the good life here on Earth. However, beyond their aim and earthly focus, these views had little in common.

In his book *The Romans*, historian R. H. Barrows comments " . . . the Romans were natural Stoics long before they heard of Stoicism."[1] He notes that Roman philosophers generally did not attempt to create their own original theories, but rather had a tendency to shop around established doctrines, selecting and assembling points that appealed to them:

> No Roman adopted the whole of any philosophy; some parts did not interest him, other parts he adapted to his own instinctive beliefs and found in them a statement of what he had never clearly articulated for himself. It may perhaps be an exaggeration to say that the Roman adopted only what suited his Roman ideals, for undoubtedly philosophical studies did influence the conduct and outlook of many. But certainly the Roman was not greatly interested in the coherence of a system, or in pursuing the fundamental questions of metaphysics. He was interested primarily in action and its springs and justification. Hence Roman philosophy is largely eclectic, and it is concerned chiefly with morals.[2]

Given the Roman habit of evaluating moral theories by the "gut-check" method, it may help us to understand their attraction to stoicism if we begin with a sketch of the Roman character. In a valuable summary of Roman ideals, Barrows highlights the qualities of character that the Romans most admired:

> *Gravitas* means a 'sense of the importance of the matters in hand', a sense of responsibility and earnestness. It is a term to apply at all levels—to a statesman or a general as he shows appreciation of his responsibilities, to a citizen as he casts his vote with

---

*This piece is an excerpt from Chapter Three of Shannon E. French's book, *The Code of the Warrior: Exploring Warrior Values, Past and Present* (Rowman and Littlefield Publishers, 2003).

[1] R. H. Barrows, *The Romans* (1949; reprint, New York: Penguin, 1986), 151.

[2] Barrows, *The Romans*, 152.

consciousness of its importance, to a friend who gives his advice based on his experience and on regard for your welfare; ... It is the opposite of *levitas*, a quality the Romans despised, which means trifling when you should be serious, flippancy, instability. *Gravitas* is often joined with *constantia*, firmness of purpose, or with *firmitas*, tenacity; it may be seasoned with *comitas*, which means the relief given to overseriousness by ease of manner, good humour, and humour. *Disciplina* is the training which provides steadiness of character; *industria* is hard work; *virtus* is manliness and energy; *clementia* the willingness to forgo one's rights; *frugalitas*, simple tastes.[3]

Keeping in mind this construction of Roman virtues, let us now consider the central points of stoic ethics as they may have appealed to the Roman mind.

The philosopher Zeno, who taught in Athens until his death in 263 B.C, is credited with having founded the stoic school of philosophy. Zeno had a habit of lecturing to his students from the front porch, or *stoa*, and it is from this that the word "stoicism" is derived. By all accounts Zeno was a charismatic teacher who persuaded many followers to adopt stoic principles. Along with his most famous pupil, Cleanthes, Zeno was viewed by his contemporaries as something of a political or religious activist.[4]

Zeno challenged the conventional wisdom, which held that any human being's hope for happiness rested entirely in the hands of fate or some other unyielding and amoral divine power. It was thought by some that Zeus himself, ruler of all the Olympic gods, whimsically parceled out unequal portions of joy and sorrow to all the poor, helpless mortals who were his playthings.

While Zeno did believe firmly in fate, he did not believe it, or the callousness of any imagined deity, to be responsible for an individual's happiness. He told his students that only they themselves could be held responsible for their emotional experience of life. He taught that the key to both happiness and virtue lay in understanding the nature of control.

As rational creatures, we naturally want to control our lives and organize our world so that we get to enjoy as many pleasant moments as possible while avoiding that which is unpleasant. Generally speaking, we all want to possess the same sorts of goods, such as health, love, friendship, success, wealth and respect, while avoiding the evils of pain, illness, death, loneliness, failure, poverty and ignominy. We tend to envy the man or woman who is presented with the many opportunities for delight and pity the one who seems overburdened with occasions for sorrow.

Unfortunately, we seem to exercise little control over the goods that we crave. I can try to take care of my physical self, but no amount of healthy living will shield me from all injury or disease, and my eventual death is a certainty. Nor can I protect the people that I love from illness, injury, or death. I cannot be certain that those I love will love me back or that I will ever find a mate with whom I am well matched. The ultimate success of many of my projects depends on factors beyond my control, as does my financial status. Even my reputation can be ruined through no fault of my own if my enemies choose to malign me.

Rather than concluding from these dismal points that my happiness is completely out of my own hands, Zeno would have urged me to reconsider where and on what I have chosen to pin my happiness. The central tenet of stoicism is that it is in your power to direct whether or not you live a good life, a life worth living. But first you must recognize that what constitutes a good life is not acquisition of those goods found outside yourself which you cannot control. Instead, living the good life depends on cultivating those internal goods that can be made immune to the unfeeling machinations of fate.

According to stoicism, the one thing we always control is our own will. And it is by our will that we decide how to respond to events in our lives. We have exclusive rule over our inner selves: our mind and emotions. None of us can control whether or not our bodies will remain healthy. However, Zeno assured his students, if we are struck down by an ailment, the way we allow that fate to affect our inner selves is entirely up to us.

---

[3] Barrows, *The Romans*, 22–23.

[4] Philip P. Hallie, "Epictetus," in the *Encyclopedia of Philosophy*, Vol. 3, New York: Macmillan Publishing Co., Inc., 1967, p. 1.

Stoics hold the view that nothing that happens to you has the power to make you unhappy unless you choose to let it. Unhappiness comes from wanting what you cannot have. Happiness is achieved by willing yourself to be satisfied with whatever external goods destiny decides to bestow upon you while focusing your energy on improving your inner self so that you can always enjoy those internal goods which no external force can strip away from you, such as your own good character, honor and integrity.

Stoic philosophy was already an influence in Rome in the last days of the Republic, before the rise of Julius Caesar and his nephew Augustus. One of the most prominent early proponents of stoicism was the influential orator and statesman, Marcus Tullius Cicero. As a man dedicated to public life (he was a successful lawyer who argued many famous cases before the Roman senate, he held the high office of Consul of Rome in 63 B.C. and put down an attempted *coup d'etat*, and he served from 51 to 50 B.C. as governor of the Roman province of Cilicia in Asia Minor), Cicero was attracted to the stoic emphasis on order, duty, and self-discipline. He wrote several essays promoting stoic ethics, including *On Duties, The Boundaries of Good and Evil, On Friendship, On the Laws* and *On the State*.

Building on the lessons of the early stoa, Cicero's essays emphasize that the greatest loss that anyone can suffer is the loss of character, the lapse of virtue. Since there is no life after death in the stoic vision, the time that we have on earth is our only opportunity to define ourselves. We can choose to exhibit nobility of character by always doing our moral duty and behaving honorably towards our fellow human beings, or else we can abandon the pursuit of virtue, undercut others to achieve our own selfish ends, and expend our energies on the pursuit of meaningless pleasures and transient goods. Not only is the noble life morally superior, it is ultimately more satisfying, since the goals we set for our own character development are within our power to realize whereas, as Zeno taught, our quests for externally-based happiness are likely to be frustrated by twists of fate.

Explaining, "How to Make the Right Decisions," Cicero wrote:

> . . . a man who wrongs another for his own benefit either imagines, presumably, that he is not doing anything unnatural, or he does not agree that death, destitution, pain, the loss of children, relations, friends, are less deplorable than doing wrong to another person. But if he sees nothing unnatural in wronging a fellow-being, how can you argue with him? —he is taking away from man all that makes him man. If, however, he concedes that this ought to be avoided, but regards death, destitution, and pain as even more undesirable, he is mistaken in believing that *any* damage, either to his person or his property, is worse than moral failure.[5]

This passage of Cicero's highlights the demanding nature of stoic ethics. The dedicated stoic must be so committed to his moral duty that no physical or emotional distractions will be able to sway him or her from doing what is right. Not even the loss of a loved one can be claimed as justification for the most minor moral infraction.

There were several other well-known stoics of the Roman era who attempted to embody the principles they advocated despite serious personal trials. Three of the most famous of these were Seneca the Younger, Epictetus, and the Emperor Marcus Aurelius. Some have questioned Seneca's successful adherence to his stoic beliefs, but most historians agree that the lives of Epictetus and Aurelius did in fact do justice to the ideals they embraced.

Seneca the Younger lived from 4 B.C. to 65 A.D. and served as an advisor to two of the Julio-Claudian emperors: Claudius and Nero. In Nero's youth, Seneca also played the role of a tutor, although he apparently failed to indoctrinate the future tyrant with any stoic principles. Seneca was a prolific writer, producing numerous tragic plays, essays, letters, and epigrams, many of which have been preserved. All of his works, whether fiction or non-fiction, feature stoic themes.

Seneca paid particular attention in his stoic writing to the subjects of fate and death. He drew an analogy between a human life in the hands of fate and a dog being walked on a leash. The dog on the leash has two choices. He is going to be taken from point A to point B, that much is unavoidable. But

---

[5]Michael Grant, *Latin Literature, An Anthology*, New York: Penguin Books, 1958, p.35.

it is up to him whether he fights the leash and forces his trainer to drag him from A to B, protesting painfully all along the way or whether he chooses instead to cover the distance in a calm and orderly fashion, accepting without question his trainer's view that to go from A to B is the most rational thing for him to do. For humans, fate is the leash; and God, understood as the source of all logic and order in the universe, is the trainer.

When Seneca writes about death, he insists that death is not an evil. If there is no life-after-death, then the state of being dead is simply the state of non-existence. There is no self left after death to experience anything at all, negative or positive. Suffering requires existence; therefore death is freedom from suffering. Those who fear death, Seneca contends, are deluded by their mistaken belief in an afterlife.

Seneca staunchly supported the stoic belief that taking one's own life is warranted if the only alternative is to fail one's moral duty in some way. Since death is nothing, a true stoic should never allow the fear of death to drive him or her to act dishonorably. Suicide should always be preferable to forsaking one's commitments, even under duress. In the following two excerpts from his *Letters*, Seneca first explains why a virtuous person should be willing to commit suicide and then provides examples of two suicides undertaken with the proper attitude, reflecting the prioritizing of one's character above one's physical self:

> ... I will not allow any wound to penetrate through the body to the real me. My body is that part of me which can be injured; but within this fragile dwelling-place lives a soul which is free. And never will that flesh drive me to fear, never to a role which is unworthy of a good man. Never will I tell lies for the sake of this silly little body. Whenever it seems the right time, I will end my partnership with the body.[6]

> Recently, ... a German who was destined to be one of the wild animal fighters at a public entertainment was preparing for the morning show. He withdrew from the rest for a moment to relieve himself (he was given no other opportunity to withdraw without a guard.). There, in the toilet area, he found a wooden stick with a sponge attached to the end (it was used for wiping away the excrement). He stuffed the whole thing down his throat and choked to death... Though apparently without any resources, he devised both a method and a means of death. From his example you, too, can learn that the only thing which makes us hesitate to die is the lack of will.... Recently, again, when a man was being carted off under close guard to the morning show, he pretended to nod his head in sleep. Then he lowered his head until he had stuck it between the spokes of the cartwheel, and remained calmly in his seat until his neck was broken by the turning wheel. And so, he used the very vehicle which was carrying him to punishment to escape it.[7]

The examples that he gives are both of men opting to die at their own hands rather than permit their deaths to become base spectacles for the entertainment of the Roman masses. The German to whom he refers is no doubt a prisoner of war from the Roman campaigns against the Germanic tribes. Romans would commonly parade their captured opponents before their own cheering populations as an act of pro-war propaganda and then force them to fight wild beasts (or each other) in an arena, surrounded by jeering crowds baying for their blood.

The stoic wisdom Seneca presented was the view that nothing or no one can do damage to your character without your complicity. Fate controls many things, but not your own possession, or lack, of virtue (*virtus*). In his *Essay about Constancy*, Seneca writes:

> The wise man cannot suffer injury or loss, because he keeps all his "valuables" within himself and trusts nothing to Fortune. He has "goods" which are safe and secure because he finds satisfaction in *virtus*, which does not depend on chance occurrences.... Fortune does not snatch away what she herself has not given; and certainly

---

[6]Shelton, *As the Romans Did*, 435.

[7]Shelton, *As the Romans Did*, 350.

> *virtus* is not a gift of Fortune, so it cannot be taken from us. *Virtus* is free, inviolable, immovable, unshaken, and so steeled against the blows of chance that it cannot be bent, much less toppled. It looks straight at the instruments of torture and does not flinch; its expression never changes, whether adversity or prosperity comes into its view. The wise man therefore will lose nothing which he might perceive as a loss, for his only possession is *virtus*, and he can never be separated from it. He treats everything else like someone else's property, and who can be distressed by the loss of things which are not yours? And so it follows that if injury can do no harm to the things which truly belong to the wise man, and if his things are secure because his *virtus* is secure, then injury cannot harm the wise man. . . .[8]

Seneca's position is that we are ourselves fully responsible for developing the excellence of our own characters. This should be the central project of our lives. And he advises in his *Letters* that we take advantage of the few peaceful moments we are given to prepare for the more demanding ones:

> The soul should use times of security to prepare itself for harsh circumstances. It should fortify itself, when enjoying the blessings of Fortune, against the blows of Fortune. A soldier practices maneuvers during peacetime and constructs defensive ramparts, although no enemy is near, and wearies himself with nonessential exertion so that he can be ready for necessary exertion. If you don't want someone to panic in a crisis, you must train him before the crisis. And people who simulate poverty every month and come close to real need are following the same plan as the soldier, so that they will never panic at what they have often learned to deal with.[9]

The great stoic writer Epictetus, who was born about five or ten years before Seneca's death (historians dispute the date of his birth), actually appears to have employed all of the demanding precepts of stoicism in the direction of his own life, despite (or perhaps because of) the many challenges he faced. Born the son of a slave mother, Epictetus was himself enslaved to a freedman by the name of Epaphroditus who may have been the Emperor Nero's administrative secretary. It was Epaphroditus who sent Epictetus to be educated by a philosopher named C. Musonius Rufus who was a powerful proponent of the stoic perspective.[10] Ironically, Epaphroditus then provided the first testing grounds for Epictetus' stoicism by punishing and even torturing him for no reason other than his own sadistic pleasure. According to one story, Epictetus was left with a permanently lame leg after Epaphroditus twisted it with such force that he destroyed the knee joint.

It is not clear how or why, but Epictetus was eventually set free and became a teacher of stoic philosophy in Rome. He taught there for many years until he was expelled in 89 A.D. by the Emperor Domitian who viewed stoicism as subversive. (Epictetus got off lightly; in 93 A.D. Domitian had several "subversive" stoics executed.) Epictetus then moved to Nicopolis in northwest Greece, where he set up his own school of philosophy and continued to advance stoic views. He never wrote down any of his teachings, but notes from his lectures taken by one of his students were preserved and are now known as the *Enchiridion*, or Handbook, of Epictetus. Providing a portrait of the philosopher for the Encyclopedia of Philosophy, Philip Hallie describes Epictetus as " . . . a man of great sweetness, as well as personal simplicity, who was humble, charitable, and especially loving towards children, but he was also possessed of great moral and religious intensity."[11]

[*Note:* **For details of Epictetus's stoic thought, please see the readings in this volume from his *Enchiridion*.**]

Epictetus is believed to have died in 120 A.D., one year before the birth of the future emperor, Marcus Aurelius, whose commitment to stoicism would be equally sincere. Marcus certainly had many opportunities to apply Epictetus' advice regarding proper stoic attitudes of emotional detachment. He

---

[8]Shelton, *As the Romans Did*, 434.

[9]*As the Romans Did*, 424.

[10]Hallie, "Epictetus," 3:1.

[11]Hallie, "Epictetus," 3:1.

was orphaned at a young age, plagued by health problems all of his life, and was pre-deceased not only by his beloved (though by some accounts repeatedly unfaithful) wife but also by four of his five sons. During his reign Marcus had to contend with barbarian invasions along the northeastern and eastern borders of the empire, a devastating plague that swept through the entire empire, and an attempted *coup d'etat* led by Avidius Cassius, commander of the Roman forces in Asia (which was brought to a swift end when Avidius was murdered by his own subordinate officers). Duty required Marcus to be a general to his legions and an administrator of the most complex bureaucracy the world had ever seen, when he would rather have lived a quiet life of contemplation and philosophical reflection. He is described as "By nature a saint and a sage, by profession a ruler and a warrior."[12]

Marcus Aurelius spelled out his understanding of stoicism in his private *Meditations*. Much of the work was written while Marcus was on military campaign, defending the Roman frontiers. As historian R.H. Barrows explains:

> ... all unwillingly he [Marcus] shouldered his duty to turn himself from a meditative student into the commander of an army defending the Northern frontier of the Roman Empire. And conscientiously and successfully he did it. But at times he withdrew into himself; and, as he fought with some of his problems in the melancholy places of his mind, he jotted down his musings and wrestlings and resolutions, and by some queer accident his jottings have come down to us.[13]

Some of the emperor's reflections deal expressly with the value of stoicism for those in the profession of arms and many of his points are illustrated with martial examples.

It should not be difficult to see why stoic philosophy appealed to those charged with the training and leadership of Rome's legions. As Maxwell Staniforth notes in the introduction to his translation of Marcus Aurelius' *Meditations*, stoicism offered the ideal code for the Roman warrior: "A code which was manly, rational, and temperate, a code which insisted on just and virtuous dealing, self-discipline, unflinching fortitude, and complete freedom from the storms of passion was admirably suited to the Roman character."[14] Emperor Aurelius' interpretation of stoicism is especially practical, as he explains how to apply the "no excuses" ethic in daily life.

Book seven of his *Meditations* contains some of Marcus's most powerful statements of the stoic doctrine of self-regulation. The ideal stoic warrior, according to the emperor, would not allow any disturbance in life to provoke him to act in a way that would taint his personal honor. Again, even the experience of evil cannot supply sufficient warrant for any moral transgression, however slight.

Marcus suggests that the way to make oneself immune to the influence of evil is to see it as both commonplace and impermanent. Encounters with evil should not rattle our composure, because they are an expected feature of our earthly existence. Why should we be thrown if, for example, we find ourselves betrayed? Acts of betrayal are hardly rare in human history. And the effects of that betrayal, whatever they may be, cannot last forever. So Marcus writes:

> 1. What is evil? A thing you have seen times out of number. Likewise with every other sort of occurrence also, be prompt to remind yourself that this, too, you have witnessed many times before. For everywhere, above and below, you will find nothing but the selfsame things; they fill the pages of all history, ancient, modern, and contemporary; and they fill our cities and homes today. There is no such thing as novelty; all is as trite as it is transitory.[15]

When events do threaten to disturb our composure, Marcus advises us to employ mental discipline to remind ourselves how we felt before the world conspired to alter our perspective. Then we are, as a sheer act of will, to cause ourselves to return to our former way of viewing matters, as though nothing unpleasant had occurred. To the person who can master this technique, Marcus proclaims, "A

---

[12] Maxwell Staniforth, Marcus Aurelius' *Meditations*, New York: Penguin Books, 1964, p. 22.

[13] Barrow, *The Romans*, 156.

[14] Barrow, *The Romans*, 10.

[15] Staniforth, trans., Marcus Aurelius' *Meditations*, Book Seven, p. 105 (Penguin 1964).

new life lies within your grasp. You have only to see things once more in the light of your first and earlier vision, and life begins anew."[16]

Consider how this advice might have played out in the life of a Roman legionnaire. The Roman Empire maintained a fighting force of approximately 30 legions[17], with 5,400 men in each legion. These men were stationed in all corners of the empire, from Egypt to Gaul, often thousands of miles away from where they were recruited. Their training was intense, and they were committed to serve for at least 25 years. They fought in tight formations, where they learned to rely on one another implicitly. The ancient historian Josephus recorded his observations of the Roman army in his *History of the Jewish War*:

> If you study very carefully the organization of the Roman army, you will realize that they possess their great empire as a reward for valor, not as a gift of fortune. For the Romans, the wielding of arms does not begin with the outbreak of war, nor do they sit idly by in peacetime and move their hands only during times of need. Quite the opposite! As if born for the sole purpose of wielding arms, they never take a break from training, never wait for a situation requiring arms. Their practice sessions are no less strenuous than real battles. Each soldier trains every day with all his energy as if in war. And therefore they bear the stress of battle with the greatest ease. No confusion causes them to break from their accustomed formation, no fear causes them to shrink back, no exertion tires them. Certain victory always attends them since their opponents are never equal to them. And so it would not be wrong to call their practice sessions bloodless battles and their battles bloody practice sessions.[18]

Given the harsh conditions of their service and the fact that they were often torn from the familiar and asked to defend Roman interests in what must have seemed to them disturbingly alien lands, it is not surprising that the legionnaires carved out some sense of stability in their lives by forging close-knit friendships with their comrades-in-arms. Though clearly beneficial, these bonds among the empire's professional soldiers also left them open to the pain of constant bereavement. The Roman legions were a superior fighting force, seldom failing to meet their objectives. But their successes came at a price. The average Roman soldier had less than a 50% chance of surviving until retirement.[19]

The wisdom of Marcus Aurelius and his stoic predecessors would have directed the legionaries not to allow the loss of their comrades to affect their performance in battle or the completion of any of their professional obligations. Seeing your best friend skewered by a Gaulish spear would be no excuse for falling out of step in an advance or failing to keep an adequate watch the following day. According to the stoics, it is your choice whether or not to let any experience distract you from your duties. Emotional disturbance is just an indication of weakness of will.

Marcus provides a useful image for aspiring stoics to hold in their minds to keep them focused on maintaining their personal honor and behaving virtuously (performing their moral duty) no matter what occurs around or to them and with no regard for the reactions of others:

> 15. Whatever the world may say or do, my part is to keep myself good; just as . . . an emerald . . . insists perpetually, "Whatever the world may say or do, my part is to remain and emerald and keep my color true."[20]

---

[16] Op. Cit.

[17] There were 28 legions under the Emperor Augustus, after Varus' defeat by the German "barbarians" the number of legions fell to 25, and up to 33 for a brief time under the Emperor Septimus Severus. In the reign of Marcus Aurelius, there were 30 legions. (For a very clear account of the fluctuation of legion numbers please see Colin Wells' *The Roman Empire*, Chapter VI: "The Army and the Provinces in the First Century AD," Stanford: Stanford University Press, 1984.),

[18] Josephus, *A History of the Jewish War*, 3.71–97, Shelton p. 260.

[19] Those who did make it to retirement, however, were given a fairly generous pension, their own parcel of land on which to live and farm, and of course all the benefits of Roman citizenship. Please see again Colin Wells, *The Roman Empire*, pps. 136–140.

[20] Staniforth, trans., Marcus Aurelius' *Meditations*, Book Seven, p. 107 (Penguin 1964).

Nor does the emperor hesitate to spell out the military applications of this stoic instruction to remain true in all situations and at all costs. Always he returns to the theme that to endure physical pain and/or death is vastly preferable to suffering the loss of honor that must necessarily attend any act of cowardice or dereliction of duty. For example, to emphasize the need for the stoic warrior to set aside any concern for his own well being when military objects are at stake, he draws on a quote from the Greek philosopher Plato, whom he holds in high regard for recognizing the moral importance of service to the state:

> 45. For thus it is, men of Athens, in truth: wherever a man has placed himself thinking it the best place for him, or has been placed by a commander, there in my opinion he ought to stay and to abide the hazard, taking nothing into the reckoning, either death or anything else, before the baseness of deserting his post.[21]

Marcus' reflections on the tolerance of pain also imply that he would expect the stoic warrior to resist the mental effects of torture if captured by an enemy. A true stoic would not permit the threat or actual infliction of physical agony to compel him to compromise his integrity. Marcus comments:

> 33. Of Pain. If it is past bearing, it makes an end of us; if it lasts, it can be borne. The mind, holding itself aloof from the body, retains its calm, and the master-reason remains unaffected. As for the parts injured by the pain, let them, if they can, declare their own grief.[22]

In other words, the emperor's position on pain is that it must fall into one of two categories: either it kills you (and recall the stoic view that death is nothing to fear) or it can, technically, be tolerated. He does not entertain the possibility that there could be some form of non-lethal but intolerable pain which no human will could resist. He trusts that strict mental discipline will fortify the stoic against all assaults on his physical self. So he advises,

> 68. Live out your days in untroubled serenity, refusing to be coerced though the whole world deafen you with its demands, and though wild beasts rend piecemeal this poor envelope of clay.[23]

Marcus' philosophy certainly is unforgiving towards the stoic himself, but it is important to note that another recurring message in his *Meditations* is that the dedicated stoic should never assume a haughty or superior attitude or find fault with those who are not yet persuaded of the wisdom of pursuing the stoic lifestyle. Humility is a stoic virtue, both because excessive pride can mar the performance of one's duties (remember how hubris led to the downfall of the Homeric heroes) and because lauding one's achievements over others is a misdirection of energies that are better turned to more honorable pursuits.

The emperor is at pains to point out that dwelling on the weaknesses of others and parceling out blame for failures advances no worthwhile cause. Several of his meditations express the idea that identifying others' faults is a useless distraction from the project of correcting our own. A stoic should only be interested in the sub-par performance of others if he can help to improve it for some greater good. Consider the following remarks:

> 17. ... All thoughts of blame are out of place. If you can, correct the offender; if not, correct the offence; if that too is impossible, what is the point of recriminations? Nothing is worth doing pointlessly.[24]

> 20. Leave another's wrongdoing where it lies."[25]

---

[21]Marcus Aurelius, *Meditations*, in *Ancient Philosophy*, ed. Walter Kaufman and Forrest E. Baird (Englewood Cliffs, N.J.: Prentiss Hall, 1994), 450.

[22]Staniforth, trans., Marcus Aurelius' *Meditations*, Book Seven, pps.110–111 (Penguin 1964).

[23]Ibid p. 117.

[24]Staniforth, trans., Book Eight, p. 124.

[25]Ibid, Book Nine, p. 142.

4. If a man makes a slip, admonish him gently and show him his mistake. If you fail to convince him, blame yourself, or else blame nobody.[26]

30. When another's fault offends you, turn to yourself and consider what similar shortcomings are found in you. Do you, too, find your good in riches, pleasure, reputation, or such like? Think of this, and your anger will soon be forgotten in the reflection that he is only acting under pressure; what else could he do? Alternatively, if you are able, contrive his release from that pressure.[27]

Marcus reminds us that stoics view their lives as opportunities to play out to the best of their abilities the roles that fate has assigned them. It is the actual accomplishment of tasks that matters, not pride or credit or fame. He addresses the importance of the frank acknowledgment of one's own limitations, the unimportance of reputation, and the practical value of stoic humility in a sequence of three aphorisms:

5. Is my understanding equal to this task, or not? If it is, I apply it to the work as a tool presented to me by Nature. If not, then either I make way—if my duty permits it—for someone more capable of doing the business, or else I do the best I can with the help of some assistant, who will avail himself of my inspiration to achieve what is timely and serviceable for the community. For everything I do, whether by myself or with another, must have as its sole aim the service and harmony of all.

6. How many whose praise used once to be sung so loudly are now relegated to oblivion; and how many of the singers themselves have long since passed from our sight!

7. Think it no shame to be helped. Your business is to do your appointed duty, like a soldier in the breach. How, then, if you are lame, and unable to scale the battlements yourself, but could do it if you had the aid of a comrade?[28]

Expounding further on the subject of legacy and reputation, the emperor appeals to the fundamental principal that whatever is beyond the stoic's control should not concern him. Our reputations are formed by the opinions and expressions of others and so are clearly beyond our control. However well we behave, it is always possible for others to think and/or speak ill of us, either from misapprehension or malice. And even if we achieve some positive fame while living, we cannot control how future generations will regard us. No one can ever guarantee that his or her complete and accurate life history will be remembered forever.

Beyond the practical consideration that such matters cannot be controlled, Marcus makes the additional argument that it is beneath the dignity of a stoic to crave praise, reputation or fame. He asks, "When you have done a good action, and another has had the benefit of it, why crave for yet more in addition—applause for your kindness, or some favour in return—as the foolish do?"[29] And he comments further on the same subject in books seven and eight:

34. Of Fame. Take a look at the minds of her suitors, their ambitions and their aversions. Furthermore, reflect how speedily in this life the things of today are buried under those of tomorrow, even as one layer of drifting sand is quickly covered by the next.[30]

---

[26]Ibid, Book Ten, p. 152.

[27]Ibid, Book Ten, p. 160.

[28]Ibid, Book Seven, p. 106.

[29]Ibid, aphorism 73, Book Seven, p. 118.

[30]Ibid, Book Seven, p. 111.

44. Make the best of today. Those who aim instead at tomorrow's plaudits fail to remember that future generations will be nowise different from the contemporaries who so try their patience now, and nowise less mortal. In any case, can it matter to you how the tongues of posterity may wag, or what views of yourself it may entertain?[31]

In considering what benefits stoicism has to offer prospective followers, certainly one answer is not immortality of any kind. The stoic was not meant to look towards a life-after-death, either in the literal sense or in terms of some lasting legacy. The stoic focus is always on the present moment; on making the best of the here and now.

But if there is no final reward for virtuous stoics—no Heaven or Valhalla or even earthbound celebrity—then how are they motivated to maintain their stoic discipline? How do they find any meaning in their efforts if everything in life is just fleeting experience with no ties to the eternal? Why should they struggle to uphold high moral standards when the only inescapable judgment they must face is the one that they pass on themselves?

The stoic emperor embraces the conclusion that if life on earth, from birth to death, is all that there is, then the only possible way to make that life worth living is to pursue virtue. Exposing his own humanity, he explains that this conclusion was not self-evident to him from the start. Rather, he found empirical support for the superiority of the stoic path by exploring its weaker alternatives first hand. In the first meditation of Book Eight, Marcus admits that he was not always such a committed stoic. As if to reinforce his own decision to adopt the stoic perspective, he reminds himself,

1. Up to now, all your wanderings in search of the good life have been unsuccessful; it was not to be found in the casuistries of logic, nor in wealth, celebrity, worldly pleasures, or anything else. Where, then, lies the secret? In doing what man's nature seeks. How so? By adopting strict principles for the regulation of impulse and action. Such as? Principles regarding what is good or bad for us: thus, for example, that nothing can be good for a man unless it helps to make him just, self-disciplined, courageous, and independent; and nothing bad unless it has the contrary effect.[32]

Marcus' motivation to be a stoic springs from his conviction that no other style of life can ultimately offer any peace or satisfaction. The good life must include a sense of purpose. The pursuit of pleasure may seem appealing, especially when it carries no threat of post mortem punishment, but its lure loses force over time as the law of diminishing returns ensures that pinnacles of delight will become harder and harder to achieve or sustain. Costs will accrue for which no commensurate benefits are received. At the end of the day, Marcus reflects, those who live for pleasure alone will be left with nothing but regrets, whereas those who chose a more disciplined path can make even a brief life count.

Following the stoic perception that individual human lives pass quickly, end completely, and are then forgotten, the emperor argues that the only fulfilling life must be one which aims at the good for humanity as a whole. Even with no expectation of life-after-death, a stoic can find meaning in life by making a positive contribution to his society, by being part of something larger than himself. In Book Eight, Marcus asks, ". . . For what task, then, were you created? For pleasure? Can such a thought be tolerated?"[33] And in Books Eleven and Twelve he spells out the direction that a meaningful life should take:

21. If a man's life has no consistent and uniform aim, it cannot itself remain consistent or uniform. Yet that statement does not go far enough unless you can also add something of what the aim should be. Now, it is not upon the whole range of things which are generally assumed to be good that we find uniformity of opinion to exist,

---

[31]Ibid, Book Eight, pps. 130–131.
[32]Ibid, Book Eight, p. 121.
[33]Ibid, aphorism 19, Book Eight, p. 125.

but only upon things of a certain kind: namely, those which affect the welfare of society. Accordingly, the aim we should propose to ourselves must be the benefit of our fellows and the community. Whoso directs his every effort to this will be imparting a uniformity to all his actions, and so will achieve consistency with himself.[34]

20. Firstly, avoid all actions that are haphazard or purposeless; and secondly, let every action aim solely at the common good."[35]

A stoic legionnaire facing likely death in battle could not find consolation in dreams of lasting glory or eternity in the Elysian Fields. But he could take solace instead in the thought that his life will serve some useful purpose. If death finds him, he will fall fighting beside his comrades, in the service of his state.

The final goal of stoic dedication is to become a "sage"—the perfect embodiment of stoic ideals. In *The Romans*, Barrows paints a vivid picture for us of the stoic sage:

Neither trouble nor tribulation distresses the sage. He is superior to riches and poverty, to opinion critical and friendly; he does all for conscience's sake. He is kind to friends and to enemies merciful, and his forgiveness outstrips requests for it. His neighbours, whether in city or state or the world, he respects, and he does nothing to reduce their liberty. He will depart this world with the consciousness that in independence of spirit he has borne alike its joys and its sorrows and that death holds no terrors.[36]

The soldiers of the Roman army, it has been noted, were not all stoic sages, and nor were they immune to vice. The life of a soldier did not provide the same opportunities for hedonistic indulgence as the life of wealthy patrician. But practical constraints, rather than moral convictions, may have been all that held some legionnaires back from imitating the dissolute behavior of their civilian counterparts. Yet regardless of how many members of the legions might have strayed far from the ideals of Zeno, Cicero, Seneca, Epictetus, and Marcus Aurelius, the image of the Roman army which has survived to inspire soldiers through the centuries is her clearly her stoic face.

The lasting impression that history holds of the Roman legionnaire is that of an efficient and committed professional warrior, unwavering in his focus as he marches with his comrades in perfect formation, winning battle after battle. It does not matter that sometimes the Romans were defeated by foes they considered mere "barbarians," that discipline was often maintained only by the fear of draconian punishments, or even that violent mutinies were not unheard of in the Roman legions. What is remembered is the strength of the stoic face of Rome and the awe that it invoked in others. Consider these further observations by the Jewish historian Josephus, describing the Roman army in the reign of the emperor Vespasian:

All their duties are performed with the same discipline, the same safety precautions: gathering wood, securing food if supplies are low, hauling water—all these are done in turn by each unit. Nor does each man eat breakfast or dinner whenever he feels like it; they all eat together. Trumpets signal the hours for sleep, guard duty, and waking. Nothing is done except by command. . . . Absolute obedience to the officers create an army which is well behaved in peacetime and which moves as a single body when in battle—so cohesive are the ranks, so correct are the turns, so quick are the soldiers' ears for orders, eyes for signals, and hands for action. . . . One might rightfully say that the people who created the Roman Empire are greater than the Empire itself.[37]

---

[34]Ibid, Book Eleven, p. 176.

[35]Ibid, Book Twelve, p.183.

[36]Barrow, *The Romans*, 159.

[37]Josephus, *A History of the Jewish War*, 3.71–97, 104, 105, 107, 108, Shelton pps. 260–261.

A Vietnam Experience, Duty
VADM James B. Stockdale, USN

Address to the Class of 1983, United States Military Academy, West Point, July 13, 1979

The subject tonight is *duty* and I'm going to begin with a Naval reference. Don't be alarmed at the navy film, and now a naval story. Duty for us all is the same; moreover, half the philosophy students you saw in my class at the War College are Army officers. I was in prison with Marine, Air Force and Army officers and I am not parochial.

So I would like to begin this discussion of duty with a well-known British admiral's flag hoist signal to his fleet as they closed on the enemy to commence the Royal Navy's most famous battle. It was, of course, Lord Horatio Nelson's signal before the Battle of Trafalgar, a signal that history has shown to be the beginning of the end of Napoleon's hope to dominate Europe by force. The ultimately victorious Admiral Nelson had ordered the hoisting of flags which said simply this:

England expects that every man will do his duty.

That signal is a short but complete lesson in the fundamental and necessary concept of *duty*, a lesson I hope you cadets will long remember. One important thing to remember is that it was given by a man in uniform.

Take note of the important word "expects" in Nelson's signal. The idea of expectation is very much a part of the concept of duty.

The old Greeks understood this notion of expectation. They had a word for what we call virtue or moral excellence: *arete*. To the Greeks a good man was a man who did what was *expected* of him depending upon his particular station in the world.

A good cobbler was one who was expected to produce well-made shoes. And this was reiterated by Plato and Aristotle.

One should expect a good navigator, then and now, to guide his ship safely over the sea and into harbor.

A good soldier, as Aristotle in particular emphasized, was one who could be expected to display certain characteristics on the battlefield: courage, obedience, loyalty, steadfastness, resourcefulness. Courage (or the Greek word *andreia,* a synonym for manliness) was the first virtue of a man as well as of a soldier. Plato defines courage as endurance of the soul. The Greeks stressed endurance; they admired a man who could give the quick thrust, the audacious dash, but they reserved their highest praise for the man who "hung in there" battling against the terrible odds, the nearly certain defeat. A more modern military leader, Frederick the Great, took as his personal watchword the command "Stand fast!"

Finally, a good man to the old Greeks was a man who could be expected to display those virtues of character proper to a human as a human, not just his occupational standards as a cobbler, or as a navigator, or as a soldier.

The stoic philosophers, most of whom lived right after the heyday of the Greeks, illustrated this idea of expectation in the moral life by the metaphor of the actor, of the stage, of the drama. Men and women are called on in life to play a part. The part may be a big one or a small one, but once the part is given to us on life's stage, it is expected of us to play it well. In Hanoi's prisons at times when I was so depressed that military virtue seemed almost at the point of irrelevance, I was comforted and strengthened by remembering Epictetus' admonition (and I *did* remember it—as I've written, Epictetus' *Enchiridion* was one of my prize memories). "Remember that you are an actor in a drama of such sort as the Author chooses—if short, then in a short one; if long, then in a long one. If it be his pleasure that you should enact a poor man, or a cripple, or a ruler, or a private citizen, see that you act it well. For this is your business—to act well the given part, but to choose it belongs to another." You young men and women of this West Point Class of 1983 have been given a part and your part is that of a military officer. Your duty is doing what is expected of you. You are expected to play your part well.

So was it for Lord Nelson, and he played his part well. Courage? He lost an eye in one action and his right arm in another and still fought his greatest battles in the seven years that followed. Nelson's first thought was for his men. Wounded in action as a young officer, he refused to let the surgeons tend to him first. He was famous for saying that he would take his turn with, as he said, "his brave fellows."

Robert Southey published his *Life of Nelson* just eight years after Nelson had been mortally wounded and died in his flagship HMS *Victory* as he turned the tide at that Battle of Trafalgar, the same day he hoisted that signal about expecting Englishmen to do their duty. Southey tells us:

> Never was any commander more beloved. He governed his men by their reason and their affections; they knew that he was incapable of caprice or tyranny; and they obeyed him with alacrity and joy because he possessed their confidence as well as their love. "Our Nel" they used to say "is as brave as a lion, and as gentle as a lamb." Severe discipline he detested, although he had been bred in a severe school (he went to sea when he was 12); he never inflicted corporal punishment if it were possible to avoid it. And when compelled to enforce it, he, who was familiar with wounds and death, suffered like a woman. In his whole life, Nelson was never known to act unkindly towards an officer. In Nelson there was more than easiness and humanity of a happy nature: he did not merely abstain from injury; his was an active and watchful benevolence, ever desirous not only to render justice, but to do good.

Such was the character of the officer who expected every man under his command to do his duty, as he himself surely always did.

The concept of duty is not popular today in some circles. We live in a world of social turmoil and shifting values, a world where people insist on their rights but often ignore their duties. So great is the concern for rights today that people will invoke the total machinery and full power of the law to secure those rights. As a result, our nation in my view is choked with legalism, a situation that even a distinguished persecuted foreigner thought dangerous enough to bring to our attention. Of course I'm talking about Solzhenitsyn in his Harvard commencement address a year ago, when he warned that our nation had shifted the focus from the substance of the good to the rule book of rights. In this he saw the beginning of the decline of our nation's strength and national will. And I agree with him. Unless we are willing to balance each of the rights we claim with a correlative duty, we'll be as a nation like the man who wants a dollar's pay but is not willing to do a dollar's worth of work to get it. Rights incur obligations.

You of the military profession, although just initiated, will soon feel the weight of this responsibility and must lead the way in awareness of the crucial importance of duty. You must be not only leaders to your men but examples to the nation of the truth that for any position of responsibility in society, whether it be in the family government, business, or military, there is a corresponding obligation to carry out the assigned task.

Where did this idea of duty come from? What are its historical roots? In his book *Essay on Human Understanding*, the 17th-century philosopher John Locke discussed the simple question, "Why a man must keep his word." He found three different answers to this question; answers that I believe are as applicable today as they were then.

First, said Locke, a Christian man will say, "Because God who has the power of eternal life and death requires it of me that I keep my word."

Secondly, said Locke, if one takes the Hobbesian view of life, he will say, "Because society requires it and the state will punish you if you don't." (Hobbes was a very practical kind of hard-nosed guy.)

And thirdly, John Locke observed that had one of the old Greek philosophers been asked why a man should keep his word, the latter would say, "Because not to keep your word is dishonest, below the dignity of man, the opposite to virtue *(arete)*."

Two of the answers Locke cites, the Christian and the Hobbesist, seem to derive duty from the command of law, external law, the law of God in one case, the law of the state in the other. But the third answer, the Greek answer, shows that duty can be understood without reference to external law or to compulsion, divine or human. We share this understanding whenever, having made a promise, taken an oath, contracted a debt of duty—as you cadets have recently done—we feel an obligation to discharge it, even if no superior commands the act. Duty in this perspective has an absolute character. Duty is its own justification. It does not have to be propped up by anything outside itself, particularly in the line of reward or punishment. This was the teaching of Socrates who urged that men should obey the law, pay their debts, discharge their obligations, not to avoid the pain of censure or punishment but simply because they ought to.

Closer to our own time, the great German philosopher Immanuel Kant said much the same thing. Kant was a very bright man. Known generally as a moral philosopher, he gave us much more. He really explained the function of the human mind. Moral obligation in his view rests on an internal conviction of duty, the law we give to ourselves from ourselves, conscience if you like, and not the law that pressures us from the outside government. To Kant carrying out our moral obligation is obeying the law we set each for ourselves. He happened to be a religious man but he was very careful in his instruction never to rely on religion as a justification for any of his ideas; he relied only on what he called pure reason. The law we set for ourselves is *free* in contrast with the external law which is *compelling*, said Kant. The argument echoes the age-old irony of the necessity of discipline for freedom. That internal law we may call the voice of conscience. It is the inner awareness of what our duty is, and it rests on no foundation but itself. The obligation to do our duty is *unconditional.* That is, we must do it for the sake of duty, because it is the right thing to do, not because it will profit us psychologically or socially or economically, not because if we don't do it and get caught we'll be punished. The categorical imperative was Kant's name for this inbred, self-imposed restraint, for the command of conscience within that tells us that the only true moral act is done from a pure sense of duty. So you can't ask what benefits will accrue from performing your duty. You must do your duty because it *is* your duty. Period. Simply put, that is the concept I want to leave with you and there's plenty of intellectual background to support it.

I hope this posits the concept of duty in historical perspective. But Locke and Kant may seem a long way from the officer (very likely many of you), soon perhaps to be standing in front of a platoon or leading a group of men in harm's way, into this very peculiar enterprise we call war. On the battlefield, you very well may find yourselves in new decision-making territory where all previous bets are off, where the rational, managerial approach of many of our fathers is no longer valid. I am describing that duty of arms that Clausewitz described as "a special profession. . . . However general its relation may be and even if all the male population of a country capable of bearing arms were able to practice it, war would still continue to be different and separate from any other activity which occupies the life of man." Another old warrior, William Tecumseh Sherman (just 143 classes ahead of you plebes here tonight), said "War is cruelty and you can't refine it."

The duty of uniformed men has a long, colorful, and frequently bloody history, and it will be no different in the future. Those who think that we've seen our last war are, in my opinion, dead wrong. I make a Pascal wager that a general war will blight this planet, probably before the end of the century. Pascal, of course, advised us all to wager on that outcome by which one would stand to lose the least in case we were wrong. The trends as I read them make war the safer wager.

A lot needs to be said about the kind of education most appropriate for the professional soldiers you have chosen to be in these times of impending peril. You must aspire to a strength, a compassion, and a conviction several octaves above that of the man on the street. You can never settle for the lifestyle that Joseph Conrad characterized as "... skimming over the years of existence to sink gently into a placid grave, ignorant of life to the last, without ever having been made to see all it may contain of perfidy, of violence, and of terror...."

How to avoid ignorance of perfidy, of violence, of terror? Your education must include those intense emotional experiences of the sort common here. You will leave this place with more than a diploma. You will likely leave it with a highly developed conscience. It's almost impossible to graduate from an institution like this without it. You will have undergone an irreversible process which will never again allow you the comfort of self-satisfaction while being glib or shallow. You will likely forever carry the burdens, and they are burdens, of loyalty, commitment, passion and idealism. You will have undergone an education of the sort people refer to when they say that education is what's left over after you've forgotten all the facts you learned. And that which is left over, that conscience, that sentiment is indispensable to that capability for which the graduates of this institution are known. And that capability is leadership.

Here at West Point, you'll learn the range of responsibility that a commitment to duty demands. Some of the things a good leader with a strong sense of duty is expected to do may surprise you. I'd like to examine some of the seldom mentioned obligations of an officer.

I say it's your duty to be a *moralist*. I define the moralist not as one who sententiously exhorts men to be good, but *one who elucidates what the good is*. (Under the press of circumstance this is sometimes unclear—perhaps in a prison camp.) The disciplined life *here* will encourage you to be men and women of integrity committed to a code of conduct and from these good habits a strength of character and resolve will grow. This is the solid foundation from which you elucidate the good, by your example, your actions and your proud tradition. A moralist can make conscious what lies unconscious among his followers, lifting them out of their everyday selves, into their better selves. The German poet Goethe once said that you limit a man's potential by appealing to what he is; rather, you must appeal to what he might be.

Secondly, there are times when you'll have to act as a jurist, when the *decisions you'll make will be based solely on your ideas of right and wrong, your knowledge of the people who will be affected, and your strength of conviction*. There won't be a textbook or school solution to go by. I'm talking about hard decisions when you'll be the one with a problem that has seemingly endless complications—when you'll have to think it through on your own. As a jurist, you may be writing the law, or at least regulations, and that's a weighty responsibility. When you're in the hot seat, you'll need the courage to withstand the inclination to duck a problem or hand it off; you've got to take it head on.

One word of caution: Many of your laws will be unpopular. You'll have to learn to live with that. But your laws should never be unjust. Moreover, you must never cross that fatal line of writing a law that cannot be obeyed. You must be positive and clear and not lapse into a bureaucratic welter of relativism that will have others asking what you really mean or trying to respond in the most politically acceptable way.

And you'll find it's going to be your duty to be a *teacher*. Every great leader I've known has been a great teacher, *able to give those around him a sense of perspective and to set the moral, social and particularly the motivational climate among them*. You must have the sensitivity to perceive philosophic disarray in your charges and to put things in order. A good starting point is to put some time in on that old injunction, "Know thyself."

Here at West Point you will follow in the footsteps of greatness. During your years here, I challenge you to leave those same clear footprints of greatness. During your years here, I challenge you to leave those same clear footprints for future generations to follow. In John Ruskin's words such a process is "painful, continual and difficult . . . to be done by kindness, by waiting, by warning, by precept, by praise, but above all by example." Teachership (in my view) is indispensable to leadership and an integral part of duty.

Fourth, you must be willing to be a *steward*. By that I mean *you must tend the flock as well as crack the whip*; you have to be compassionate and realize that all men are not products of the same mold. The old Civil War historian Douglas Southall Freeman described his formula for stewardship at my school, the Naval War College, thirty years ago last month. He said you have to know your stuff, to be a man, and to take care of your men. There are flocks outside these walls that will require your attention and test your stewardship. They're not all West Pointers out there.

The final duty is that you must be able to act as *philosophers* in your careers in order to *explain and understand the lack of a moral economy in this universe.* Many people have a great deal of difficulty with the fact that virtue is not always rewarded nor is evil always punished. To handle tragedy may indeed be the mark of an educated man, for one of the principal goals of education is to prepare us for failure. When it happens, you have to stand up and cope with it, not lash out at scapegoats or go into your shell. *The test of character is not "hanging in there" when you expect a light at the end of the tunnel, but performance of duty and persistence of example when you know that no light is coming. Believe me, I've been there.*

Admiral James B. Stockdale's Leadership Model
Col. Paul E. Roush, USMC (Retired)

I begin these remarks with reference to a speech given to the Class of 1983, United States Military Academy, West Point, July 13, 1979 by Vice Admiral James B. Stockdale on the topic of duty. In the latter part of that speech Admiral Stockdale described the obligations of an officer in terms of five roles each of us must play well if we are to fulfill our duty. Those roles are *moralist, jurist, teacher, steward,* and *philosopher.* The words are not those one normally thinks of when contemplating military virtue. They seem especially incongruous with respect to the requirements levied on an officer in periods of combat. Yet the five roles represent the distillation of years of reflection on how a group of heroic individuals were able to endure in combat and then as prisoners of war for a combined total of up to ten years. If ever a concept for military excellence was forged exclusively in and for the fires of combat it is this one.

As we review the leadership model, we can have the utmost confidence that it represents a way of life that sustains in the most trying of times and that it represents the essence of the nation's expectations for its military leaders—those in whom it reposes special trust and confidence for the care of its sons and daughters who serve in the armed forces. Our task is to internalize (hence, act in accordance with) the roles of moralist, jurist, teacher, steward, and philosopher. By so doing we will have been faithful to our oath of office and will have honored those who paid the terrible price by which this lofty expression of the military ethic was forged.

The Stockdale Leadership Model: Value-and Principle-Centered Leadership

Leaders are imbued with national and professional values

- Their reference point for national values is the Constitution
- Their professional values find practical expression in our armed forces' core values
- Their leadership principles derive from the core values
- They know and celebrate their national and professional heritage
- They incorporate the national and professional values in leader roles of *moralist, teacher, jurist, steward,* and *philosopher*

Leaders in the role of MORALIST make plain the good by the way they live their lives
Leaders are people of honor

- They practice and promote truthfulness in all its nuances
- They practice and promote fairness in all their human interactions

- They practice and promote respect for persons by always treating others as ends rather than means
- They practice and promote the keeping of commitments

Leaders demonstrate a disciplined lifestyle for emulation by their followers

- They pursue intellectual development over a lifetime
- They acquire and enhance physical skills which extend the limits of their capabilities
- They strive in the small daily choices to increase their commitment to the military and moral ethos of military service
- They continually shape their own character as they habituate the doing of that which is right

Leaders exercise a priority of loyalties

- Their primary loyalty in their capacity as members of the armed services of the United States is to the supreme law of the land as defined in the Constitution, then to mission, then to service, then to unit or ship, then to shipmate or comrade, then to self
- They resolve loyalty conflicts in favor of higher elements in the priority of loyalties
- They reserve their loyalty for entities and persons engaged in ethical behavior
- They prepare themselves and their followers for mission accomplishment

Leaders in the role of JURIST make decisions, rules, and policies based on the strength of their character

Leaders hold themselves accountable for all their units do or fail to do

- They answer for the actions of self and followers
- They assess and enhance their own performance
- They improve constantly the processes under their purview
- They are willing to change and they count the costs of the status quo

Leaders enhance their followers' ability to know what is right and do what is right

- They make clear their expectations and help their followers meet those expectations
- They seek from their followers a commitment that goes beyond mere compliance with the minimum standard of the law to the idealized expression of the law's intent
- They allow followers the freedom to make mistakes while finding ways to promote their success
- They punish when necessary for the purpose of teaching their followers better ways of behaving

Leaders act upon well-placed conviction

- They make the best decisions with available facts
- They make clear, impartial, and achievable policy
- They make decisions based on their understanding of right and wrong, their knowledge of people, and their values
- They understand there are many situations for which there is no one right answer

Leaders in the role of TEACHER provide a sense of perspective and set the moral, social, and particularly the motivational climate in the unit

Leaders appeal to their followers' highest aspirations

- They provide the vision which inspires their followers to adopt as their own the goals of the unit

- They inspire their followers to do more than they think they are capable of doing
- They exercise the charisma which calls followers from "their everyday selves into their better selves"

Leaders create confidence in their followers by serving as the catalyst for decisive, forceful action when appropriate

- They confront problems head-on
- They willingly accept the consequences of their decisions and actions
- They seek responsibility and strive toward task completion
- They tolerate hardships, frustrations, and interpersonal stress for the sake of mission accomplishment
- They persist in the vigorous pursuit of goals
- They know when the situation calls for personal intervention and act accordingly

Leaders create a climate which promotes unit cohesion

- They articulate, demonstrate, and enforce appropriate unit norms
- They move their followers from dependence to independence to interdependence
- They obtain input from those likely to be affected by decisions
- They create an attitude of win-win in their unit

Leaders enhance unit effectiveness

- They provide direction; they motivate; they implement
- They eliminate artificial barriers to follower effectiveness
- They provide followers the rationale for decisions when possible
- They emphasize the teaching-learning process whether the issue is training, education, or mentoring

Leaders in the role of STEWARD invest their lives in the lives of their followers

Leaders view themselves as servants

- They serve the nation subject to the primacy of civilian authority
- Their priority of service is "service, comrades, self"
- They consider their followers a sacred trust for whose care they are answerable
- They are committed to their followers' development and well-being

Leaders guard the fundamental dignity of their followers

- They promote self-esteem in followers by respecting them and by holding them accountable for high standards
- They suffer hardship along with their followers in meeting professional obligations
- They are intolerant of formal or informal norms which diminish the dignity of their followers

Leaders understand human nature and value individual differences

- They understand individual differences as strengths which can enhance the overall functioning of the group
- They understand the values of their followers and appeal to those values and associated needs to bring out the motivation which already exists in their followers
- They know themselves—including vulnerabilities as well as strengths
- They have significant involvement with their followers while maintaining appropriate rank distinctions

Leaders in the role of PHILOSOPHER persevere "when virtue is not rewarded and evil is not punished"
Leaders do their duty because it is their duty

- They are moved to action by a sense of obligation
- They choose the harder right rather than the easier wrong
- They respond to intrinsic motivation rather than extrinsic rewards
- They know that cost-benefit analyses neither count all the costs nor know all the benefits

Leaders know how to deal with uncertainty and adapt to change

- They subordinate personal ambition to a higher cause
- They look beyond individual events to focus on patterns of behavior and ultimately on underlying systemic structure
- They forge new strength by learning from and coping with failure

Part VII
Epilogue

The Hiding Places of Memory
George R. Lucas, Jr. *

> "Prague, in Franz Kafka's novels, is a city without memory. It has even forgotten its name. Nobody there remembers anything, nobody recalls anything... Time [for Kafka]... is the time of a humanity that no longer knows anything nor remembers anything, that lives in nameless cities with nameless streets or streets with names different from the ones they had yesterday, *because a name means continuity with the past and people without a past are people without a name.*"

—Milan Kundera, *The Book of Laugher and Forgetting*[1]

A few years after the end of the Cold War, following the final collapse of the Berlin Wall in the fall of 1989, a retired former Soviet army officer, Colonel Vladimir Malinin, related how he and his wife, Yevgenia, first learned for themselves of how the Soviet state dealt with political prisoners:

> Yevgenia, an archivist for the state prison system, accidentally discovered a secret report written to Soviet leader Nikita Khrushchev by the director of the camps administration in the Far East. The report recited a litany of horrors that shocked the couple out of their previous unquestioning devotion to the Soviet state. According to Mr. Malinin, the report recounted that 17.5 million people had been imprisoned in a sprawling network of labor camps for political prisoners in the Kolyma River valley north of Magadan between 1933 and 1952. Of those, the report said, 16.3 million had died of exhaustion or illness and another 85,877 had been shot to death. Having stumbled upon such forbidden knowledge, Mr. Malinin said, he agonized over it for months, then finally shared his secret and sought advice from a friend named Ivan Chistiakov, who held a high position in the Magadan regional administration. "He told me, 'It's better you forget all about it,' " Mr. Malinin said.[2]

Likewise, at the relatively advanced age of seventy-six, a former Japanese army physician decided to break a longstanding code of silence and denial and speak out for the record concerning Japanese military atrocities during World War II. At issue were allegations, never acknowledged by the Japanese,

---

*Portions of these remarks are drawn from the author's essay, "Recollection, Forgetting, and the Hermeneutics of History," in *Hegel, History, and Interpretation*, ed. Shaun Gallagher (Albany, NY: State University of New York Press, 1994), 97–115.

[1]Trans. Michael Henry Heim (London: Penguin Books, 1983), p. 157; my emphasis.

[2]*Newsday* article on Soviet prison camps, reprinted in *The Baltimore Sun*, "Photo helps ex-Soviet officer recall U. S. prisoner." Monday, September 20, 1993, p. 3A.

463

that army doctors had conducted a variety of cruel and scientifically unwarranted experiments on Chinese and Korean prisoners of war. The physician, Dr. Ken Yuasa, remarked:

> I must confess, with embarrassment for myself and the country, because I strongly believe everyone should know the truth. If I don't tell my story, what the Japanese military has done will be *erased from history*.[3]

It is the business of the historian to remember, but the tendency in history itself is to forget. In most instances the forgetting, the fading of immediacy and the loss of intricate detail, is inadvertent, and seems unavoidable. Memories, it seems, have a way of hiding from us.

Forgetting, however, is not always simply unavoidable or inescapable, it is often intentional. The forgetting of history in many instances occurs as the end result of deliberate actions. What Col. Malinin uncovered was a strategy to *suppress* the deeds of an evil regime. Dr. Yuasa, by comparison, finally refused to condone *repression*, a conspiracy of silence and denial concerning medical and pseudoscientific activities that violated established, universal codes of decency and humanity. The motivations in both instances are literally to cast these black deeds into historical oblivion, to have them (and the guilt of their perpetrators) *erased from history*. Both strategies very nearly succeeded, save that someone, somewhere, chose instead to "remember." Remembering is an act of caring.

Attempts like these to force people to forget all about the historical record constitute the stock in trade of tyranny. Indeed, Czech author Milan Kundera maintains, in his *Book of Laughter and Forgetting*, that "the only reason people want to be masters of the future is to change the past."[4] Tyrants often see the need for this social or cultural amnesia, this need "to change the past" either to hide what they have done or to disguise or distort what they propose to do. They threaten to succeed only when the rest of us cease to care.

Warriors themselves must always care, and must never become complicit in these ongoing acts of loss and forgetfulness. This volume of readings is, in reality, an act of remembering. It represents a voyage of discovery, seeking to uncover the moral foundations of military leadership, and to ground these, in turn, in the history of moral reflection and the moral resources that arise in western culture.

We have sought to link military service to its high moral purpose in the defense of liberty, the protection of human rights, and the establishment of justice and the rule of law. The high moral purpose of military service is not simply the "defense of the homeland," but the never-ending struggle against the abuse of power by tyrants and criminals, as portrayed in these examples, and the protection of the vulnerable rights and liberties of their prospective victims. Warriors, as distinct from tyrants and criminals, use force reluctantly, and only when necessary, in order to protect the well-being of others, and never simply to harm them.

Warriors have not always remembered these things. Warriors discover in these readings, however, and especially in the experiences of CWO Hugh Thompson, VADM James Stockdale, and Bishop Desmond Tutu, that it is vital that they do so.

Why then, one might ask, did we not begin "at the beginning," and work our way through history to our present day? How is it we considered contemporary views, like Relativism, at the beginning of this book, or discussed the grounding of military service in a document, the Constitution, that is barely two hundred years old? Why is it only now, with the assistance of Vice Admiral Stockdale, that we finally arrive, in conclusion, at the views and the sources of ideals from centuries and even millennia before, that formed the historical basis of those founding documents, and that uphold the moral basis of the military profession itself?

Historians naturally prefer to begin at the beginning, and to conclude by showing how we have arrived at the point at which we now dwell. We, however, were not in fact privileged to stand at the beginnings of things, because the memories we seek lie hidden from us, forgotten, buried beneath the accrued traditions, the familiar formulae, the vacant rituals, and the easy assumptions of an uncaring present.

---

[3] Dr. Yuasa, imprisoned for three years as a war criminal at the conclusion of WWII, as quoted in an Associated Press article on Japanese war crimes: *The Baltimore Sun*, Tuesday, September 7, 1993, p. 7A. My emphasis.

[4] Milan Kundera, *The Book of Laughter and Forgetting*, p. 22.

Instead, we are compelled, like archaeologists, or like forensic scientists, to begin by surveying the world around us. We evaluate the evidence and the circumstances as we find them, and then patiently begin the process of teasing out their "myths of origin." By carefully examining the world of the present, we begin to discern within, behind, and beneath those forms, habits, and structures of contemporary life the patterns and habits of earlier times, the forms of prior beliefs, and the structures of earlier, founding values and moral commitments. Slowly, carefully, we work our way backward through the traditions of the profession, and the variety of proposals advanced throughout history concerning how we ought to fashion our lives. Only at the end of our voyage of discovery do we arrive, at long last, at the conclusion of that voyage, at the very foundations of our civilization, and of the military profession itself.

Such a task is not simply an idle, academic or philosophical exercise. It is an important, and ongoing, responsibility. We have *a duty to remember*. Something about our commonly-held views of the past makes it conceivable to the tyrant that a forgetting of history is possible, that the past can be undone and cast into oblivion. Kundera suggests that actual loss of the past is possible partly because of the sheer welter of detail, the relentless and oppressive weight of subsequent events crowding out the prior ones, perhaps aided by exhaustion, complacency, or despair. Events and crises come and go, or repeat themselves (as in Haiti, or Iraq, or Liberia) with wearisome frequency. Personnel come and go, leaders and their policies come and go, everything keeps changing and yet somehow remains the same, until soon it seems as if nothing matters, and everyone is fungible, interchangeable, and no one any longer cares:

> The bloody massacre in Bangladesh quickly covered over the memory of the Russian invasion of Czechoslovakia, the assassination of [Chilean President Salvadore] Allende drowned out the groans of Bangladesh, the war in the Sinai Desert made people forget Allende, the Cambodian massacre made people forget Sinai, and so on and so forth *until ultimately everyone lets everything be forgotten*.[5]

Where do these memories then hide? From its very beginnings, philosophy has concerned itself centrally with this question. The ancient Pythagoreans engaged in daily rituals of recollection, believing that by developing a perfect capacity to hold everything in mind all at once, they would thereby attain immortality. The "Myth of Er" at the end of Plato's *Republic* describes forgetting, or a sense of forgetfulness, as endemic to the human condition. The philosopher is one who responds to that absence, recognizing and lamenting this perceived loss, and willing to spend a lifetime in the quest to recover those lost, primordial memories. For St. Augustine, at the end of a lengthy autobiography narrating his own spiritual journey, a final, strange meditation on time and memory seems to suggest that divine experience itself consists in the ability to hold all memories together, timelessly, steadily, in an everlasting present, without loss. We who are mortal, finite, and imperfect find that we cannot do this, and so we forget, and cease any longer to care.

The German philosopher, George W. F. Hegel, figures prominently in this Western philosophical preoccupation with the problem of memory and forgetting. Like Plato, Hegel suggests in the *Phenomenology of Spirit* (1807) that it is *forgetting*, rather than memory, that is the true engine of history: humankind forgets what it has learned, and is forced to retrace its painful steps on the road to the recovery of the forgotten truth. The loss, the absence, and the sense of "something missing" or incomplete impels human consciousness on its historical journey—on what Hegel, in the Preface to this great work, dramatically terms "its highway of despair."

The cessation of suffering, the metaphorical end of humanity's journey on this historical highway, the rectification of injustice, and the antidote to tyranny—all, by contrast, lie in recollection. At the conclusion of the *Phenomenology*—on its final page, in fact—Hegel suggests that philosophical wisdom lies less in the attainment of some new, undiscovered truth than in the recollection and full retention of all that consciousness has already learned—as with Augustine and the ancient Pythagoreans, holding everything together simultaneously. This full internalization of what Hegel terms philosophical or "comprehended" history is characterized, in an interesting and poignant metaphor, as "the Calvary of

---

[5]Kundera, p. 7.

absolute Spirit"—an image, like that of James Stockdale, of an end that is also a beginning, of suffering that is also a victory over suffering.[6]

The recent confessions and acknowledgements of Japanese war crimes, delving the depths of Stalinist terror in the recently-accessible archives of the former Soviet Union, and the opening of the United States Holocaust Museum in Washington, DC in 1993, all likewise signified that historical recollection—recovering, retaining, and reweaving what was forgotten, suppressed, and distorted by the tyrant into a full and complete narrative of the present—alone can bring us to closure, to reconciliation, and, as Bishop Tutu's example suggests, to some measure of peace with ourselves. All of these memorial events are instances of moral closure, of ends that are simultaneously beginnings, of narrations of suffering that represent simultaneously a *victory over* suffering.

It is always our prior awareness of absence, of loss, of incompleteness, even in the absence of a full knowledge of the forgotten details, that prompts and prods historical consciousness, in these instances, not to rest, not to remain content with indifference and forgetting, but to "press on" along the highway of despair toward a full understanding of what actually occurred and why. The very sense of absence that forgetting induces in us—that gap or space in the airbrushed image of our collective experience—awakens a sense of discomfort, a vague sense of absence or of loss. This kindles what Plato then called *eros*, the desire to know, to recover that which was lost. The kingdom of heaven, a rural rabbi once taught his disciples, is like a lost sheep or a lost coin: when the owners discover their loss, they set aside everything else, and seek after it, and will not rest until what was lost has been fully restored.

Our loss of cultural memory amounts to a loss of self. Our lost or partial selves, like Plato's philosopher, or like the rabbi's owner of the sheep or the coin, wander the earth in search of the hiding places of memory.

In our present age, we have all but forgotten who we are, what we believe, and what we are expected to do. Those who read these words, however, are, or will soon be, warriors. And warriors, above all, must always remember who they are, for remembering is an act of caring.

Having remembered, resolve never to forget.

---

[6]Cf. *Hegel's Phenomenology*, trans. A.V. Miller (Oxford: The Clarendon Press, 1977), pp. 492–493.